Lecture Notes in Networks and Systems

Volume 1307

The series "Lecture Notes in Networks and Systems" publishes the latest developments in Networks and Systems—quickly, informally and with high quality. Original research reported in proceedings and post-proceedings represents the core of LNNS.

Volumes published in LNNS embrace all aspects and subfields of, as well as new challenges in, Networks and Systems.

The series contains proceedings and edited volumes in systems and networks, spanning the areas of Cyber-Physical Systems, Autonomous Systems, Sensor Networks, Control Systems, Energy Systems, Automotive Systems, Biological Systems, Vehicular Networking and Connected Vehicles, Aerospace Systems, Automation, Manufacturing, Smart Grids, Nonlinear Systems, Power Systems, Robotics, Social Systems, Economic Systems and other. Of particular value to both the contributors and the readership are the short publication timeframe and the world-wide distribution and exposure which enable both a wide and rapid dissemination of research output.

The series covers the theory, applications, and perspectives on the state of the art and future developments relevant to systems and networks, decision making, control, complex processes and related areas, as embedded in the fields of interdisciplinary and applied sciences, engineering, computer science, physics, economics, social, and life sciences, as well as the paradigms and methodologies behind them.

Indexed by SCOPUS, EI Compendex, INSPEC, WTI Frankfurt eG, zbMATH, SCImago.

All books published in the series are submitted for consideration in Web of Science.

For proposals from Asia please contact Aninda Bose (aninda.bose@springer.com).

Saroj Kumar Biswas · Sivaji Bandyopadhyay ·
Yoichi Hayashi · Valentina Emilia Balas
Editors

Intelligent Computing Systems and Applications

Select Proceedings of the International Conference, ICICSA 2022

Springer

Editors
Saroj Kumar Biswas
Department of Computer Science
and Engineering
National Institute of Technology Silchar
Silchar, Assam, India

Yoichi Hayashi
Department of Computer Science
Meiji University
Kawasaki, Japan

Sivaji Bandyopadhyay
National Institute of Technology Silchar
Silchar, Assam, India

Valentina Emilia Balas
Aurel Vlaicu University of Arad
Arad, Romania

ISSN 2367-3370 ISSN 2367-3389 (electronic)
Lecture Notes in Networks and Systems
ISBN 978-981-96-3859-8 ISBN 978-981-96-3860-4 (eBook)
https://doi.org/10.1007/978-981-96-3860-4

This Springer imprint is published by the registered company Springer Nature Singapore Pte Ltd.
The registered company address is: 152 Beach Road, #21-01/04 Gateway East, Singapore 189721, Singapore

Contents

Data Science, Big Data and Computational Intelligence

Comparative Analysis of Neural Machine Translation Models for Low Resource English–Assamese Language Pair

Basab Nath and Sunita Sarkar

Abstract Language translation makes it possible for people to communicate, share information, and build strong relations. Because it converts text from one language to another, Neural Machine Translation contributes to the development of the performances. For the English–Assamese language pair, this paper analyses four Neural Machine Translation models based on distinct Techniques. This paper presents four models of Sequence Architecture, which include: (1) Bidirectional Gated recurrent units (GRUs) (2) Bidirectional Long Short-Term Memory (BI-LSTM) (3) Bidirectional Gated recurrent units (GRUs) +BPE (4) Bidirectional Long Short-Term Memory (BI-LSTM) +BPE, with attention mechanism. The results of the comparison revealed that GRU with BPE outperforms Bi-LSTM in terms of performance.

Keywords Neural machine translation · Natural language processing · GRU · Bi-LSTM · Attention mechanism

1 Introduction

Humans communicate with one another using a mutually understandable language. Communication is essential for sharing knowledge and socializing. But as there are roughly more than 6500 different languages across the globe, language can be a barrier to communication. This is where machine translation enters the picture. With the help of rich data sets accompanied by appropriate methodologies, machine translation will be able to generate quality translations that have the potential to surpass human translators. However, to achieve such high levels of accuracy, several criteria must be taken into account, including the availability of vocabulary in both the source and target languages, a grasp of the semantic qualities of both the source and target languages, and many more. The viable models for machine translation are rule-based, statistical-based, and neural network-based [1]. The rule-based methods are based on

B. Nath (✉) · S. Sarkar
Department of Computer Science and Engineering, Assam University, Silchar, India
e-mail: basabnath@gmail.com

S. K. Biswas et al. (eds.), *Intelligent Computing Systems and Applications*, Lecture Notes in Networks and Systems 1307, https://doi.org/10.1007/978-981-96-3860-4_1

maintaining huge dictionaries containing predefined semantic and syntactical rules of the concerned languages. The statistical-based models use probabilistic functions for computing statistical tables, which comprise the semantic rules learned from the bilingual corpus. The statistical tables may be syntax trees or phrase translation models. The advent of neural machine translation models eliminates the need for feature engineering tasks such as building parse trees, phrase translation models, rule-based dictionaries, etc. In the case of Indian languages, the linguistic resources are limited, especially for the languages spoken in the northeast region of India. Our survey focuses on exploring the various techniques in Neural Machine Translation, which can be used to deal with issues found in low-resource languages, such as Out Of Vocabulary (OOV) words, rare words, and many more [3]. With the aid of LSTMs (Long short-term memory) and GRUs (Gated Recurrent Units), as well as an attention layer, we were able to create four unique neural machine translation models for the English–Assamese language pair. We used the BLEU metric, which stands for Bi-lingual Evaluation Understudy, to compare the performance of various models.

2 Literature Survey

In the previous few years, a huge number of Neural Machine Translation (NMT) research projects have been conducted, including a number of excursions. Despite the fact that the vocabulary size was fairly small, using the backpropagation neural approach, Allen [3] released a paper in 1987 in which he attempted to translate English to Spanish. Sennrich et al. [4] proposed a novel character-aware encoder–decoder structure. This structure had an increased encoder to produce better source word portrayals and an expanded decoder with a second consideration to process source side characters in order to produce better translations. Sutskever et al. [5] presented an encoder–decoder system with an LTSM neural network that initially processes the source sentence by the encoder and acquires a fixed vector representation. The decoder part then produces outputs at each time step by processing this vector representation. Activity stops when an extraordinarily nary token demonstrating the part of the agreement is delivered. Johnson et al. [6] created a method that uses a single neural machine translation to translate many languages by inserting a token at the start of the source sentence and translating it to another sentence. Because of shared word pieces, it is possible to interpret several languages using a single model language. They also revealed a problem with zero shots. Agrawal and Jain [7] in this paper illustrated two procedures to build a device called transliteration that translates the En-San language pair. The first approach uses a keyboard, and the second approach uses a virtual keyboard. The translator uses Unicode to translate the sentences. It takes Sanskrit or Hindi as input (in English) and then, using transliteration, the sentence is converted into Sanskrit. Later, it translates it into Hindi. According to Narayan et al. [8], quantum machine learning may be used to comprehend the structure of corpora by extracting part-speech information from each word. Chandola and Mahalanobis [9] presented a tool working as an automatic language translator that is based on neural networks. For Hindi–English translation, neural networks are employed to find some set of ordered rules that are often effective for other Indo-European languages if the dictionary is modified. Despite the fact that [14] used SMT-based techniques to achieve a BLEU score of 5.02 and 9.72 for English-to-Assamese translation and vice-versa, MT on the English–Assamese language pair is still in its infancy. The current focus of this paper is on improving current machine translation work on one of the Northeast language pairs using the neural machine translation system.

3 Methodology

Neural machine Translation (NMT) is a translation system that uses neural networks and the probability of translating phrases from the source language [2] to the target language to achieve translation accuracy. In the next parts, we'll talk about how our suggested translator will use a sequence-to-sequence architecture and an attention model.

3.1 Sequence to Sequence Architecture

Sequence-to-sequence architecture is the most common method for generating responses. It's used in MT models to figure out how two different language pairings are related to one another. An encoder and a decoder make up the system. The encoder and decoder receive input from the source; the encoder generates an encoded vector and words, and the decoder outputs a decoded version of the encoded vector and words. It's a fair bet that A and B are generated from each other. Using conditional probability, the decoder sends each word of the encoded source phrase out one at a time a1, a2, a3, and an. The encoder then transforms it into a vector of the same dimensions. The letters in the equation represent each encoded vector of a specified size. Equation (1) is converted to Eq. (2) using the chain rule [5].

$$P(C/D) = P(C|D1, D2, D3, ..., DN) \tag{1}$$

$$P(C|D) = P(ci|c0, c1, c2, \ldots, ci1; d1, d2, d3, \ldots, dn) \tag{2}$$

During decoding, the very next word is determined by using the previously predicted word vectors and the source sentence vectors from Eq. (1). With a softmax over all the terms in the vocabulary, each term in a distribution is represented as a single term in another distribution.

3.2 Gated Recurrent Units (GRUs)

Simplified LSTMs (Long short-term memory), or GRUs (Gated Recurrent Units), are a kind of Long Short-Term Memory. GRUs have only two gates, Reset and Update, unlike Long Short-Term Memory, which has a single hidden layer and two gates. LSTMs require more time and resources to develop than other types of neural networks [10]. Means the time and resources needed to properly train and prepare a dataset for practical use is substantial. Linear layers in each cell need a large memory bandwidth, which is often lacking in the system. Thus, hardware-wise, LSTMs become rather inefficient. It is also easier to train GRUs since they are less complex in terms of the number of gates and states they have, whereas LSTMs are more complex. The reset gate determines how much of the previous hidden state data will be used to compute the current hidden state in order to distinguish between critical and non-critical memory. Using an update gate, you may regulate how much data from a prior hidden state can be transmitted to the current hidden state [11].

3.3 Word Embedding

When words with the same meaning are encoded on a vector space, word embedding is a way of encoding them in which the vector representations of the words are similar. In this game, each word from the lexicon has hundreds of different dimensions. Ordinarily, pre-trained word embeddings are employed in conjunction with transfer learning to translate words from vocabulary to vectors [12]. The word2vec vectors we employed in our model were used to turn the combined English and Assamese vocabulary into a 300-dimensional vector. A BLEU score of 6.74 was achieved after training the model with the same layers, optimization approach, attention, and regularization as we used in the previous training.

3.4 Encoder–Decoder Architecture with Attention Mechanism

The bottleneck in a conventional encoder–decoder system is caused by the compressed nature of the whole input sequence information into a fixed-length vector called a context vector [15]. This context vector is used by the decoder to predict the decoding sequence. This fixed-length context vector is unable to retain longer sequences, which corresponds to poor translation. Hence, an additional layer in the encoder–decoder architecture, called the attention layer, comes into the picture. With the help of this attention layer, the context vector will have access to the full state history of the encoder for each decoding time step. This allows the decoder to have access to a rich representation of the input sequence. The attention layer helps the decoder to pay different amounts of attention to different input words at different decoding steps. Let $a = [a1, b2, b3, \ldots, bn]$ be the input sequence and $c = [c1, c2, c3, \ldots, cm]$ the output sequence and $t = f(x_t, h_{t-1})$ denote encoder's hidden state, for $t = 1, 2, \ldots, n$.

Let the decoder hidden state, $d_t = f(d_t, y_{t-1}, c_t)$ for the output word at position $t = 1, 2, \ldots, m$. Here, f can be any instance of a recurrent neural network. The context vector ct is the weighted sum of the hidden states of the encoder. The following is the context vector c_t:

$$c_t = \sum_{i=1}^{n} \alpha_{t,i} h_i \tag{3}$$

for the output y_t.

In the attention model proposed by Bahdanau et al. [12], the variable sized alignment vector, $\alpha_{t,i}$ given as

$$\alpha_{t,i} = \frac{exp(e_t i)}{\sum k = 1^i exp(e_t k)} \tag{4}$$

where $e_t i = a(d_{t-1}, h_i)$ Using the alignment model, the pair of input and output at position I and position t is given an alignment score $\alpha_{t,i}$ depending on how well they match each other. The weights in $\alpha_{t,i}$ specify how much of each encoder's hidden states should be taken into account when generating an output. According to [12], all the components of the proposed system are trained together, including the alignment model 'a,' which is parameterized as a feed-forward neural network.

3.5 Byte Pair Encoding

Byte Pair Encoding is a lossless data compression approach in information theory in which the most frequently occurring pair of successive bytes is replaced with a byte that is not present in the data. Byte Pair Encoding (BPE) is a sub-word tokenization approach used in natural language processing. A variation of this technique is used to tokenize word sequences in natural language processing models such as Transformers and GPT-2. A word is segmented into distinct characters in BPE. A count is kept of all the vocabulary's unique terms. Consider the vocabulary terms "clever," "smart," "oldest," and "wildest," which occur three times, four times, six times, and three times, respectively [12]. In BPE, these words will be split into characters, and the vocabulary will be formed along with their occurrence count as

's m a r t </w>': 3, 's m a r t e r </w>': 4, 'o l d e s t </w>': 6, 'w i l d e s t </w>': 3

A fixed number of iterations will be run, and at every iteration, the frequency of each consecutive byte pair will be counted. The pair with the highest count will be merged as a single token. In this example, the pairs “e” and “s” are the most frequent, with 9 occurrences. Hence , after merging the pair, the vocabulary changes to [12],

‘s m a r t </w>’: 3, ‘s m a r t e r </w>’: 4, ‘o l d es t </w>’: 6, ‘w i l d es t </w>’: 3

The number of merge operations, i.e., iterations, is set accordingly. One of the prime advantages of BPE is that it solves the issue of out-of-vocabulary to an extent because the vocabulary prepared by BPE consists of sub-word tokens instead of whole words. A word which is unknown to the vocabulary can be split up using the sub-word tokens in the vocabulary, hence eliminating the out-of-vocabulary issue. Also, by manipulating the number of merge operations, the size of the vocabulary can also be obtained as desired.

4 Translator

Machine translations for Indian languages [16] have been the subject of a lot of research. We have found that there are very few machine translations for North-Eastern languages. Our neural machine translation models are capable of translating between English and Assamese. We employ GRUs and BPE in the encoder and decoder, as well as the attention mechanism developed by Cho et al. [12], which follows the Encoder–Decoder architecture. We have used TensorFlow and Keras as frameworks to build our models, so we can make them. In order to get the word embeddings of the vocabulary, we used the Skip-gram algorithm from Word2Vec (Fig. 1).

5 Results and Discussions

In this work, we have constructed four distinct models and computed a differentiation between them using their BLEU ratings.

5.1 Evaluation Metrics

The BLEU assessment metric was used to assess the effectiveness of our implemented models. In order to evaluate an MT model’s [13] performance, researchers employ the Bi-lingual Evaluation Understudy (BLEU) method. For scoring purposes, it compares the output sentences of the model as a candidate to the human-translated words as a reference. In this section, the adjusted n-gram precision of candidate sentences is determined in relation to the reference sentence accuracy. Additionally, a brevity penalty has been created, which is meant to penalize translators who provide shorter

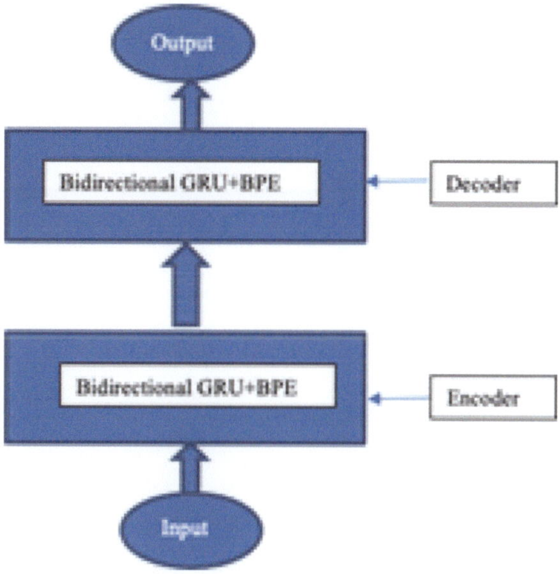

Fig. 1 Neural machine translation based Encoder–Decoder architecture

translations. The BLEU score can be expressed as a number between 0 and 1, but it can also be expressed as a percentage. The BLEU measure is one of the most often used metrics in machine translation [12].

5.2 *Corpus Statistics*

The diagram below shows the detailed statistics of the corpus that we have used (Table 1).

5.3 *Pre-processing of Data*

Each English and Assamese sentence in the original bilingual corpus was stored in a separate file at first. Files for each language were created by dividing the sentences into two separate folders. The English sentences were written in lowercase characters. There was no deletion of punctuation or numerical values. White space is added before punctuation marks to show that they are separate characters. We made four new symbols: '<sos>', '<eos>', '<unk>', '<pad>'. The tokens <sos> and <eos> are used to show when a sentence starts and ends. Use the token to show a word in the dictionary that isn't there with the word. All source and target phrases must be

Table 1 Corpus details

Corpus containing parallel sentences	61,138	
English vocabulary count	29,339	
Assamese vocabulary count	63,904	
Training set with parallel sentences	58,089	
Testing set with parallel sentences	3,058	
Minimum length of the sentence (no. of words)	English language	1
	Assamese language	1
Maximum length of the sentence (no. of words)	English language	21
	Assamese language	21

the same length in order to train the model in batches; tokens are used to fill in gaps in the shorter sentences. The target sentence and the source sentence may or may not be the same length.

5.4 System and Training Details

The Google Colab platform has been used to deploy our models. The settings of the system on which our model was trained and evaluated are summarized in the Tables 2 and 3.

Table 2 System and training details

Processor	1x single core hyper threaded Xeon Processors @2.3 GHz
Processor base frequency	2.3 GHz
Ram	25 GB
GPU	1x Tesla P100-PCIE, having 2496 CUDA cores and 16 GB GDDR5 VRAM
Disk space	68 GB

Table 3 Low-resource English–Assamese translation system's training and testing accuracy

Sl. no	No. of layers	Encoder	Decoder	Testing accuracy (%)	Training accuracy (%)
1	1	Bidirectional GRU	Bidirectional GRU	28.0271	29.2125
2	1	Bidirectional LSTM	Bidirectional LSTM	32.9858	34.9876
3	1	Bidirectional GRU+BPE	Bidirectional GRU+BPE	36.1738	36.1996
4	2	Bidirectional LSTM+BPE	Bidirectional LSTM+BPE	29.961	30.8765

6 Result

Models like LSTMs and GRUs, which have attention mechanisms in their structure, are among the models we designed on our own. Unidirectional RNNs, unlike bidirectional RNNs, only do one search of the input data. It is a different story when it comes to bidirectional RNNs, which search for input in both directions at once. Bidirectional RNNs can be quite helpful when trying to estimate the semantics of a sequence's midwords. We used at least 512 dimensions in all of our models. Diagrams depicting our architectures and a table showing the BLEU scores for each model are shown in Table 4.

7 Analysis

Because deep-layered neural networks outperform other methods in prediction, they were used in this paper. Because they have more layers and learning units per layer, they can predict better than unilayered neural networks. However, there is a risk,

Table 4 Low-resource English–Assamese translation system's training and testing accuracy

Sl. no	No. of layers	Encoder	Decoder	BLEU score (%)
1	1	Bidirectional GRU	Bidirectional GRU	29.0371
2	1	Bidirectional LSTM	Bidirectional LSTM	34.9878
3	1	Bidirectional GRU+BPE	Bidirectional GRU+BPE	36.1938
4	2	Bidirectional LSTM+BPE	Bidirectional LSTM+BPE	30.8765

Table 5 The sample input output of translated NMT system

Input	Actual Output	Predicted Output (With Transliteration)
This is primarily grown in green house.	ইয়াক প্ৰধানকৈ গ্ৰীণ হাউচত উৎপন্ন কৰা হয়।	ইয়াক প্ৰধানকৈ গ্ৰীণ হাউচ উৎপন্ন কৰা হয় ।
Kamlesh is a maths teacher by profession.	কমলেশ পেছাত এজন গণিত শিক্ষক ।	কমলেশ এজন গণিত শিক্ষক হয় ।
Rangoli is made in festive occasions.	ৰংগোলী উৎসৱৰ অনুষ্ঠানত তৈয়াৰ কৰা হয়।	ৰংগলী উৎসৱৰ অনুষ্ঠান তৈয়াৰ কৰা হয়।
Geetakumari is a wrestler from Haryana.	গীতাকুমাৰী হাৰিয়ানাৰ এগৰাকী মল্লযোদ্ধা ।	গীতাকুমাৰী হাৰিয়ানা মল্লযোদ্ধা ।
Amjad Khan was a great Indian actor.	আমজাদ খান এজন মহান ভাৰতীয় অভিনেতা আছিল ।	আমজদ খান এজন মহান ভাৰতীয় অভিনেতা ।

which is the complexity of deep neural network training. It's hard for these kinds of models to agree with each other, and if they don't, the model won't be able to predict well for data that we don't know about yet. The model with bidirectional GRU on both sides, i.e., encoder and decoder, as well as BPE and attention mechanism, proved to be the best among the four, receiving an approximate BLEU score of 36%. An example of our implemented models' translated sentences is shown in Table 5.

8 Conclusion

We used NMT to improve one of the North East language pairings in this study, and the results were promising (English–Assamese). We demonstrated that when applied to Indian languages and variations, NMT outperforms sophisticated translation approaches using pre-trained word embedding and Byte Pair Encoding. In this research, we primarily focused on developing an NM translator for the English–Assamese language pair, and we were successful in implementing four different NMT models in the process. We can also conclude from the test that our model is successful in overcoming the Out of Vocabulary (OOV) problem. Future work will include experimenting with alternative neural networks other than RNNs to develop NMT models, among other things. When compared to RNNs, feed-forward neural networks require significantly less training time and are more capable of supporting parallelization.

References

1. Koehn P, Knowles R (2017) Six challenges for neural machine translation. In: Proceedings of the first workshop on neural machine translation. https://doi.org/10.18653/v1/w17-3204
2. Chen H, Huang S, Chiang D, Dai X, Chen J (2018) Combining character and word information in neural machine translation using a multi-level attention. In: Proceedings of NAACL-HLT, vol 2018, pp 1284–1293
3. Allen RB (1987) Several studies on natural language and back-propagation. In: Proceedings of the IEEE first international conference on neural networks, vol 2. IEEE Piscataway, NJ, pp 335–341
4. Sennrich R, Haddow B, Birch A (2016) Neural machine translation of rare words with subword units. In: Proceedings of the 54th annual meeting of the association for computational linguistics (Volume 1: Long papers)
5. Sutskever I, Vinyals O, Le QV (2019) Sequence to sequence learning with neural networks. In: Advances in neural information processing systems, vol 27, pp 3104–3112. http://papers.nips.cc/paper/5346-sequence-to-sequence-learning-with-neural-networks.pdf
6. Johnson M, Schuster M, Le Q, Krikun M, Wu Y, Chen Z, Thorat N, Viégas F, Wattenberg M, Corrado G, Hughes M, Dean J (2019) Google's multilingual neural machine translation system: enabling zero-shot translation. arXiv:1611.04558
7. Agrawal P, Jain L (2018) Anuvaadika: implementation of Sanskrit to Hindi translation tool using rule-based approach. Recent Patents Comput Sci 12. https://doi.org/10.2174/2213275912666181226155829
8. Narayan R, Chakraverty S, Singh VP (2016) Quantum neural network based machine translator for Hindi to English. Appl Soft Comput 1060–1075
9. Chandola A, Mahalanobis A (1994) Ordered rules for full sentence translation: a neural network realization and a case study for Hindi and English. Pattern Recogn 515–521
10. Revanuru K, Turlapaty K, Rao S (2017) Neural machine translation of Indian languages. In: Proceedings of the 10th annual ACM India compute conference on ZZZ. ACM, pp 11–20
11. Rubino R, Marie B, Dabre R, Fujita A, Utiyama M, Sumita E (2020) Extremely low-resource neural machine translation for Asian languages. Mach Transl 34(4):347–382. https://doi.org/10.1007/s10590-020-09258-6
12. Cho K, Merrienboer BV, Bahdanau D,Bengio Y (2014) On the properties of neural machine translation: encoder-decoder approaches. In: Proceedings of SSST-8, eighth workshop on syntax, semantics and structure in statistical translation. https://doi.org/10.3115/v1/w14-4012
13. Papneni K, Roukos S, Ward T, Zhu W (2002) BLEU: a method for automatic evaluation for machine translation. In: Proceedings of the 40th annual meeting of the association for computational linguistics (ACL)
14. Kanchana Baruaj K, Das P, Hannan A, Sarma SK (2014) Assamese–English bilingual machine translation. Int J Nat Lang Comput (IJNLC) 3. https://doi.org/10.5121/ijnlc.2014.3307
15. Hassan A, Alom F, Chowdhury AS, Khan N (2019) Neural machine translation for Bengala-English language pair. In: 22nd international conference of computer and information technology (ICCIT)
16. Shah P, Bakarola V (201) Neural machine translation system of Indic language-attention based approach. In: Second international conference on advanced computational and communication paradigms. https://doi.org/10.1109/ICACCP.2019.8882969

Data Balancing in Darknet Traffic Classification Using Deep Learning

M. Mahalakshmi, P. Vijay Kumar, S. Navin Kumar, S. Meenakshi Sundaram, M. P. Ramkumar, and G. S. R. Emil Selvan

Abstract Darknet, apart from the surface web, is mostly used for privacy and illicit activities over the internet. These are networks that are only accessible to a small number of people rather than the entire internet public and are only accessible through authorization, specific software, and configurations. This includes both benign sites such as academic databases and corporate websites, as well as sites that deal with shadier issues such as black markets, fetish communities, and hacking and piracy. This study proposes a generic machine learning-based technique for identifying Darknet visits, as well as examines and performs statistical preprocessing on the dataset, which provides significant information on Darknet users. Following that, this research looks at various function selection approaches to see which ones are better for detecting and classifying Darknet users. This study employs tuned machine learning (ML) approaches such as Decision Tree (DT), Naive Bayes (NB), Random Forest (RF), and Logistic Regression (LR) based on identified capabilities. Our proposed technique of Deep learning (DL)-based oversampling utilizing Variational Autoencoders (VAE) and Generative Adversarial Networks (GAN) known as Adversarial Autoencoders (AAE) that aids in handling class imbalance is now able to attain 100% accuracy in almost every ML classifier model. To cut down on time, the entire classification model is built on the cloud computing resource, Apache Spark.

M. Mahalakshmi · P. Vijay Kumar (✉) · S. Navin Kumar · S. Meenakshi Sundaram · M. P. Ramkumar · G. S. R. Emil Selvan
Department of Computer Science and Engineering, Thiagarajar College of Engineering, Madurai, India
e-mail: vijaykumarptce@student.tce.edu

S. Navin Kumar
e-mail: navinkumar@student.tce.edu

S. Meenakshi Sundaram
e-mail: meenakshisundarams@student.tce.edu

M. P. Ramkumar
e-mail: ramkumar@tce.edu

G. S. R. Emil Selvan
e-mail: emil@tce.edu

S. K. Biswas et al. (eds.), *Intelligent Computing Systems and Applications*, Lecture Notes in Networks and Systems 1307, https://doi.org/10.1007/978-981-96-3860-4_2

Keywords Darknet classification · Artificial intelligence · Machine learning · Deep learning · Variational autoencoder · Generative adversarial autoencoders · Class imbalance · Apache spark

1 Introduction

The Darknet is an internet overlay network that can only be accessed with specialist software, configurations, and special authorizations, and which frequently employs non-standard communication protocols to be intentionally unavailable to the internet. In recent years, there has been a slew of cybersecurity vulnerabilities that have wreaked havoc on society. The modern technological advancement on the internet has immeasurably enhanced people's lives, and yet the Darknet has its dark side. There are more than 52% of active dark websites around the world, 29.7% are illicit in total, and these dark webs are comparatively more dangerous than surface web. In this paper, the standard Canadian Institute for Cybersecurity (CIC) datasets have been used that contain instances of both Benign and Darknet Traffic. Using CIC-Darknet 2020 dataset, an artificial-intelligence-based traffic classification system is proposed. Both the subsets of AI say ML and DL are used to serve the purpose. Autoencoders (AE) of DL algorithm are used for dimensionality reduction that aids in feature extraction and the class imbalance is then handled by performing oversampling using AAE. The classification of the traffic type is then done using several ML classifiers including DT, RF, NB, and LR. This pipeline performs with an accuracy of 100% with almost every ML classifier model for traffic classification.

2 Literature Survey

Various machine learning (ML)- and deep learning (DL)-based algorithms were currently in usage for Darknet classification aiding in different applications. While some authors use publicly available datasets, some prefer the simulation of the Darknet dataset. One such way of simulation is collecting traffic probes in the real Darknet environment was proposed [1]. The author proposed a three-layer hierarchical classifier and considered six ML classifiers and two DL classifiers for the training of local classifiers of which the XGBoost algorithm performs the best in this Darknet scenario, whereas the DL algorithms fail to do. This is highly contradicting the well-known fact of DL which is superior to ML. Thus, a careful analysis is required of those preprocessing steps to analyze their impact on ML's performance. In the case of classification with the publicly available dataset, the new dataset CIC-Darknet 2020 plays a vital role [2]. Here common machine learning classification algorithms are employed to identify Darknet traffic and the classifiers were trained to both binary and multiclass classification and proved an average accuracy of 98% using Random Forest. Though the author attained remarkable results, CIC-Darknet

2020 holds a high level of class imbalance problem which is ignored. One of the other essential preprocessing steps is feature extraction. Researchers on the field of Darknet classification inhibited different approaches on different datasets, say, the coefficient score approach [3] and chi-squared statistical score [4] on the CIC-Darknet 2020 dataset. And the class imbalance problem is handled by using the Generative Adversarial Networks (GAN) as an oversampling method [5] on three different NSL-KDD, UNSW-NB15, and CICIDS-2017 datasets. The GAN-based minority class oversampling approach is purely dependent on random input. This can be handled by the use of VAE, which learn on individual class distribution in the dataset for oversampling. Further, the autoencoder model serves the purpose of feature extraction, by compressing the data using the encoder.

On considering the importance of preprocessing steps and drawbacks of the proposed approaches, an AE model combined with AAE is proposed for feature extraction and for handling class imbalance problems.

3 Proposed Methodology

The entire proposed system to classify Darknet is provided as a flow diagram in Fig. 1. After the training and testing dataset creation, the significant features of the training dataset are extracted using the AE model. The minority class training dataset is further oversampled using the AAE. Finally, the generated minority class oversampled dataset is augmented with the original training dataset. Thereby the class imbalance problem is resolved. Thus, the classification of ML using well-balanced dataset results is observed. Figure 1 shows the flow of execution.

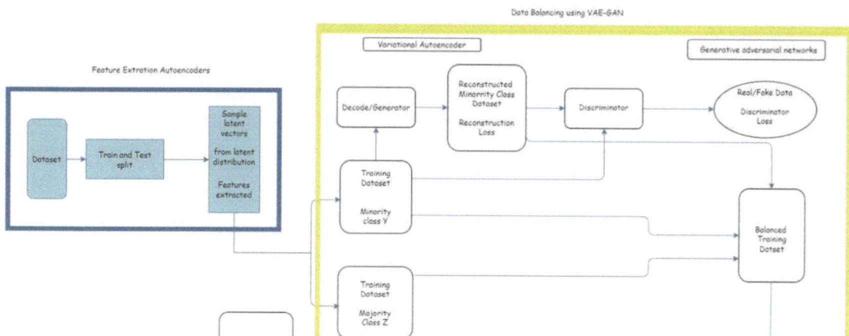

Fig. 1 Flow of execution

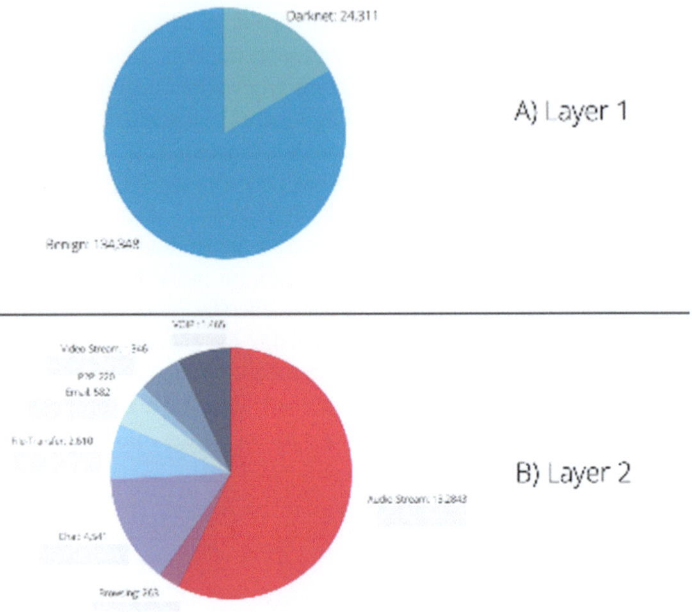

Fig. 2 Dataset pie chart

3.1 Dataset

At the first layer, a two-layered technique is employed to produce benign and Darknet traffic using the CICDarknet2020 dataset. The second layer of the Darknet traffic is made up of Audio-Stream, Browsing, Chat, Email, P2P, Transfer, Video-Stream, and VOIP. The dataset providers merged the respective VPN and Tor traffic in corresponding Darknet categories to build the representative dataset by combining our previously developed datasets, namely, ISCXTor2016 and ISC VPN 2016 and combining the respective VPN and Tor traffic in corresponding Darknet categories [6]. Figure 2 shows the dataset in pie chart. Figure 3 shows the different types of Darknet traffic and the programs that are used to produce it.

3.2 Data Preprocessing

The dataset is preprocessed to check missing values, noisy data, and different inconsistencies before executing it to the algorithm. Whereas label encoding converts the labels into a numeric value to be able to convert them into machine-readable. ML algorithms can then determine in a higher manner how the ones labels should be operated. Grouping in Benign and Darknet, before everything there have been four unique values which include Non-Tor, Non-VPN, Tor, and VPN have been grouped

Traffic Category	Applications Used
Audio-Stream	Vimeo and Youtube
Browsing	Firefox and Chrome
Chat	ICQ, AIM, Skype, Facebook and Hangouts
Email	SMTPS, POP3S and IMAPS
P2P	uTorrent and Transmission (BitTorrent)
Transfer	Skype, FTP over SSH (SFTP) and FTP over SSL (FTPS) using Filezilla and an external service
Video-Stream	Vimeo and Youtube
VOIP	Facebook, Skype and Hangouts voice calls

Fig. 3 Dataset traffic category

into two subclasses as benign and Darknet. The first two classes, namely, non-tor and NordVPN as benign and subsequent two classes tor and VPN as Darknet [6].

3.3 *Dimensionality Reduction*

Dimensionality reduction reduces the number of dimensions in data, by removing less valuable elements (feature selection) or by translating the data into lower dimensions (feature extraction). Overfitting is avoided by dimensionality reduction. For this purpose, AE has a dimensionality reduction approach [7]. AE is an unsupervised artificial neural network that compresses data into lower dimensions (bottleneck layer or code) and then decodes the data to recreate the original input. The compressed representation of the input data is stored in the bottleneck layer (or code). The data should be scaled between 0 and 1 before being sent into the AE, using MinMaxScaler. MinMaxScaler divides by the range after subtracting the feature's minimal value as shown in Eq. (1):

$$x_{\text{scaled}} = \frac{x - x_{\min}}{x_{\max} - x_{\min}} \tag{1}$$

The entire process is observed as feature extraction [8].

3.4 Minority Oversampling Using Adversarial Autoencoders

AAE blends the AE architecture with the adversarial loss concept introduced by GAN. The encoded value is used as an input to both the GAN discriminator model and the decoder. AAE works in two phases: reconstruction phase and regularization phase. This study starts with training the VAE (encoder and decoder) in the reconstruction phase of the learning process [9]. The discriminator further classifies the real data and the fake data. The ultimate aim of AAE model is to reconstruct minority class data points more similar to that of original data using the VAE (encoder and decoder). This obviously leads to a decreased reconstruction loss in the reconstruction phase. This reconstruction is regularized by means of GAN (discriminator), such as the discriminator fails to classify the fake data from real data. This in turn leads to increased discriminator loss, proving that the minority class data points created for oversampling is more similar to the original data points.

3.5 Traffic Classification

The term “traffic classification” refers to the process of analyzing and identifying the type of network traffic class. DT, NB, RF, and LR were among the machine learning methods employed [10]. For better performance, Apache spark has been used. Apache Spark is a free and open-source unified analytics engine for processing huge amounts of data. Spark is a programming interface for clusters that has implicit data parallelism and fault tolerance [11].

4 Results and Discussion

4.1 Performance Metrics

The evaluation metrics considered are based on the confusion matrix as tabulated in Table 1 based on confusion matrix.

Confusion Matrix: A $n \times n$ square matrix represents the number of instances correctly or incorrectly classified in n classes.

- *True positives (TP)*: The number of normal class instances that are correctly classified as normal by the classifier.
- *False Negatives (FN)*: The number of normal class instances that are incorrectly classified as an attack by the classifier.
- *False Positives (TP)*: The number of attack class instances that are incorrectly classified as a normal class by the classifier.

Table 1 Evaluation metrics

Evaluation metrics	Formula
Accuracy	$\frac{TP+TN}{TP+FP+TN+FN}$
Precision	$\frac{TP}{TP+FP}$
Recall	$\frac{TP}{TP+FN}$
F1-score	$2 * \frac{Precision \cdot Recall}{Precision+Recall}$
False Positive Rate (FPR)	$\frac{FP}{TN+FP}$
False Negative Rate (FNR)	$\frac{FN}{TP+FN}$
False Alarm Rate (FAR)	$\frac{FPR+FNR}{2}$

- *True Negatives (TN)*: The number of attack class instances that are correctly classified as an attack class by the classifier.

4.2 Experimental Setup

The proposed system is built on Anaconda 3 using Jupyter notebook. Pandas- and Sklearn-based libraries serve the purpose of preprocessing. Keras- and PyTorch-based autoencoder and AAE model are built for feature extraction and oversampling. Added to that the PyTorch is enabled using the CuDNN. CuDNN is a high-performance GPU accelerator used by deep learning researchers and framework developers all over the world. Instead of spending time on low-level GPU performance optimization, they may focus on training neural networks and designing software applications. And at last, the ML classification on the cloud computing resource Apache spark. This provides the MLlib a package to carry out ML classification tasks.

4.3 Evaluation on Dimensionality Reduction

The phase is to evaluate the performance of a machine learning classifier while working with all the 82 features of the original dataset versus the significant features extracted using autoencoders.

The dataset is processed through the autoencoder to train with minimal loss. Then the encoder model is alone executed for a single epoch to extract significant features, which is then converted as a dataframe. The autoencoder's model for feature extraction is built with three layers of neurons on each layer as 32,16,7. The encoder output is the significant features extracted. Further to evaluate its performance impact on classification, different machine learning classifiers and their metrics are tabulated in Table 2. Time is the predominant factor that portrays the impact of significant feature processing. From the table, it is observed that on using the significant feature extract time consumption is constantly reduced on using feature extraction. Table 2

Table 2 Classification with and without feature extraction

Algorithm	With all 82 features					With significant feature extract				
	Accuracy	Pre	Recall	F1-score	Time (s)	Accuracy	Pre	Recall	F1-score	Time (s)
DT	0.93	0.90	0.86	0.88	8.0	**0.82**	0.41	0.50	0.45	**3.6**
RF	0.80	0.0	0.0	0.45	8.4	**0.97**	0.41	0.50	0.45	**3.6**
NB	0.58	0.58	0.64	0.53	8.2	**0.82**	0.41	0.50	0.45	**3.6**
LR	0.82	0.41	0.50	0.45	7.5	0.82	0.41	0.50	0.45	**3.6**

shows the accuracy improved to 0.97 on Random Forest comparatively to the without feature extraction model accuracy (0.80).

4.4 Evaluation on Adversarial Autoencoder-Based Class Imbalance Solution

The term "generative models" refers to a type of statistical model that can create new data instances. The learning algorithms prefer to represent the underlying patterns or distribution of the data points, whereas the generative models focus on the distribution of specific classes in a dataset. These models employ the concept of joint probability to generate situations in which a given feature (x) or input and the intended output or label (y). One such model is AAE. The AAE model was created to resolve class imbalance issues and generate minority class data points. The model is fed only to the train dataset with the minority class Darknet as input. Figure 4 describes the class distribution is balanced after oversampling using AAE.

The encoder model is constructed using three fully connected layers, with the following number of neurons per layer: "8,4,2". The decoder network is designed to symmetrically replicate the encoder network's architecture by inverting the encoder network layers "2,4,8" layer by layer. The encoder and the decoder models are built with the three layers having a sigmoid activation function at the last layer. The

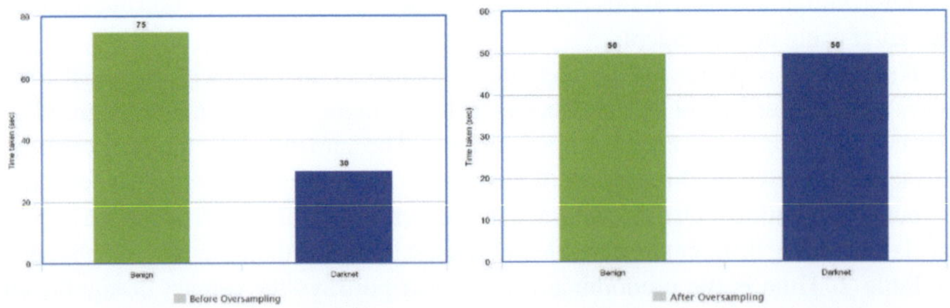

Fig. 4 Before and after oversampling

challenge of negotiating a negative gradient neuron is resolved by each layer with the activation function of LeakyRelu at first two layers. LeakyRelu with a negative slope of 0.5 is used. Finally, the discriminator has four fully connected layers with neurons "8,4,2", respectively.

Thus, the built AAE model is trained using the reconstruction and regularization phases. The encoder is trained with the latent distribution of minority class Darknet dataset samples during the reconstruction phase. The encoder's output is then fed as input to the decoder to reconstruct the data with minimum reconstruction error. The mean-squared reconstruction error is used to calculate the error (MSE). In the regularization phase, the discriminator and encoder model is trained. From the original minority class data, a latent space is generated using random sampling from the Gaussian distribution. The sampled input is fed to the encoder, which generates fake data. Finally, the discriminator is trained to classify the real and fake data, such that the discriminator loss is high. For this binary cross entropy is used. The entire training process was carried out using epoch 100 and batch size 127.

After the training phase, to collect the reconstructed oversampled minority class dataset a single epoch and batch size equal to the entire dataset is used. The class distribution on the dataset obtained using AAE is now well balanced when compared to the imbalanced original train dataset as shown in Fig. 4.

While evaluating class imbalance using four machine learning classifiers, Table 3 shows that the majority of class instances have been correctly classified as benign, whereas all the Darknet class instances have been wrongly classified as benign. This naturally leads to increased accuracy and decreased F1-score, precision-recall rate as shown in Table 4. Thus, the accuracy is not always the right metric, in the case of class imbalance problem is hence proved here. This misclassification of Darknet as benign has led to an increased FPR rate of 1.0 for the benign class and an increased FNR rate of 1.0 for the Darknet class. This in turn led to a 0.5 FAR rate, which is to be closer to 0.0. On the other hand, while evaluating without class imbalance all the three misclassification rates FPR, FNR, and FAR are exactly zero, which is the optimal required value.

From Table 4, it shows that the Accuracy, Pre, Recall, and F1-score for the classification without class imbalance rates are maxed to 1.0 which is 100% compared to the classification with class imbalance.

4.5 *Evaluation on Traffic Classification Using Apache Spark*

The traffic classification using Apache spark big data frameworks for parallel computing that offers a combination of in-memory processing, speed, and ease of programming is represented. ML being more iterative, Apache Spark. Spark can store large datasets in cluster memory, paging from the disk as needed, and run a variety of machine learning algorithms without having to sync multiple times to the disk, resulting in 100 times faster performance. The results of using Apache spark for classification are shown in Fig. 5 with respect to time.

Table 3 Misclassification with and without class imbalance

Algorithm	Class label	Misclassification with class imbalance			Misclassification without class imbalance		
		FPR	FNR	FAR	FPR	FAR	FPR
DT	0 (Benign)	1	0	0.5	0	0	0
	1 (Darknet)	0	1	0.5	0	0	0
RF	0	1	0	0.5	0	0	0
	1	0	1	0.5	0	0	0
NB	0	1	0	0.5	0	0	0
	1	0	1	0.5	0	0	0
LR	0	1	0	0.5	0	0	0
	1	0	1	0.5	0	0	0

Table 4 Classification with and without class imbalance

Algorithm	Classification with class imbalance				Classification without class imbalance			
	Accuracy	Pre	Recall	F1-score	Accuracy	Pre	Recall	F1-score
DT	0.82	0.41	0.50	0.45	1.0	1.0	1.0	1.0
RF	0.80	0.41	0.50	0.45	1.0	1.0	1.0	1.0
NB	0.82	0.41	0.50	0.45	1.0	1.0	1.0	1.0
LR	0.82	0.41	0.50	0.45	1.0	1.0	1.0	1.0

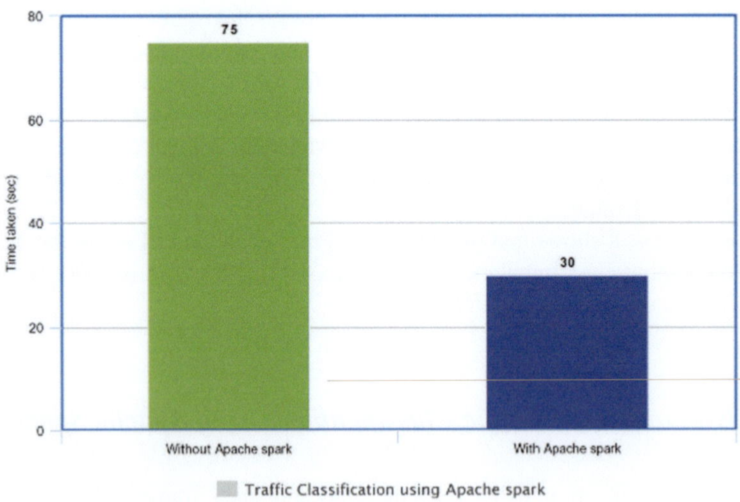

Fig. 5 Classification with and without Apache spark

From Fig. 5, it shows that time consumption is less on using an Apache spark framework when compared to the classification without using Apache spark.

5 Conclusion and Future Work

The AI model proposed for Darknet classification using the privileges of both ML and DL is examined. The proposed method has solved the problem of target variable distribution imbalance, and a suitable subset of features has been chosen. In terms of evaluation measures, the model described here combining VAE and GAN, DL-based model, and ML for classification performed well. As a future work, the proposed AI model will be implemented for a real-time application. To further encounter the real-time challenges, we aim to make the system more efficient.

References

1. Hu Y, Zou F, Li L, Yi P (2020) Traffic classification of user behaviors in Tor, I2P, ZeroNet, Freenet. In: 2020 IEEE 19th international conference on trust, security and privacy in computing and communications (TrustCom), pp 418–424
2. Iliadis LA, Kaifas T (2021) Darknet traffic classification using machine learning techniques. In: 2021 10th International conference on modern circuits and systems technologies (MOCAST), pp 1–4
3. Al-Haija A, Qasem MK, Elhaija WA (2022) Machine-learning-based darknet traffic detection system for IoT applications. Electronics 11(4):556. https://doi.org/10.3390/electronics11040556
4. Sridhar S, Sanagavarapu S (2021) DarkNet traffic classification pipeline with feature selection and conditional GAN-based class balancing. In: 2021 IEEE 20th international symposium on network computing and applications (NCA), 2021, pp 1–4
5. Liu X, Li T, Zhang R, Wu D, Liu Y, Yang Z (2021) A GAN and feature selection-based oversampling technique for intrusion detection. Secur Commun Netw 2021:15 p, Article ID 9947059
6. Dataset Homepage: https://www.unb.ca/cic/datasets/darknet2020.html
7. Chapaneri R, Shah S (2022) Enhanced detection of imbalanced malicious network traffic with regularized generative adversarial networks. J Netw Comput Appl 103368
8. Lan J, Liu X, Li B, Li Y, Geng T (2022) DarknetSec: a novel self-attentive deep learning method for darknet traffic classification and application identification. Comput Secur 116:102663
9. Makhzani A, Shlens J, Jaitly N, Goodfellow I, Frey B (2015) Adversarial autoencoders. arXiv preprint arXiv:1511.05644
10. Nguyen TTT, Armitage G (2008) A survey of techniques for internet traffic classification using machine learning. IEEE Commun Surv Tutor 10(4):56–76. https://doi.org/10.1109/SURV.2008.080406
11. Biswas R, Fidalgo E, Alegre E (2017) Recognition of service domains on TOR dark net using perceptual hashing and image classification techniques. In: 8th International conference on imaging for crime detection and prevention (ICDP 2017). IET, pp 7–12
12. Guan J, Cai J, Bai H, You I (2021) Deep transfer learning-based network traffic classification for scarce dataset in 5G IoT systems. Int J Mach Learn Cybern 12(11):3351–3365

Using Stacking Ensemble Method for Rental Bike Prediction

S. Akashdeep, Akshith Nettar Mahalinga, R. Harshvardhan, Sunil Chinnahalli KomariGowda, and Nagamma Patil

Abstract Rental bike platforms that improve mobility comfort are on the rise in major cities worldwide. One of the essential requirements for these rental bike systems is that bikes are available to end users at the specified time, reducing waiting time. Increased waiting time indicates that movement has been halted, implying that more efficiency can be gained. As a result, the city's main priority is ensuring a steady supply of bicycles. It's crucial to be able to forecast the number of bikes needed at each hour for this. This work look at alternative models for forecasting the bike count per hour needed to maintain a steady supply of bikes. Weather data (Temperature, Humidity, Wind speed, Dew point), the quantity of bikes hired every hour, and time information are all used to train the models. Filtering can also be used to exclude non-predictive parameters and rank features based on how well they predict outcomes. The effectiveness of the regression model was assessed using a testing set after they had been trained using repeated cross-validation. For the model Gradient Boosting Machine, the optimum $R2$ value is 0.96. The most significant predictors are also determined, as well as their relationships. Bike-sharing demand, data mining, predictive analytics, public bikes, regression.

Keywords Bike sharing · Data mining · Predictive analytics · Public bikes · Regression

S. Akashdeep · A. N. Mahalinga · R. Harshvardhan · S. Chinnahalli KomariGowda (✉) · N. Patil
Information Technology Department, National Institute of Technology Karnataka, Surathkal, India
e-mail: sunilchinnahalli@gmail.com

N. Patil
e-mail: nagammapatil@nitk.edu.in

S. Chinnahalli KomariGowda
Computer Science and Engineering, Indian Institute of Information Technology Dharwad, Dharwad, India

S. K. Biswas et al. (eds.), *Intelligent Computing Systems and Applications*, Lecture Notes in Networks and Systems 1307, https://doi.org/10.1007/978-981-96-3860-4_3

1 Introduction

Bike-sharing networks work on the principle that customers can rent a bike from one network station and then return it to another station for a small charge. Due to the rise of initiatives to reduce pollution and raise public awareness about health, bike sharing has gotten much attention in recent years. Bike-sharing encourages people to ride their bikes, reduces traffic, and helps to build a cycling community.

After efficient tracking strategies were devised in the 1960s, the boom in bike-sharing systems occurred [1]. The fundamental advantage of bike-sharing over bike-renting is that a user can return a bike to another station after renting it from one station. This improves movement while also assisting other users. To make use of the benefits of a bike-sharing platform, one must first sign up as a member and then utilize the bikes for a fee. Many of the most modern bike-sharing programs employ smartphones to automate the process.

This work aims to use machine learning models to forecast the number of bikes needed at any given hour based on past data. There has been some research into the impacts and correlations between weather and climate conditions and cycling. Pucher et al. [2] found that most US cities with mild winters and little rain had high cycling rates when compared to places with excessive heat and moisture. Precipitation and temperature levels were statistically relevant in Buehler's study [3], and were linked to reduced cycling rates. This shows how important weather data is in affecting riding patterns.

2 Related Work

Many bike-sharing platforms have embraced Multiple Linear Regression as one of their approaches. Random Forest algorithms have also been used to forecast bike demand on an hourly basis [4]. In [5], historical cycling data were combined with a spatiotemporal mobility model and employed a filter model with a data-driven Graph Convolutional Neural Network. It successfully predicted per station demand by learning pairwise hidden correlations between stations.

Zhang et al. [6] used historical bicycle data to anticipate demand using spatiotemporal approaches. Spatial-temporal data are used to generate more relevant information. A sub-hour roughness model was built at the station level. Data analysis and visualization systems were also developed to demonstrate the concept and improve mobility prediction. This was used to estimate how many bikes would be needed per docking station in a large metropolis. To increase the efficacy and training components of the prediction task, the regressor may recognize external factors that influence the process and factor in some data that tackles imbalance concerns and also helps capture any non-linearity of the spatiotemporal information. This is performed by utilizing expected rent demands. To anticipate bike rent and return in different regions of the city, a model is created utilizing past data, time-related data, and weather-related information [7]. Satishkumar et al. [8] used weather information to determine hourly bike demand. The authors applied simple machine learning algorithms; however, the authors did not consider the impact of previous bike demand.

To create a data-based network, a community-based detection method is utilized, and the most significant demand for shared bikes is discovered in two communities. A two-layer Long-Short-Term Memory (LSTM) model is utilized to forecast bike rent and return, employing an LSTM gating mechanism and the ability to handle RNN-based sequence information, using information collected from stations in both communities as a data set. The use of meteorological data for hourly bike demand prediction was proposed by Sathishkumar et al. [8]. Simple machine learning methods were deployed, but no prior bike demand data was used.

3 Proposed Methodology

The proposed approach combines two datasets of weather and rented bike count data; it is described in Fig. 1. Applied data preprocessing, feature engineering techniques, and feature selection methods, and trained the stacking ensemble model to predict the rented bike count on a given day.

3.1 Dataset

This work uses two datasets in our project, namely the rented bike count dataset and the respective weather information dataset. The first dataset consists of the count of rented bikes in the city of Seoul, and the frequency is hourly count, and the data is collected for a year from December 2017 to December 2018. The dataset is open to access and can be downloaded from the Seoul Public Data Park website of South Korea.

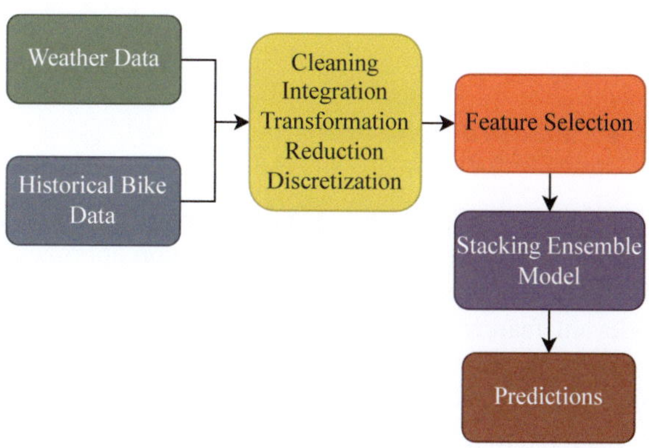

Fig. 1 Flow diagram

The second dataset contains weather information for the first dataset corresponding days, including weather information such as temperature, visibility, etc. The dataset's data points are gathered at an hourly rate. The training set has 65710 variables and 16 variables, while the testing set has 21890 variables and 16 variables.

3.2 Data Preparation

The rented bike count and weather information datasets are merged based on the timestamp of the data points in both datasets. The continuous features in the dataset are normalized to the range [0, 1], while the categorical characteristics are stored as a single hot vector. The outliers in the rented bike count are found using the interquartile range and removed from the dataset. To get additional information out of the timestamp, applied feature engineering approaches and further derived the week status, i.e., is the day a weekend or weekday, and the day as well. Based on the timestamp, the data on holidays in South Korea is gathered and added to the dataset. We also add the season data, i.e., Autumn, Spring, Summer, or Winter, which is derived from the timestamp. The functional and nonfunctional days of the rental bikes are also added to the dataset.

3.3 Exploratory Data Analysis

The data were subjected to exploratory data analysis, providing some patterns. Temperature, wind, visibility, dew, and sunshine all have a positive link with the number of bikes. The count has a negative correlation with humidity, rain, and snow, respectively. This shows that the number of bikes rented during rain or snowfall decreases. We also found that the average rented bike count is higher in the summer when compared to the winter season. The average rented bike count is higher for the time between 8 and 18 h on the weekdays in all seasons, as this time is the peak hours in the city.

3.4 Feature Selection

The dataset was left with 32 features after data preprocessing and feature engineering. Used a Zero Importance Feature Selection approach to discover the best features. Features that aren't important are deleted using this strategy. In order to find the relevant features, a tree-based model was employed using the gradient boosting concept that conducts regression tasks on the data and determines the relevance of the features. The features that are not used to split any of the nodes in the tree are considered to be irrelevant and are eliminated. The rest of the features are selected.

3.5 Ensemble Model

The proposed approach uses a stacking ensemble model to predict the rental bike count based on the given features. Using an ensemble model can avoid the model becoming stuck in local optima. The ensemble model is shown in Fig. 2. Because local searches are done from different points, having a large number of models reduces the chances of this happening. A combination of hypotheses is also provided via ensemble learning, which increases the possibility of presenting a decent approximation of the actual hypotheses. As base layer models, employed Random Forests, Neural Networks, LSTM, and Holt-Winters technique, and as an ensemble, this work used an extreme gradient boosting model. The ensembler is fed predictions from the base layer models, which will be the inputs.

Random Forests. Random forests are ensemble models that combine the predictions of many decision trees using the bagging strategy. The outputs of all the weak learners are averaged to get the ensemble output, i.e., the final output. All decision trees are given the same weight. It employs bootstrap sampling to choose training samples and the features to train the weak learners in random forests. The best hyper-parameters for the random forest were discovered using Bayesian optimization, and they are listed below (Table 1).

Neural Networks. An artificial neural network with three layers is used. The ANN's input layer contains 16 neurons, each corresponding to one of the dataset's 16 features. The ReLu activation function is used in the hidden layers, while linear activation is used in the output layer. The buried layer of a neural network assists in the learning of more abstract input representations. It simulates non-linear dependencies between the input and output better than a shallow neural network. On the other hand, deep neural networks are prone to overfitting and poor generalization. To tackle this problem, we use Bayesian Optimisation. Given a set of values for

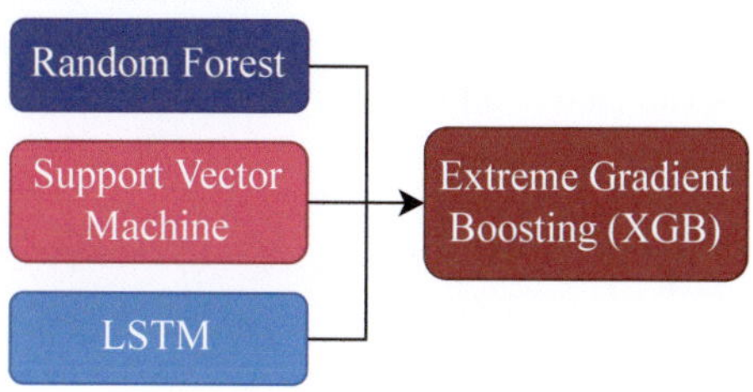

Fig. 2 Ensemble model

Table 1 Parameters for Random Forests

Parameter	Value
Estimators	320
Max leaf nodes	230
Min samples split	3
Min samples leaf	3

the model's hyper-parameters, Bayesian Optimisation attempts to discover values for these hyper-parameters that minimize the ANN model's losses. It is an iterative process in which the optimizer comes closer to the optimum values for the hyper-parameters with each iteration. By selecting a random sample of hyper-parameters, the Bayesian optimizer trains the model. We build a Gaussian regressor with the validation error as the target using a random sampling of hyper-parameters. To choose the next group of samples for the next iteration, a Gaussian regressor-based acquisition function is generated. It finds the sample that is most likely to produce the best outcome using the regressor's mean and variance. The acquisition function's global optima are utilized to generate the next set of hyper-parameter values to train the model. This process is done iteratively until the optimal hyper-parameters are discovered.

LSTM. LSTM is one of the base layer models we use. LSTM is distinguished from an RNN, a normal machine learning model, or a neural network by the concept of memory. The model is made up of an LSTM input layer followed by another LSTM layer with 30 cells and a dense network (Table 2).

Holt-Winters Method. To check the outcomes with simply the timestamp feature, the Holt-Winters time series forecasting method was used. The Holt-Winters approach uses exponential smoothing methods to account for seasonality and trends in the data. The data is assumed to be level in the Simple Exponential Smoothing (SES) update. The trend in the data is accounted for via Holt's Exponential Smoothing (HES). The Winter's Exponential Smoothing (WES) method adjusts for the data's seasonality.

Extreme Gradient Boosting (XGB). To create an ensemble of trees, XGB employs gradient descent. To limit the risk of overfitting, XGB employs a more regularized model. Dropout regularization for regression trees (DART) is implemented in XGB [8]. As a result, XGB is able to generalize well and has a low bias and variation. XGB can be parallelized to use a multi-core processor or a GPU to drastically improve the computation speed. It also makes use of the processor cache for better performance. Using Bayesian optimization, we have found the optimal hyper-parameters for the model, and they are as follows (Table 3).

Table 2 Parameters for LSTM

LSTM input layer	22 cells
LSTM layer 1	30 cells
Dense layer	16 nodes
Output layer	1 node
Learning rate	0.01
Optimizer	Adam

Table 3 Parameters for XGB

Estimators	220
Learning rate	0.05
Max depth	5
Min node weight	0.004
Min node samples	27
Num leaves	42

4 Results

4.1 Metrics

Coefficient of Determination (R2), Root Mean Square Error (RMSE), Mean Absolute Error (MAE), and Coefficient of Variance (CV) are the metrics utilized to analyze the results of our model. The higher the coefficient of determination, the closer the prediction is to the truth. It has a range of 0–1. Equation (1) is the formula for the coefficient of determination. Root Mean Square Error is the standard deviation between the anticipated value and the ground truth. It's a scale-dependent statistic that determines how far apart the expected and actual values are. The formula for RMSE is shown in Eq. (2).

$$R^2 = 1 - \sum_{t=1}^{n} \frac{e_t^2}{|y_t + \bar{y}_t|} \tag{1}$$

$$\text{RMSE} = \sqrt{\frac{1}{n} \sum_{t=1}^{n} e_t^2} \tag{2}$$

The average difference between the anticipated outcome and the ground truth, as indicated in Eq. (3), is the mean absolute error. The relative dispersion of the prediction around the mean is indicated by the coefficient of variation. The formula for CV is given in Eq. (4).

$$\text{MAE} = \frac{1}{n} \sum_{t=1}^{n} |e_t| \tag{3}$$

$$\text{CV} = \frac{\text{RMSE}}{\bar{y}_t} * 100. \tag{4}$$

4.2 Analysis

As seen in Fig. 3, the RMSE is compared with the model's results when the weather data is not used. The RMSE value is lesser for both the Artificial Neural Network and the Random Forest, which proves that the weather data plays an important role in the prediction. A similar analysis was done to check if feature engineering would improve our results. It has a significant difference as all the models perform better with a lower RMSE score, as seen in Fig. 4.

The time window is the period that the model uses to predict the currently rented bike count. If X is the time window, then the model gets the rented bike counts based on the last X hours. To decide the right time window, analyze with the base layer models as well as the Ensemble model. As seen in Fig. 5, the RMSE value decreases as we increase the time window. It plateaus after the time window = 6 h, and this was the value that we chose.

This work compared the outcomes of the Artificial Neural Network (ANN), Random Forest model, Long Short-Term Memory model, and Ensemble model (XGB) as constructed to the results of the base paper models. As basic models, ANN and LSTM outperform Linear Model, Support Vector Machine, and Gradient Boosting Machine. It creates a higher Coefficient of Determination score, indicating that it is closer to the ground truth, as well as lower RMSE, MAE, and CV, indicating that the discrepancy between predicted and ground truth values is smaller.

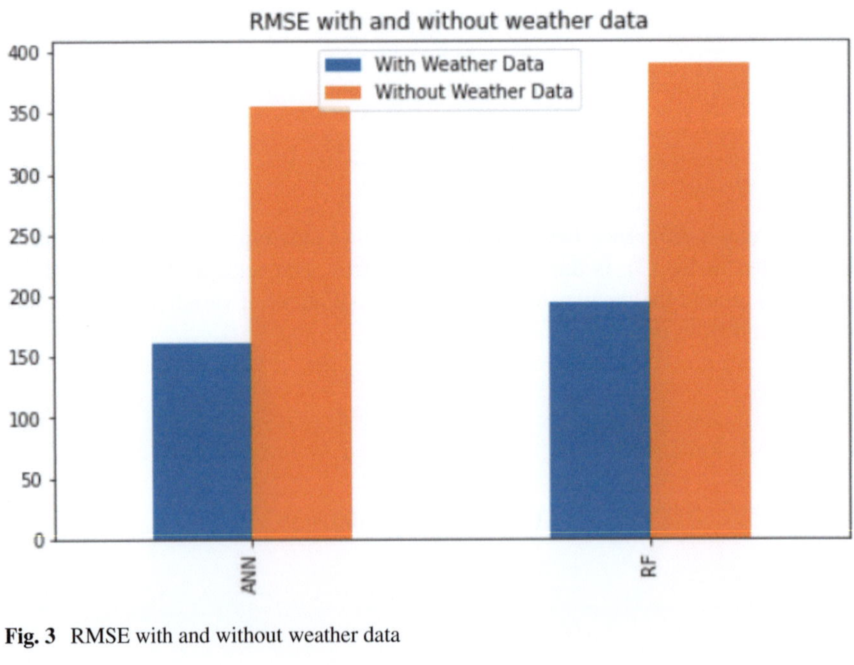

Fig. 3 RMSE with and without weather data

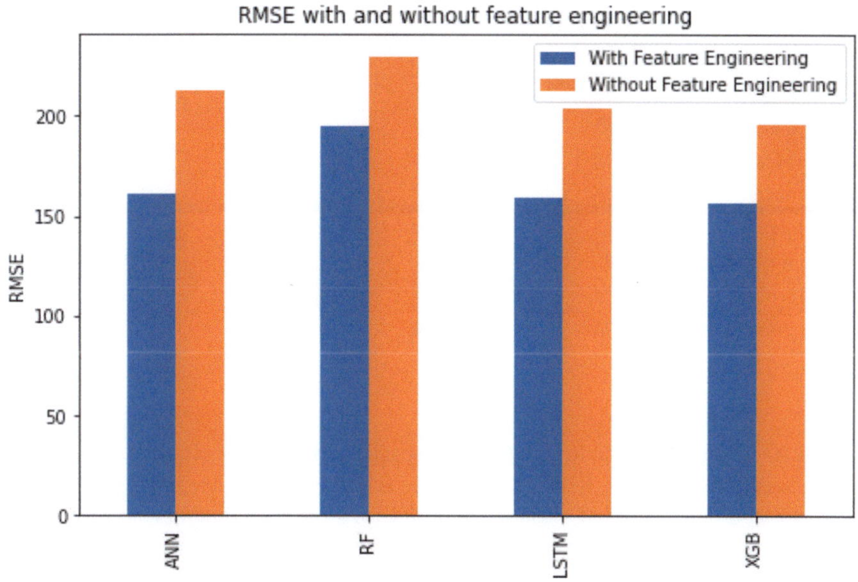

Fig. 4 RMSE with and without feature engineering

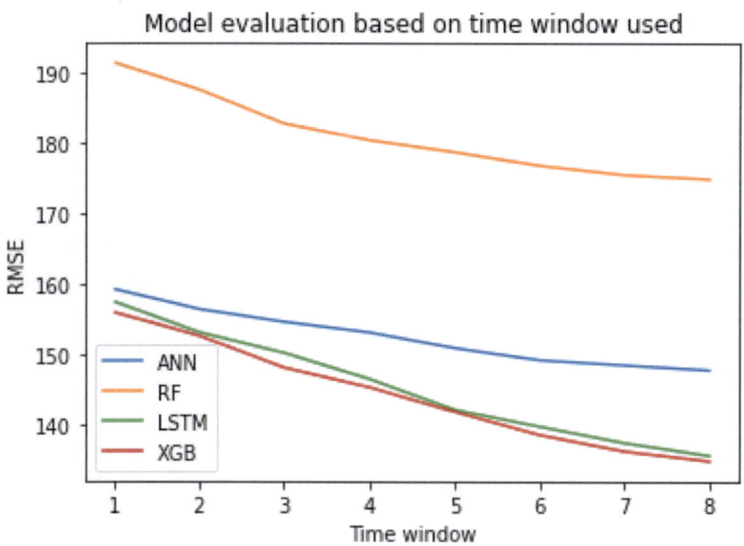

Fig. 5 Model evaluation based on time window used

Table 4 Metrics comparison for various methods

	R^2	RMSE	MAE	CV
LM (base paper)	0.55	427.71	322.32	61.03
SVM (base paper)	0.85	241.94	151.21	34.52
GBM (base paper)	0.92	172.73	109.78	24.64
Holt winters	0.56	421.52	318.29	59.73
ANN (ours)	0.93	161.29	102.14	22.74
Random Forest (ours)	0.9	194.61	123.52	25.97
LSTM	0.94	159.17	99.27	22.13
XGB (ensemble)	**0.94**	**156.82**	**97.15**	**21.74**

The XGB Ensemble model provides the best results with a Coefficient of Determination score (R2) of 0.94, Root Mean Square Error of 156.82, Mean Absolute Error of 97.15, and Coefficient of Variance of 21.74 for a time window of 6 h (Table 4).

5 Conclusions

The proposed work merged bike rental and weather data in this project because our analysis suggested that weather data was important. This work used zero importance-based feature selection to construct significant features and Bayesian optimization to fine-tune hyper-parameters. The Artificial Neural Network, Random Forest, and LSTM were trained as the ensemble's base layer models. Finally, choose the XGB model for our Ensemble model.

The Ensemble model outperformed all other models, indicating that it is the most effective. The linear dependence instances were located using random forests. The non-linear dependence cases were successfully modeled using the Neural Network. Long-term memory remembered the series' pattern. As a result, each model addresses the flaws of the others while also complementing them. The XGB model accurately integrates the predictions of the weak learners by assigning different priorities to each base layer model depending on the input. In comparison to methods in the literature on the same dataset, the presented solutions produce much superior outcomes.

Future work could include evaluating the proposed approach's performance in a variety of domains, such as agriculture commodity forecasting, stock price forecasting, and so on. In the future, an experimental investigation could use a hierarchical ensemble technique.

References

1. DeMaio P (2009) Bike-sharing: history, impacts, models of provision, and future. J Pub Transp 12(4):3
2. Pucher J, Komanoff C, Schimek P (1999) Bicycling renaissance in North America?: recent trends and alternative policies to promote bicycling. Transp Res Part A: Policy Pract 33(7–8):625–654
3. Pucher J, Buehler R (2006) Why Canadians cycle more than Americans: a comparative analysis of bicycling trends and policies. Transp Policy 13(3):265–279
4. Feng Y, Wang SA (2017) forecast for bicycle rental demand based on random forests and multiple linear regression. In: 2017 IEEE/ACIS 16th international conference on computer and information science (ICIS). IEEE, pp 101–105
5. Yang Z, Hu J, Shu Y, Cheng P, Chen J, Moscibroda T (2016) Mobility modeling and prediction in bike-sharing systems. In: Proceedings of the 14th annual international conference on mobile systems, applications, and services, pp 165–178
6. Zhang L, Tang S, Yang Z, Hu J, Shu Y, Cheng P, Chen J (2016) Data analysis and visualization in bike-sharing systems. In: Proceedings of the 14th annual international conference on mobile systems, applications, and services companion, pp 128
7. Pan Y, Zheng RC, Zhang J, Yao X (2019) Predicting bike sharing demand using recurrent neural networks. Procedia Comput Sci 147:562–566
8. Sathishkumar VE, Park J, Cho Y (2020) Using data mining techniques for bike sharing demand prediction in metropolitan city. Comput Commun 153:353–366

Efficacy of Recurrent Neural Networks in Rainfall Forecasting: A Case Study of Shillong Weather Data

Sudipta Mandal, Saroj Kumar Biswas, and Biswajit Purkayastha

Abstract Rainfall prediction is a noteworthy and challenging task in today's world. Due to the nonlinear and dynamic behavior, advanced computer modeling is required to predict rainfall accurately, in lieu of unreliable statistical techniques. In recent days, Machine Learning (ML) such as Deep Learning (DL) and Artificial Neural Network (ANN) enables self-learning to build a data-driven module for a time-series dataset. This paper proposes a DL model called "Short-term Data-Driven Rainfall Forecasting Model" (SDDRFM) which uses Long Short-Term Memory (LSTM)-based rainfall prediction model for weather and rainfall forecasting in the region of Shillong, a city in Meghalaya. In this paper, a rainfall forecasting model SDDRFM is proposed with LSTM algorithm using the rainfall dataset of Shillong collected from NASA power grid data, and is compared with other ANN models such as Gated Recurrent Unit (GRU), Convolutional Neural Network (CNN), and Feedforward Neural Networks (FFNN). The model uses dominant features such as humidity, temperature, wind speed and has achieved an accuracy of 85.73% using LSTM which is better than other ML models.

Keywords Rainfall prediction · ANN · DL · LSTM

1 Introduction

Rainfall remains one of the most influential meteorological parameters in many aspects of our day-to-day life. Rainfall is a form of precipitation where the water is dispersed within environment, river, oceans, and earth's surface. It holds the majority

---

S. Mandal (✉) · S. K. Biswas · B. Purkayastha
Computer Science and Engineering, National Institute of Technology, Silchar, Assam, India
e-mail: sudipta_pg@cse.nits.ac.in

S. K. Biswas
e-mail: saroj@cse.nits.ac.in

B. Purkayastha
e-mail: biswajit@nits.ac.in

S. K. Biswas et al. (eds.), *Intelligent Computing Systems and Applications*, Lecture Notes in Networks and Systems 1307, https://doi.org/10.1007/978-981-96-3860-4_4

of the world's freshwater for plants and animals. Water evaporates as vapor from the earth surface's water bodies into the atmosphere, which is carrying heat from the surface with it. The water vapor rises into atmosphere, and after condensation it forms cloud and releases rainfall [1]. As India produces and contributes a maximum amount of crops, rainfall estimation has become a great priority. Prior knowledge of rainfall is needed to save and grow crops, to fight against flood and drought.

On a global scale, numerous attempts have been undertaken by various researchers to reliably estimate rainfall using various methodologies which is starting from numerical weather prediction (NWP) models and statistical models. But these models had several drawbacks in every cases. Because of the dynamic nature of climate and the nonlinear nature of rainfall data, numerical and statistical approaches cannot be used to forecast long-term rainfall [2]. Artificial Neural Network (ANN) becomes an appealing inductive approach in rainfall prediction as it owns highly nonlinearity, flexibility, and data-driven learning in building models using prior knowledge about catchment behavior and flow processes. That is why ANNs have been successfully used in recent days in various aspects of science and engineering. Deep Learning (DL) technologies differ from neural networks in that the models are altered in response to the growth of hidden layers, resulting in improved performance.

This paper has focused on Shillong [3], a city of the state Meghalaya of India, which is considered famous for climate changing with its altitude. The Khasi and Jaintia Hills' climate is very bracing and pleasant which is neither too cold in winter season nor too warm in summer season, but the climate of the plains of Garo Hills is very humid and warm. The sky in Meghalaya is rarely cloudless, as its name suggests. The average annual rainfall is approximately about 3636 mm. May to October is the time of year where the majority of Shillong's annual precipitation occurs. So, prediction of rainfall in Shillong has been a motivation of this paper, so that it will be easy to understand the trend of Shillong rainfall. The objective of this paper is to implement rainfall forecasting using LSTM and make a comparison with different ML models on the basis of their performances and also highlights some special observations during the implementation of the proposed model SDDRFM. This paper has several sections. Section 2 discusses the Literature Survey work, Sect. 3 explains the proposed methodology of the models which is used in this paper, Sect. 4 contains detailed results and discussion, Sect. 5 contains the comparison between baseline model and proposed model, and at last Sect. 6 concludes the paper.

2 Literature Survey

The convolutions of hydrological equations for rainfall prediction have increased the reliability on statistical models, but with the evolvement of neural networks, the use of learning models has geared up efficiently. Many approaches have been proposed for accurately predicting rainfall in order to save the situation of drought and flood. Table 1 shows some ML models which are proposed by several researchers and accuracy measures are presented in format of Mean Absolute Percentage Error (MAPE), Root Mean Squared Error (RMSE), Mean Squared Error (MSE), Mean Absolute Error (MAE), and R-Squared Error (R^2) score.

3 Proposed Methodology

Rainfall forecasting algorithms can be divided into two categories: traditional statistical techniques and Machine Learning (ML) techniques. Traditional statistical techniques are the most commonly used algorithms for rainfall prediction. Statistical methods are appropriate for linear applications, while ML algorithms are appropriate for nonlinear applications like rainfall forecasting. Neural networks engaged in FFNN structure are not adequate for the rainfall prediction since it doesn't take previous state into account. On the other hand, Recurrent Neural Network (RNN) has a smaller memory to store the rainfall history's previous status. Hence, we consider the LSTM which comprises a large number of previous dataset, by alleviating the vanishing gradient problem, thus it improves the accuracy in predicting rainfall. Therefore, this paper proposes a rainfall forecasting model named as "Short-Term Data-Driven Rainfall Forecasting model (SDDRFM)" which predicts the shilling rainfall based on stacked LSTM. SDDRFM architecture is descripted in Fig. 1.

3.1 Dataset

The Shillong rainfall dataset between January 1, 2016 and December 31, 2020 is obtained from NASA power grid data [15]. In the raw weather dataset, there are some important parameters like temperature, wind speed, and humidity. All numerical data are considered here for experimentation.

Detailed description of the parameters is described in Table 2.

Next step is to try each and every feature selection methods mounted with LSTM model to find out the best method to predict rainfall.

3.2 Feature Selection

The reducing process of input variables when training a predictive model is known as feature selection [19]. In a linear regression test, filter-based feature selection approaches rate the correlation or dependence between independent (input) variables and dependent (output) variables using statistical measures, which may be filtered to find the most relevant feature. To discover the best characteristics, wrapper approaches use a specific ML algorithm. Embedded methods embed features at the

Table 1 Literature survey on different rainfall prediction papers

Authors	Dataset used	Model used	Parameters used	Accuracy measure
Kanchan et al. [4]	Karnataka and Maharashtra	Back propagation neural network	Rainfall	MAPE = 109.6% RMSE = 162.5%
Yadav et al. [5]	Udham Singh Nagar district of Uttarakhand	Back propagation neural network	Maximum temperature, maximum humidity, minimum temperature, minimum humidity, and wind speed	MSE = 0.00547823
Rasel et al. [6]	6 years (Chittagong Bangladesh)	Back propagation neural network	Rainfall, temperature	RMSE = 31.97 MAE = 10.87
Hasnul et al. [7]	Malaysia	Radial basis function neural network	Temperature (°C), humidity (%), air pressure (mbar), dam wind speed (km/h)	RMSE = 49% and the percentage accuracy = 57%
Chai et al. [7]	Kuching, Sarawak, Malaysia	Radial basis function neural network	Mean relative humidity (%), maximum temperature (°C), minimum temperature (°C), mean temperature (°C), mean sea level pressure (hPa), mean wind speed (m/s), and mean precipitation	MSE = 0.2206 $R^2 = 0.8191$
Lee et al. [8]	Daily rainfall of 367 locations in Switzerland	Radial basis function neural network	Rainfall data of 4 regions	RMSE = 78.65 Relative error = −0.46 Absolute Error = −55.9
Aswin et al. [9]	July 1979 to January 2018 of GPCP	Convolutional neural networks	Global precipitation data	RMSE = 2.44
Zhang et al. [10]	Shenzhen, China, in the year of 2014 to 2016 from March to September	Deep convolutional neural networks	Gauge rainfall Doppler radar echo maps	RMSE = 9.29
Narejo et al. [11]	Meteo weather station	Convolutional neural network	Rainfall, humidity, pressure, temperature	MSE = 4.59E−05, RMSE = 0.0068 and R = 0.94

(continued)

Table 1 (continued)

Authors	Dataset used	Model used	Parameters used	Accuracy measure
Kaneko et al. [12]	Kyushu region in Japan	Two-layer stacked long short-term memory	Wind direction, precipitation, temperature, pressure, relative humidity, and wind velocity	MSM = 2.07 RMSE = 2.44
Poornima et al. [13]	Hyderabad region dataset from 1980 to 2014	Intensified long short-term memory	Maximum relative humidity, maximum temperature, wind speed, minimum temperature, minimum relative humidity, sunshine and evapotranspiration, rainfall	Accuracy = 87.99%
Pranolo et al. [14]	Tenggarong East Kalimantan-Indonesia within year of 1986 to 2008	Long short-term memory	Rainfall data	RMSE = 0.2367

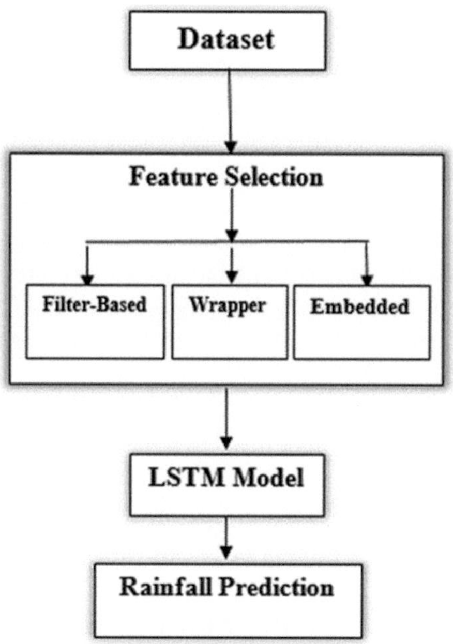

Fig. 1 SDDRFM architecture

Table 2 Description of dataset attributes

Attributes	Description
T2M	Temperature at 2 m (C)
T2M_MAX	Maximum temperature at 2 m (C)
T2M_MIN	Minimum temperature at 2 m (C)
T2M_RANGE	Range temperature at 2 m (C)
QV2M	Specific humidity at 2 m (g/kg)
RH2M	Relative humidity at 2 m (%)
WS2M	Wind speed at 2 m (m/s)
WS2M_MAX	Maximum wind speed at 2 m (m/s)
WS2M_MIN	Minimum wind speed at 2 m (m/s)
WS2M_RANGE	Range wind speed at 2 m (m/s)
PRECTOTCORR	Precipitation corrected (mm/day)

time of process of build model and feature selection is done by analyzing each iterations of training phase of model.

The Pearson correlation approach is the most widely used filter-based feature selection method for numerical variables; its range is between −1 and 1, with 0 representing no correlation, 1 representing total positive correlation, and −1 representing total negative correlation. High correlated features are more linearly dependent and approximately having the same influence on the dependent variable. When two features have a high association, one of them can be eliminated. In this experimentation, co-relation threshold values of 0.9, 0.8, and 0.75 are used to select the features as input variables of the proposed model. A detailed description of results using Pearson co-relation co-efficient is given in Table 3.

Selected features are fed into the proposed LSTM model. Resulting accuracies are noted down in results and discussion section for every threshold value of the Pearson co-efficient method. Analysis of variance (ANOVA) is another filter-based statistical method and is used to check the means of two or more groups that are significantly

Table 3 Pearson co-relation co-efficient results

Threshold value	Features removed	Remaining features considered as input variables
0.9	QV2M, T2M_MIN, T2M_MAX, WS2M_MAX	T2M, T2M_RANGE, RH2M, WS2M, WS2M_MIN, WS2M_RANGE (6 input features)
0.8	QV2M, T2M_MAX, T2M_MIN, WS2M_MAX, WS2M_MIN, WS2M_RANGE	T2M, T2M_RANGE, RH2M, WS2M (4 input features)
0.75	QV2M, T2M_MAX, T2M_MIN, WS2M_MAX, WS2M_MIN, WS2M_RANGE, RH2M	T2M, T2M_RANGE, WS2M (3 input features)

different from each other. Variance is the measure of the spread between numbers in a variable. If the variance is low, it implies that there is less impact of that feature on target variable. A detailed description of feature relevance using ANOVA is given in Table 4.

Among these 10 features, firstly, RH2M is removed with least feature relevance, then T2M_MAX is removed and finally WS2M_RANGE is removed. Results of R^2 score using LSTM for remaining features are described in Sect. 4.

Forward selection, backward elimination, and bi-directional elimination (stepwise selection) are wrapper approaches for feature selection. In forward selection, it starts with a null model and then starts fitting the model with each individual feature one at a time and select with minimum *p*-value. *P*-value is a statistical measurement used to validate a hypothesis against observed data. In backward elimination, it starts with a full model, then removes the insignificant features with highest *p*-value. The process repeats until final set of features is achieved. Bi-directional elimination is similar to forward selection but the difference is while a new feature is added, it also checks the existing feature set, if finds any of the already selected features insignificant, the particular feature will be removed. A detailed description of feature selection using wrapper method is given in Table 5.

In every methods, while eight features are considered, WS2M_RANGE is removed, and while seven features are considered then RH2M is also removed. Results of LSTM model using these selected features are described in Sect. 4.

L1 regularization or least absolute shrinkage and selection operator (LASSO) method is one of the embedding feature selection methods and is a regression analysis method that performs variable selection and regularization to enhance the prediction accuracy and interpretability of the resulting statistical model. It can be understood as adding a penalty against complexity to reduce the degree of over-fitting or variance of a model by adding more bias. Alpha is a tuning parameter to find relevant features. Feature relevance value less than the alpha value will be removed from the feature set. Results using L1 method are described in Table 6. Results of LSTM model using these selected features are described in Sect. 4.

Table 4 ANOVA feature relevance results

Feature name	Feature relevance
T2M	2.7867
T2M_MAX	1.7151
T2M_MIN	3.5623
T2M_RANGE	4.3073
QV2M	5.4254
RH2M	1.3418
WS2M	6.8451
WS2M_MAX	2.6888
WS2M_MIN	6.2562
WS2M_RANGE	1.8282

Table 5 Wrapper feature relevance results

Backward elimination			Forward selection			Stepwise selection		
9 Features	8 Features	7 Features	9 Features	8 Features	7 Features	9 Features	8 Features	7 Features
T2M	T2M	T2M	T2M	T2M	T2M	T2M	T2M	T2M
T2M_MAX	T2M_MAX	T2M_MAX	T2M_MAX	T2M_MAX	T2M_MAX	T2M_MAX	T2M_MAX	T2M_MAX
T2M_MIN	T2M_MIN	–	T2M_MIN	T2M_MIN	T2M_MIN	T2M_MIN	T2M_MIN	T2M_MIN
T2M_RANGE	–	–	T2M_RANGE	T2M_RANGE	T2M_RANGE	T2M_RANGE	T2M_RANGE	T2M_RANGE
–	–	–	QV2M	QV2M	–	QV2M	QV2M	QV2M
RH2M	RH2M	RH2M	RH2M	RH2M	RH2M	RH2M	RH2M	–
WS2M	WS2M	WS2M	WS2M	WS2M	WS2M	WS2M	–	–
WS2M_MAX	WS2M_MAX	WS2M_MAX	WS2M_MAX	–	–	WS2M_MAX	WS2M_MAX	WS2M_MAX
WS2M_MIN	WS2M_MIN	WS2M_MIN	–	–	–	–	WS2M_MIN	WS2M_IN
WS2M_RANGE	WS2M_RANGE	WS2M_RANGE	WS2M_RANGE	WS2M_RANGE	WS2M_RANGE	WS2M_RANGE	–	–

Table 6 Results using LASSO method

Alpha value	Removed feature	Remaining features
ALPHA 0.2	QV2M, WS2M, WS2M_RANGE	T2M, T2M_MAX, T2M_MIN, T2M_RANGE, RH2M, WS2M_MAX, WS2M_MIN
ALPHA 0.18	QV2M, WS2M, WS2M_RANGE	T2M, T2M_MAX, T2M_MIN, T2M_RANGE, RH2M, WS2M_MAX, WS2M_MIN

3.3 Long Short-Term Memory (LSTM)

LSTM is a distinct architecture of RNN, which was deliberately designed for handling time-related sequences. LSTM possesses a long-range dependence which distinguishes LSTM from traditional RNN architecture. The primary LSTM structure is a memory cell to memorize and propagate output accurately in several timestamps. To remember the temporal context information, the LSTM memory cell constructs the cell state concept. It consists of an input gate, output gate, and forget gate for controlling information flow within various time steps. This work presents LSTM depending on neural networks which is selected for forecasting from time-series weather data. Activation function which generally works in LSTM cell is hyperbolic tangent (tanh) and sigmoid. In LSTM architecture, forget and input gates are implemented with sigmoid function and candidate vector is with tanh function which is used to update the cell state vector. These LSTM activation functions are calculated for input gate I_t, hidden state h_t, output gate O_t, forget gate F_t, candidate vector C'_t, and cell state C_t, with the use of the following formulae:

$$I_t = \text{sigmoid}(W_i[X(t), h_{t-1}] + b_i)$$

$$F_t = \text{sigmoid}(W_f[h_{t-1}, X(t)] + b_f)$$

$$O_t = \text{sigmoid}(W_o[h_{t-1}, X(t)] + b_o)$$

$$C'_t = \tanh(W_c[h_{t-1}, X(t)] + b_c)$$

$$C_t = F_t * C_{t-1} + I_t * C'_t$$

$$h_t = O_t * \tanh(C_t)$$

Here, input vector is $X(t)$, previous state hidden vector is h_{t-1}, weight W and each gate bias is b. The basic structure of LSTM network is shown in Fig. 2.

Here, current input $X(t)$ goes into the LSTM memory cell, previous hidden state h_{t-1}, and previous memory state C_{t-1}. This is a neural network with single layer where input multiplied with sigmoid function acts as input gate activation function and another activation function is tanh multiplied with input. Using forget gate, current input influence on the previous state is analyzed and makes a decision whether storing the current input in cell state is required or not. Sigmoid function produces

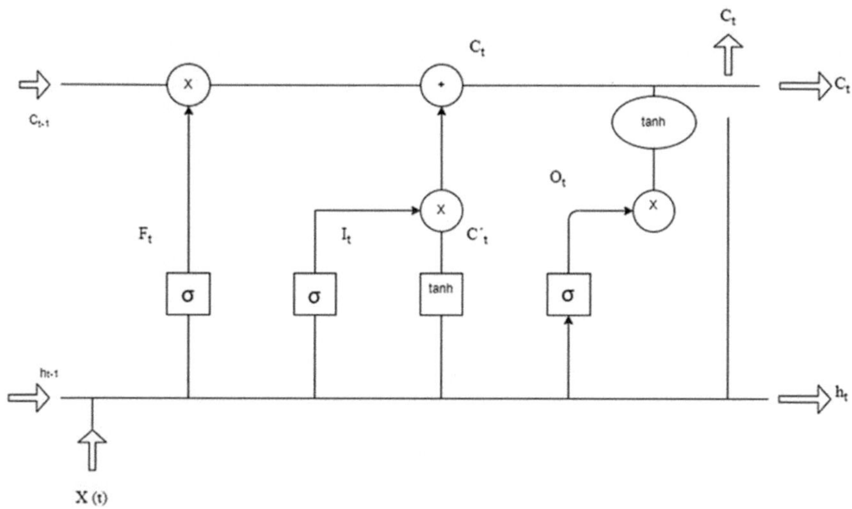

Fig. 2 Basic LSTM architecture

0 if not required, then forget gate is deactivated, it produces 1 to activate the gate if required. Depending upon this decision, on certain input, LSTM gates get activated or deactivated.

LSTM time-series forecast is a modern approach for building a rainfall forecasting model. As neural network is ready to learn complex nonlinear relationships, LSTM is leading successful neural network that is directly supporting multivariate sequence prediction problem. LSTM produces RMSE lesser than other models for testing. The dominance in difference type of accuracy measures in contrast with other forecasting models demonstrates that the LSTM model is efficient in predicting rainfall.

Stacked LSTM Model Architecture

Stacked LSTM model is considered as deep learning model, Deep learning is the deepest part of neural networks which is in generally assigned to the succeeding approaches on a broad range of challenging prediction problems. Stacked LSTM with hidden layers produces model deeper and more accurate. Due to the introduction of several layer architectures in the hierarchy the successful use of deep neural network is attributed. Some part of the task is processed by each layer and passed it to the next layer. In this point of view, the deep neural network can be considered as a pipeline process, where a part of the task is processed by each layer before passing it on to the next, until output is provided by the last layer. Stacked LSTMs are stable technique for sequence prediction problems. We can describe a stacked LSTM architecture as an LSTM model consisting of multiple LSTM layers. First, LSTM produces an output sequence rather than an output with single value to the second LSTM layer. A two-layer stacked LSTM layer is used in this experimentation. A simple outline of the model is shown in Fig. 3.

In SDDRFM, this two-layer stacked LSTM model is used with different combinations of feature selection methods.

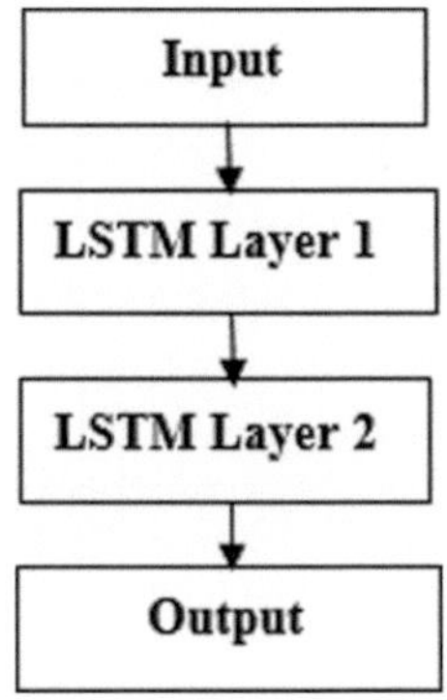

Fig. 3 Two-layer stacked LSTM model

4 Results and Discussions

The SDDRFM model trains with a Python package, namely, keras of Tensorflow backend. In this model, dataset is divided into two parts: training dataset and testing dataset. Training dataset takes the rainfall data between 2016 and 2019 to train the proposed model. After the training, this trained model takes for test the model with the rainfall dataset of the year of 2020. The features used in experiment are T2M, T2M_MAX, T2M_RANGE, RH2M, WS2M, WS2M_MAX, WS2M_MIN, WS2M_RANGE and the outcome variable is PRECTOTCORR.

In this paper, a baseline model [13] is implemented using this dataset [15]. Obtained results are included in Table 7. Further, comparative analysis is performed using three other predictive algorithms, namely, FFNN, CNN, and GRU.

Other than the input and output layers, there are two hidden layers utilized to build the proposed model, where each layer is having 50 neurons. The MSE is used as the loss function and Adam is used as the optimizer. The loss function's error value is defined by the discrepancy between actual and forecast rainfall. Errors are steadily decreased when the proposed model achieves stability as resulting in minimum losses. Accuracies of models using R^2 score, MSE, and RMSE are recorded in Table 8. For Pearson feature selection, best R^2 accuracy is achieved in the LSTM model with 6 input features using Pearson co-efficient of 0.9. For this particular model, in Fig. 4, the losses are drastically decreased in the early epochs and to remain within limited boundaries in later epochs. The global minimum calculates 0.0031 at roughly 123 epochs and remains within 0.0031 until the 150th epoch, which demonstrates the proposed model's stability.

Figure 5 depicts the actual rainfall as well as the forecasted rainfall. The forecasted rainfall levels are represented for the final 366 days of the year of 2020, using prior data from the years of 2016–2019 as training dataset.

To evaluate the model's ability to track the real graph, this plot is compared to a plot of actual rainfall of the same year. Resulting accuracy of LSTM using ANOVA, wrapper, and embedded method feature selection is described in Table 9.

Best accuracy of R^2 score 0.857 is achieved using LSTM with WRAPPER feature selection method (Table 10). Corresponding training loss and actual versus predicted rainfall are shown in Figs. 6 and 7.

Table 7 Results using baseline method

Model name	Result		
Baseline model	R^2	MSE	RMSE
	0.7115	Training MSE: 0.00314 Testing MSE: 0.0021	Training RMSE: 0.06 Testing RMSE: 0.05

Table 8 Accuracy of LSTM models using different input parameters

Model name	Results		
	R^2 score	MSE	RMSE
LSTM (Pearson correlation threshold 0.9) (6 input features)	0.7302	Training MSE: 0.0029 Testing MSE: 0.0019	Training RMSE: 0.05 Testing RMSE: 0.04
LSTM (Pearson correlation threshold 0.8) (4 input features)	0.7157	Training MSE: 0.0030 Testing MSE: 0.0020	Training RMSE: 0.06 Testing RMSE: 0.05
LSTM (Pearson correlation threshold 0.75) (3 input features)	0.7129	Training MSE: 0.0029 Testing MSE: 0.0019	Training RMSE: 0.05 Testing RMSE: 0.04

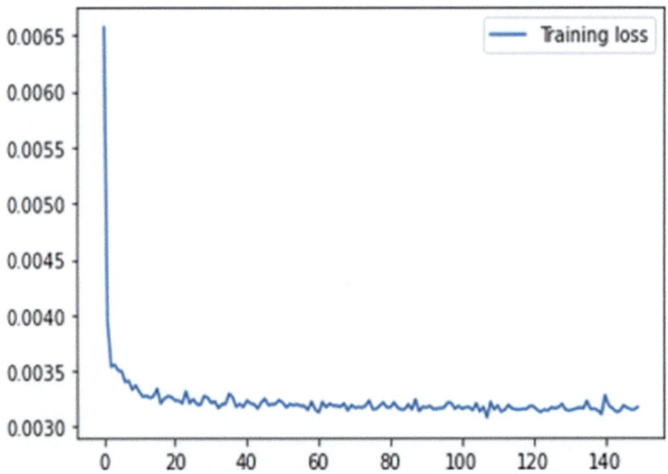

Fig. 4 Loss function values against number of epochs for 6 input values

5 Comparison with Baseline Model

The baseline model has achieved a R^2 score of 0.711. However, the proposed model using feature selection methods in this paper using the same dataset has obtained a highest R^2 score of 0.8573. From this observation, it can be concluded that the proposed method has over-powered the baseline model by 14.63 using the feature selection process. The importance of input features with respect to the rainfall is also discussed. Feature selection has brought more dynamicity in architecture. Comparison in actual versus predicted rainfall in baseline model and best model achieved in proposed SDDRFM is shown in Fig. 8. In the actual versus predicted rainfall

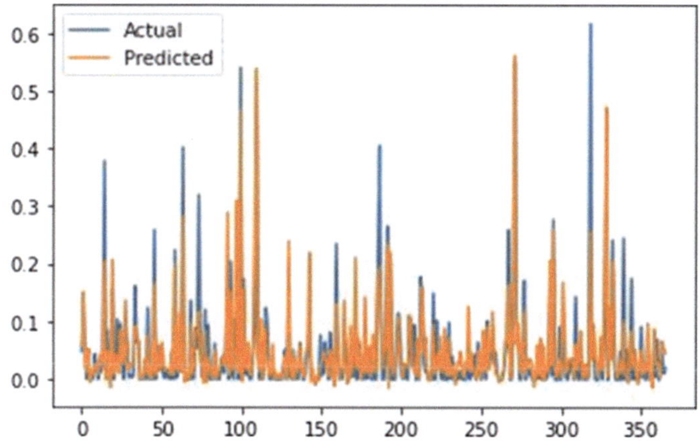

Fig. 5 Actual versus predicted rainfall of the year 2020 for Pearson correlation threshold 0.9

Table 9 Results of LSTM model using different feature selection methods

Feature selection method	Parameters	R^2 score	MSE score	RMSE score
LASSO	Alpha 0.2	0.7272	Training MSE: 0.0030 Testing MSE: 0.0020	Training RMSE: 0.05 Testing RMSE: 0.04
	Alpha 0.18	0.7334	Training MSE: 0.0029 Testing MSE: 0.0019	Training RMSE: 0.05 Testing RMSE: 0.04
ANOVA	9 Input features	0.733	Training MSE: 0.0029 Testing MSE: 0.0019	Training RMSE: 0.05 Testing RMSE: 0.04
	8 Input features	0.727	Training MSE: 0.0029 Testing MSE: 0.0020	Training RMSE: 0.05 Testing RMSE: 0.04
	7 Input features	0.754	Training MSE: 0.0030 Testing MSE: 0.0018	Training RMSE: 0.06 Testing RMSE: 0.04
	6 Input features	0.717	Training MSE: 0.0030 Testing MSE: 0.0020	Training RMSE: 0.06 Testing RMSE: 0.05
WRAPPER	9 Input features	0.661	Training MSE: 0.1619 Testing MSE: 0.1655	Training RMSE: 0.40 Testing RMSE: 0.41
	8 Input features	0.857	Training MSE: 0.0324 Test Score: 0.0341	Training RMSE: 0.18 Test Score: 0.18
	7 Input features	0.416	Training MSE: 0.4281 Test Score: 0.5040	Training RMSE: 0.65 Test Score: 0.71

graph, it can be noticed that only some peak points are missed or failed to predict by this proposed model. However, throughout a year rainfall value cannot be similar as maximum time in summer and winter season there is no rainfall, in the monsoon season maximum of rainfall happens. As due to some natural calamity rainfall can happen in odd times, so predicting those rainfall points is a constraint in rainfall prediction.

Table 10 Accuracy measures of rainfall prediction using 7 parameters with different ANN models

Model name	Results		
	R^2 score	MSE	RMSE
FFNN	0.4570	Training MSE: 0.0024 Testing MSE: 0.0055	Training RMSE: 0.05 Testing RMSE: 0.07
CNN	0.5321	Training MSE: 0.0016 Testing MSE: 0.0047	Training RMSE: 0.04 Testing RMSE: 0.07
GRU	0.7321	Training MSE: 0.0030 Testing MSE: 0.00197	Training RMSE: 0.06 Testing RMSE: 0.04

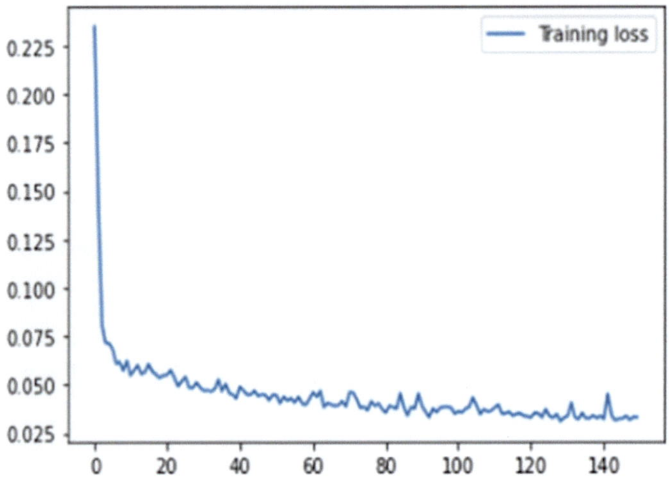

Fig. 6 Loss function value against number of epochs for Wrapper 8 input feature

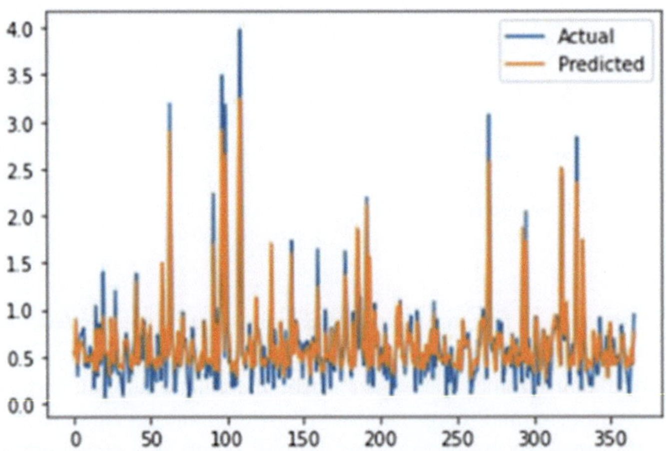

Fig. 7 Actual versus predicted rainfall of using Wrapper 8 input feature approach

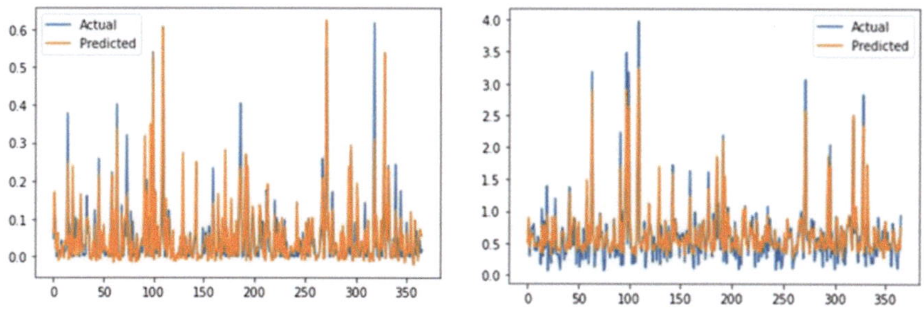

Fig. 8 Baseline model versus proposed model resulting graph comparison

6 Conclusion

Nowadays, many research works are going on different regions of our country. But analyzing the northeast region data is quite rare. Regardless of the application, data analytics and prediction of northeast region can give insights for better decision-making. The DL method is used in this research paper to predict rainfall. This strategy is compared to other approaches such as FFNN, CNN, and GRU in order to show that the suggested approach improves rainfall prediction. Weather parameters such as relative humidity, wind speed, and temperature are used as important parameters for predicting rainfall. In future works, the rainfall forecast dataset can aid in the hydrological situation of drought and flood in order to save the crops. Therefore, this research will be carried on by analyzing the negative and positive consequences based on rainfall forecasting data, which will aid in prescriptive analytics.

References

1. Rain, Society. https://www.nationalgeographic.org/encyclopedia/rain. Accessed 1 Apr 2022
2. Gado Djibo A, Karambiri H, Seidou O, Sittichok K, Philippon N, Paturel JE, Moussa Saley H (2015) Linear and non-linear approaches for statistical seasonal rainfall forecast in the Sirba watershed region (SAHEL). Climate 3(3):727–752
3. Climate Shillong (India), climate-data.org. https://en.climate-data.org/asia/india/meghalaya/shillong-24618. Accessed Jan 2022
4. Kanchan P, Shardoor NKB (2021) Rainfall analysis and forecasting using deep learning technique. J Inform Electr Electron Eng (JIEEE) 2(2):1–11
5. Yadav P, Sagar A (2019) Rainfall prediction using artificial neural network (ANN) for Tarai region of Uttarakhand. Curr J Appl Sci Technol 33:1–7
6. Rasel RI, Sultana N, Meesad P (2017) An application of data mining and machine learning for weather forecasting. In: International conference on computing and information technology. Springer, Cham, pp 169–178
7. Auzani H, Has-Yun KS, Nazri FAM (2021) Development of trees management system using radial basis function neural network for rain forecast. Comput Water Energy Environ Eng 11(1):1–10

8. Chai SS, Wong WK, Goh KL, Wang HH, Wang YC (2019) Radial basis function (RBF) neural network: effect of hidden neuron number, training data size, and input variables on rainfall intensity forecasting. Int J Adv Sci Eng Inf Technol 9(6):1921–1926
9. Lee S, Cho S, Wong PM (1998) Rainfall prediction using artificial neural networks. J Geogr Inf Decis Anal 2(2):233–242
10. Aswin S, Geetha P, Vinayakumar (2018) Deep learning models for the prediction of rainfall. In: International conference on communication and signal processing
11. Zhang CJ, Wang HY, Zeng J, Ma LM, Guan L (2020) Tiny-rainnet: a deep convolutional neural network with bi-directional long short-term memory model for short-term rainfall prediction. Meteorol Appl 27(5)
12. Narejo S, Jawaid MM, Talpur S, Baloch R, Pasero EGA (2021) Multi-step rainfall forecasting using deep learning approach. PeerJ Comput Sci 7:e514
13. Kaneko R, Nakayoshi M, Onomura S (2019) Rainfall prediction by a recurrent neural network algorithm LSTM learning surface observation data. In: AGU fall meeting abstracts, vol 2019, pp GC43D-1354
14. Poornima S, Pushpalatha M (2019) Prediction of rainfall using intensified LSTM based recurrent neural network with weighted linear units. Atmosphere 10(11):668
15. Feature selection techniques in machine learning, Analytics Vidya. https://www.analyticsvidhya.com/blog/2020/10/feature-selection-techniques-in-machine-learning/. Accessed on 4 Jan 2022
16. Understanding architecture of LSTM, Analytics Vidya. http://analyticsvidhya.com/blog/2021/01/understanding-architecture-of-lstm. Accessed on 4 Jan 2022
17. Power Grid Data. https://power.larc.nasa.gov/data-access-viewer/. Accessed 4 Jan 2022

Analysis of AprioriAll Sequence Mining Algorithm Using Hash and TRIE with Database Scan Reduction

Nayanjyoti Mazumdar and Pankaj Kumar Deva Sarma

Abstract Advancement of sequential mining algorithms is seen toward multiple dimensions. Alongside, the hardware technologies are also advanced in parallel. But due to the exponential growth of data, even terabytes of main memory become insufficient for handling the huge data into the memory for computing. The problem of the sequential mining algorithms lies there. Loading the whole data into the memory reduces computing ability. When computing is to be enhanced, data is to be decreased. Loading data in chunks in to the memory may skip important features of the data and loss meaningful inference rules. Reading dataset in multiple passes increases computing overhead. So finding an algorithm that manages an optimum tradeoff between the memory consumption and computing efficiency is really a great challenge for the researchers today. Use of proper data structures in algorithms with database scan reduction techniques can elevate the sequence mining process up to manifolds. In this paper, the AprioriAll sequence mining algorithm and its variants are implemented by using Hash and TRIE structures with database scan reduction techniques. We consider Nifty 50 index data for studying the algorithms. A thorough study is presented to compare the generated results against different performance parameters like support, confidence, dataset size, scan reductions, and space requirements. It is observed that the AprioriAll with TRIE structures outperforms the AprioriAll with Hashing in respect of processing time whereas AprioriAll with Hashing is a better choice to make in respect of memory consumptions.

Keywords Sequence mining · AprioriAll · Database scan · Hash · TRIE

1 Introduction

Sequential mining [1] permeated into diversified domains due to its application flexibility. A sequence is an ordered list of events generated in some span of time. These events can be found in different areas not limited to audio, videos, web logs,

N. Mazumdar (✉) · P. K. D. Sarma
Department of Computer Science, Assam University, Silchar, Assam, India
e-mail: nayan.mazumdar@gmail.com

S. K. Biswas et al. (eds.), *Intelligent Computing Systems and Applications*, Lecture Notes in Networks and Systems 1307, https://doi.org/10.1007/978-981-96-3860-4_5

social media contents, financial transactions, trading portfolios, educational datasets, weather data, bioinformatics, genome sequences, etc. The broad task of sequential mining is to uncover and visualize the meaningful sequential behaviors that are hidden in the datasets. Sequential algorithms [2–4] support a host of applications to analyze important sequences from the datasets in multi-ways. In recent times, a number of sequential algorithms are evolved not only to produce frequent sequential rules, but also to help in investigating outliers, negative rules, and gray point's identification.

One of the objectives of sequential mining algorithms is to enquire the time–space tradeoffs in evaluating the sequence behaviors in data. Dealing with variable datasets which are available either in horizontal or vertical formats in heterogeneous sources needs considerable efforts to bring them into a common format so that sequential algorithms could be applied. Database scanning while mining a bigger dataset is a tedious process that consumes a lot of processing time. An intermediate task of a sequential mining process is that it requires management of a large number of candidate sequences in the memory.

A host of data structures are proposed to handle these candidates for ease of storage, modification, searching, and deletions. When the data structure is efficient enough, it minimizes memory requirement and provides faster processing. So, the success of a sequential mining algorithm is solely depending on the selection of a proper data structure and use of database scan reduction techniques. Hash Tree and TRIE [5] are advanced data structures known to have lesser time complexity and better memory optimization. AprioriAll is a sequence mining algorithm that supports a range of applications. For this study, we consider AprioriAll algorithm to implement with Hashing and TRIE structures for investigating time and space complexities.

1.1 Measures of Interestingness

In sequence mining, an itemset is a non-empty ordered subset of items present in a transaction sequence. If an itemset contains n-number of items in it, then it is called an n-itemset. Support count of an itemset refers to the numbers of transactions which contains that particular itemset. Support refers to the total number of transactions that possesses a particular itemset. If we have two disjoint itemset, say A and B, then support finds the frequency of how often A and B occur together in the dataset of size N as given in Eq. (1). Confidence determines how frequently itemsets A and B occur together in some transactions where the itemset A occurs as given in Eq. (2) [6]. Mathematically, the support and confidence are derived as follows:

$$\text{Suport,}\ \ s(A \rightarrow B) = \frac{\sigma(A \cup B)}{N} \tag{1}$$

$$\text{Confidence},\ c(A \rightarrow B) = \frac{\sigma(A \cup B)}{\sigma(A)} \tag{2}$$

In a sequence mining process, support and confidence are considered to be the most widely used measures of interestingness. Some itemset with very low support and confidence may accidentally occur in the dataset by chance and these itemsets do not attract any interestingness for supporting the business process. Itemset with high support counts or high confidence infers strong rules. The support and confidence frequency are tested against some user-specified minimum support. If an itemset satisfies the min_sup then it is called a *frequent itemset*. An itemset is said to be most frequent if it qualifies for the highest min_sup. Finding the frequent sequences is the desired objective of most of the sequence mining activities.

1.2 Sequence Mining and Stock Index Data

A time is involved with the occurrence of an event. The occurrence of event(s) may affect the occurrence of some other event(s) and these events may be mutually inclusive or exclusive. NSE, the National Stock Exchange Limited, manages a large number of capital market segment indices such as Nifty 500, Nifty Midcap 150, and Nifty 50 in India [7]. A huge volume of stock buy-sale index prices for the above segments are available both in offline and online modes in terms of per tick, per second, per minute, per day, weekly, quarterly, or yearly basis. It is a good example of sequence data that can be considered for mining purposes. We have considered here the last 25 years' Daily Trading Data of *NIFTY 50 Index* for this study as shown in Fig. 1.

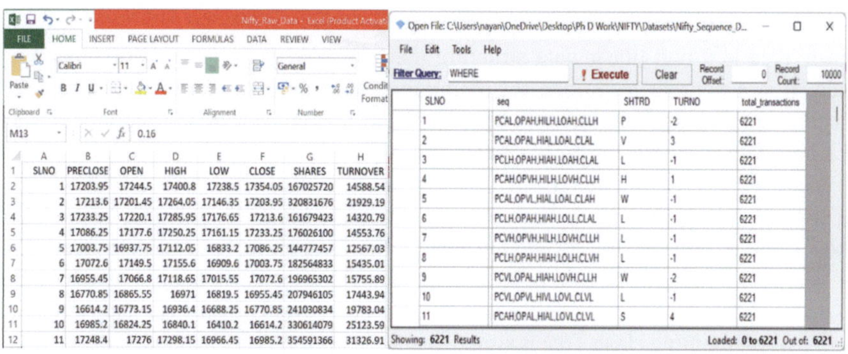

(a) Nifty 50 Indecx Raw Data for last 25 Years - 6221 Records

(b) Nifty 50 Index Raw Data Converted into Sequence Dataset - 6221 Records

Fig. 1 Sample of Nifty 50 index data

2 Data Structures and Data Preprocessing

Data refers to a wide variety of information that are collected and stored by many offline institutions and online applications for domain-specific consumptions. The data may be structured, semi-structured, or unstructured and these may either be qualitative or quantitative data. Different data structures are required for handling different data. Data structures define the way to organize and manage data in memory so that these can be efficiently accessed. Data structures may be primitive or non-primitive. Some basic data structures *like int, char, float, double, pointer,* etc. that can hold a single value are called the primitive data structures. Non-primitive data structures are either linear or non-Linear. Data structures that arrange data in a continuous manner such as *arrays, stacks, queues, and linked-lists* are called linear structures. Here, one element of the data is connected only to another element of the data. In non-linear structures, an element of the data can be connected to any number of elements. Examples include different *tree structures, graphs,* etc. Data structures may also be static or dynamic in nature. In static structures, memory allocation is required at compile time, so, it is difficult to predict beforehand the actual memory requirements for a program. Whereas in dynamic structures, memory allocation is flexible as it has to be done at the run-time only, therefore optimum memory requirement is possible. Static structures are easy to implement but dynamic structures become costly in this regard. Searching, sorting, insertion, updating, and deletion are the operations that can be performed in any data structures. Selection of a proper data structure for handling a certain type of data is important for enhancing the time–space efficiency, re-usability, and data abstraction process.

2.1 Hash Tree and TRIE Data Structures

Hash Tree is a non-linear data structure that can hold large amount of sequential data. It is a popular technique for mapping some arbitrary size of data into fixed sized data using hashing function (also called as message digests function) (Fig. 2). It stores data as key–value pairs into small tables, called the hash tables. It is capable of identifying a specific item from a huge pool of similar itemset in a very short span of time. As shown in Fig. 3, each data is assigned a unique hash value or index by the Hash function for storage and retrieval of the data. Here, hash value is needed to be searched instead of searching the actual data. Once the hash value is matched, the associated data could easily be located. Hashing can evenly distribute data across an array. Implementing Hash Tree is simple but it may possess difficulties like collisions and predefined memory requirements. Collisions occur when two keys are assigned similar hash values and the later become unable to find a space to store the key–value pair in the hash table. However, these drawbacks could be overcome with proper hash functions. Linear probing and double hashing are such solutions that help in finding next empty space in the array. The time complexity of creating the hash table is $O(n)$, where n is the length of the input key.

TRIE is also a non-linear data structure that stores set of strings in their child nodes as shown in Fig. 4. It is also called the digital tree or the prefix tree structures. Position of a node in the tree determines the key with which it is associated. It is difficult to implement as it has to maintain large numbers of pointers for storage and retrieval of elements in the tree. These pointers consume extra memory management overheads. In worst-case scenarios, creating a TRIE has also a time complexity of $O(n)$, where n is the length of the input sequence. In sequential mining on bigger datasets, TRIE may yield good results in respect of space complexity because lots of common subsequences are normally generated and the duplicates need not be stored again and again, only the occurrence of these subsequences is to be maintained.

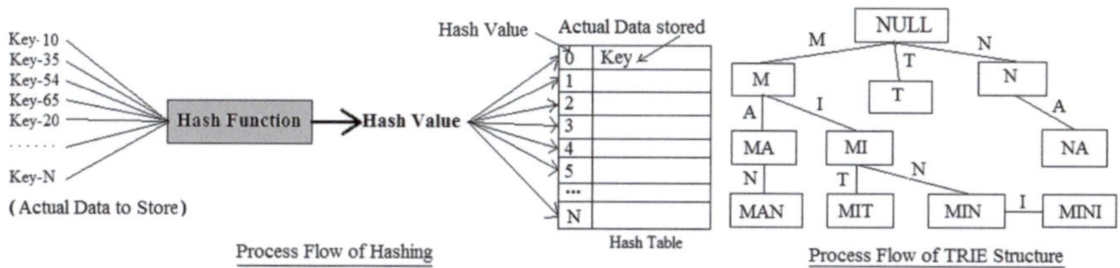

Fig. 2 Process flow of Hashing and TRIE structures

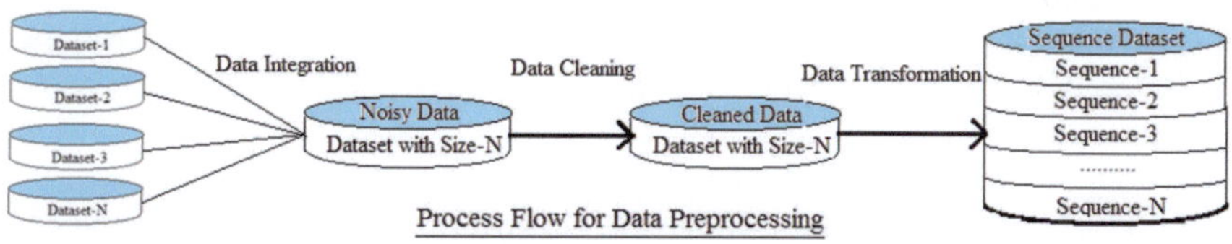

Fig. 3 Process flow for data preprocessing and cleaning

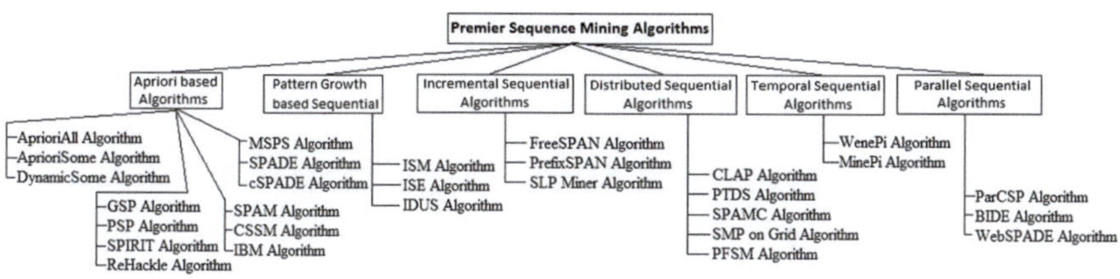

Fig. 4 Taxonomy of the premier sequence mining algorithms

2.2 *Preprocessing of the Dataset*

Dataset preprocessing is one of the necessary steps required for the sequence mining process. It involves the data integration, data cleaning, and data transformation phases. It is not necessary that the data considered for sequence mining should reside in a single flat file but it is always necessary to have it in a common structure. After the data is set to a common format, data cleaning activities such as noise removal and missing value substitutions are required. The noise-free data is then converted to a sequence dataset so that further mining activities could be performed.

3 Premier Sequential Mining Algorithms

In the last two decades, a number of sequential mining algorithms [2, 3, 8] were designed and implemented to support a wide range of applications. Among these, the Apriori-based or the candidate generation and test algorithms, breath-first search or horizontal format-based algorithms, depth-first search or vertical format-based algorithms, and sequential closed pattern-based algorithms were the premier algorithms proposed. The main concern of these algorithms was the methods of generating the candidate item sets and counting the support of the candidate itemset against some user-specified minimum support threshold. New approaches such as pattern growth-based algorithms, temporal, parallel, incremental and distributed sequential mining algorithms, multi-dimensional sequential rules, maximal sequential patterns, constraint-based sequential patterns, and approximate sequential patterns have also been developed in last few years. The following section gives an overview of the algorithms developed in recent times.

Some applications demand that massive serial processing of the data sequences is required for discovering the important hidden meanings in it. Parallel sequential mining algorithms become unable to cope up with such situations. However, parallelization of different activities in different levels like data integration, preprocessing, etc. could be eased with the help of parallel processing algorithms.

4 Implementing Sequential Mining Algorithms

In recent times, a host of sequential mining algorithms was evolved and tested for verifying the performances in respect of processing efficacy, data structure improvements, database handling, and memory managements. Different application domains demand different type of algorithms, to uncover the hidden important sequential inference rules from a dataset. For implementations, Windows Environment (Intel Core-i7, 16 GB RAMs with windows-11) and Linux Environment (Core-i7, 8 GB RAMs,

Ubuntu Open source) and the tools—Python distributions, Pandas with Jupyter notebook, Pandas Profiler, Jason editor, MS-Excel, Parquet viewer, etc. are used. AprioriAll algorithm and its variants are tried with the use of two data structures—Hashing and TRIE. AprioriAll algorithm using TRIE structures with database scan reductions is also implemented.

4.1 AprioriAll Algorithm Using Hash Tree with Multiple Database Scans

AprioriAll algorithm [9] process flow as shown in Fig. 5a requires that multiple database passes are required for generating candidates of different lengths. For example, if we consider five features of a dataset (i.e., maximum sequence length is 5), we require at least five database scans for generating the subsequences called the candidate sequences. In the first pass, length-1 candidates are generated. In second pass, length-2 candidates are generated and so on. Hash Tree is used to handle these candidates for storage and retrievals.

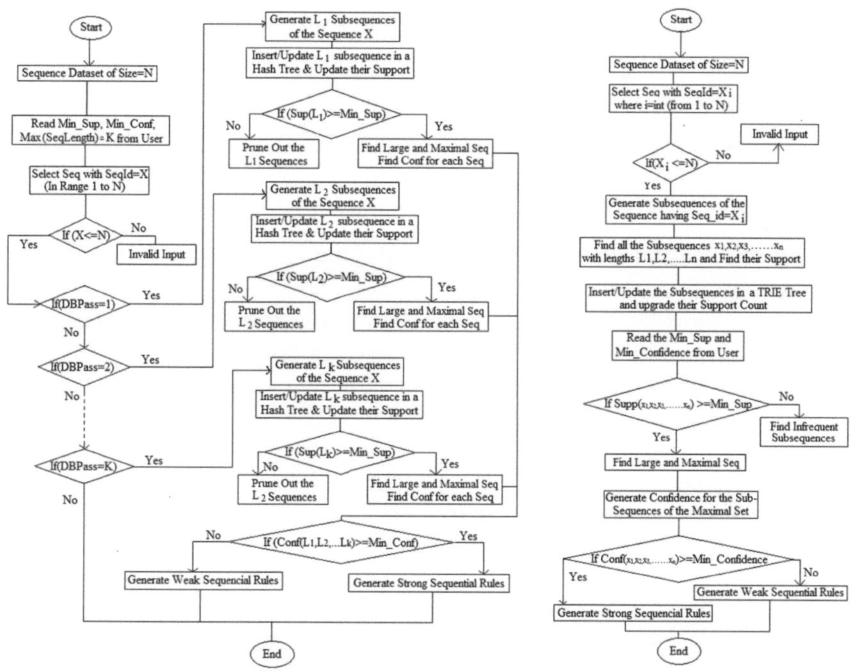

(a) AprioriAll Algorithm Flowchart with Hash Tree and Multiple DB Scan (b) AprioriAll Algorithm Flowchart with TRIE and Single DB Scan

Fig. 5 Data flow diagrams for AprioriAll algorithm variants

It is essentially a candidate generation and pruning strategy. User specifies a certain minimum support threshold (*min_sup*) for testing the frequentness of subsequence(s). If the occurrence frequency of a candidate is higher or equal to that *min_sup*, the candidate is said to be frequent and they are retained in the database. If the occurrence frequency of the candidate does not satisfy the min_sup threshold it is pruned out (deleted) from the database. This process continues till all the frequent candidates with higher lengths are generated. Frequent candidates having length k are said as Large-K (L_k) sequences. Maximal candidates are those which satisfy a certain level of min_sup and occur in candidate sets up to a certain length but it does not satisfy to occur in higher length candidate sets. Confidence of each of the maximal candidates of different lengths is calculated and tested against some user-defined minimum confidence (*min_conf*). If these candidates have higher or equal confidence than the *min_conf* then we get the strong rules of the candidates with some support and the confidence percentage. Weak rules identify the candidates that do not satisfy the *min_conf*. The algorithm stops here.

Advantage of this algorithm is that during subsequent passes over the dataset, some infrequent candidates are pruned out due to which search space complexity is reduced. In real-time transactions, repetition of sequences is very common. In such cases, repeated sequences need not be stored again which saves a lot of memory. Another advantage of this algorithm is that the mining process can be stopped in any stage of the database pass after getting some desired results. For smaller datasets having shorter sequence lengths, this algorithm can outperform other related algorithms.

4.2 AprioriAll Using TRIE Structure with Single Database Scan

To overcome the difficulties posed by the AprioriAll algorithm such as candidate sequence management and multiple database scans, we tried to implement a variant of the AprioriAll algorithm using the TRIE or the prefix tree structures with database scan reduction technique. The process flow of this algorithm is shown in Fig. 5b. It requires only a single pass over the database. Each sequence records are picked up and the corresponding subsequences called the candidate set are generated. After generating the candidate set of all sequences in the dataset, they are inserted into a single big TRIE tree and the occurrence frequency of each of the candidates is recorded. Whenever any candidate is repeated, it is not required to insert it in the tree again. Instead the frequency of occurrence is updated.

User-specified *min_sup* and the *min_conf* are taken beforehand when the algorithm starts. Supports for each of the candidates are tested against the *min_sup*, like in the AprioriAll algorithm. Large sequences and maximal sequences are searched in the tree. No candidates are pruned out here. Frequent sequences are nothing but some sub-tree of the main TRIE, where frequency of occurrence of each of the nodes in the tree satisfies the min_sup. Here also, the confidence of the maximal sequences is found out, to test against the min_conf supplied by the user. This algorithm ends with the generation of sequential rules. If a candidate sequence satisfies itself for the min_conf, it generates the strong rule(s). Otherwise, some weak rules are found. Advantage of using a TRIE instead of a Hash Tree is that, it can handle a large number of candidates, and the memory requirement for storing these candidates need not be pre-specified. Single pass over the database can considerably save processing times. The same tree can be reused for all levels of candidates storing and searching.

5 Result Analysis and Discussion

Algorithms become unable to address all the user requirements—both in respect of processing time and space consumption. Some algorithms need very negligible amount of time to get the desired result but consume a very high amount of memory. In some others, take a little longer processing time but can save considerable amount of memory. It is very difficult for the algorithms to find an optimum time–space tradeoff. It is up to the user and applications in which way to make any adjustments.

5.1 Candidate Generation and Database Scans

AprioriAll algorithm using Hash Tree incorporates some difficulties—a huge number of candidate sequences are generated. For example, in worst-case scenarios, if there is n-number of sequences in a dataset, a total of $2^n - 1$ candidates are generated as shown in Fig. 6. Storing and managing these candidates in main memory is a big challenge even if advanced structures like Hash Tree are used. Moreover, hashing requires a huge amount of memory to be pre-assigned considering all candidates to be as disjoint, and presuming the optimum memory requirement during program implementation is not feasible.

Main disadvantage of AprioriAll algorithm using TRIE includes construction cost of the TRIE tree both in case of implementation complexity and pointer managements. Additional memory gets consumed for maintaining such link pointers which is equal to total number of candidate items generated.

5.2 Processing Time Analysis

Scenario (1) Fixed datasets against variable min_sup—The processing time consumed by AprioriAll algorithm using Hash and TRIE in different situations is shown in Fig. 7. In case of AprioriAll with Hash Tree structure Fig. 7a, a considerable time is consumed in finding the large frequent sequences. For each database pass

Generation of Number of Large Sequences against variable Dataset Size with Fixed Min_Sup=0%											
Large Seq	Min_Sup	623	1245	1867	2489	3111	3733	4355	4977	5599	6221
L1	0	30	30	30	30	30	30	30	30	30	30
L2	0	349	352	355	359	360	360	360	360	360	360
L3	0	1268	1414	1514	1620	1715	1341	1861	1887	1946	1978
L4	0	1302	1653	1938	2243	2499	2780	2866	2950	3128	3277
L5	0	381	542	676	833	965	1098	1159	1208	1296	1383
Total	0	3330	3991	4513	5085	5569	6109	6276	6435	6750	7028

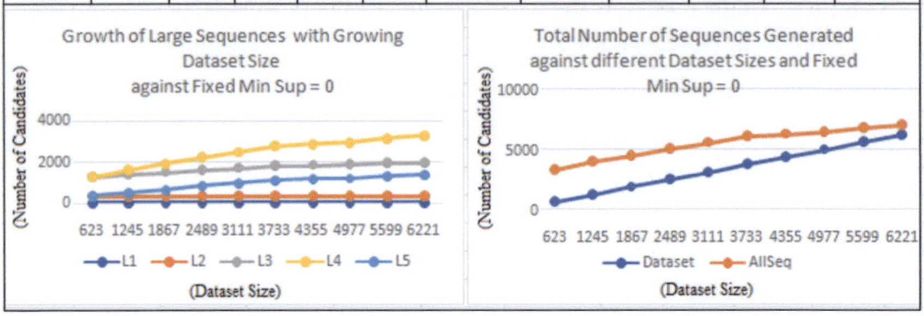

Fig. 6 Candidate sequences against variable dataset size

in generating an L_k sequence, Hash Tree needs to be constructed. Initialization of variables and arrays for handling candidates and construction of the Hash Tree again and again leads to more computational overheads. More the length of the sequences more is the number of database passes required. In case of AprioriAll using TRIE with single database scan Fig. 7b, the TRIE tree is constructed only once for the whole dataset. Initially, both memory and time are consumed for generating the tree, but in subsequent phases, when the min_sup varies, a very little amount of time is taken for counting the support of candidates. Only the searching time is spent to get the large frequent sequences.

Scenario (2) Fixed Min_sup against variable Datasets—Experiments show that, when we change the number of records in the dataset, keeping the min_sup fixed, AprioriAll using TRIE with single database scan outperforms the AprioriAll using Hash as shown in Fig. 8. In worst-case scenarios when the min_sup tends to zero, TRIE starts consuming time exponentially after growth of the dataset from a certain limit.

5.3 Memory Consumption

The main problem of sequential mining algorithms is that most of them consume a high volume of main memory. When the size of the dataset increases, candidate generation is also increased. From the experimental results as shown in Fig. 9, it is clear that for a linear growth in size of the dataset, there is a manifold increase in consumption of memory for storing the candidates. Therefore, it is desirable to load

(a) Processing Time (in seconds) for variable Min_Sup vs Fixed Datasets using AprioriAll with Hash Tree Structures

Datasets	0	5	10	15	20	25	30	35
623	77.78	50.11	52.53	49.79	51.05	50.26	50.99	49.61
1245	102.12	70.26	69.76	69.74	69.75	70.15	70.511	70.4
1867	111.56	99.28	101.24	99.42	99.26	100.86	100.06	99.74
2489	177.84	130.33	130.64	130.4	129.65	129.2	129.22	130.24
3111	92.77	70.46	69.16	68.79	69.27	68.19	72.14	70.02
3733	115.198	85.49	85.62	85.54	85.29	86.11	86.26	85.03
4355	349.77	273.45	271.62	268	278.88	287.08	274.82	280.25
4977	393.65	309.21	357.01	324.89	325.35	330.5	331.42	316.304
5599	358.14	280.12	228.36	198.23	124.83	121.12	120.82	122.98
6221	171.63	134.68	140.28	136.84	136.29	139.38	141.77	147.91

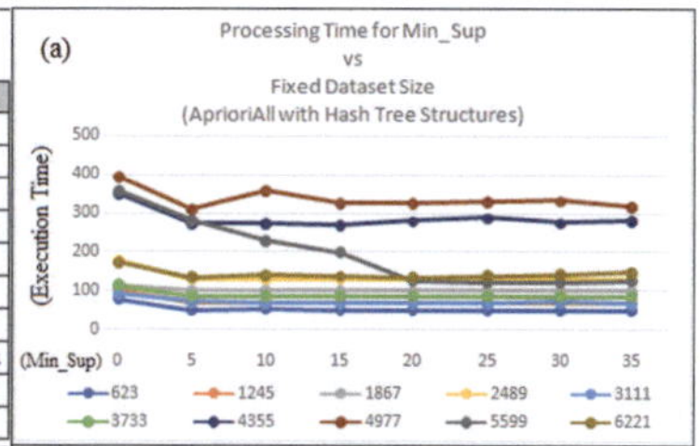

(b) Processing Time (in seconds) for variable Min_Sup vs Fixed Datasets using AprioriAll Algorithm with TRIE and Database Scan Reduction

Datasets	0	5	10	15	20	25	30	35
623	78.38	1.487	0.368	0.198	0.091	0.032	0	0
1245	107.6	1.955	0.365	0.124	0.113	0.089	0.018	0.009
1867	118.672	1.817	0.513	0.131	0.109	0.087	0.015	0
2489	155.252	2.015	0.499	0.167	0.113	0.099	0.011	0.011
3111	179.156	1.588	0.439	0.148	0.115	0.066	0.018	0.018
3733	195.202	1.504	0.54	0.54	0.54	0.156	0.107	0.019
4355	213.047	1.842	0.528	0.139	0.19	0.022	0	0
4977	326.109	5.729	1.246	0.677	0.265	0.051	0.051	0.051
5599	575.313	5.932	1.574	0.827	0.304	0.054	0.054	0.054
6221	579.175	5.874	1.716	0.888	0.296	0.022	0.022	0.022

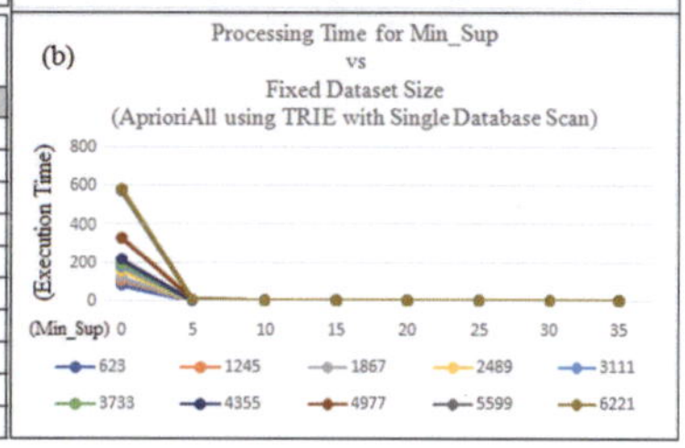

Fig. 7 Processing time analysis of AprioriAll with variable Min_Sup

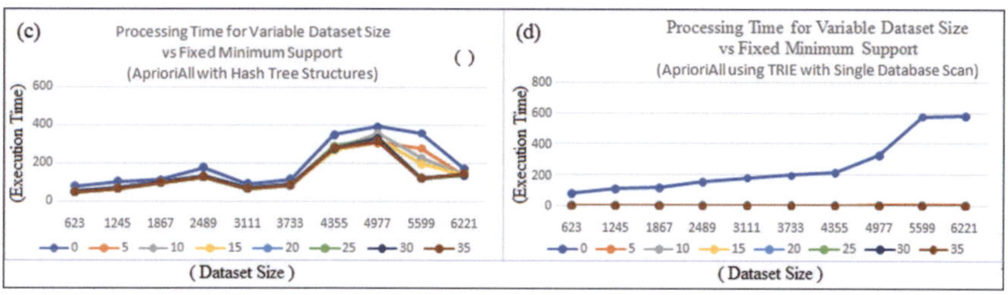

Fig. 8 Processing time analysis of AprioriAll with variable dataset size

a dataset in main memory of optimum size for performing sequence mining. Finding this optimum size is really a great challenge for the researchers. Some real world datasets like the market basket data, students passing grades, etc. contain repetition of events. This type of applications generates a large number of common candidates.

When the TRIE structure is used for storage and retrieval of candidates in such situations, TRIE do not grow exponentially. Even when more common candidates start generating, a situation may come when the size of the input dataset and size of the candidate set become closely equal. This is a best case scenario and is valid for huge data volumes only. From the following table, it is seen that, after the growth of the dataset up to a certain limit, generation of disjoint candidates decreases. Therefore, memory for the dataset and the candidates becomes adjacent. When more common candidates are generated, TRIE become a good choice, when more disjoint candidates are generated, TRIE become computationally expensive.

No of Records (Size)	623	1245	1867	2489	3111	3733	4355	4977	5599	6221
Size of Dataset (KB)	13	19	24	31	37	43	50	55	61	67
Hash tree size (KB)	30	35	39	44	48	52	54	56	58	60
Prefix Tree Size (KB)	52	62	71	79	87	94	98	101	106	110

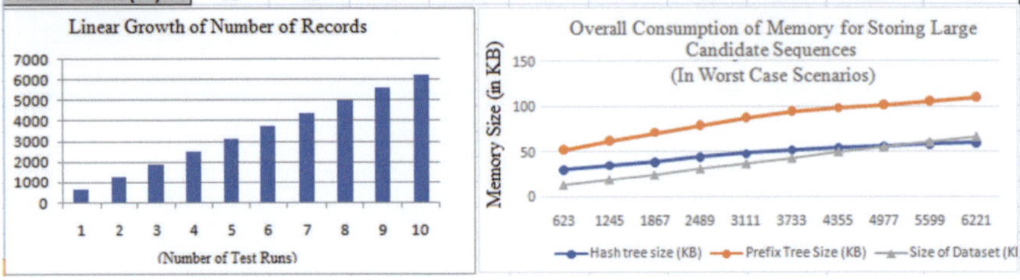

Fig. 9 Overall space requirement analysis

6 Applicability of Sequence Mining Algorithms

Sequence mining is a forever evolving paradigm. In recent times, applications of sequence mining can be seen in a vast range of domains. In the field of medical sciences like medical diagnosis and disease prediction, drug design, cause and effect of medical treatments, etc. have a tremendous use of sequence mining algorithms [10–12]. Educational sequence mining is relatively a new area where sequential mining algorithms are used [13]. In tourism industry, route clustering [14] and travel diary analysis [15] play important role in proposing and discovering new tourism packages.

Identification of areas for business diversification [16] is supported by the sequential mining algorithms. Due to extensive use of social networks, a huge volume of multimedia data is generated. Multimedia feature selection [17] is relatively a new field where sequence mining can contribute a lot. For cybertracking and user authentications [18], sequential algorithms with the use of machine learning are successfully used. A lot of research articles related to the recent COVID-19 pandemic situations for drug discovery as well as pandemic control mechanisms are supported by the sequential mining algorithms.

7 Conclusion and Future Works

Data is multiplying in an exponential pace with time. It is equally important to manage data efficiently before any meaning is derived from it. Finding some sequence rules from a huge pool of sequence data requires use of efficient data structures to be used with the algorithms for storage and retrieval of the data. Moreover, database scan reduction technique plays an important role in decreasing computing time of the mining process. Implementation of AprioriAll sequence mining algorithms with variations of data structures and database scan reductions is carried out in this study and an investigation is presented to see the time–space tradeoffs among them by varying the characteristic features like support thresholds, dataset size, etc. There is an ample opportunity for analysis, modification, and development of new sequential mining algorithms for optimizations in different levels of computing, memory requirements, and database handlings.

References

1. Agrawal R, Srikant R (1995) Mining sequential patterns. In: Proceedings of the international conference on data engineering (ICDE'95), Taipei, Taiwan, pp 3–14
2. Mooney CH, Roddick JF (2013) Sequential pattern mining-approaches and algorithms. ACM N, M 20YY:1–46
3. Motegaonkaret VS et al (2014) A survey on sequential pattern mining algorithms. Int J Comput Sci Inf Technol 5(2)
4. Fournier-Viger P, Lin JC-W, Kiran RU, Koh YS (2017) A survey of sequential pattern mining. Data Sci Pattern Recogn Ubiq Int 1(1)
5. Silberschatz, Korth, Sudarshan (2006) Indexing and hashing, chap 12. In: Database system concepts, 5th edn
6. Tan, Steinbach, Kumar (2004) Association analysis: basic concepts and algorithms, chap 6. In: Introduction to data mining
7. NSE India Limited (2022) Methodology document for equity indices. www.niftyindices.com
8. Kour A (2017) Sequential rule mining, methods and techniques: a review. Int J Comput Intell Res 13. ISSN 0973-1873
9. Agrawal R, Srikant R (1994) Fast algorithms for mining association rules. In: Proceedings of the 20th VLDB conference, Santiago, Chile
10. Rjeily CB, Badr G, Hajjarm El Hassani A, Andres E (2018) Medical data mining for heart diseases and the future of sequential mining. In: Machine learning paradigms, vol 149. Springer Link, pp 71–99
11. Kaur I, Doja MN, Ahmad T (2020) Time-range based sequential mining for survival prediction in prostate cancer. J Biomed Inform Elsevier 110
12. Estiri H, Strasser ZH et al (2020) Transitive sequencing of medical records for mining predictive and interpretable temporal representations. Patterns ScienceDirect 1(Issue-4)
13. Lee G, Kim D, Lee C (2020) A sequential pattern mining approach to identifying potential areas for business diversification. Asian J Technol Innov 28(Issue-I)
14. Lee GH, Han HS (2019) Clustering of tourist routes for individual tourists using sequential pattern mining. J Supercomput Springer 76:5364–5381
15. Choong SS, Wong L-P, Low MYH, Chong CS (2020) A bee colony optimization algorithm with sequential pattern mining based pruning strategy for the travelling salesman problem. Int J Bio-Inspired Comput 15(04):239–253
16. Wang X, Yu X, Guo L, Liu G, Xu L (2020) Student performance prediction with short-term sequential campus behaviours. MDPI J 11(Issue-4). https://doi.org/10.3390/info11040201
17. Pushpalatha K, Ananthanarayana VS (2020) Multimedia document mining using sequential multimedia feature patterns. In: IEEE sixth international conference on multimedia big data. https://doi.org/10.1109/BigMM50055.2020.00040, ISBN INSPEC Accession No-20075532, September
18. Levi M, Hazan I (2020) Deep learning based sequential mining for user authentication in web applications. In: International workshop on emerging technologies for authorization and authentication, ETAA-2020. Springer, pp 1–15

A Review on FPGA-Based CNN Accelerators

Anish M. George and Shajimon K. John

Abstract Convolutional Neural Networks (CNNs) have turned out to be the most sought-after algorithm for implementing many computer-vision tasks. But its huge number of computations and memory bandwidth requirement led to the need of hardware accelerators like GPUs and FPGAs. Due to the flexibility and energy efficiency offered by FPGAs over the other platforms, CNN inference accelerators based on FPGAs have now become a hot research area, especially in embedded and mobile applications where the resources are limited. So, this paper abstracts a review on the various state-of-the-art FPGA-based CNN inference accelerators. Also, a comparison is performed based on their performance and resource utilization.

Keywords FPGA · CNN · Inference · Acceleration

1 Introduction

Machine vision or Computer Vision is the ability of computer systems to intelligently act like human visual systems and thereby automate many different applications like classification, detection, segmentation, and video surveillance [1]. CNNs [2], one of the classes of Deep Neural Networks (DNNs), have become one of the prominent techniques used for deploying machine vision tasks [3, 4]. But the number of computations in CNNs is enormously high so that they cannot be efficiently deployed in General-Purpose Processors (GPPs) and mobile devices where the resources are limited. This calls for a high demand for various hardware accelerator platforms like Graphics Processing Units (GPUs), Field Programmable Gate Arrays (FPGAs), and even Application-Specific Integrated Circuits (ASICs) for implementing CNNs.

A. M. George (✉)
Saintgits College of Engineering, Kottayam, Kerala, India

Research Scholar, APJ Abdul Kalam Technological University, Thiruvananthapuram, Kerala, India
e-mail: anish.mg@saintgits.org

S. K. John
Muthoot Institute of Technology and Science, Kochi, Kerala, India

S. K. Biswas et al. (eds.), *Intelligent Computing Systems and Applications*, Lecture Notes in Networks and Systems 1307, https://doi.org/10.1007/978-981-96-3860-4_6

One of the most sought-after and computationally efficient hardware accelerators for CNNs is GPU [5] due to its support for massive parallelism in operations and high throughput. But the major hurdle in using GPU-based accelerators is the lack of energy efficiency, especially in low power-embedded platforms. Due to the aforementioned problems of GPUs, FPGAs have come into the role and currently, a lot of research is happening on FPGA-based CNN accelerators [6–9]. The major reasons for the increasing trend in using FPGAs for CNN acceleration are, they offer flexibility for the designers because of their reconfigurability and more importantly they are highly energy efficient. Moreover, modern state-of-the-art FPGAs do provide high computational throughput also and therefore they have become a de facto standard for CNN acceleration, especially in embedded applications [10, 11].

CNNs have two computation phases, namely, training and inference. During training, the network is trained with the huge amount of training data samples and using back propagation algorithm, the parameters are all iteratively updated so that the model performs the best. After training, the next stage is inference where the trained model will be input with a new set of testing data samples and the model classifies or predicts the output. The training process of CNNs is quite compute-intensive and memory bound and hence usually this is done using GPUs. Since the inference process has to take place for every new input data sample, it is worth focusing on its accelerating mechanisms. Also, the huge number of computations and memory requirements of CNN usually burdens FPGA-based CNN implementations. Therefore, CNN inference can be accelerated by different methods of optimizations. Even though similar papers have been published earlier [11–15], this paper specifically tries to perform a detailed review and a comparison of some of the recent studies on FPGA-based CNN inference accelerators using both hardware and software optimization techniques like model compression (pruning & data quantization).

The rest of the paper is structured as follows: Sect. 2 provides a brief review on the main features of CNNs, Sect. 3 discusses on the different approaches being currently used in accelerating CNN inference and some of the recent studies pertaining to the mentioned approaches.

2 Review of CNNs

CNNs are one of the types of Deep Neural Networks (DNNs) which are nowadays widely being used for image classification, object detection etc. A typical CNN has a combination of different layers which varies according to the selected CNN model. The combination of layers will be performing two basic functions namely feature extraction and image classification where feature extraction is done by the convolution block and classification is done by the fully connected block a shown in Fig. 1.

The layers in a typical CNN model are convolution layers, activation layers usually following the convolution layers, pooling layers and fully connected layers. A brief description of these layers is given below.

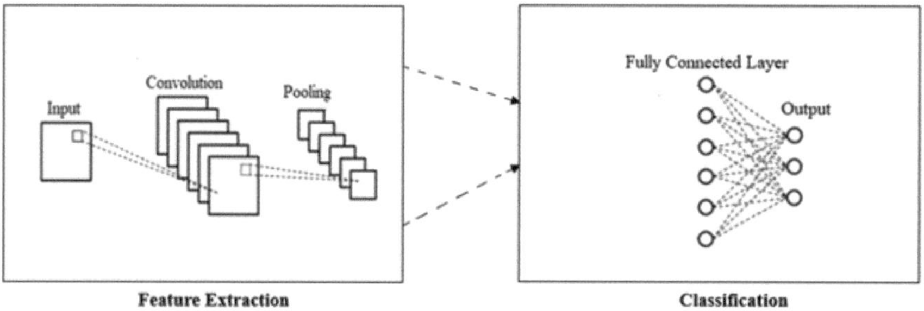

Fig. 1 Black box model diagram

Convolution (CONV) Layers: This is the first layer which extracts the input image features such as corners and edges by convolving the kernel over the input pixels of the image with a specified window size and stride value. The weighted sum output generated from this layer is termed as feature map which is given as input to the succeeding layer.

Activation Layers: Usually, convolution layers are immediately followed by activation layers. The activation functions will introduce non-linearity to the weighted sum outputs of the previous convolution layer and transforms it into a number that lies within a lower and an upper limits. The most widely used activation function in CNNs is ReLU (Rectified Linear Unit) which transforms all the negative inputs to 0 and positive values to same values.

Pooling Layers: Pooling layers are typically used following the convolution layers. They shrink the dimensions or sizes of the output feature maps from the convolution layer by sliding a specified filter over them. In other words, this layer performs downsampling of the image and thus reduces the complexity in computation. Also, it makes the features, it detects, more robust. The commonly used pooling functions are Max Pooling and Average Pooling.

Fully Connected (FC) Layer: The classification layer consists of a fully connected layer which is a feed-forward neural network. This can be multi-layer as well. The input given to the FC layer is usually the output from the final pooling layer which are flattened to a single-dimensional vector. Here, the number of weights will be enormously large which accounts for its high memory demand. The classification layer gets normally terminated with softmax activation function which gives the probabilities of the inputs for being classified into one among the different output classes.

3 Different Approaches of CNN Inference Acceleration

Some of the popular and commonly discussed CNN models suitable for image classification are AlexNet [17], VGG-16 [18], VGG-19 [18], GoogleNet [19], ResNet-50 [20], ResNet-101 [20], and ResNet-152 [20]. The computational workload including both CONV and FC layers of these models are given in Table 1 [15]. The accuracy of classification of these CNN models have been increasing gradually since the inception of AlexNet in 2012. But this was always a challenge for the designers as the computational workload was also increasing. Specifically, there was an imbalance in the mapping between the computational workload and the memory requirements. This was due to the fact that more than 90% of the MAC operations happen in the CONV layers which make them compute bound and most of the parameters (weights) are included in the FC layers which make them memory bound. So, due to this issue, the FPGA-based CNN inference accelerators make use of different optimization strategies when implementing the CONV and FC layers. The two important points to be considered while designing CNN accelerators are high speed and high energy efficiency. On analysis of these two aspects, we understand that CNN accelerators can be designed by optimizing either the CNN model or the hardware architecture.

Table 1 Workload and parameters of popular CNN models [15]

Model	AlexNet	VGG-16	VGG-19	GoogleNet	ResNet-50	ResNet-101	ResNet-152
Top1 err	42.9%	28.1%	27.3%	31.3%	24.7%	23.6%	23.0%
Top5 err	19.80%	9.90%	9.00%	10.07%	7.8%	7.1%	6.7%
Conv layers	5	13	16	57	53	104	155
Conv parameters	2.33 M	14.7 M	20 M	5.97 M	23.5 M	42.4 M	58 M
Conv workload							
(MACs)	666 M	15.3 G	19.5 G	1.58 G	3.86 G	7.57 G	11.3 G
Activation							
Layers	ReLU	ReLU	ReLU	ReLU	ReLU	ReLU	ReLU
Pooling layers	3	5	5	14	2	2	2
FC Layers	3	3	3	1	1	1	1
FC parameters	58.6 M	124 M	124 M	1.02 M	2.05 M	2.05 M	2.05 M
Conv workload							
(MACs)	58.6 M	124 M	124 M	1.02 M	2.05 M	2.05 M	2.05 M
Total workload							
(MACs)	724 M	15.5 G	19.6 G	1.58 G	3.86 G	7.57 G	11.3 G
Total parameters	61 M	138 M	144 M	6.99 M	25.5 M	44.4 M	60 M

So, the major optimization techniques used for CNN inference acceleration are algorithmic/architectural optimization and CNN model optimization. Presently, most of the FPGA-based CNN inference accelerators do focus on heterogenous approaches where a combination of these strategies is incorporated.

4 Algorithmic and Architectural Optimization

As stated above, hardware-level or architectural optimizations in CNN accelerator can result in better performance and better energy efficiency. There are many hardware-level techniques, of which efficient and careful design of computation unit can result in high performance and energy efficiency. Convolution operation is the most significant computation in CNNs. Therefore, optimizing and reducing the convolution operations can bring a remarkable advantage in CNN inference. Several computational transforms like General Element Matrix Multiplication (GEMM), Winograd minimal filtering and FFT are used to reduce the total number of arithmetic operations in the inference process. So, this section provides a detailed study on various recently carried out architectural optimizations for accelerating CNN inference using FPGAs.

Even though many successful implementations of conventional convolution algorithm in CNNs are present, more efficiency can be achieved if the convolution algorithm itself is optimized. So, Lu et al. [21] have proposed a novel architecture which used the Winograd algorithm for convolution operation and eventually obtained better performance of CNNs on FPGAs. In Winograd algorithm, making use of its structural symmetry, elements in the output feature maps are generated simultaneously thereby reducing the number of computations. Though it was proved that, Winograd algorithm is most efficient for convolution operation when the kernel size is small [22], current state-of-the-art CNN models are all having deeper layers with small filter sizes. The authors designed a line buffer memory structure to store the feature maps which allowed different tiles to reuse the data during the convolution operation.

CNN inference can be accelerated by using fast convolution techniques like FFT when compared with conventional direct convolution technique. Kala et al. [23] proposed such an accelerator design for AlexNet CNN model using a radix-4 butterfly FFT-based architecture. Unlike conventional radix-4 architectures, it has a control signal for the selection of output. The availability of the control signal eliminates the need for output reordering. Also, the authors have performed the complexity analysis of different convolution techniques. The analysis was validated by implementing the model in Virtex-7 FPGA device using both the convolution techniques and proved that FFT-based convolution gives better performance when the filter size is comparatively large.

Podili et al. [24] have proposed a convolution engine which is also based on Winograd minimal filtering in order to accelerate the convolution layers and to achieve maximum resource efficiency. Thus, brings down the count of multiplications and improves the throughput per multiplier by almost 3.7 times compared with the other state-of-the-art implementations. This implementation marks about 1.2 times improvement in throughput by using approximately 3 times lesser number of multipliers and 2 times lesser on-chip memory without affecting the classification accuracy. The engine is highly parallelized and pipelined which eventually lead to higher working frequency and this call for higher memory bandwidth requirements. Considering this, the authors have implemented a novel data layout also which can reduce the memory bandwidth by almost 50%. The proposed engine was implemented on VGG-16 network using HARP platform which houses an Intel Xeon E5-2600 v2 processor and Altera Stratix-V FPGA.

Most of the previous studies on algorithmic optimizations for accelerating CNNs, use separate computing elements for the convolution layers as well as the fully connected layers. But in the work done by Kala et al. [25, 26], they proposed a hybrid approach which is named as UNIWIG-Unified Winograd-GEMM architecture where the resources utilized for both the convolution layers and fully connected layers are the same. This CNN accelerator architecture is based on blocked Winograd-GEMM Algorithm and was implemented both in 32-bit floating point and 16-bit fixed point precision. AlexNet and VGG-16 CNN models were implemented using the proposed architecture on Virtex-7 FPGA (Xilinx XC7VX690T) board.

In another work by Aydonat et al. [27], they have designed a novel accelerator architecture based on OpenCL which addressed the memory bandwidth issue and boosted data reuse. They have also made use of Winograd transform and the memory bandwidth issue is resolved by incorporating on-chip buffers for storing all intermediate feature maps in CONV and FC layers.

Usually in an FPGA, two integer 8-bit multiplications are implemented on a single DSP element and it does only multiplications. Because accumulation operations require a lot of LUT accesses which incur high energy consumption. Hence, Hu et al. [28] proposed an efficient DSP Processing Element with an efficient DSP optimization which can perform Multiply and Accumulate (MAC) operation reducing the LUT access overhead. Also, a dynamic precision quantization is done to avoid the accuracy loss due to the overflow of partial sums. Incorporating the above-mentioned elements, they proposed a CNN accelerator named Tiny Neural Network Accelerator (TiNNA). They suggested an efficient DSP optimization approach named Signed Magnitude Format-INT8 (SMF-INT8) which can perform two signed 8-bit MAC operations wherein all inputs and weights are in sign-magnitude format. The SOC, housing the proposed accelerator was implemented on Xilinx ZC706 FPGA Board using Xilinx Vivado 2018.3.

According to Yang et al. [29], most of the previous accelerators focus on CNN models like AlexNet and VGG-16 which contain highly memory and compute-intensive operations while modern state-of-the-art CNN models achieve the same level of accuracy with lesser memory requirement and fewer operations. Moreover, accelerators of these models achieve higher throughput in terms of GOPS, but ends up with low inference latency in terms of Frames Per Second (FPS). They also point out that FPS is considered to be a more essential metric of accelerator efficiency.

So, the authors implemented a CNN accelerator called "Synetgy" and a new CNN model called DiracDeltaNet which are specifically tailored for embedded FPGAs. The proposed CNN model was based on one of the newer and efficient CNN models, ShuffleNetV2, which is relatively less complex in terms of size, FLOP count and gives decent and comparable accuracy. The authors have optimized this model by replacing all the 3*3 convolutions with 1*1 convolutions and shift operations. Here, the filter functions can be viewed as two-dimensional DiracDelta functions and so the name. Also, the weights of the proposed CNN model were quantized to 4-bits to reduce the computational requirements. This accelerator was implemented on the Ultra96 development board with Xilinx Zynq UltraScale+ MPSoC.

Also, owing to the flexibility in design and the reduced design time, most of the conventional FPGA-based CNN inference accelerators use High-Level Synthesis tools like Xilinx Vivado HLS. But they do suffer from on-chip memory shortage, lower hardware utilization, and memory bandwidth bottleneck. While on the other side, RTL-based CNN accelerators could be customized for specific CNN models which will eventually result in better performance and better hardware utilization. Unfortunately, most software designers find it very difficult to write RTL code and thus it incurs a long delay in the development. So, Ma et al. [30] had proposed an RTL compiler, ALAMO: Acceleration of deep Learning Algorithms with a MOdularized RTL compiler, which combines the advantages of both HLS and RTL implementations. The authors have introduced customized RTL templates for each CNN layer, namely, Scalable convolution module (CONV), Pooling module (POOL), Normalization module (NORM), and Inner-product module (FC). The proposed compiler would automatically configure these modules and connections between them would also be made accordingly by analyzing the given input CNN structure and the amount of parallelism required. The ALAMO RTL compiler is proposed to accelerate the CNN inference in FPGA platforms. Therefore, the proposed system was validated by implementing the CNN models AlexNet and NiN on a DE5-Net board with Altera Stratix-V GXA7 FPGA.

Other architectural optimizations include loop optimizations like unrolling [8, 32], loop tiling and interchange [31], and cross-layer scheduling. But since our scope of discussion focuses only on computation unit level optimizations, this paper does not include the review of the same.

5 CNN Model Optimization

Optimization of CNN models can also result in higher performance and power efficient neural network accelerators. Usually, larger the CNN model, greater is its accuracy. But then, there is a trade-off between the model accuracy and hardware speed and energy efficiency. So, compression of CNN models is one of the methods to achieve this trade-off. The prominent model compression techniques are data quantization and pruning.

5.1 Data Quantization

One among the common methods of CNN model compression is data quantization of model weights and activations. Usually, CNN models are deployed with the same numerical precision in which they were trained, which is IEEE754 Floating Point Standard. Attempts to replace this by low-bit fixed point formats will reduce the memory requirements as well as reduce the cost of hardware. Also, several studies have proved that reducing the operand precision does not harm the accuracy of the model considerably. Most of the previous studies were focusing on accelerating CONV layers which are compute-intensive. But as already mentioned, the FC layers are memory bound as they need millions of parameters to be loaded from the external memory. So, addressing the memory bandwidth issue is of great relevance. Here, some of the latest works which demonstrate CNN inference with reduced data precision are reviewed.

Qiu et al. [33] has introduced a quantization method which is dynamic and also a convolver design that works well for all types of layers in CNN to rectify the memory bandwidth constraint and reduce resource utilization. Here, the proposed quantization method works in two stages, namely, weight quantization phase and data quantization phase. The former one tries to find an optimal length for the weights in each layer and the latter one determines an optimal length for the feature maps between two layers of CNN. They had put forward the design of a heterogenous CNN accelerator which has two major parts namely Programmable Logic (PL) and Processing System (PS). The PL part is the reconfigurable FPGA which includes all the computing elements called Processing Elements (PE). The PE houses the convolver and the proposed dynamic quantization units. The PS part has got CPU and external memory for storage of all model parameters, input and output data, programs, etc. The proposed CNN accelerator was implemented in Xilinx Zynq ZC706.

In the work by Guo et al. [34], they have proposed a run-time configurable CNN accelerator named Angel-Eye, which also incorporates a dynamic data quantization strategy as well as a compiler which efficiently maps the given CNN network onto the flexible hardware. Here for performing data quantization as part of model compression, a greedy method of finding the radix position of each layer in the CNN is used which is shown in the figure. After data quantization is done on all the layers, accuracy is tested and fine tuning is performed to improve the accuracy, if needed. The proposed quantization scheme was evaluated on the state-of-the-art CNN models like GoogleNet, VGG-16, SqueezeNet, and VGG-CNF. It was proved that the 16-bit and 8-bit fixed point format gave decent accuracy without significant performance loss. Another major contribution of this work is that they proposed an efficient compiler to map the given CNN model onto the run-time configurable hardware architecture. There are three steps included in the suggested compilation process and they are block partitioning, memory mapping, and dependency check. This was evaluated on Xilinx XC7Z045 platform and XC7Z020.

Wang et al. [35] proposed a CNN accelerator for edge devices where resources are limited and power is also a stringent constraint. The proposed work suggested a design process for speeding up extremely low-bit-width neural network. The striking aspect in their work was it enabled different bit-width for different layers in CNN. They used 8-bit precision for the first convolution layer and the last fully connected layer. They have also developed an automation tool named AccELB which will automatically generate the best possible FPGA implementation for a given low-bit CNN model. The evaluation of this implementation was done on a Xilinx ZC706 evaluation platform with Zynq XC7Z045 FPGA and obtained the most energy-efficient solution compared to GPU and other FPGA implementations.

5.2 *Pruning*

In addition to cutting down the bit-width of weights and activations, yet another method for reducing the computations in the inference process is weight pruning. Weight pruning is the process where the less influential weights are made to zero, thereby increasing the sparsity of the network. As CNNs are over parameterized, this can significantly reduce the number of parameters while keeping the accuracy of classification at a decent level. So, this section reviews few of the recent works on CNN inference accelerator employing pruning technique.

Zhang et al. [36] proposed a novel mechanism of CNN model compression which was implemented on AlexNet model. The novelty of the proposed compression mechanism was that they have combined both reverse pruning and peak pruning techniques and had achieved the same classification accuracy. Reverse pruning does its job from the last layer of the CNN model to the first layer and peak pruning starts from the layer with highest number of parameters to the smallest. And as already mentioned, since the CONV layers are compute-intensive and FC layers are memory-intensive, pruning is more aggressively done in the FC layers. The strategy they adopted is a three-step pipeline process wherein the first stage adjusts the weights by traditional method of network training. The second stage removes the low weight connections which do not make a significant contribution to the network output, leading into a sparse network. The final stage is retraining the remaining weights to get adjusted with the pruned weights. This process was done iteratively for each layer in the CNN. The proposed pruning technique could achieve 13x better compressibility for AlexNet when compared to the originally inspired Han's network without any accuracy loss. Even though pruning was the major strategy of model compression used here, another compression was also done on the pruned AlexNet model by data quantization where the bit-width of the weights were reduced to 8-bits from 32-bit floating point. Linear maximum quantization was the technique used for the same. One of the major advantages of performing the quantization on top of the pruned model is that CNN model becomes even more sparse. They also proposed an efficient storage technique considering the characteristics of CNN and FC layers so that cache overhead could be reduced to a great extent. Finally, the hardware architecture of FPGA accelerator incorporating the proposed pruning and data quantization strategy was implemented and performance was evaluated on Xilinx ZCU104 evaluation board. So, this can be considered as a heterogenous approach of CNN inference acceleration.

Qiu et al. [33] used Singular Value Decomposition (SVD) technique to the weight matrix of the first FC layer and achieved a memory footprint reduction of about 85% by removing sufficient number of weights. Also, it could compress the largest FC layer of the VGG network to about 36% of its original size with only 0.04% accuracy loss.

Liu et al. [37] introduced Sparse Convolution Neural Networks (SCNN) by reducing the redundancy of parameters using sparse decomposition which eventually cut down the computation cost drastically. This procedure of obtaining maximum sparsity was done on AlexNet CNN model and could remove 90% of the weights with an accuracy drop <1%. The authors also proposed a CPU-based sparse matrix multiplication for SCNN models.

Han et al. [38] proposed a three-stage pipelined deep compression technique which combined pruning, quantization and Huffman coding to reduce the memory demand without loss of accuracy. Here weight pruning was done directly by removing weights which are below a certain threshold value. Then the remaining sparsely connected network is trained with the training data. According to them, this has reduced the parameters by 9 and 13 times for AlexNet and VGG-16 CNN models. This was followed by quantization and weight sharing which further compressed the network model. Altogether, this technique reduced the memory required by AlexNet by 35% and VGG-16 by 49%.

Aimar et al. [39] adopted the technique of utilizing the sparsity of activations in CNNs to reduce the memory constraints and to speed up the computations. The proposed accelerator architecture can efficiently work with kernel sizes up to 7*7 and could process up to 128 input and 128 output feature maps per layer in one pass. Also, the architecture was validated in all the state-of-the-art CNN models and achieved lesser memory bandwidth and lesser computation time.

6 Evaluation

In this section, we have tabulated the performance of the different CNN accelerators discussed in Sects. 4 and 5. Because of the heterogeneity of the techniques used and the target FPGA devices, it is actually unfair to do a comparison between them. But still, it can give some insights into the recent trends and future directions in the area of FPGA-based CNN inference accelerators. The designs used for comparison are listed in Table 2. The major parameters used for comparison are precision, logic utilization, DSP utilization, BRAM utilization, and throughput.

7 Conclusion

In this paper, we reviewed some of the recent works on the design of CNN inference accelerators and discussed in brief on the optimization strategies used in these works. We classified the different types of CNN accelerators based on the technique used for optimization. Mainly, we categorized the works based on whether the optimization is done in the architectural or algorithmic level or in the CNN model. Also, the different myriad of possibilities for accelerating CNN inference, give immense opportunities for the new researchers in this area. The results of these works tabulated in Table 2 clearly show that FPGA-based CNN accelerators give pretty decent performance in terms of throughput and resource utilization.

Table 2 FPGA-based CNN accelerators employing different optimization strategies

References	CNN model	FPGA platform	Freq (MHz)	Precision (bit fixed)	Logic usage	DSP usage	BRAM usage	Throughput (GOPS)
[26]	AlexNet	Virtex-7	200	16	468 K (67%)	1436 (39.8%)	1465 (99%)	433.63
[26]	VGG-16	Virtex-7	200	16	468 K (67%)	1436 (39.8%)	1465 (99%)	407.23
[24]	VGG-16	HARP	200	32	196 K (83.6%)	256 (100%)	–	229.2
[28]	ResNet-50	Zync-ZC706	200	8	78.98K	448	168	241.2
[29]	Dirac DeltaNet	Zync-ZU3EG	250	4	51 K (73.4%)	360 (100%)	159 (73.6%)	–
[31]	AlexNet	Stratix-V	100	8–16	121 K (52%)	256 (100%)	1552 (61%)	114.5
[31]	NiN	Stratix-V	100	8–16	121 K (52%)	256 (100%)	23390 (91%)	117.3
[33]	VGG-16	Zync-XC7Z045	150	16	182 K (83.5%)	780 (89.2%)	486 (86.7%)	136.97
[34]	VGG-16	Zync-XC7Z045	150	16	182 K (83.5%)	780 (89.2%)	486 (86.7%)	187.8
[34]	VGG-16	Zync-XC7Z020	214	8	29 K (56%)	190 (86.4%)	85.5 (61%)	84.3
[36]	AlexNet	Zync-XCZU7EV	300	8	101 K (44.25%)	696 (40.28%)	1998.5 (63.62%)	14.11
[39]	VGG-16	Zync-ZC7100	60	16	229 K (83%)	128 (6.3%)	386 (51.1%)	368.47

References

1. Huang T (1996) Computer vision: evolution and promise
2. Lecun Y, Bottou L, Bengio Y, Haffner P (1998) Gradient-based learning applied to document recognition. Proc IEEE 86(11):2278–2324
3. Farabet C, Poulet C, Han JY, LeCun Y (2009) CNP: an FPGA-based processor for convolutional networks. In: 2009 international conference on field programmable logic and applications. IEEE
4. Ji S, Xu W, Yang M, Yu K (2012) 3D convolutional neural networks for human action recognition. IEEE Trans Pattern Anal Mach Intell 35(1):221–231
5. Nurvitadhi E, Venkatesh G, Sim J, Marr D, Huang R, Ong Gee Hock J, Boudoukh G (2017) Can FPGAs beat GPUs in accelerating next-generation deep neural networks? In: Proceedings of the 2017 ACM/SIGDA international symposium on field-programmable gate arrays, pp 5–14
6. Farabet C, Martini B, Akselrod P, Talay S, LeCun Y, Culurciello E (2010) Hardware accelerated convolutional neural networks for synthetic vision systems. In: Proceedings of 2010 IEEE international symposium on circuits and systems. IEEE, pp 257–260
7. Farabet C, Martini B, Corda B, Akselrod P, Culurciello E, LeCun Y (2011) Neuflow: a runtime reconfigurable dataflow processor for vision. In: CVPR 2011 workshops. IEEE, pp 109–116
8. Zhang C, Li P, Sun G, Guan Y, Xiao B, Cong J (2015) Optimizing FPGA-based accelerator design for deep convolutional neural networks. In: Proceedings of the 2015 ACM/SIGDA international symposium on field-programmable gate arrays, pp 161–170
9. Zhang C, Wu D, Sun J, Sun G, Luo G, Cong J (2016) Energy-efficient CNN implementation on a deeply pipelined FPGA cluster. In: Proceedings of the 2016 international symposium on low power electronics and design, pp 326–331
10. Pérez I, Figueroa M (2021) A heterogeneous hardware accelerator for image classification in embedded systems. Sensors 21(8):2637
11. Blaiech AG, Khalifa KB, Valderrama C, Fernandes MA, Bedoui MH (2019) A survey and taxonomy of FPGA-based deep learning accelerators. J Syst Archit 98:331–345
12. Li Z, Wang Y, Zhi T, Chen T (2017) A survey of neural network accelerators. Front Comput Sci 11(5):746–761
13. Lacey G, Taylor GW, Areibi S (2016) Deep learning on FPGAs: past, present, and future. arXiv preprint arXiv:1602.04283
14. Guo K, Zeng S, Yu J, Wang Y, Yang H (2017) A survey of FPGA-based neural network accelerator. arXiv preprint arXiv:1712.08934
15. Abdelouahab K, Pelcat M, Serot J, Berry F (2018) Accelerating CNN inference on FPGAs: a survey. arXiv preprint arXiv:1806.01683
16. Mittal S (2020) A survey of FPGA-based accelerators for convolutional neural networks. Neural Comput Appl 32(4):1109–1139
17. Hinton GE, Krizhevsky A, Sutskever I (2012) Imagenet classification with deep convolutional neural networks. Adv Neural Inform Process Syst 25(1106–1114)
18. Simonyan K, Zisserman A (2014) Very deep convolutional networks for large-scale image recognition. arXiv preprint arXiv:1409.1556
19. Szegedy C, Liu W, Jia Y, Sermanet P, Reed S, Anguelov D, Rabinovich A (2015) Going deeper with convolutions. In: Proceedings of the IEEE conference on computer vision and pattern recognition, pp 1–9
20. He K, Zhang X, Ren S, Sun J (2016) Deep residual learning for image recognition. In: Proceedings of the IEEE conference on computer vision and pattern recognition, pp 770–778
21. Lu L, Liang Y, Xiao Q, Yan S (2017) Evaluating fast algorithms for convolutional neural networks on FPGAs. In: 2017 IEEE 25th annual international symposium on field-programmable custom computing machines (FCCM). IEEE, pp 101–108
22. Lavin A, Gray S (2016) Fast algorithms for convolutional neural networks. In: Proceedings of the IEEE conference on computer vision and pattern recognition, pp 4013–4021
23. Kala S, Jose BR, Paul D, Mathew J (2018) A hardware accelerator for convolutional neural network using fast Fourier transform. In: International symposium on vlsI design and test. Springer, Singapore, pp 28–36

24. Podili A, Zhang C, Prasanna V (2017) Fast and efficient implementation of convolutional neural networks on FPGA. In :2017 IEEE 28th international conference on application-specific systems, architectures and processors (ASAP). IEEE, pp 11–18
25. Kala S, Jose BR, Mathew J, Nalesh S (2019) High-performance CNN accelerator on FPGA using unified winograd-GEMM architecture. IEEE Trans Large Scale Integr (VLSI) systems, vol 27, issue 12, pp 2816–2828
26. Kala S, Jose BR, Mathew J, Nalesh S (2019) Efficient hardware acceleration of convolutional neural networks. In: 2019 32nd IEEE international system-on-chip conference (SOCC). IEEE, pp 191–192
27. Aydonat U, O'Connell S, Capalija D, Ling AC, Chiu GR (2017) An opencl™ deep learning accelerator on Arria 10. In: Proceedings of the 2017 ACM/SIGDA international symposium on field-programmable gate arrays, pp 55–64
28. Hu X, Li X, Huang H, Zheng X, Xiong X (2022) TiNNA: a tiny accelerator for neural networks with efficient dsp optimization. IEEE Trans Circ Syst II: Express Briefs 69(4):2301–2305
29. Yang Y, Huang Q, Wu B, Zhang T, Ma L, Gambardella G, Keutzer K (2019) Synetgy: algorithm-hardware co-design for convnet accelerators on embedded FPGAs. In: Proceedings of the 2019 ACM/SIGDA international symposium on field-programmable gate arrays, pp 23–32
30. Ma Y, Suda N, Cao Y, Vrudhula S, Seo JS (2018) ALAMO: FPGA acceleration of deep learning algorithms with a modularized RTL compiler. Integration 62:14–23
31. Ma Y, Cao Y, Vrudhula S, Seo JS (2017) Optimizing loop operation and dataflow in FPGA acceleration of deep convolutional neural networks. In: Proceedings of the 2017 ACM/SIGDA international symposium on field-programmable gate arrays, pp 45–54
32. Zhang C, Prasanna V (2017) Frequency domain acceleration of convolutional neural networks on CPU-FPGA shared memory system. In: Proceedings of the 2017 ACM/SIGDA international symposium on field-programmable gate arrays, pp 35–44
33. Qiu J, Wang J, Yao S, Guo K, Li B, Zhou E, ... Yang H (2016) Going deeper with embedded FPGA platform for convolutional neural network. In: Proceedings of the 2016 ACM/SIGDA international symposium on field-programmable gate arrays, pp 26–35
34. Guo K, Sui L, Qiu J, Yu J, Wang J, Yao S, ... Yang H (2017) Angel-eye: a complete design flow for mapping CNN onto embedded FPGA. IEEE Trans Comput-Aided Des Integr Circ Syst 37(1):35–47
35. Wang J, Lou Q, Zhang X, Zhu C, Lin Y, Chen D (2018) Design flow of accelerating hybrid extremely low bit-width neural network in embedded FPGA. In: 2018 28th international conference on field programmable logic and applications (FPL). IEEE, pp. 163–1636
36. Zhang M, Li L, Wang H, Liu Y, Qin H, Zhao W (2019) Optimized compression for implementing convolutional neural networks on FPGA. Electronics 8(3):295
37. Liu B, Wang M, Foroosh H, Tappen M, Pensky M (2015) Sparse convolutional neural networks. In Proceedings of the IEEE conference on computer vision and pattern recognition, pp 806–814
38. Han S, Mao H, Dally WJ (2015) Deep compression: compressing deep neural networks with pruning, trained quantization and Huffman coding. arXiv preprint arXiv:1510.00149
39. Aimar A, Mostafa H, Calabrese E, Rios-Navarro A, Tapiador-Morales R, Lungu IA, Delbruck T (2018) NullHop: a flexible convolutional neural network accelerator based on sparse representations of feature maps. IEEE Trans Neural Netw Learn Syst 30(3):644–656

Automatic Cloud Detection from Satellite Images Using U-Net Based Convolutional Neural Network

S. R. Surya and M. Abdul Rahiman

Abstract Cloud detection is an important and difficult task in the processing of satellite remote sensing data. Clouds are one of the most major noise presents in the satellite image. Presence of clouds hide the ground information. So, it is not possible to predict the values of pixels affected by clouds. Satellite images are having wide variety of applications. Because of the presence of clouds, the satellite images are partially unusable. To use the information from satellite images it is necessary to detect the presence of clouds. Accurate detection of clouds is particularly important before using satellite images for applications like land use classification, object detection, forest fire detection, climate change analysis, etc. In this paper a novel method for automatic cloud detection based on U-Net convolutional neural network is proposed. Qualitative and quantitative evaluations are conducted on publicly available cloud validation datasets, which indicates that our proposed method is more effective and accurate than the state-of-the-art methods and can accurately detect clouds under different conditions.

Keywords Cloud detection · Remote sensing images · U-Net architecture · Deep learning · Convolutional neural network

S. R. Surya (✉)
LBS Centre for Science and Technology, University of Kerala, Thiruvananthapuram, Kerala, India
e-mail: sooryasr07@gmail.com

Department of CSE, College of Engineering Perumon, Kollam, India

M. Abdul Rahiman
Director, LBS Centre for Science and Technology, Govt of Kerala, Thiruvananthapuram, Kerala, India
e-mail: rehmanpaika@gmail.com

S. K. Biswas et al. (eds.), *Intelligent Computing Systems and Applications*, Lecture Notes in Networks and Systems 1307, https://doi.org/10.1007/978-981-96-3860-4_7

1 Introduction

Optical remote sensing images are important data source available for monitoring changes in nature. But cloud cover present in Earth's surface in is more than 50% most of the area around land and sea were covered by clouds [1]. Cloud cover present in optical satellite images may help in detecting climate change analysis [2, 3]. But one of the after effect of cloud is it hides the ground information, so cloud covered regions are present in satellite images as high intensity values. It is difficult to get land information under the cloud covered regions. This necessitates the development of accurate cloud detection algorithms. Many researchers are working in the fields of cloud detection and removal areas. In the literature several cloud detection methods have been proposed and which can be grouped into three categories: threshold-based methods, machine learning based methods and deep learning-based methods. Threshold based method is one of the basic method available for cloud detection. The function of mask (Fmask) algorithm proposed by Zhu et al. [4], which used a decision tree and multiple thresholds to identify cloud and non-cloud pixels. In [5] improvements were made on Fmask algorithm. For cloud detection in Landsat images an automatic cloud cover assessment algorithm was designed in [6]. Tan et al. [7] proposed machine learning (ML)-based multilayer cloud detection algorithm for the passive Advanced Himawari Imager (AHI) aboard the geostationary Himawari-8 satellite. The extreme learning machine, support vector machine, and multilayer perceptron machine learning techniques are used for the detection of clouds in the method proposed by Tian et al. [8]. C. Williams et al. developed supervised accurate and fast annotation technique for cloud detection [9]. Zang et al. [10] proposed an encoder–decoder neural network architecture for cloud detection with new activation function ActivateOrNot and attention mechanism. It is based on U-Net architecture with attention mechanism. Image-to-image domain translation network (DTNet) is used for detecting clouds without using any paired and pixel-level annotation training data proposed by Guo et al. [11]. Peng et al. [12] analyzed the role of receptive field on the cloud segmentation results. They explored the relationship between the receptive field size in the performance of a cloud detection network. And they found that the network with a small receptive field is good for cloud detection of Landsat 8 OLI imagery and for increasing the performance.

For detection of all variants of clouds an encoder-decoder architecture which uses a deformable convolutional block at the encoder side and incorporate skip-connection mechanisms into the decoder for precise cloud detection results proposed by Liu et al. [13]. Luotamo et al. [14] proposed two cascaded CNN models for the detection of clouds. This method reduced the computational complexity by successively processing under sampled and full-resolution images.

A deep learning technique for cloud detection on RGB data proposed by Jeppesen et al. [15]. They combined SLIC, CRF and PCA Net used for cloud detection in [16] and SVM classifier is used to predict the pixel as cloud or non-cloud.

Because of the powerful representation learning capability and yield significant performance now a days, deep convolutional neural networks (CNNs) have been widely used in cloud detection. Since clouds are having variable geometry and complexity and under heterogeneous landscape, most of the existing CNN based methods are having challenges.

The structure of this paper is as follows: the U-Net architecture used in this paper is described in Sect. 2. Subsequently, the dataset, experiment environment, results and experimental analysis are given in Sect. 3. Finally, Sect. 4 summarizes the work of this paper.

2 Proposed Method

In the proposed automatic cloud detection method, convolution neural network based on U-Net is used to segment each pixel as cloud or non-cloud. The original cloudy satellite image is shown in Fig. 1 and corresponding ground truth is shown in Fig. 2. The block diagram of the proposed model is shown in Fig. 3. Initially the input cloudy satellite image is preprocessed. And in the preprocessing step image pixel values are normalized between 0 and 1. After preprocessing the given data set is divided into training set and test set. 70% of the data is used for training and 30% for testing and out of 70% training data 10% is used for validation.

The U-Net was developed by Ronneberger et al. [17] for Bio Medical image segmentation. The proposed architecture is inspired from the idea proposed in [17]. It contains two paths. First path is the contraction path which is down sampler or encoder. The down sampler is stack of convolution and max pooling layers. It is used to extract the context in the image. The second path is the symmetric expanding path which is up-sampler or decoder. The up-sampler blocks are used to precisely predict

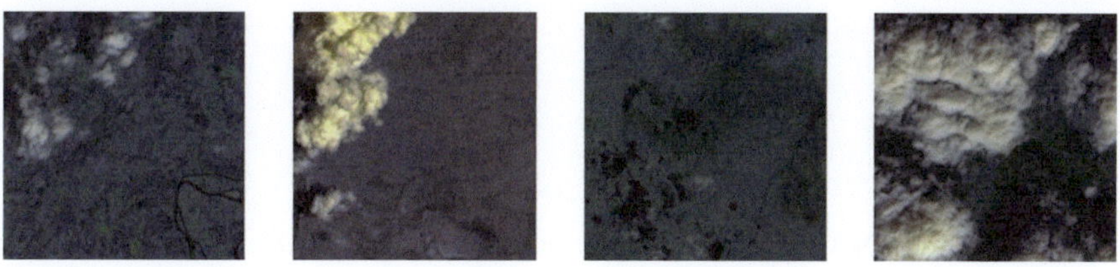

Fig. 1 Cloudy input image

Fig. 2 Ground truth cloud mask image

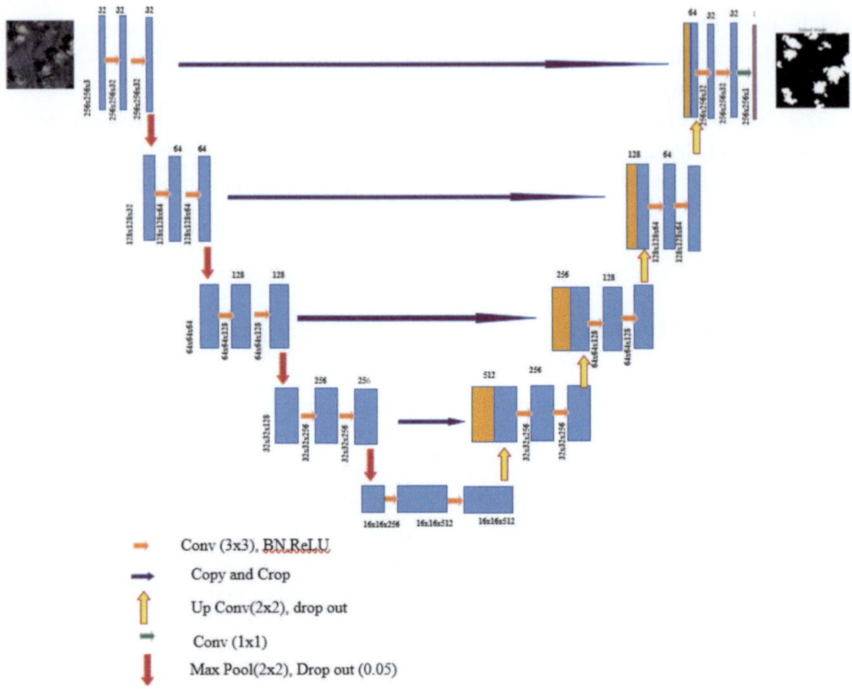

Fig. 3 U-Net CNN for cloud detection (based on U-Net [17])

cloud pixels using transposed convolution. In between encoder and decoder there is a bottleneck layer. The bottleneck layer is used to extract more features corresponding to cloud.

2.1 Encoder

In the proposed U-Net model for cloud detection the encoder consists of stack of four blocks. Each block has two layers of convolution, batch normalization and ReLU activation. Batch normalization is used to speed up training. Then some random numbers of neurons are dropped out to prevent overfitting and down sampled using max pooling. The down sampled data given to the next block. Initially the number of convolutional filters used is 32. At each block number of convolutional filters are increased and max pooling is used to reduce the dimensionality of features and to reduce the computational costs. The output of the convolution layer is then passed through a non-linear activation function the rectified linear unit (ReLU).

$$\mathrm{ReLU}(Zi) = \max(0, Zi) \tag{1}$$

2.2 Decoder

The decoder consists of stack of four blocks. The decoder blocks up-samples the cloud features back to the original image size. At each up-sampling level, the output of the corresponding encoder block is concatenated with up sampled values before forwarding to the next decoder block. The up-sampling operation is performed by transposed convolutions which helps to localize the cloud pixels. Finally in the last stage 1 × 1 convolution is performed with sigmoid activation function which will generate an output image O^{out} with same size as input image. Thresholding is applied to output mage to generate final binary cloud mask in which cloud pixels are represented as white and non-cloud pixels are represented in black color.

$$\mathrm{Predicted}_{\mathrm{Mask}i,j} = \begin{cases} 1 \text{ if } \left(O_{i,j}^{\mathrm{out}} \geq 0.5\right) \\ 0 \quad \text{otherwise} \end{cases} \tag{2}$$

2.3 Training

After building the model and forming training and test set, the model undergo training to evaluate the validation error. We have used Nesterov-accelerated Adaptive Moment Estimation (Nadam) optimizer with initial learning rate of 10^{-4}, loss function used is dice loss function and metrics used are Dice coefficient, Precision and Recall, with batch size 32 and epochs is set to 300.

3 Dataset and Experimental Environment

3.1 Dataset

The dataset that is used in the experiment is created by Joseph Hughes [18], Oregon State University. It was derived manually from Pre-Collection Landsat 8 scenes. It contains 80, 1000 × 1000-pixel cloud validation data. We extracted several slightly overlapping and non over lapping 256 × 256 patches from these images and generated total 936 images. Out of 936 images 655 images are used for training and 281 images are used for testing.

3.2 Experimental Environment

In this paper, all the experiments are programmed and implemented with Keras framework on Ubuntu 20.04.3 LTS and trained with NVIDIA Corporation GK208B [GeForce GT 710]. The result of cloud detection of various 256 × 256 patches extracted from SPARCS validation dataset [18] is shown in Fig. 4. Both qualitative and quantitative analysis shows that our approach with dice loss function gives better performance. With this approach thin clouds also detected effectively. Due to consistent color distributions and similar local texture patterns cloud and snow share similar low-level features. But discrimination between cloud and snow is not addressed in this paper. The cloud masks obtained with this approach is close to original ground truth. Various quantitative measures of the images shown in Fig. 4 are shown in Table 1. From Table 1 our approach gives good result in almost all images.

The Result of cloud detection in various scenes in the dataset is shown in Fig. 5. In Fig. 5 first column represents the original cloudy input image, ground truth is shown in column 2, cloud detected by our approach is in column 3 and last column represents the result of cloud detection by U-Net with ResNet-34 Backbone. From Fig. 5 visual quality of the proposed cloud detection method is better compared to cloud detected by using U-Net with ResNet-34 Backbone. In second row of Fig. 5 the red circle represents the small thin cloud detection result, the result obtained using

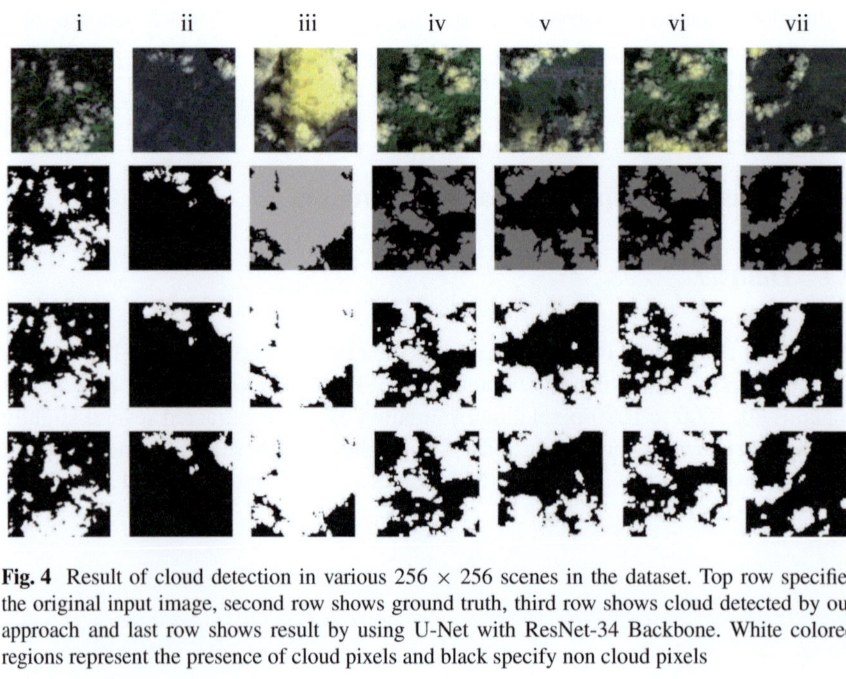

Fig. 4 Result of cloud detection in various 256 × 256 scenes in the dataset. Top row specifies the original input image, second row shows ground truth, third row shows cloud detected by our approach and last row shows result by using U-Net with ResNet-34 Backbone. White colored regions represent the presence of cloud pixels and black specify non cloud pixels

Table 1 Values of various performance metric obtained for the images shown in the Fig. 4

	Model	Input image						
		i	ii	iii	iv	v	vi	vii
Accuracy	ResNet-34	0.9701	0.9819	**0.9605**	0.9643	0.9802	0.9625	**0.9796**
	Our approach	**0.9745**	**0.9834**	0.9578	**0.9667**	**0.9807**	**0.9652**	0.9784
F-Score	ResNet-34	0.9319	0.9832	**0.9543**	0.8960	0.9806	0.9362	**0.8992**
	Our approach	**0.9592**	**0.9845**	0.9518	**0.9029**	**0.9813**	**0.9409**	0.8981
Precision	ResNet-34	0.9560	0.9906	**0.9637**	0.9303	**0.9894**	0.9242	**0.9207**
	Our approach	**0.9623**	**0.9927**	0.9484	**0.9391**	0.9846	**0.9274**	0.8749
Recall	ResNet-34	0.9090	**0.9759**	0.945	0.8641	0.9719	0.9486	0.8786
	Our approach	**0.9424**	0.9765	**0.9551**	**0.8693**	**0.9781**	**0.9547**	**0.9227**
Jaccard index	ResNet-34	0.8726	0.9669	**0.9125**	0.8117	0.9619	0.8801	**0.8168**
	Our approach	**0.8910**	**0.9695**	0.9079	**0.8231**	**0.9633**	**0.8884**	0.8151

our approach is close to ground truth but in segmentation using U-Net with ResNet-34 Backbone the thin cloud as well as the nearby non-cloud pixels also misclassified as cloud pixels. Our approach reduced the number of misclassifications.

3.3 *Evaluation Metrics*

The accuracy, Dice, Precision, Recall, and Jaccard Index to portray the performance of the algorithm. All the metrics are denoted as following:

$$\text{Accuracy} = \frac{(\text{TP} + \text{TN})}{(\text{TP} + \text{TN} + \text{FP} + \text{FN})} \tag{3}$$

$$\text{Dice} = \frac{2 * (\text{Precision} * \text{Recall})}{(\text{Precision} + \text{Recall})} \tag{4}$$

$$\text{Precision} = \frac{\text{TP}}{\text{TP} + \text{FP}} \tag{5}$$

$$\text{Recall} = \frac{\text{TP}}{\text{TP} + \text{FN}} \tag{6}$$

$$\text{Jaccard Index} = \frac{\text{TP}}{(\text{TP} + \text{FP} + \text{FN})} \tag{7}$$

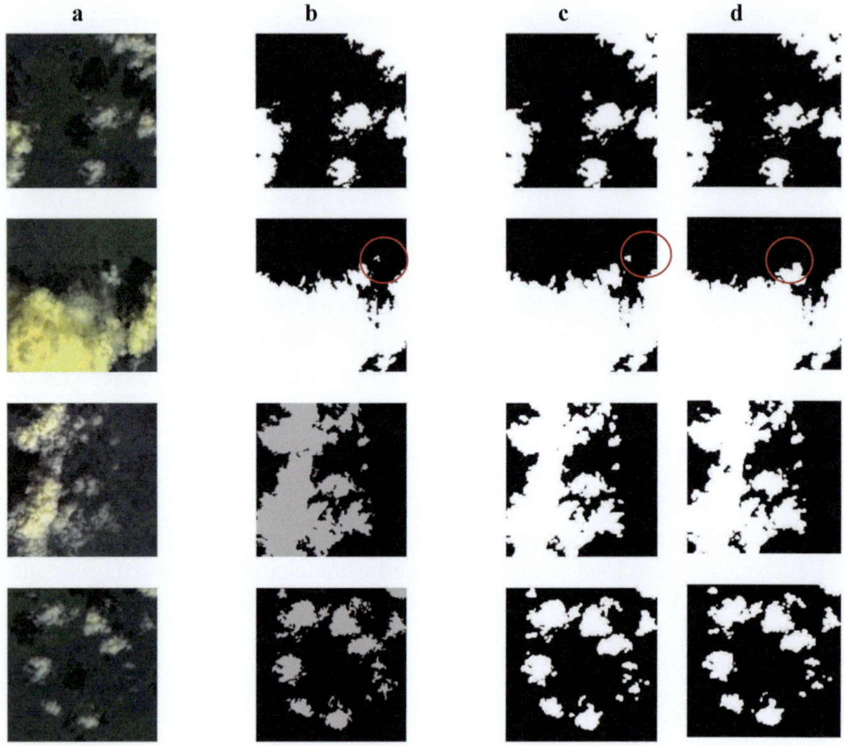

Fig. 5 Result of cloud detection in various scenes in the dataset. **a** Original input image, **b** ground truth, **c** cloud detected by our approach and **d** cloud detected by using U-Net with ResNet-34 Backbone. White colored regions represent the presence of cloud pixels and black colored regions specify non cloud pixels

In Table 2 the pixelwise semantic segmentation method proposed is compared with U-Net with ResNet-32, ResNet-18 and InceptionV3 backbone defined in Keras segmentation models. From Table 2 our approach gives better result in dice coefficient and recall compared to other models. The learning curve is shown in Fig. 6. From Fig. 6 as the number of epochs increases dice coefficient, precision and recall increases and the loss decreases.

Table 2 Comparison with other methods

Model	Dice coefficient	Precision	Recall
ResNet-32	0.9210	**0.9585**	0.8924
Our approach	**0.9240**	0.9457	**0.9094**
ResNet-18	0.9207	0.9583	0.8904
Inception V3	0.9226	0.9574	0.9004

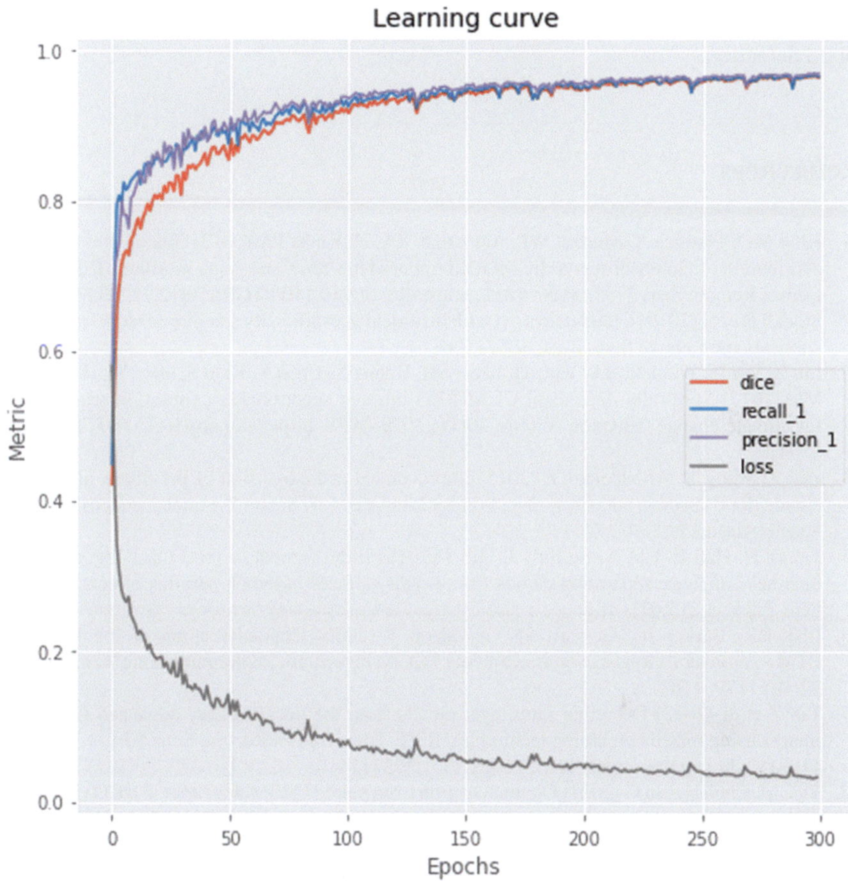

Fig. 6 Training metric

4 Conclusion

This paper presents an automatic cloud detection based on U-Net convolutional neural networks. Cloud detection is important preprocessing steps in the processing of satellite images. Nowadays deep learning plays an important role in image segmentation and processing. In this paper clouds are automatically detected by using U-Net architecture. The model is trained on labelled images, where it learns to classify cloud covered and non cloud covered regions at pixel level. The U-Net convolutional neural network with dice loss function can be able to detect both thin and thick clouds most effectively minimizing the false positives and false negatives. Discrimination between cloud and snow is not addressed in this paper. Experimental results proved that the proposed cloud detection is better than the existing methods. In future we will optimize this network and will apply the same technique for detecting clouds in satellite images produced by different sensors. Also, we will consider other bands in cloud detection.

References

1. King MD, Platnick S, Menzel WP, Ackerman SA, Hubanks PA (2013) Spatial and temporal distribution of clouds observed by MODIS onboard the Terra and Aqua satellites. IEEE Trans Geosci Remote Sens 51(7):3826–3852. https://doi.org/10.1109/TGRS.2012.2227333
2. Stöckli R et al (2019) Cloud detection with historical geostationary satellite sensors for climate applications. Remote Sens 11(9):1052–1063
3. Liu X, Wu W, Wielicki BA, Yang Q, Kizer SH, Huang X, Chen X, Kato S, Shea YL, Mlynczak MG (2017) Spectrally dependent CLARREO infrared spectrometer calibration requirement for climate change detection. J Clim 30(11):3979–3998. https://doi.org/10.1175/JCLI-D-16-0704.1
4. Zhu Z, Wang S, Woodcock CE (2015) Improvement and expansion of the fmask algorithm: cloud, cloud shadow, and snow detection for Landsats 4–7, 8, and Sentinel 2 images. Remote Sens Environ 159:269–277
5. Frantz D, Haß E, Uhl A, Stoffels J, Hill J (2018) Improvement of the Fmask algorithm for Sentinel-2 images: separating clouds from bright surfaces based on parallax effects. Remote Sens Environ 215:471–481
6. Irish RR, Barker JL, Goward SN, Arvidson T (2006) Characterization of the landsat-7 ETM+ automated cloud-cover assessment (ACCA) algorithm. Photogramm Eng Remote Sens 72(10):1179–1188
7. Tan Z et al (2022) Detecting multilayer clouds from the geostationary advanced Himawari imager using machine learning techniques. IEEE Trans Geosci Remote Sens 60:1–12, Art no. 4103112. https://doi.org/10.1109/TGRS.2021.3087714
8. Tian M, Chen H, Liu G (2021) Cloud detection from paired CrIS water vapor and CO_2 channels using machine learning techniques. IEEE Trans Geosci Remote Sens 59(4):2781–2793. https://doi.org/10.1109/TGRS.2020.3020120
9. Williams C, Dagobert T, De Franchis C, Morel JM, Hessel C (2021) Fast accurate supervised cloud annotation. In: 2021 IEEE international geoscience and remote sensing symposium IGARSS, 2021, pp 3237–3240. https://doi.org/10.1109/IGARSS47720.2021.9554213
10. Zhang L, Sun J, Yang X, Jiang R, Ye Q (2022) Improving deep learning-based cloud detection for satellite images with attention mechanism. IEEE Geosci Remote Sens Lett 19:1–5, Art no. 6005505. https://doi.org/10.1109/LGRS.2021.3133872
11. Guo J, Yang J, Yue H, Chen Y, Hou C, Li K (2022) Cloud detection from remote sensing imagery based on domain translation network. IEEE Geosci Remote Sens Lett 19:1–5, 2022, Art no. 5000805. https://doi.org/10.1109/LGRS.2021.3061569
12. Peng L, Chen X, Chen J, Zhao W, Cao X (2022) Understanding the role of receptive field of convolutional neural network for cloud detection in Landsat 8 OLI imagery. IEEE Trans Geosci Remote Sens 60:1–17, Art no. 5407317. https://doi.org/10.1109/TGRS.2022.3150083
13. Liu Y, Wang W, Li Q, Min M, Yao Z (2022) DCNet: a deformable convolutional cloud detection network for remote sensing imagery. IEEE Geosci Remote Sens Lett 19:1–5, Art no. 8013305. https://doi.org/10.1109/LGRS.2021.3086584
14. Luotamo M, Metsämäki S, Klami A (2021) Multiscale cloud detection in remote sensing images using a dual convolutional neural network. IEEE Trans Geosci Remote Sens 59(6):4972–4983. https://doi.org/10.1109/TGRS.2020.3015272
15. Jeppesen JH, Jacobsen RH, Inceoglu F, Toftegaard TS (2019) A cloud detection algorithm for satellite imagery based on deep learning. Remote Sens Environ 229:247–259
16. Zi Y, Xie F, Jiang Z (2018) A cloud detection method for Landsat 8 images based on PCANet. Remote Sens 10(6):877

17. Ronneberger O, Fischer P, Brox T (2015) U-Net: convolutional networks for biomedical image segmentation. In: Navab N, Hornegger J, Wells W, Frangi A (eds) Medical image computing and computer-assisted intervention—MICCAI 2015. MICCAI 2015. Lecture notes in computer science, vol 9351. Springer, Cham. https://doi.org/10.1007/978-3-319-24574-4_28
18. Hughes MJ, Hayes DJ (2014) Automated detection of cloud and cloud shadow in single-date Landsat imagery using neural networks and spatial post-processing. Remote Sens 6(6):4907–4926. https://doi.org/10.3390/rs6064907

A Fuzzy FUCOM-TOPSIS-Based Optimal Framework Towards Cloud Service Selection

Satyabrata Nath, Purnendu Das, and Pradip Debnath

Abstract With the advancement and expansion of cloud computing services, selecting a proper cloud service is regarded as a challenging endeavor, since there is fierce competition in the industry for delivering a quality experience to its customers. For addressing this challenge, we introduced a hybrid framework consisting of the Full Consistency Method (FUCOM) and Technique for Order Preference by Similarity to Ideal Solution (TOPSIS) in a fuzzy environment to determine the optimal cloud service. Here fuzzy FUCOM is used to generate the criteria weights which will be utilized by fuzzy TOPSIS to predict the rank of the alternatives. Comparative study and sensitivity analysis are done to check the efficacy and robustness of our model. The outcomes show that the proposed model involves fewer pairwise comparisons while providing a more consistent result than existing approaches.

Keywords Fuzzy TOPSIS · Fuzzy FUCOM · Cloud service selection · MCDM

1 Introduction

One of the most evolving platforms based on the internet computational paradigm that transforms the way people compute, store data, and provide services is cloud computing [1]. This computing paradigm delivers significant benefits to enterprises connected to IT ecosystem, thus allowing them to start with limited resources and scale them up as needed, freeing up more time for innovating and developing

S. Nath (✉) · P. Das
Department of Computer Science, Assam University, Silchar, Assam, India
e-mail: satyabrata.nath@aus.ac.in

P. Das
e-mail: purnen1982@gmail.com

P. Debnath
Department of Mathematical Sciences, Tezpur University, Tezpur, Assam, India
e-mail: debnath.pradip@yahoo.com

S. K. Biswas et al. (eds.), *Intelligent Computing Systems and Applications*, Lecture Notes in Networks and Systems 1307, https://doi.org/10.1007/978-981-96-3860-4_8

economic interest [2]. Cloud computing is a significant service-oriented paradigm that provides a range of online services such as—SaaS, PaaS, and IaaS [3].

The rapid expansion of cloud providers, along with the numerous services offered by them at an attractive and varying pricing scheme, has made it difficult to compare cloud providers in context of quality and cost of service, leading to complexity in cloud service selection (CSS) [4]. Furthermore, for a non-expert with minimal knowledge of the cloud ecosystem, selecting a cloud service is a time-consuming and complicated process. The primary parameters for selecting and ranking different cloud service providers (CSP) from an identical group are QoS (Quality of Service) criteria which can be both functional and non-functional in nature. These QoS criteria are used by the cloud customer for cloud service evaluation of different types. CSS is a common instance in which a client or decision-maker selects the best cloud service depending upon several conflicting QoS criteria [5, 6]. As there are various aspects involved in CSS, such as various QoS criteria, number of alternatives, and decision-makers, thus the issue of selecting optimal service comes under Multi-criteria Decision-Making (MCDM) problem [7]. Therefore, a more effective MCDM method is needed to cope with client's specifications and evaluate cloud services depending on their capability.

Several recent research attempts have been done in the field of CSS [7–9]. In perspective of their environment, uses, and constraints, they have certain advantages and disadvantages. Recent studies show the use of various MCDM methodologies for evaluating, ranking, and selecting trustworthy cloud services [10]. Numerous MCDM techniques, such as TOPSIS [11], PROMETHEE [12], AHP [13], BWM [14], and others, have proved their use in resolving the CSS problem. Thus, service selection is a critical subject in cloud computing, and its significance is projected to grow in future as cloud computing expands. With this in mind, we suggested a hybrid architecture to address the challenges of CSS, stating the primary contribution as follows:

1. Our model integrates fuzzy FUCOM (F-FUCOM) and fuzzy TOPSIS (F-TOPSIS). In this case, F-FUCOM is employed to calculate criteria weights, which then become input for F-TOPSIS to establish the final alternative rankings.
2. Since our model is based on fuzzy data, it will be more efficient in dealing with inconsistencies and vagueness.
3. The proposed model's robustness and efficiency are validated by sensitivity and comparative analysis.

The remaining paper is structured as: Sect. 2 summarizes related works on the topic of CSS. Some fundamental concepts are described in Sect. 3 followed by the proposed model and a case study with real-life dataset, to support the proposed framework in Sect. 4. The concluding remarks are addressed in Sect. 5.

2 Related Works

Over the last few years, several research attempts have been made to address the issue of CSS using QoS criteria. In paper [15], the authors generate a series of mapping rules and introduce a new method to calculate end-to-end QoS metrics in the cloud. They calculated the QoS weights with AHP. The authors of [16] presented a Service Management Index (SMI) based SMICLOUD framework for comparison and ranking of cloud services (IaaS). Here also AHP method was employed for SMI criteria weights calculation according to user preferences and compared the IaaS cloud services using the generated weights. A fuzzy MCDM model consisting of Analytic Network Process (ANP) and interpretive structure modeling has been presented in [17] for reducing the interactive importance among the pairs of criteria. In article [18], a fuzzy user-focused paradigm was proposed. The AHP approach was employed in the paper to generate the weights of criteria based on QoS, while fuzzy TOPSIS assessed the cloud alternatives and gave final rankings. Nawaz et al. [14] suggested a hybrid model for acceptable cloud selection that combines the Markov chain and the Best Worst Method (BWM) to deal with changing user preferences. The Markov chain method is used to establish a pattern of changing user priorities, which is then utilized to determine criteria weights. Using an expanded version of Grey TOPSIS, Jatoth et al. [19] offered a hybrid MCDM model for choosing cloud services. In this work, AHP determines criteria weights that are combined with Grey TOPSIS for alternative rankings. TOPSIS includes grey values to account for the uncertainty that comes with cloud services. The authors of [7] proposed CCS-OSSR, a novel hybrid CSS model which incorporates BWM and TOPSIS for ranking and prioritizing the cloud services based on QoS factors.

3 Fundamental Concepts

This section highlights certain principles and basic notions related to our work. This is needed to comprehend the proposed model.

3.1 Triangular Fuzzy Numbers (TFNs)

The linguistic variables in the domain of Fuzzy Set Theory are assigned with certain membership degrees that are taken into consideration for the probability distribution. To accomplish this, fuzzy numbers are used. For our study, we used triangular fuzzy numbers as it's very popular and easy to interpret.

Definition 1 A TFN can be expressed as pair of numbers in the form $G = (G^L, G^M, G^U)$ such that $L \leq M \leq U$, where *L*, *M*, *U* represent the lower bound, middle and upper bound of a fuzzy number, respectively. The triangular membership function of G is given as

$$\mu_G(x) = \begin{cases} 0, & x < L \\ (x - L)/(M - L), & L \le x \le M \\ (U - x)/(U - M), & M \le x \le U \\ 0, & x > U \end{cases} \quad (1)$$

Let $G_1 = (G_1^L, G_1^M, G_1^U)$ and $G_2 = (G_2^L, G_2^M, G_2^U)$ be two TFNs. Following are the key operations of the respective two fuzzy numbers:

$$\left(G_1^L, G_1^M, G_1^U\right) \oplus \left(G_2^L, G_2^M, G_2^U\right) = \left(G_1^L + G_2^L, G_1^M + G_2^M, G_1^U + G_2^U\right) \quad (2)$$

$$\left(G_1^L, G_1^M, G_1^U\right) \otimes \left(G_2^L, G_2^M, G_2^U\right) = \left(G_1^L G_2^L, G_1^M G_2^M, G_1^U G_2^U\right) \quad (3)$$

$$\left(G_1^L, G_1^M, G_1^U\right) / \left(G_2^L, G_2^M, G_2^U\right) = \left(G_1^L / G_2^L, G_1^M / G_2^M, G_1^U / G_2^U\right), \ (G_i^L, G_i^M, G_i^U) > 0 \quad (4)$$

$$\left(G_i^L, G_i^M, G_i^U\right)^{-1} \approx \left(\frac{1}{G_i^L}, \frac{1}{G_i^M}, \frac{1}{G_i^U}\right), (G_i^L, G_i^M, G_i^U) > 0 \quad (5)$$

Definition 2 ([20]) The conversion of a TFN $s_j = (L_j, M_j, U_j)$ to a crisp number R can be formulated as

$$R(s_j) = \frac{L_j + 4M_j + U_j}{6}. \quad (6)$$

3.2 *Fuzzy FUCOM*

Pamučar et al. [21] introduced the novel MCDM subjective model FUCOM in the year 2018. It is based on the concept of pairwise comparison of the criteria and performs deviation from maximum consistency (DMC) for result validation. FUCOM is a much simpler method compared to the existing techniques for defining criteria weight, requiring only a few binary comparisons to estimate the weights [22].

For any MCDM problem, there must be n evaluation criteria represented as w_j, $j = 1, 2, \ldots, n$ as well as the calculation of their weight coefficients. Decision-makers must identify the scope of impact of criterion i over criterion j using subjective models in order to establish weights based on pairwise comparisons of criteria. The comparison value (a_{ij}) represents the level of effect of criterion i over criterion j. Since the a_{ij} comparison values are based on subjective assessments rather than accurate measurements, any existing uncertainties can be stated in fuzzy numbers. As a result, the preferences of decision-makers (DMs) in F-FUCOM are based on a fuzzy linguistic scale given in Table 1.

Table 1 Fuzzy linguistic scale [20]

Linguistic terms	Membership function
Equally Significant (ES)	(1, 1, 1)
Weakly Significant (WS)	(0.67, 1, 1.50)
Fairly Significant (FS)	(1.50, 2, 2.50)
Very Significant (VS)	(2.50, 3, 3.5)
Absolutely Significant (AS)	(3.5, 4, 4.5)

The algorithmic steps of F-FUCOM are listed below.

Step 1: Establishing decision criteria. Assume there are n ($j = 1, 2,.., n$) criteria for evaluation denoted by the set $\overline{C}_j = \{C_1, C_2, \ldots, C_n\}$.

Step 2: Ranking of decision criteria. The criteria are ranked by the DMs based upon their preference of importance. The criterion with highest weight coefficient is assigned the first rank and so on till the criterion with the least weight coefficient receives the last rank. The criteria ranking obtained is $\overline{C}_{j(i)} > \overline{C}_{j(2)} > \ldots > \overline{C}_{j(q)}$, where q is the observed criterion rank. The equality operator "=" is used instead of ">" if two or more criteria are the same.

Step 3: Criteria comparisons with TFNs. Using the linguistic expressions from Table 1, all the criteria are compared with one another. The comparison is conducted considering the criteria with first rank (most significant). For all of the criteria given in Step 2, fuzzy criteria significance ($\overline{w}_{C_{j(q)}}$) is acquired. As the first criterion is self-compared ($\overline{w}_{C_{j(q)}} = ES$), the remaining $n - 1$ criteria must be compared. Therefore, the fuzzy comparative significance ($\overline{\varphi}_{q/(q+1)}$) is accomplished by using Eq. (7).

$$\overline{\varphi}_{q/(q+1)} = \frac{\overline{w}_{C_{j(q+1)}}}{\overline{w}_{C_{j(q)}}} = \frac{\left(w^L_{C_{j(q+1)}}, w^M_{C_{j(q+1)}}, w^U_{C_{j(q+1)}}\right)}{\left(w^L_{C_{j(q)}}, w^M_{C_{j(q)}}, w^U_{C_{j(q)}}\right)} \tag{7}$$

Thus, utilizing Eq. (8), a fuzzy vector with the comparative importance of evaluation criteria is achieved.

$$\overline{\Phi} = (\overline{\varphi}_{1/2}, \overline{\varphi}_{2/3}, \ldots, \overline{\varphi}_{q/(q+1)}) \tag{8}$$

where $\overline{\varphi}_{q/(q+1)}$ signifies the importance of $C_{j(q)}$ rank with respect to $C_{j(q+1)}$ rank.

Step 4. Optimal fuzzy weights calculation. Here the final values of fuzzy criteria weight coefficients are generated $(\overline{w}_1, \overline{w}_2, \ldots, \overline{w}_n)^T$. The final weight coefficient values must fulfill two conditions given in Eqs. (9) and (10).

$$\frac{\overline{w}_q}{\overline{w}_{q+1}} = \overline{\varphi}_{q/(q+1)} \tag{9}$$

$$\frac{\overline{w}_q}{\overline{w}_{q+2}} = \overline{\varphi}_{q/(q+1)} \otimes \overline{\varphi}_{q/(q+1)(q+2)} \tag{10}$$

To fulfill the conditions, the weight coefficients $(\overline{w}_1, \overline{w}_2, \ldots, \overline{w}_n)^T$ must satisfy $\left|\frac{\overline{w}_q}{\overline{w}_{q+1}} = \overline{\varphi}_{q/(q+1)}\right| \leq \chi$ and $\left|\frac{\overline{w}_q}{\overline{w}_{q+2}} = \overline{\varphi}_{q/(q+1)} \otimes \overline{\varphi}_{q/(q+1)(q+2)}\right| \leq \chi$ states to minimize the value of χ. With the defined requirements, Eq. (11) can be used to generate the optimal fuzzy values of the weight coefficients $(\overline{w}_1, \overline{w}_2, \ldots, \overline{w}_n)^T$ of the evaluation criteria.

$$\begin{aligned} &\min\ \chi \\ &\text{s.t.} \\ &\begin{cases} \left|\frac{\overline{w}_q}{\overline{w}_{q+1}} = \overline{\varphi}_{q/(q+1)}\right| \leq \chi, \forall_j \\ \left|\frac{\overline{w}_q}{\overline{w}_{q+2}} = \overline{\varphi}_{q/(q+1)} \otimes \overline{\varphi}_{q/(q+1)(q+2)}\right| \leq \chi, \forall_j \\ \sum_{j=1}^{n} w_j = 1, \forall_j, \\ w_j^L \leq w_j^M \leq w_j^U, \\ w_j^L \geq 0, \forall_j \\ j = 1, 2, \ldots, n \end{cases} \end{aligned} \tag{11}$$

where $\overline{w}_j = w_j^L, w_j^M, w_j^U$ and $\overline{\varphi}_{q/(q+1)} = \varphi^L_{q/(q+1)}, \varphi^M_{q/(q+1)}, \varphi^U_{q/(q+1)}$.

3.3 *Fuzzy TOPSIS*

The TOPSIS method, proposed by Hwang and Yoon in 1980, is among the most popular MCDM methods present today [23]. The TOPSIS works on the notion that the best alternative is closest to the positive-ideal solution while being distant from the negative-ideal solution [24]. Many changes have been made to TOPSIS over time, one of which is the incorporation of fuzzy set theory for dealing with imprecise and fuzzy data. The algorithmic steps for fuzzy TOPSIS are given below:

Step 1: Select a group of decision-makers and explain to them the criteria for selecting alternatives.

Step 2: The next step is to compute the average weights of each criterion so that each criterion can be allocated a single value. This step was skipped in our case because the weights were estimated using the F-FUCOM method.

Step 3: In this step, appropriate linguistic variables are selected so that the significance rating of each choice in relation to each and every criterion may be determined. For K decision-makers, calculate the aggregated $\widetilde{x}_{ij}$ such that

$$\tilde{x}_{ij} = \frac{1}{k}(\tilde{x}_{ij}^{1} + \tilde{x}_{ij}^{2} + \cdots + \tilde{x}_{ij}^{k}) \tag{12}$$

where $\tilde{x}_{ij}^{k}$ is the rating of the kth decision-maker for the ith alternative and jth criterion.

Step 4: Following the successful conversion of linguistic variables into fuzzy triangular numbers, a fuzzy decision matrix is built.

$$\tilde{G} = \begin{bmatrix} \tilde{x}_{11} & \tilde{x}_{12} & \dots & \tilde{x}_{1q} \\ \tilde{x}_{21} & \tilde{x}_{22} & \dots & \tilde{x}_{2q} \\ . & . & \dots & . \\ . & . & \dots & . \\ \tilde{x}_{p1} & \tilde{x}_{p2} & \dots & \tilde{x}_{pq} \end{bmatrix} \tag{13}$$

where $\tilde{x}_{ij}, \forall i, j$ are linguistic variables defined by TFNs $\tilde{x}_{ij} = (p_{ij}, q_{ij}, r_{ij})$ and p denotes the number of alternatives and q is the number of criteria and $\tilde{x}_{ij}$ signifies the triangular fuzzy number.

Step 5: Normalized fuzzy decision matrix $\tilde{Q}$ for benefit (B) and cost (C) criteria are constructed:

$$\tilde{Q} = \left[\widetilde{Q_{ij}}\right]_{m \times n} \tag{14}$$

$$\tilde{Q}_{ij} = \left(\frac{p_{ij}}{r_j^*}, \frac{q_{ij}}{r_j^*}, \frac{r_{ij}}{r_j^*}\right), j \in B, \tag{15}$$

$$\tilde{Q}_{ij} = \left(\frac{p_j^-}{r_{ij}}, \frac{p_j^-}{q_{ij}}, \frac{p_j^-}{p_{ij}}\right), j \in C, \tag{16}$$

$$r_j^* = \max_i c_{ij}, j \in B, \tag{17}$$

$$p_j^- = \min_i a_{ij}, j \in C, \tag{18}$$

Step 6: Create the weighted normalized fuzzy decision matrix $\tilde{V} = [\tilde{V}_{ij}]_{m \times n}$, where $\tilde{V}_{ij} = \tilde{s}_{ij} \,.\, \tilde{w}_j$ represent normalized TFNs in the range [0, 1].

Step 7: Compute the fuzzy positive-ideal solution (G*) and fuzzy negative-ideal solution (G^-), $G^* = (\tilde{v}_1^*, \tilde{v}_2^*, \ldots, \tilde{v}_n^*)$, and $G^- = \left(\tilde{v}_1^-, \tilde{v}_2^-, \ldots, \tilde{v}_n^-\right)$.

Step 8: Find each alternative's distance from G* and G^- with the vertex method:

$$D_i^* = \sum_{j=1}^{n} D\left(\widetilde{V}_{ij}, \widetilde{V}_j^*\right), \forall i, \tag{19}$$

$$D_i^- = \sum_{j=1}^{n} D\left(\widetilde{V}_{ij}, \widetilde{V}_j^-\right), \forall i, \tag{20}$$

where D denotes the distance between two fuzzy numbers and is determined by the formula:

$$D\left(\tilde{r}_{ij}, \tilde{s}_{ij}\right) = \sqrt{\frac{1}{3}\left[(r_1 - s_1)^2 + (r_2 - s_2)^2 + (r_3 - s_3)^2\right]} \tag{21}$$

Step 9: Using the following formula, evaluate the closeness coefficient of each alternative:

$$CC = \frac{D_i^-}{D_i^- + D_i^*} \forall i, \tag{22}$$

Step 10: Ranking of alternatives with decreasing order of $CC_{i.}$

4 Proposed Methodology

This section introduces the suggested hybrid F-FUCOM + F-TOPSIS method for ranking and prioritizing alternatives of cloud service. The proposed schematic diagram is depicted in Fig. 1 that comprises three major phases, which are as follows:

Phase A: Establishing a pairwise decision matrix and determining the criteria for evaluation that will be used in the service selection step.
Phase B: The F-FUCOM approach estimates the fuzzy weights of decision criteria provided by a group of decision-makers. It generates a non-linear programming model which is solved in LINGO 18.0 software.
Phase C: Utilize the optimum weights from the previous phase and generate the final service rankings of the alternatives using the F-TOPSIS method.

4.1 Case Study

For implementing the proposed approach, a real cloud service dataset QWS [25] is used for this experiment using LINGO 18.0 software on a system having Intel core i5 processor at 1.8 GHz along with 16 GB RAM and Windows 10 Operating System. For our experiment, we examined 6 QoS parameters (Response Time (CR1),

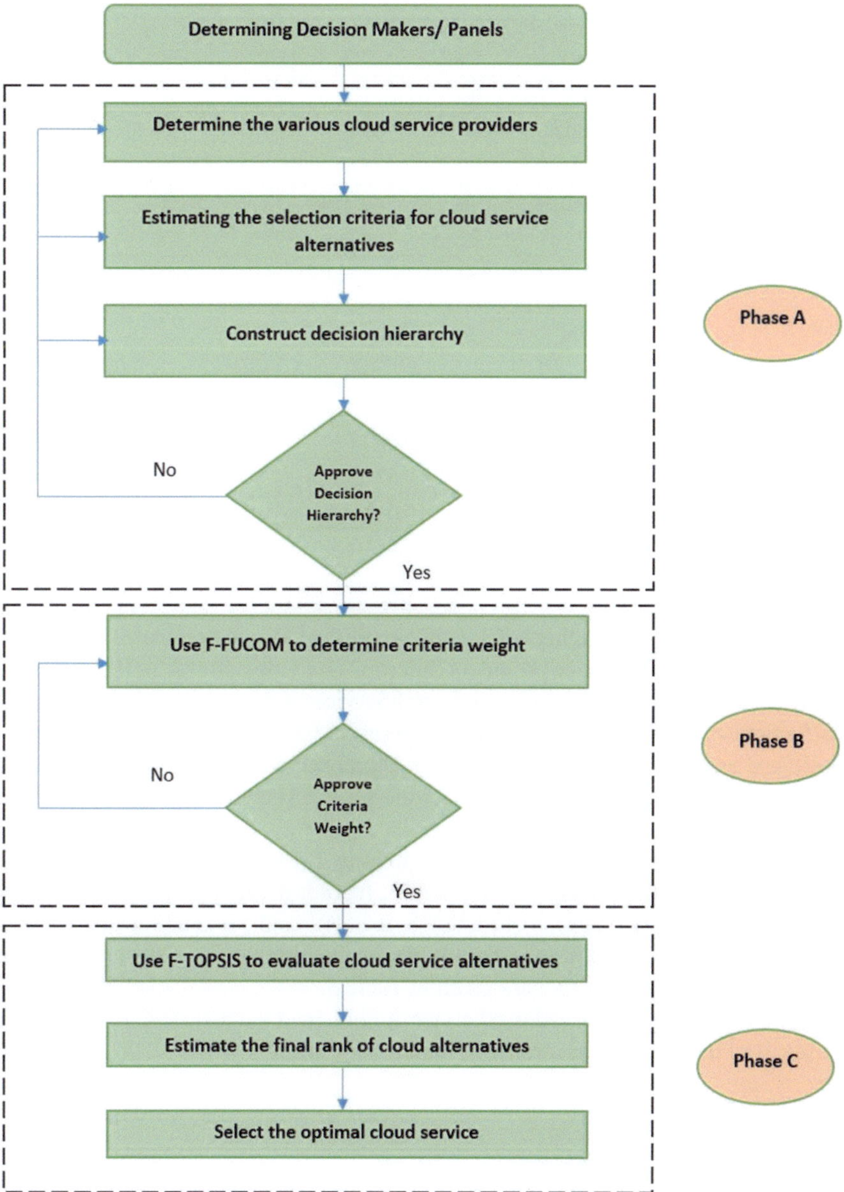

Fig. 1 Flow diagram of various phases of the proposed framework

Table 2 Decision matrix of Cloud Service Providers (CSP) and QoS criteria

CSP	CR1	CR2	CR3	CR4	CR5	CR6
Interop2	107.57	80	1.7	67	78	18.21
Skynode	80.55	94	13.6	60	89	26.19
IP2Geo	130.25	79	2.4	73	78	93.75
MyAmazonSecure	107.57	89	2.2	73	100	5.71
GoogleSearchService	172.67	87	6.8	73	78	14

Table 3 Final result and ranking

CSP	D_i^*	D_i^-	CC_i	Rank
Interop2	0.501	0.676	0.574	5
Skynode	0.341	0.665	0.661	3
IP2Geo	0.431	0.671	0.609	4
MyAmazonSecure	0.311	0.649	0.676	2
GoogleSearchService	0.130	0.654	0.834	1

Availability (CR2), Throughput (CR3), Reliability (CR4), Compliance (CR5) and Latency (CR6)) and 5 service providers (interop2 (A1), skynode (A2), IP2Geo (A3), MyAmazonSecure (A4), and GoogleSearchService (A5)), while picking the best CSP amongst the varied alternatives. The criteria CR3 is considered as the most significant and CR1 as the least significant. Moreover, we did not evaluate any sub-criteria in our case study. Table 2 depicts the service providers' decision matrix in relation to the criteria.

The criteria weights obtained after applying F-FUCOM algorithm are $\overline{w}_j$= ((0.076, 0.108, 0.108), (0.147, 0.204, 0.208), (0.183, 0.252, 0.252), (0.067, 0.106, 0.134), (0.100, 0.156, 0.174), (0.142, 0.218, 0.244))^{T,} and $\chi = 0.023$ which are then used by F-TOPSIS to output the final ranks of the alternatives. The overall cloud services ranking based on CC_i are given in Table 3.

From Table 3, it is observed that service A5 obtained the first rank, while service A1 received the last rank according to the decision-makers.

4.2 *Comparative Analysis*

We compared the proposed framework with other MCDM approaches like BWM-TOPSIS [7] and Fuzzy AHP-TOPSIS [26] to analyze its consistency. An identical dataset was utilized for comparison, and ranking similarity was observed, with the ranking of BWM-TOPSIS being considerably closer to the suggested technique, justifying the proposed method's consistency. The results can be observed in Fig. 2.

Fig. 2 Comparative ranking of CSP with other MCDM methods

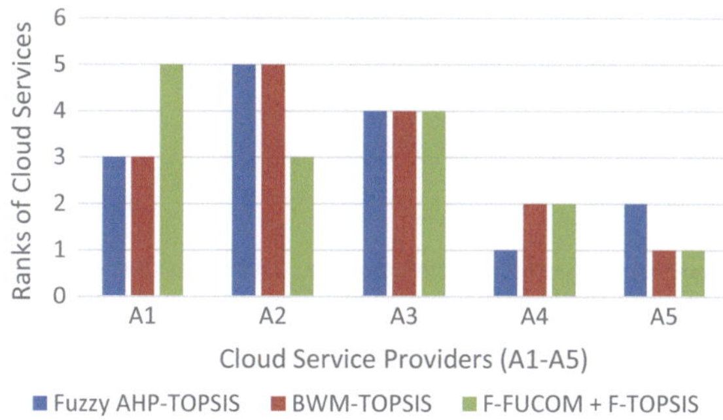

FUCOM has an advantage in comparison to the pairwise matrix calculation as it requires only $n - 1$ comparisons, whereas BWM requires $2n - 3$ and AHP requires $n(n - 1)/2$ comparisons, thus making it more feasible.

4.3 *Sensitivity Analysis*

By conducting sensitivity analysis, we can analyze the impact on the overall result of ranking various CSPs by QoS criteria weights and ensure the ranking solution's robustness. For this, we altered the level of weights based on QoS attribute to examine how the final outcomes changed. The system is considered robust if the ranking of CSPs appears unaltered in the majority of cases or else it's sensitive. For our experiment, we exchanged the criteria weights one at a time and conducted 15 test runs to observe the results where we got different CC_i values. For instance, CR1–CR2 represents the weight exchange between CR1 and CR2. The finding of the sensitivity analysis is depicted in Fig. 3, where it can be noted that A5 is the best CSP followed by A4. Thus, it can be said that our proposed framework is not sensitive to changing criteria weights.

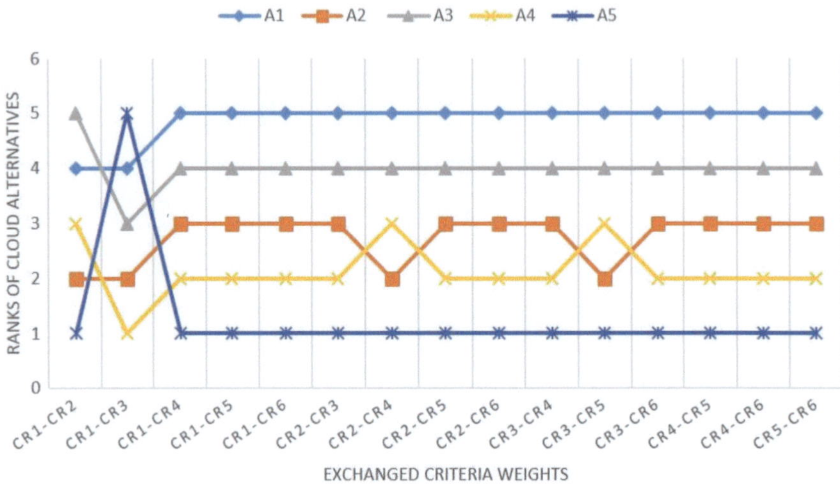

Fig. 3 Assessment of sensitivity for different rankings of cloud alternatives

5 Conclusion

An efficient hybrid decision-making model based on fuzzy data has been proposed in this paper for evaluating and selecting the best CSP. We combined fuzzy FUCOM which is a new and better method to calculate the subjective weights with fuzzy TOPSIS which outputs the final rankings of the cloud alternatives. A real-life cloud dataset QWS is chosen to illustrate the efficacy of the proposed hybrid approach by picking the best cloud database server and conducting a detailed analysis. Furthermore, comprehensive experiments revealed that the suggested methodology is suitable for complicated MCDM problems and for dealing with ambiguity throughout the decision-making phase. However, some limitations remain, such as biases and subjectivity during pairwise comparisons because human judgments are involved. In future works, the proposed model will be analyzed more rigorously under diverse circumstances.

References

1. Kumar RR, Mishra S, Kumar C (2018) A novel framework for cloud service evaluation and selection using hybrid MCDM methods. Arab J Sci Eng 43(12):7015–7030
2. Al-Faifi A, Song B, Hassan MM, Alamri A, Gumaei A (2019) A hybrid multi criteria decision method for cloud service selection from Smart data. Future Gener Comput Syst 93:43–57
3. Sun L, Dong H, Hussain FK, Hussain OK, Chang E (2014) Cloud service selection: state-of-the-art and future research directions. J Netw Comput Appl 45:134–150
4. ur Rehman Z, Hussain FK, Hussain OK (2011) Towards multi-criteria cloud service selection. In: 2011 fifth international conference on innovative mobile and internet services in ubiquitous computing. IEEE, pp 44–48

5. Khanam R, Kumar RR, Kumar C (2018a) Qos based cloud servicecomposition with optimal set of services using PSO. In: 2018 4th international conference on recent advances in informationtechnology (RAIT). IEEE, pp 1–6
6. Khanam R, Kumar RR, Kumari B (2018b) A novel approach for cloud service composition ensuring global qos constraints optimization. In: 2018 international conference on advances in computing, communications and informatics (ICACCI). IEEE, pp 1695–1701
7. Kumar RR, Kumari B, Kumar C (2021) CCS-OSSR: a framework based on hybrid MCDM for optimal service selection and ranking of cloud computing services. Clust Comput 24(2):867–883
8. Sun L (2019) An influence diagram based cloud service selection approach in dynamic cloud marketplaces. Cluster Comput 22:7369
9. Tiwari RK, Kumar R (2021) G-TOPSIS: a cloud service selection framework using Gaussian TOPSIS for rank reversal problem. J Supercomput 77(1):523–562
10. Sun L, Dong H, Hussain FK, Hussain OK, Chang E (2014) Cloud service selection: state-ofthe-art and future research directions. J Netw Comput Appl 45:134–150
11. Singh S, Sidhu J (2017) Compliance-based multi-dimensional trust evaluation system for determining trustworthiness of cloud service providers, vol 67, pp 109–132
12. Sidhu J, Singh S (2017) Design and comparative analysis of MCDM-based multidimensional trust evaluation schemes for determining trustworthiness of cloud service providers. J Grid Comput 15(2):197–218
13. Kumar RR, Mishra S, Kumar C (2017) Prioritizing the solution of cloud service selection using integrated MCDM methods under Fuzzy environment. J Supercomput 73(11):4652–4682
14. Nawaz F, Asadabadi MR, Janjua NK, Hussain OK, Chang E, Saberi M (2018) An MCDM method for cloud service selection using a Markov chain and the best-worst method. Knowl Based Syst 159:120–131
15. Karim R, Ding C, Miri A (2013) An end-to-end QoS mapping approach for cloud service selection. In: 2013 IEEE ninth world congress on services. IEEE, pp 341–348
16. Garg SK, Versteeg S, Buyya R (2013) A framework for ranking of cloud computing services. Future Gener Comput Syst 29(4):1012–1023
17. Le S, Dong H, Hussain FK, Hussain OK, Ma J, Zhang Y (2014) Multicriteria decision making with fuzziness and criteria interdependence in cloud service selection. In: 2014 IEEE international conference on fuzzy systems (FUZZ-IEEE). IEEE, pp 1929–1936
18. Sun L, Ma J, Zhang Y, Dong H, Hussain FK (2016) Cloud-fuser: fuzzy ontology and MCDM based cloud service selection. Future Gener Comput Syst 57:42–55
19. Jatoth C, Gangadharan GR, Fiore U, Buyya R (2019) SELCLOUD: a hybrid multi-criteria decision-making model for selection of cloud services. Soft Comput 23(13):4701–4715
20. Pamucar D, Ecer F (2020) Prioritizing the weights of the evaluation criteria under fuzziness: the fuzzy full consistency method–FUCOM-F. Facta Universitatis, Series: Mechanical engineering, vol 18, issue 3, pp 419–437
21. Pamučar D, Stević Ž, Sremac S (2018) A new model for determining weight coefficients of criteria in mcdm models: full consistency method (fucom). Symmetry 10(9):393
22. Pamucar D, Ecer F, Deveci M (2021) Assessment of alternative fuel vehicles for sustainable road transportation of United States using integrated fuzzy FUCOM and neutrosophic fuzzy MARCOS methodology. Sci Total Environ 788:147763
23. Hwang CL, Paidy SR, Yoon K, Masud ASM (1980) Mathematical programming with multiple objectives: a tutorial. Comput Oper Res 7(1–2):5–31
24. Gupta H (2018) Assessing organizations performance on the basis of GHRM practices using BWM and Fuzzy TOPSIS. J Environ Manage 226:201–216
25. Al-Masri E, Mahmoud QH (2008) Investigating web services on the world wide web. In: Proceedings of the 17th international conference on World Wide Web, pp 795–804
26. Kumar RR, Shameem M, Kumar C (2022) A computational framework for ranking prediction of cloud services under fuzzy environment. Enterp Inform Syst 16(1):167–187

A Novel Convolutional Neural Network Architecture for Gender and Age Prediction

Neeraj Kumar Sharma, Anuja Induri, M. Reddy Priya, and Priyanka Choubey

Abstract Soft biometrics is a field in which a person's gender, height, and weight are determined using a digital machine. This is due to the increasing number of real-world applications in day-to-day life with improved wireless technology. Soft biometrics generally consist of gender, age, ethnicity, height, and facial dimensions. In this paper, the authors propose a state-of-the-art Convolutional Neural Network (CNN) to classify gender and estimate age based on human face images. We also tune the hyperparameters of the CNN to compute the best result for the gender and age of a human face. Experimental results reveal that our proposed methodology achieved 82% accuracy for gender and 78% accuracy for age.

Keywords Deep learning · Convolutional neural network (CNN) · Rmsprop · Training · Testing · Accuracy · Loss

1 Introduction

In recent times, soft biometrics is the most active research area in the field of facial technology. It is mainly used to predict facial characteristics, namely, age, gender, and facial features of the human face. Most of the research has been done to determine age and gender using Deep Learning methods like ANN and CNN. The problem of face recognition has received significant recognition in the field of computer vision [1]. In general, a human's face consists of different features which are useful to identify the gender, age, emotions, etc. [2]. Using the key feature vectors, solving the age and gender classification problem is the most important problem. Further,

N. K. Sharma · A. Induri · M. Reddy Priya (✉) · P. Choubey
Department of CSE, SRM University, Amaravathi, AP, India
e-mail: priya_madupuri@srmap.edu.in

A. Induri
e-mail: anuja_induri@srmap.edu.in

S. K. Biswas et al. (eds.), *Intelligent Computing Systems and Applications*, Lecture Notes in Networks and Systems 1307, https://doi.org/10.1007/978-981-96-3860-4_9

the key feature vectors of human face are also important to solve different real-time problems such as security, surveillance, digital customer relationship, forensic, banking, human computer interaction, etc. [3].

However, several problems related to age prediction and gender classification are still open research problems. Age prediction along with the gender classification of unfiltered human faces using commercial and real-world applications must meet some basic requirements. Due to the development of computer vision techniques age and gender classification problem is not as challenging as in the past [4].

In recent years, many different methods have been proposed to minimize and solve the problem of facial classification. There are several handcrafted methods to solve age and gender classification, but the main limitation of these methods is that the performance is not satisfactory. The traditional age and gender classification methods mainly depend on the facial description and the main features of the face. They are generally unable to deal with the varying degrees of observations under these difficult imaging conditions. In addition, the images in the conventional categories have some different variations in terms of appearance, noise, orientation, and illumination.

This can affect the performance and accuracy of manually developed methods based on computer vision. Recently, deep learning-based methodologies [5, 6] have shown encouraging performance in face recognition, especially in age prediction and gender classification of unfiltered face images. The complex architecture of deep Convolution Neural Network (CNN) enables the extraction of complex and simple feature vectors to solve the age and gender classification problem. Complex and simple feature vectors improve the performance and accuracy of the model when processing an image dataset. Therefore, this is the main motivation for using CNN to solve age and gender classification problems. In this research, we propose a novel deep learning-based end-to-end classification model that predicts the age group of people along with their gender from unfiltered facial images. Moreover, in our research, we make use of the gender and classification task as the most significant classification problem. In this context, the CNN model is proposed to learn and predict the age and gender of the given face.

In short, the general idea of the proposed methodology in this paper is described as follows.

Age and gender classification based on the human face is performed using a CNN. First, the UTK face dataset is acquired from the UTK website [12]. This dataset is available as an open source for research and experimental purposes. After acquiring the face dataset, preprocessing, such as removing the blurred images, and data augmentation, is performed on the face dataset. Further, the dataset is divided into three different parts: the training part, the testing part, and the validation part. The CNN model is implemented by creating different layers such as convolution layers, pooling layers, fully connected layers, and output layers. The implemented CNN network is trained by applying a training dataset. Fine-tuning of the hyperparameters is performed by applying a validation dataset. After training and validating the design CNN, it is saved in memory. The testing part of the face dataset is applied to the saved CNN model. The different performance metrics, such as loss and accuracy, are calculated from the proposed CNN model. The key contribution of the proposed work in this paper is described as follows.

1. To preprocess the face dataset using data augmentation.
2. To implement a CNN for age and gender classification of a face image dataset.
3. To train and test, and validate the models using training, testing, and validation datasets
4. To analyze the performance of the proposed model by calculating through different performance metrics.

The rest of the sections are organized as follows. Section 2 deals with the background and literature review, Sect. 3 deals with proposed work in this paper, results, and discussion are described in Sect. 3, and finally Sect. 4 deals with conclusion and future works.

2 Literature Survey

Age and gender classification problems are mainly classified into three categories: multiclassification-based methods, regression-based methods, and hybrid methods. Recent research work on the age and gender classification problem is described below. Based on all the early gender and age classification methods, they focused on the concept of manual acquisition of facial features, which mainly involves the study of constrained images acquired under controlled constrained methodologies.

Some years ago, classification and regression methodologies were proposed and used to classify the age and gender of human facial images using these features. Beheshti [7] proposed a machine learning model for age and gender classification. In their proposed approach, the author used a Support Vector Machine (SVM) to classify the gender and age prediction. A model for age prediction was proposed by Demontis [8]. In their work, they used a super-sparse machine learning regression algorithm to predict the age. The main advantage of the proposed approach is that it drastically reduces the number of reference prototypes needed for age prediction.

Dileep [9] proposed an age prediction system. The author in their proposed work used an Artificial neural network for predicting age. They also used three-sigma control units in their proposed age prediction work. They also classified the gender in their proposed work. Belver [10] proposed an experimental-based classification of different face classification algorithms. In their comparative study, they classified the algorithm based on image features and deep features. In their research work, they work on two different datasets known as Morph II and Pal. The deep features they extracted from the images using CNN.

Agbo [12] proposes a CNN architecture for predicting the age and gender. In their proposed they implemented 5-layer CNN architecture, out of 5 layers, 2 are convolution layers, 2 are pooling layers, 1 fully connected layer, and 1 output layer. The author in their proposed work used unconstrained face images to train the proposed CNN architecture.

Sánchez-Hevia [13] proposed a voice-based age and gender recognition system for call centers. In their proposed work, the authors implemented a deep neural network for age and gender prediction. To train the proposed deep neural network, the authors used Mozilla's Common Voice dataset in their experiment. The proposed system achieved 98% accuracy for predicting gender and 20% error for predicting the age of the person.

This, in general, will be done by grouping the adjacent ages into the same age group.

3 Proposed Work

The flow chart of the proposed work for face age prediction and gender classification is illustrated in Fig. 1. First, the input dataset of facial images is collected. Then, data preprocessing is applied to the collected face image dataset. Then, the face image dataset is divided into three parts, namely, train data, validation data, and test data. The train dataset is used to train the implemented CNN algorithm. The validation dataset is used to tune the hyperparameters of the proposed system. The test part is used to test the saved model. In the end, to check the performance of the proposed CNN-based age and gender detection deep learning model, different performance metrics are calculated.

(a) **Data preprocessing**: (a) The data preprocessing involves reducing the size of an input image to a smaller size by removing the background area from the image. Further, we converted the size of all the images is 128*128. After converting the images into a uniform size, the data augmentation is performed on the face dataset. Data augmentation is a technique used to generate a greater number of images from the input image. There are multiple operations, such as flip, rotation, increasing brightness, increasing contrast, etc., performed on the face image dataset. By the data augmentation technique, a greater number of images are generated from an existing dataset image. Thus, using data augmentation,

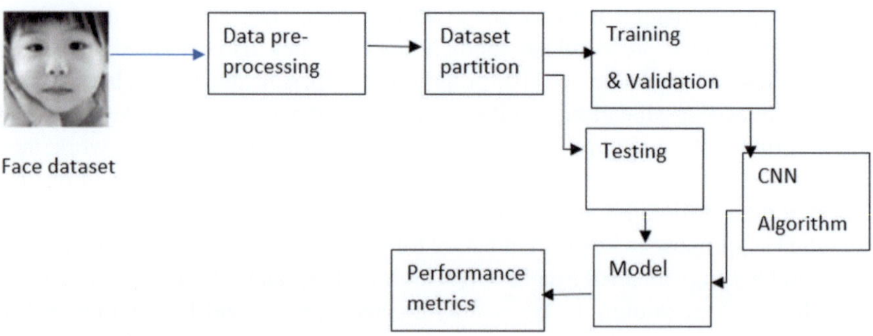

Fig. 1 Flowchart of the proposed work

we can train, test, and validate the model with a greater number of input images. After data augmentation, the blurred images are removed from the dataset. Since the acquired face dataset is labeled and there are only two types of gender images such as male and female. Further, each image is labeled by the current face images in the dataset, such as diseased images and non-diseased (fresh) images, therefore, data annotation is not performed in the proposed work.

(b) **Data set partition**: After performing the data augmentation, the face image dataset is divided into two parts, namely, 80% train data, and 20% test data. Further, the training part is divided into two parts: 87% training part and the 13% validation part. The training dataset we used to train the model. Due to low memory size, the training images are applied in the form of batches, and training is performed using batch normalization.

Here we took the image dataset and we performed the data augmentation, where we converted the single images into more images. After data augmentation, we perform the CNN model. After that, we must split the datasets into two parts: one train dataset and another test dataset. After we use optimizers to find out the accuracy for age and gender. For that, we use a total of 5 optimizers they are Adam, SGD, RMSprop, Adagard, and Adadelta. After we predicted the age and gender we estimated.

(c) **Proposed CNN Architecture**: The proposed CNN architecture for age and face classification consists of 10 different layers. These 10 layers are the combination of 4 convolution layers, 4 pooling layers, 1 dense layer, and 1 output layer. The proposed CNN architecture is a sequential deep neural network, in which alternatively convolution and pooling layers are created, then a fully connected layer and output layers are created. The four convolution layers extract the feature vectors from the input image. Dimension reduction of the input image is performed by 4 pooling layers. A total of 32 kernels or filters are applied to the first convolution layer for extracting the key feature vectors from the input image. A 3*3 size of kernels is applied to the convolution layer. Further, the pooling layer is responsible for reducing the dimension of the input image. Each kernel size is defined as 3*3. The dimension of each kernel is a defined number of filters or kernels. The feature extraction phase mainly consists of four convolutional layers, which contain the corresponding parameters, including the kernel size of each filter, the number of filters, and the step function. It also contains the convolution layer along with the activation function, such as Rectified Linear Unit (ReLU), batch normalization, and max-pooling layer. The classification layer contains two linked parameters that include the number of filters, as well as the kernel size of each filter and the step function. The detailed description of the CNN architecture is defined in Table 1.

Contains the convolutional layer together with the activation layer (rectified linear unit (ReLU)) and the batch normalization plus max-pooling layer. The classification layer contains two connected layers that can handle the classification phase. The first connected layer contains 512 neurons, followed by a ReLU. This is followed by batch normalization, and later on, it consists of a dropout layer with a dropout ratio of 0.5. The second connected layers mainly

Table 1 Proposed CNN architecture

Layer	Filter/kernels	Filter size	Feature map	Act. function
Conv2d_1	32	3*3	N*126*126*32	Relu
Max_pooling2d_1		2*2	N*63*63*32	
Conv2d_2	32	3*3	N*61* 61*64	Relu
Max_pooling2d_2		2*2	N*30*30*64	
Conv2d_3	64	3*3	N*28*28*128	Relu
Max_pooling2d_3		2*2	N*14*14*128	
Conv2d_4	64	3*3	N*12*12*256	Relu
Max_pooling2d_4		2*2	N*6*6*256	
Flatten	9216	256	N*9216	
Dense	256	256	N*256	
Dropout	256	256	N*256	

contain 512 features, which are densely mapped to 8 or 2 neurons for the classification tasks. The gender and age group classification task is a continuous deep classification problem. Therefore, we present the design of the specific parameters of the CNN architecture. To check the performance of the proposed CNN architecture there are different optimizers such as Stochastic Gradient Descent (SGD) and RMSprop are applied to the CNN. SGD is a variant of GD, in which the SGD model is trained by applying one input in an iteration. The RMSprop optimizer is the same as GD with momentum. The RMSprop optimizer restricts the oscillations in the vertical direction. Thus, we can increase the proposed learning rate in our algorithm that helps to take larger steps in the horizontal direction and converge it in a faster way.

4 Experimental Setup and Result Analysis

To check the performance of the proposed CNN architecture for predicting age and gender from the human image, an 11th Gen Intel(R) Core (TM) i5-1135G7 with 8.00 GB RAM, 1 TB Hard disk is utilized. To implement the proposed CNN architecture, we used Windows 10 system software and the Python 3.4 programming language. The other Python libraries and IDE, which are useful for machine learning and deep learning projects, such as Jupyter IDE, Numpy, Matplotlib, TensorFlow 2.1, and Keras used in our experiment (Fig. 3).

To train the proposed CNN architecture for predicting age and gender, we applied two different optimizers, such as SGD and RMSProp. In our experiment, we found that the SGD optimizer consumed more time and performed worse compared to the RMSProp optimizer. RMSprop optimizer-based loss and accuracy of the proposed CNN architecture are described in Fig. 2a and b, respectively. More than 100 epochs are used to train the proposed CNN architecture. The loss and accuracy are described for the training and validation datasets. The proposed CNN achieved 89% accuracy and 11% loss, respectively, in the case of the RMSProp optimizer.

Figure 4a and b show the accuracy and loss, respectively, for the gender classification problem in the case of Adagard optimizer. The 84% accuracy achieved by the proposed CNN architecture in the case of the Adagard optimizer. Further, approximately 26% loss was achieved by the proposed CNN in the case of gender classification.

Figure 5 shows the loss of the proposed CNN architecture in the case of age prediction. The age prediction loss is quite low, which is less than 10% of our proposed CNN architecture.

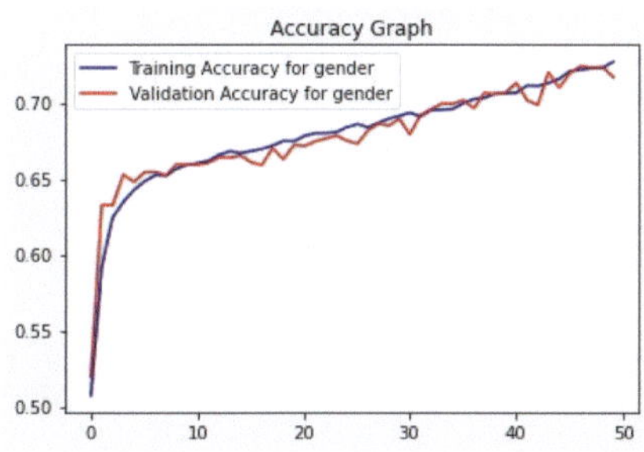

(a) Loss

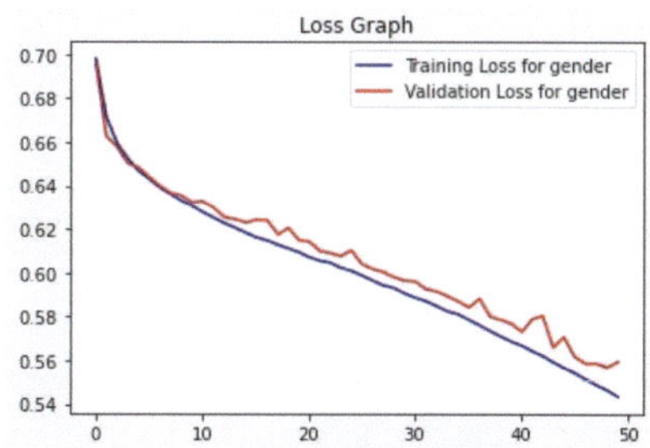

(b) Accuracy

Fig. 2 Adadelta optimizer-based Training and validation dataset **a** loss, **b** accuracy for age prediction

Fig. 3 Adadelta optimizer-based loss and accuracy of age prediction

Adagard

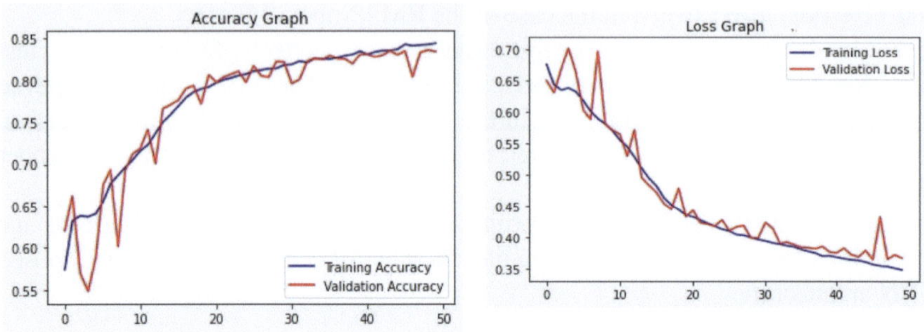

(a) Gender Accuracy **(b) Gender loss**

Fig. 4 Adagard-based Gender classification accuracy and loss

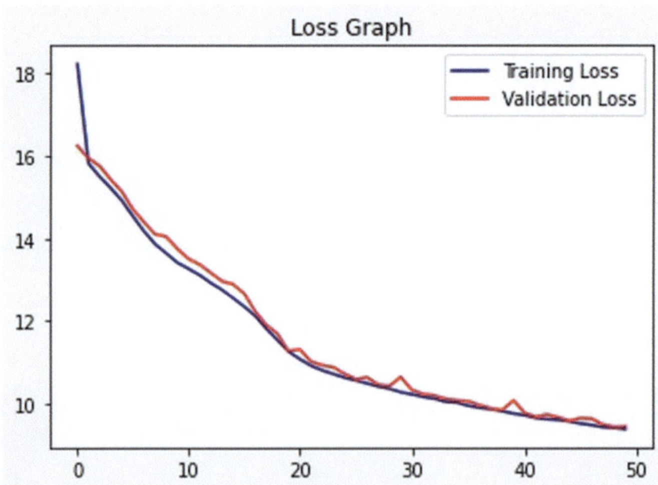

Fig. 5 Adagrad optimizer-based loss for predicting age

5 Conclusion and Future Work

This paper highlights the CNN architecture for predicting the gender and age of different human face images. The proposed CNN architecture was trained on a large UT face dataset. In our research work, we fine-tuned the hyperparameters and applied different optimizers such as SGD, RMSProp, Adadelta, and Adagrad to train the proposed CNN for predicting age and gender. The different optimizers perform differently with respect to the gender and age classification problem. The average accuracy achieved by the proposed CNN architecture in the case of gender classification is 89%, whereas in the case of age prediction, the average accuracy achieved is 92%.

References

1. Carletti V, Greco AS, Percannella G, Vento M, Fellow I (2019) Age from faces in the deep learning revolution. IEEE Trans Pattern Anal Mach Intell 1
2. Bin Gao B, Zhou HY, Wu J, Geng X (2018) Age estimation using expectation of label distribution learning. In: Proceedings of the twenty-seventh international joint conference on artificial intelligence. Stockholm, Sweden, pp 712–718
3. Rothe R, Timofte R, Van Gool L (2018) Deep expectation of real and apparent age from a single image without facial landmarks. Int J Comput Vision 126(2–4):144–157
4. Huang J, Li B, Zhu J, Chen J (2017) Age classification with deep learning face representation. Multimed Tools Appl 76(19):20231–20247
5. Cai L, Huang L, Liu C (2016) Age estimation based on improved discriminative Gaussian process latent variable model. Multimed Tools Appl 75(19):11977–11994
6. Iandola FN, Han S, Moskewicz MW, Ashraf K, Dally WJ, Keutzer K (2017) Medical image computing and computer-assisted Intervention MICCAI. ICLR 8150:1–13
7. Beheshti-nia MA, Mousavi Z (2017) A new classification method based on pairwise support vector machine (SVM) for facial age estimation. J Indus Syst Eng 10(1):91–107
8. Demontis A, Biggio B, Fumera G, Roli F (2015) "Super-sparse regression for fast age estimation from faces at test time", Image analysis and processing—ICIAP. Springer, Berlin, Germany
9. Dileep MR, Danti A (2018) Human age and gender prediction based on neural networks and three human age and gender prediction based on neural networks and three sigma control limits. Appl Artif Intell 32(3):281–292
10. Belver C, Arganda-Carrerasa I, Dornaikaa F (2017) Evaluating age estimation using deep convolutional neural nets. Electron Imaging 2017(17):100–105
11. Tan Z, Wan J, Lei Z, Zhi R, Guo G, Li SZ (2018) Efficient group-n encoding and decoding for facial age estimation. IEEE Trans Pattern Anal Mach Intell 40(11):2610–2623
12. Agbo-Ajala O, Viriri S (2019) Age group and gender classification of unconstrained faces. In: Advances in visual computing. Springer, Berlin, Germany, pp 418–429
13. Sánchez-Hevia HA, Gil-Pita R, Utrilla-Manso M et al (2022) Age group classification and gender recognition from speech with temporal convolutional neural networks. Multimed Tools Appl 81:3535–3552
14. https://susanqq.github.io/UTKFace/

An Efficient Probability Attestation Scheme for Multipath Routing and Performance Enhancement in IoT

J. V. N. Raghava Deepthi, Ajoy Kumar Khan, and Tapodhir Acharjee

Abstract The interconnection of different wireless communication devices is known as the Internet of Things (IoT). This allows for the ubiquitous accessibility of devices and their built-in intelligence capacity. The benefits of these networks are numerous and few of them are no infrastructure support, deployed spontaneously, cost-effective setup, and so on. In order to improve service quality, the Internet of Things (IoT) facilitates communication between devices that have different types of connections, which allows for the creation of interfaces and connections. Many risks impact the efficiency of the network and the data transfer between IoT devices. But still, security mechanisms are challenging. Remote Attestation (RA) is one tool that efficiently identifies a malicious entity in the network. This research offers a probability-based attestation scheme (PROB_ATT) for improving IoT network routing and performance. Depending on the estimated probability value of the journey, the attestation value is included in the network with the data. With the application of trickle time, attestation is verified and data transmission is performed. After the estimation of probability within the network, data transmission is performed based on the trust mechanism. The PROB_ATT scheme was found to decrease energy consumption, packet loss ratio, and delay in the simulation analysis. Also, the proposed PROB_ATT exhibits improved efficiency, packet delivery rate, and throughput than existing techniques.

Keywords Remote attestation · Trust mechanism · Packet delivery ratio · Throughput

J. V. N. Raghava Deepthi (✉) · T. Acharjee
Assam University, Assam State, Silchar, India
e-mail: deepthijonnalagadda1@gmail.com

A. K. Khan
Mizoram University, Aizawl, India

S. K. Biswas et al. (eds.), *Intelligent Computing Systems and Applications*, Lecture Notes in Networks and Systems 1307, https://doi.org/10.1007/978-981-96-3860-4_10

1 Introduction

The Internet of Things (IoT) links a vast array of physical devices. Internet of Things (IoT) objects can be either real or virtual; they can make their own decisions and perform their functions; they can communicate with other devices; and, because they are interoperable, they can operate interactively. In addition, the Internet of Things allows for versatile, always-on, anywhere-service interaction with objects [1]. In recent years, there has a tremendous growth in IoT. Many different types of heterogeneous Internet of Things (IoT) devices are interconnected for a wide range of uses, including health and environmental monitoring, smart city planning, building automation, and military operations. These simplify, intelligently, and automatically make life much easier. But now there are dangers, and hackers have a new cyberspace to operate in. Researchers are motivated to develop security and safety mechanisms after attacks like the Mirai botnet and the smart TV hack [2]. Users are vulnerable to a multitude of attacks due to devices' lack of security [3–7]. Many times, these types of attacks, such as man-in-the-middle [8] attacks, smart attacks [9], and Denial of Service (DoS) attacks [10], result in monetary loss or even worse consequences [11]. It is possible to detect these harmful devices by performing inexpensive attestation. However, a cheap, unskilled device-to-device remote attestation solution has issues with communication overhead and long attestation times. [12]. Plus, systems that use a cluster of dynamic technology devices, such as ITS and oil and gas search robots, do not scale [13]. New, reliable attestation solutions that are both scalable and innovative are thus required to guarantee the safe operation of IoT devices on networks [14, 15]. The devices connected to the infrastructure of IoT are attacked by attackers where the protection mechanism is easily broken [16]; these circumstances motivated the researchers to develop various protection methods that provided network security [17]. In addition to security, other challenges are also faced in IoT networks. Security solutions are a need for an expanding Internet of Things (IoT) network, but they add complexity, computational overhead, and expense as the network's node count grows. This highlights the need for the development of a scalable, secure, and lightweight protocol that can adapt to the ever-changing requirements of networks [18]. To make Internet of Things (IoT) communications more secure, this study developed a probability-based attestation scheme (PROB_ATT). Additionally, this study employs a trust mechanism to efficiently route data within the network. In the proposed PROB_ATT scheme, within the network, the node generates and sends its attestation. Once the attestation is created, the proposed PROB_ATT calculates the different paths and transmits data across the network. If the node responds with the correct attestation within the tickle time, data is transmitted in the network; otherwise, transmission is terminated. The last step in efficient data routing within the network is the application of a trust mechanism. According to the results of the simulations, the proposed PROB ATT is far superior to the existing methods.

2 Related Works

Security at every layer of the IoT-based applications is the most significant feature to be considered. In an IoT system, implementation is successful if security is concentrated more. Hence, security is addressed while designing applications based on IoT networks [18], Tournier et al. [11] and [11] and Libert et al. [19] have addressed a common approach with dual challenges, and Yousefnezhad et al. [20] provided a promising solution to address security in IoT. The studies focus on Security and its taxonomy on product lifestyle, security solutions for every stage in the lifecycle, and issues in these stages that introduce challenges, Abdul-Qawy et al. [21] attempted to provide a common and widespread taxonomy with complete classification covering almost all the major energy-saving approaches of IoT-based constrained WSNs, Aarika et al. [22] examined various solutions for IoT attacks at perception, network and application layers, Karunanithy and Velusamy [23] implemented a CEEDG protocol to enhance lifetime and throughput, Elappila et al. [24] created the survivability Aware Channel Allocation (SACA) fuzzy-based method, in which the nodes in the network could express their preferences for which channels to used. When compared to other methods, SACA significantly improved network reliability, PDR, and throughput, Lee et al. [25] reviewed the RPL's performance after making some adjustments. Step one involves disabling the timer of the vacation spot integrated outlet trickle due to its unsuitability for use in periodic topology changes, Tian et al. [26] By focusing on the node's location, the GI-RPL scheme improves upon the RPL in VANETs and makes it possible to deal with periodic changes to the topology, Martocci et al. [27] offer an analysis of RPL using the most recent revision. Examples include possibilities shown by RPL, point-to-point RBL (P2P-RBL), assessment, performance, challenges, and research resistance, Brandit et al. [28] planned Goal Function. The new feature based on fuzzy logic (OF-FL) allows RPL to advance beyond the standard objective capabilities restriction by taking into account crucial nodes.

3 Overview of Remote Attestation

To determine whether an adversary is present in a device, Remote Attestation (RA) is an appropriate technique. An "untrusted" device's integrity is confirmed by a trusted entity (Vrf), which is especially important in RA.(Prv) by checking, in the absence of the adversary, that the device is using the updated software version. Figure 1 depicts a challenge sent by *Vrf* to *Prv*. When the challenge is received, *Prv* performs the necessary operation and responds back to *Vrf*. Depending on this response, *Vrf* verifies the device's health. Even though RA efficiently validates the health of the device, implementing it on huge networks is challenging as it is a 1–1 verification approach. RA gains cost in many factors like computing, storing, communicating overhead, and power of batteries. Low-end embedded devices suffer

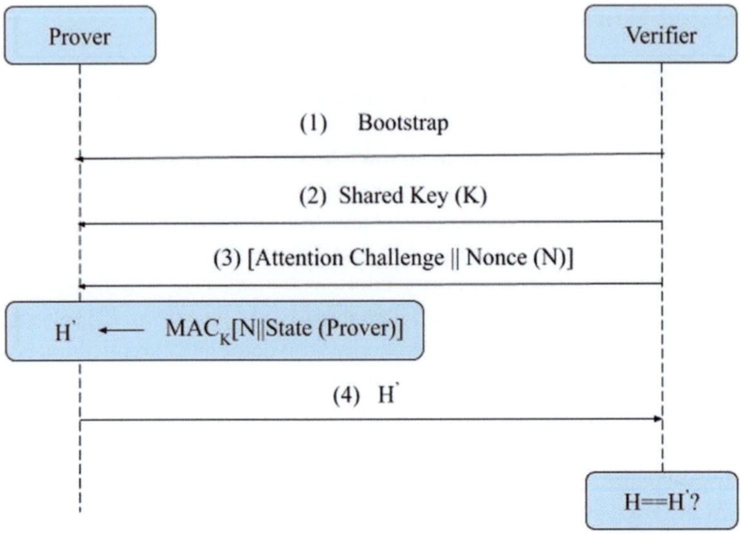

Fig. 1 Steps in remote attestation

from computational cost, hence several IoT applications do not involve high security mechanisms. To develop low-cost secure networks, RPL-integrated RA is suitable due to their interoperability. Thus with the RPL framework, the attestation process is made lightweight and efficient. In PROB_ATT, the DAO packet structure is used for communicating attestation results.

Moreover, conventional attestation methods send the attestation request through *Vrf* to the network, but here drip-time is used which automatically initiates the process of attestation. Hence, it overcomes the communication overhead and tackles several network attacks. In Fig. 1, conventional remote attestation scheme is presented.

Operating under the principle of Destination-Oriented Directed Acyclic Graph (DODAG), RPL [17] is a virtual routing topology. In DODAG, while each node can have multiple parents, the objective functions and routing metrics determine which parent each node is assigned to. The parent node forwards the data packets. The DODAG structure supports N-1 communication in RPL. Rank ID is provided to every node based on the distance from the node to the root. DODAG Information Objects (DIO) with ICMPv6 control packets create and maintain DODAG. In RPL, every node distributes DIO packets which include a link, node metrics, and an objective function which selects the preferred parent of a node among its neighbors. Node metrics include residual energy and expected transmission count (ETX).

4 Proposed Prob_Att Scheme

The RPL protocol is the current standard for Internet of Things (IoT) infrastructure implementation, serving as the network's de facto routing mechanism. The PROB_ATT design exemplifies a safe and effective means of communication. In the multipath routing method, different paths are considered for each node in the network that is meant to be a destination. Higher frequency nodes in the network necessitate a higher energy level for data transmission costs. Using the network's probabilistic value, routing decisions are observed through the incorporation of the probability model. In Eq. (1), connection probability evaluation of proposed PROB_ATT is provided.

$$G(P) = \begin{cases} 1 \; p \le p_0 \\ 0 \; p > p_0 \end{cases} \tag{1}$$

If the connection was successful, the value will be 1, and if it was unsuccessful, the value will be 0. In this, the probability of transmitting a packet is p. The communication radio environment can be better understood with the help of probability routing analysis. In order to maximize channel performance, this multipath optimization method evaluates end-to-end user communication using a game-theory approach.

The average path for the IoT communication is represented in Eq. (2),

$$f(v_i) = \frac{1}{\sigma\sqrt{2\prod}} e^{-(p_i - p_{avg})^2/2\sigma^2} \tag{2}$$

We can estimate the effective path for information communication on the IoT network using Eq. (3),

$$E(N) = \rho d \lambda \tag{3}$$

within this number of nodes, transmitted packets, and other factors are introduced.

For attestation, this approach employs RPL DAO ICMPv6 control messages with improved header fields. The updated data structures are sent in the following formats: (i) a 4-bit flag field that indicates the node ID; (ii) an 8-bit reserved field that includes a timestamp and several bits (00 for a GOOD node and 11 for a BAD node) that represent the results of the self-attestation; and (iii) a 32-bit option field that allows for the transmission of an encrypted hash of DAOcrypt in plain text along with the attestation results. To show how recent a time-bound item is, the outcome includes the hash value of the device's software and the attestation time stamp. The attestation message is as follows in Eq. (4).

$$A_k^{\text{Att}} = (\text{Hash}_{Dk} \| \text{Timestamp})\text{Root}_{pk} \tag{4}$$

$Root_{pk}$ denotes the public key of the root node. Only the root node can decrypt the attestation report, which is encrypted with this public key. Before encrypting with the root node's public key, nodes perform a hash on the modified DAO ICMPv6 message during self-attestation. An optional 32-bit field in the DAO stores both the encrypted result and the plain-text self-check report. The security of DAO messages and self-attestation reports in DAO messages relies on encryption and hashing. An adversary modifies the root path attestation report, for instance. By comparing the plain-text attestation report's hash with the decrypted hash string, changes of that kind can be found in PROB_ATT.

4.1 Encryption Using PROB_ATT

The proposed PROB_ATT is adopted for minimizing computational cost and improved security. However, the attestation result is forwarded to the root node in PROB_ATT by exploiting DAO message packets. The software's public key for root and attestation can be found in Traffic-based Energy Estimation (TEE). The proposed PROB_ATT is adopted for minimizing computational cost and improving security. However, in PROB_ATT, the root node receives the authentication result through the DAO message packet. Attestation and root software public keys are stored in Traffic Energy Estimation (TEE). Since encryption is done using its public key, decryption can only be done at the root.

Figure 2 presents the overall architecture of PROB_ATT for security improvement. Based on probability estimation and Tickle Timer data are transmitted within the network. This approach utilizes the non-storing mode of RPL as MOP2, which is ideal for devices with limited resources, as it supports computational requirements and minimum memory. Another feature of MOP2 is that all network nodes communicate with the root via control messages. At regular intervals (referred to as a "Trickle-Timer"), the DAO message in PROB_ATT transmits the device attestation report to the root. This timer monitors the rate at which beacon messages are generated [18]. With the timer set to relay DAOcrypt messages to the root, the root can learn the network's health status and use that information to redirect RoamingAdv-derived threats as in Fig. 2. The overall architecture of the proposed PROB_ATT scheme for WSN-IoT security is presented in Fig. 2.

For analyzing IoT path and the performance of data transmission the following assumptions are coined and are stated below.

PROB_ATT has Verifier (S_{rf}) and Prover *(Qrv)* as basic stakeholders. Algorithm 1 provides the pseudo code incorporated in the encryption process adopted in PROB_ATT for security improvement in WSN-IoT is provided.

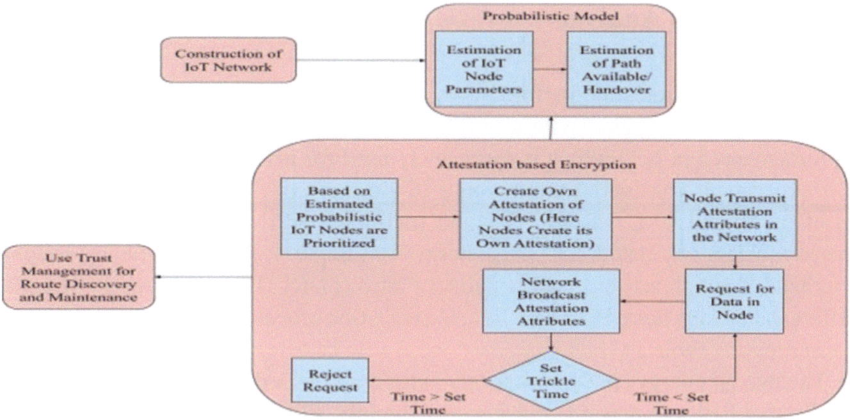

Fig. 2 Proposed model for flow of PROB_ATT

Algorithm1: Encryption by PROB_ATT

1. Initialize the Set device id as N_{id}

2. Set cryptographic details of node as E_{cry}

3. Define attestation details of each node in the network N_{att}

4. Set trickle time for attestation TT_{att}

5. Network Cryptography $D_{crypt}=0$

6. While true do

Formation();

7. If (TT_{att}==True) *then*

Perform N_{att}

D_{att} ← *Encryption of Message*

D_{crypt} ← D_{att}

Transmit the obtained D_{crypt}

else

Perform normal operation

end

end

R_{msg} = *Received attestation values*

E_{node} ← *Elimination of faulty nodes*

While True do

Check ()←flag

If check () ==0 then

Verify D_{crypt}

Perform E_{node}

else

Keep those nodes

Perform normal Operation

end

end

5 Simulation Analysis

This section illustrates the performance of PROB_ATT with the simulation results obtained and is compared with traditional RPL in various situations. PROB_ATT is implemented in NS2 which is extensively used for network deployment. Hence, this experiment considered the increase of attacker nodes related to node density as of traditional RPL. PROB_ATT and traditional RPL data are compared with the parameters calculated, loss rate, energy consumption, energy efficiency, routing overhead, throughput, and packet delivery ratio (PDR). This is necessary because different attackers harm PDR and the self-attestation process has an impact on energy consumption (Table 1).

The overall energy consumption calculated by proposed PROB_ATT scheme and Traditional RPL are presented in Fig. 3. Certainly, PROB_ATT requires minimum energy than traditional RPL. As complex hashing is utilized to provide encrypted attestation results, PROB_ATT achieves a lower energy consumption than the traditional RPL. In Fig. 4, the energy efficiency of the proposed PROB_ATT with traditional RPL is presented (Table 2).

Simulation results prove that with the increase in simulation time, energy required for PROB_ATT drastically decreases which increases energy efficiency. As malicious nodes are detected at the early stages, healthy nodes are prevented from forwarding iterative messages. In Fig. 5, packet loss rate (PLR) of proposed the PROB_ATT is comparatively examined with the traditional RPL scheme (Table 3).

From Fig. 5, the delivery probability $PTMG_{(S,D)}$ is estimated by varying the probability which is randomly defined in [10, 20, 30%] for each node. For the assigned probability value, loss rate of the packet is examined. From Fig. 4, the suggested PROB_ATT scheme outperforms the conventional RPL in terms of energy efficiency. This, in turn, decreases the packet loss rate compared with the existing traditional RPL scheme. In Fig. 6, the packet delivery rate (PDR) of the proposed PROB_ATT with the traditional RPL is presented (Table 4).

The results shown in Fig. 6 provides better PDR for the proposed PROB_ATT approach. This approach utilizes a multipath disjoint routing method and when the route discovery mechanism is considered, each trusted node with high PDR discovers a valid secure link. Hence, it is concluded that the proposed approach is better for larger networks like Mobile Ad Hoc Networks (MANET). In Fig. 7, throughput

Table 1 Energy consumption versus No. of nodes

No. of nodes	Existing (RPL)	Proposed (PROB_ATT)
0	0	0
20	14.87	9.57
40	21.85	15.67
60	28.53	21.63
80	37.83	29.50
100	38.66	30.76

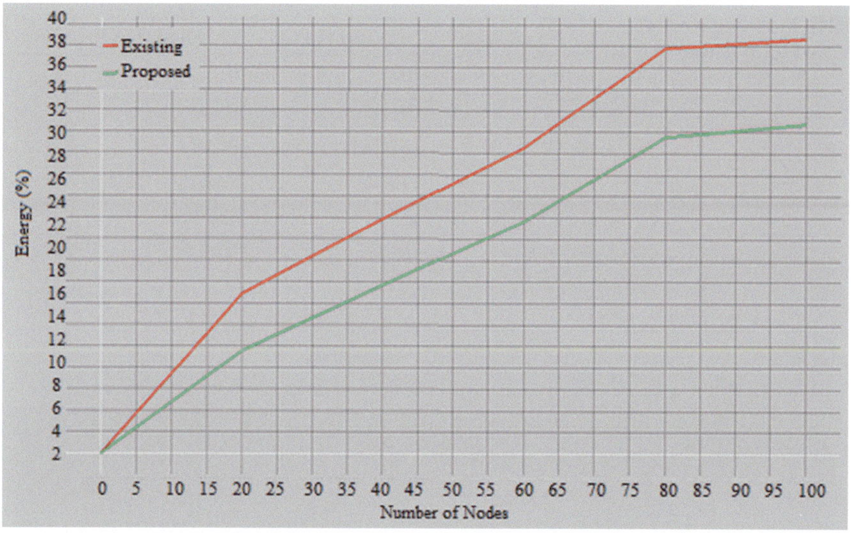

Fig. 3 Comparison of energy consumption

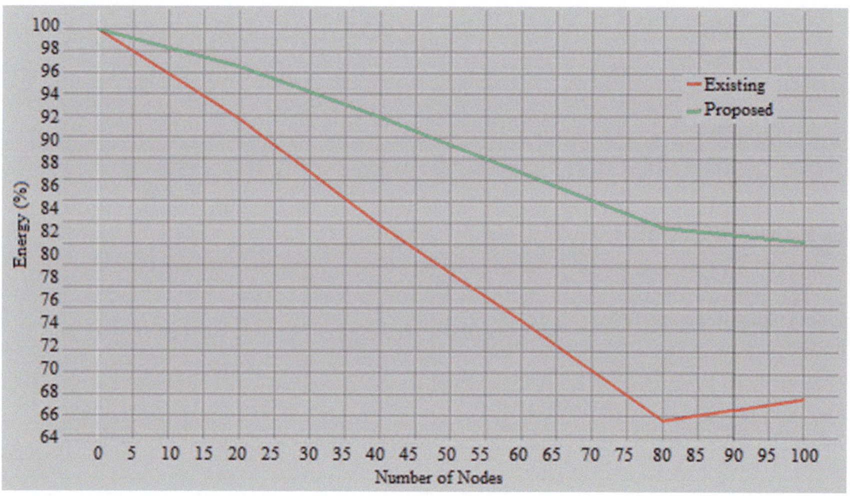

Fig. 4 Comparison of energy efficiency

measurement of the proposed PROB_ATT scheme with existing RPL scheme is provided (Table 5).

In Fig. 7, throughput measurement of the proposed PROB_ATT scheme with existing traditional RPL is presented. The analysis stated that an increase in PDR and reduced PLR increases the network throughput. In Fig. 8, delay measurements

Table 2 Energy efficiency versus No. of nodes

No. of nodes	Existing (RPL)	Proposed (PROB_ATT)
0	0	0
20	91.67	96.53
40	81.78	91.87
60	72.86	86.68
80	63.57	81.59
100	65.66	80.33

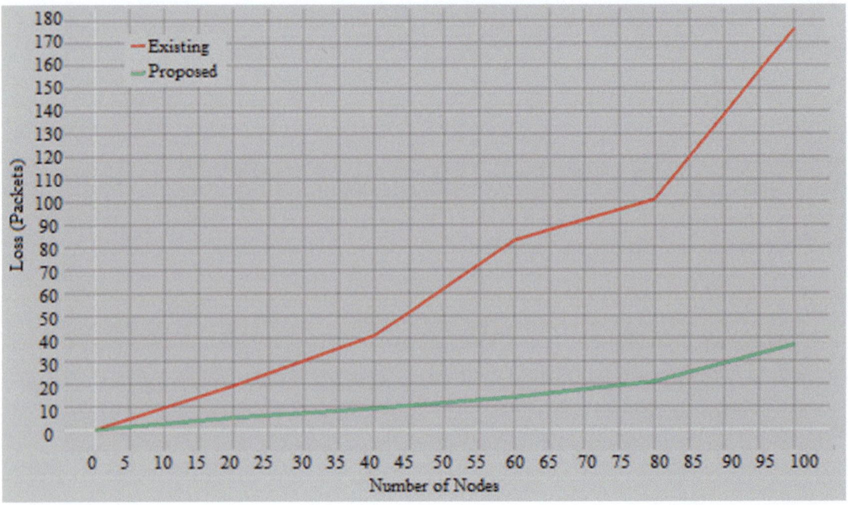

Fig. 5 Comparison of PLR

Table 3 Loss rate versus No. of nodes

No. of nodes	Existing (RPL)	Proposed (PROB_ATT)
0	0	0
20	19	5
40	41	9
60	83	14
80	101	21
100	176	37

for varying nodes are presented. In Table 6, delay measured for proposed PROB_ATT scheme for security improvement is presented.

Figure 8 shows the delay in the network. Network size and control packets play a role in delay while transmitting messages to the root because of multiple hop distances. In RPL, DAO messages maintain the downward topology of the network.

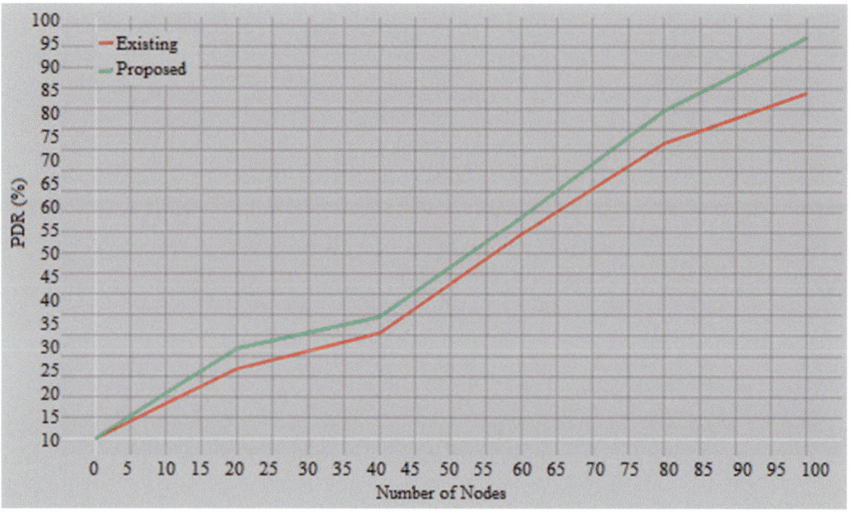

Fig. 6 Comparison of PDR

Table 4 Packet delivery ratio versus No. of nodes

No. of nodes	Existing (RPL)	Proposed (PROB_ATT)
0	0	0
20	16.89	21.87
40	25.62	29.57
60	49.57	53.67
80	71.57	79.52
100	83.78	97.23

By DAO messages, the root obtains the network view globally which controls the DODAG topology. In Fig. 9, routing overhead measurement for the proposed PROB_ATT scheme and traditional RPL is presented (Table 7).

In Fig. 9, PROB_ATT, from routing overhead exploits the DAO packet to transmit the attestation report to the root. Attestation and encryption with SHA-256 are expensive. Still, it is proved that PROB_ATT does not delay in transmitting DAO messages which is cost-effective for security. Particularly, PROB_ATT obtains network security without any additional delays and the root gets updated with the status of the network without any network delay.

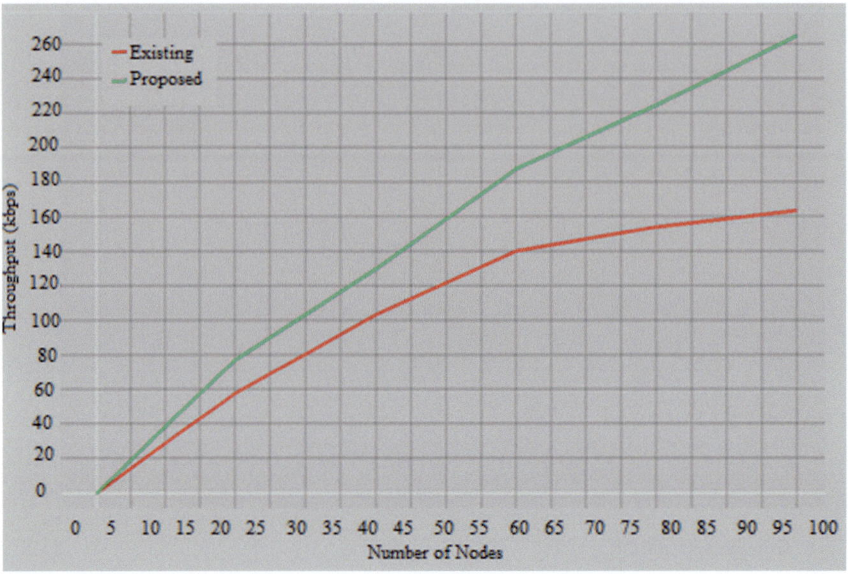

Fig. 7 Comparison of throughput

Table 5 Throughput versus No. of nodes

No. of nodes	Existing (RPL)	Proposed (PROB_ATT)
0	0	0
20	57.26	76.67
40	103.28	129.57
60	139.78	187.57
80	153.57	224.29
100	163.57	264.29

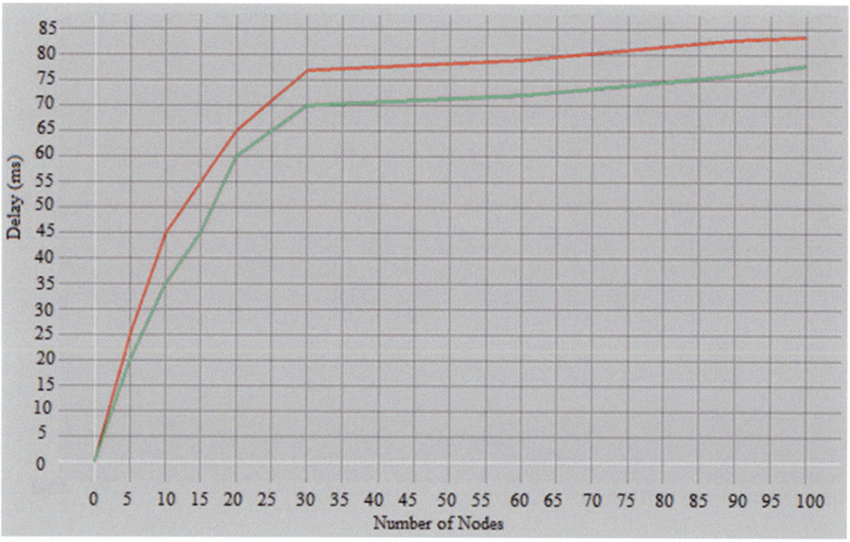

Fig. 8 Comparison of delay

Table 6 Delay versus No. of nodes

No. of nodes	Existing (RPL)	Proposed (PROB_ATT)
0	0	0
5	0.025	0.020
10	0.045	0.035
15	0.055	0.045
20	0.065	0.060
30	0.077	0.070
60	0.079	0.072
90	0.083	0.076
100	0.0835	0.078

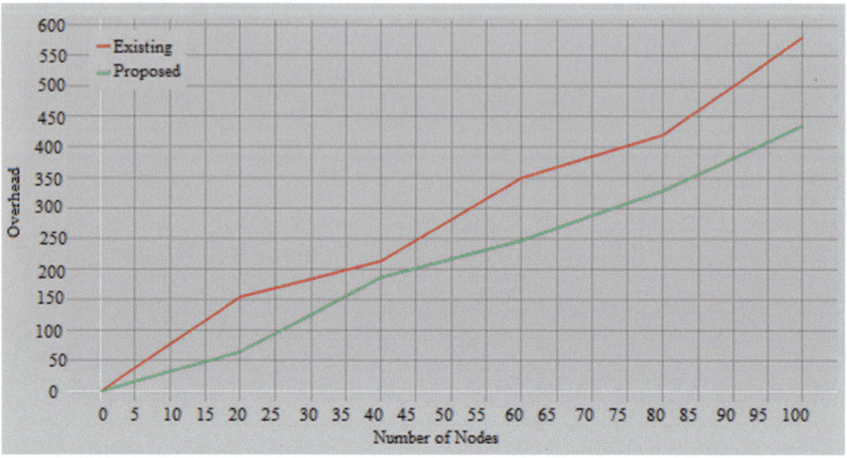

Fig. 9 Comparison of routing overhead

Table 7 Routing overhead versus No. of nodes

No. of nodes	Existing (RPL)	Proposed (PROB_ATT)
0	0	0
5	0.025	0.020
10	0.045	0.035
15	0.055	0.045
20	0.065	0.060
30	0.077	0.070
60	0.079	0.072
90	0.083	0.076
100	0.0835	0.078

6 Conclusion

IoT infrastructure is based on an ad hoc network with numerous benefits like no infrastructure, deploying spontaneously, and a chance to set up cost-effective temporary and dynamic networks. Still, few characteristics like unreliable wireless links, dynamic topology, and energy and power constraints objects, developing new protocols that ignore the security of applications for satisfying quality of service (QoS). The purpose of this study was to enhance the efficiency of Internet of Things (IoT) communication by creating a probability-based attestation scheme (PROB ATT). Also, for effective routing of data within the network this research utilizes a trust mechanism. The node in the proposed PROB_ATT scheme generates and sends its attestation across the network. The proposed PROB_ATT creates an attestation, then estimates the paths through the network and transmits data. All of the PROB_ATT evaluations include up to 100 nodes, and the suggested method yields satisfactory results. Simulation analysis stated that the PROB_ATTT scheme reduces energy consumption, packet loss ratio, and delay by 20, 10, and 5%, respectively. In terms of energy efficiency, throughput, and packet delivery rate performance improved around 20% compared to the existing method. In future, we can extend this work by implementing the variable nodes up to 500.

References

1. Airehrour D, Gutierrez JA, Ray SK (2019) SecTrust-RPL: a secure trust-aware RPL routing protocol for Internet of Things. Future Gener Comput Syst 93:860–876
2. Meneghello F, Calore M, Zucchetto D, Polese M, Zanella A (2019) IoT: Internet of threats? A survey of practical security vulnerabilities in real IoT devices. IEEE Internet Things J 6(5):8182–8201
3. Tomić I, McCann JA (2017) A survey of potential security issues in existing wireless sensor network protocols. IEEE Internet Things J 4(6):1910–1923
4. Qiu T, Chen N, Li K, Qiao D, Fu Z (2017) Heterogeneous adhoc networks: architectures, advances and challenges. AdHoc Netw 55:143–152
5. Deebak BD, Al-Turjman F (2020) A hybrid secure routing and monitoring mechanism in IoT-based wireless sensor networks. Ad Hoc Netw 97:102–122
6. Haseeb K, Islam N, Almogren A, Din IU, Almajed HN, Guizani N (2019) Secret sharing-based energy-aware and multi-hop routing protocol for IoT based WSNs. IEEE Access 7:79980–79988
7. Neshenko N, Bou-Harb E, Crichigno J, Kaddoum G, Ghani N (2019) Demystifying IoT security: an exhaustive survey on IoT vulnerabilities and a first empirical look on internet-scale IoT exploitations. IEEE Commun Surv Tutor 21(3):2702–2733
8. Conti M, Kaliyar P, Rabbani MM, Ranise S (2020) Attestation-enabled secure and scalable routing protocol for IoT networks. Ad Hoc Netw 98
9. Conti M, Kaliyar P, Rabbani MM, Ranise S (2018) SPLIT: a secure and scalable RPL routing protocol for Internet of Things. In: 14th international conference on wireless and mobile computing, networking and communications (WiMob), pp 1–8
10. Feng W, Qin Y, Zhao S, Feng D (2018) AAoT: lightweight attestation and authentication of low-resource things in IoT and CPS. Comput Netw 13(4):167–182
11. Tournier J, Lesueur F, Le Mouël F, Guyon L, Ben-Hassine HA (2020) Survey of IoT protocols and their security issues through the lens of a generic IoT stack. Internet Things 100–264
12. Yousefnezhad N, Malhi A, Främling K (2020) Security in product lifecycle of IoT devices: a survey. J. Netw Comput Appl 102–779
13. Abdul-Qawy ASH, Almurisi NMS, Tadisetty S (2020) Classification of energy saving techniques for IoT-based heterogeneous wireless nodes. Procedia Comput Sci 17(1):2590–2599
14. Aarika K, Bouhlal M, Abdelouahid RA, Elfilali S, Benlahmar E (2020) Perception layer security in the Internet of Things. Procedia Comput Sci 17(5):591–596
15. Elappila M, Chinara S, Parhi DR (2020) Survivability aware channel allocation in WSN for IoT applications. Pervasive Mob Comput 6(1):101–107
16. Karunanithy K, Velusamy B (2020) Cluster-tree based energy efficient data gathering protocol for industrial automation using WSNs and IoT. J Indus Inform Integr 100–156
17. Tao M, Li X, Yuan H, Wei W (2020) UAV-aided trustworthy data collection in federated-WSN-enabled IoT applications. Inform Sci 532:155–169
18. Rabbani MM, Vliegen J, Winderickx J, Conti M, Mentensa N (2019) Scalable heterogeneous layered attestation. IEEE Internet Things J 6(6):10240–10250

19. Libert B, Ling S, Mouhartem F, Nguyen K, Wang H (2016) Signature schemes with efficient protocols and dynamic group signatures from lattice assumption, pp 373–403
20. Yousefnezhad N, Malhi AK, Främling K (2020) Security in product lifecycle of IoT devices: a survey. J Netw Comput Appl 171:102779
21. Abdul-Qawy ASH, Nasr Musaed lmurisi S, Srinivasulu T (2020) Classification of energy saving techniques for IoT-based heterogeneous wireless nodes. Procedia Comput Sci 171:2590–2599
22. Aarika K, Bouhlal M, AitAbdelouahid R, Elfilalia S, Benlahmar E (2020) Perception layer security in the Internet of Things. Procedia Comput Sci 175:591–596
23. Karunanithy K, Bhanumathi V (2020) Cluster-tree based energy efficient data gathering protocol for industrial automation using WSNs and IoT. J Indus Inform Integr 19:100156
24. Elappila M, Chinara S, Parhi DR (2019) Survivability aware channel allocation in WSN for IoT applications. Pervasive Mob Comput 61:101107
25. Lee KC, Sudhaakar R, Dai L, Addepalli S, Gerla M (2012) RPL under mobility. In: Proceedings of 2012 IEEE consumer communications and networking conference (CCNC), Las Vegas, NV, USA
26. Tian B, Hou KM, Shi H, Liu X, Diao X (2013) Application of modified RPL under VANETWSN communication architecture. In: Fifth (ICCIS), pp 1467–1470
27. Martocci EJ, Mil PD, Riou N, Vermeylen W (2010) Building automation routing requirements in low-power and lossy networks. RFC:5867
28. Brandt A, Buron J, Porcu G (2010) Home automation routing requirements in low-power and lossy networks

Recent Trends on Low-Resource Neural Machine Translation and Research Scope for English–Mizo Pair

Vanlalmuansangi Khenglawt, Sahinur Rahman Laskar, Riyanka Manna, Partha Pakray, and Ajoy Kumar Khan

Abstract Dealing with low-resource languages is one of the challenging fields for NMT, as it struggled with restricted language resources. Hence, there exist various techniques to handle the issue of low-resource languages in NMT. Linguistic constraints of a language are one of the key considerations. Variations in the linguistic structure between the language pairs also have a remarkable impact on the translation. Therefore, we need to explore suitable techniques and address the linguistic difficulties of improving low-resource pair translation. This paper presents a survey on the low-resource English–Mizo pair's linguistic challenges and the various existing low-resource NMT approaches.

Keywords English–Mizo · Linguistic challenges · Low-resource · NMT

1 Introduction

Neural machine translation (NMT) has achieved significant progress in recent years by producing high-quality translations in terms of fluency and adequacy. In large-scale situations, NMT has accomplished remarkable outcomes, but it has struggled to make use of limited language resources. The key reason is that the output is determined by the amount of training data available. As a result, figuring out how to make the most of available data to minimize resource scarcity has become a major focus in

V. Khenglawt (✉) · A. K. Khan
Department of Computer Engineering, Mizoram University, Mizoram, India
e-mail: mzut208@mzu.edu.in

S. R. Laskar · P. Pakray
Department of Computer Science and Engineering, National Institute of Technology, Silchar, Assam, India
e-mail: partha@cse.nits.ac.in

R. Manna
Department of Computer Science and Engineering, Adamas University, Kolkata, West Bengal, India

S. K. Biswas et al. (eds.), *Intelligent Computing Systems and Applications*, Lecture Notes in Networks and Systems 1307, https://doi.org/10.1007/978-981-96-3860-4_11

neural machine translation. Additionally, the relationship between the language pairs also has a significant impact on the output of machine translation. Every language has a unique set of linguistic characters that make up its phonological structure. With machine translation techniques, closely related language pair gives better results as compared with language pair with huge structural diversity. Therefore, detailed research has been carried out in this paper for low-resource NMT as well as research gap exist between English Mizo language pair.

The following sections are classified as follows: the related work on the English–Mizo pair is discussed in Sect. 2. Sections 3 and 4 deals with low-resource NMT and techniques for handling low-resource NMT. Section 5 focuses on the research gap for the English–Mizo pair. Finally, the conclusion and future work are mentioned in Sect. 6.

2 Related Work on English–Mizo Pair

There are a few existing research works dealing with Mizo language including a study on multiword expression of Mizo language [16], defining rules for identifying named entity classes in Mizo language [2], the creation of a Mizo-to-English dictionary and a developing part-of-speech tag set for the Mizo language, which will be utilized to create an automatic POS tagger [19]. Regardless of the fact that a few NLP works related with Mizo language have been performed, only a limited amount of research has been carried out recently in the field of MT due to scarcity of resources in the English–Mizo language pair. Thihlum et al. [25], Pathak and Pakray [20] has prepared and built an English–Mizo parallel corpus. Two models, NMT and PBSMT have been tested on the prepared corpus. Automatic evaluation metrics such as BLEU, METEOR, F-measure, and human evaluation scores have been used to assess the performance of translations predicted by the trained models. Subsequently, an enhanced NMT technique is implemented on English–Mizo language pair [13]. The predicted output is evaluated based on a various aspects, including sentence length alteration, comparison to current baselines, and analysis of predicted errors. In several language pairs, NMT achieves state-of-the-art performance. However, the performance degrades when working with low-resource languages. Therefore, Lalrempuii and Soni [12] employed multiple attention-based NMT models, such as RNN, BRNN, and transformer, to analyze the English–Mizo pair.

3 Low-Resource NMT

Although NMT has struggled in the area of low-resource language, yet, various works have been done by several researchers to deal with low-resource language. The NMT methodologies can be categorized into supervised, semi-supervised, and unsupervised. A large-scale bilingual corpus is crucial for supervised NMT, which is lacking in low-resource languages. Semi-supervised techniques depend on the presence of parallel corpora in addition to monolingual corpora. The goal of unsupervised learning approaches is to build an MT system on a language pair with no parallel corpus. Both semi-supervised and unsupervised techniques deal with low-resource language pairs. There are various approaches considered improving the translation quality of limited-resource pair translation. These approaches can be categorized into two groups: the traditional approach and the knowledge-transfer approach.

Traditional Approach. It mainly deals with data scarcity, the issue of rare words, and out-of-vocabulary. To obtain better translation quality, NMT requires sufficient parallel data. However, with low-resource language, monolingual data is more accessible to obtain compared to parallel data from the viewpoint of resource availability. Therefore, the data augmentation technique is introduced by utilizing monolingual data to increase the size of the dataset. It is known as the self-training approach [9] under semi-supervised settings [4, 26]. Researchers have been investigating improving NMT by pimping monolingual data [3, 4, 21] and back-translation (BT). However, due to the presence of large-scale noise in BT, data decreases the training performance and, as a result, lowers the translation quality [3]. Although the out-of-vocabulary issue is improved in the COPY model, it increases the vocabulary only in the target data. For this reason, filtering-based approaches are introduced to refine the synthetic data obtained from BT [10, 14].

With an unsupervised setting, pre-train word-embeddings on monolingual data enhance the performance of low-resource NMT. The issues of rare words and out-of-vocabulary (OOV) are the crucial challenges of low-resource NMT [11] and byte-pair-encoding (BPE) [22] introduce to handle such problems.

Knowledge-Transfer Approach. It is mostly concerned with insufficient data for low-resource NMT through various approaches, namely, transfer learning, multitasking, zero-shot, multilingual, and multimodal-based NMT. The multitasking approaches executes multiple tasks in a limited time by employing various downstream natural language processing tasks, such as part-of-speech tagging, role-labeling, named-entity-recognition, and image caption generation [1]. The objective behind multitasking is to obtain train weights from extensive data and enhance the model's generalization capability. Furthermore, the investigation has been conducted by integrating parallel and monolingual data simultaneously to improve translation quality [6] using translation language modeling (TLM) and masked language modeling (MLM) objectives. Other approaches are discussed in the next section.

4 Techniques for Low-Resource NMT

Researchers have developed various ways to improve the limited resource language in NMT. The following strategies analyze techniques for handling low-resource language in NMT. By applying these strategies, the limitation of low-resource language may be addressed to some extent, resulting in a significant improvement for NMT.

- **Back-Translation**: In a low-resource language, the most popular and successful use for data augmentation in machine translation is back-translation. In this process, BT generates a synthetic parallel corpus from the target monolingual data by training a translation model in the backward direction. The model is then retrained using this synthetic parallel data along with the original parallel data. The purpose was to create synthetic source sentences to maximize the number of parallel training data. BT has proven to be a substantial gain in NMT [21]. Iterative back-translation (IBT) is also proposed [8] to improve the idea of BT, where both forward and backward directions of translations are utilized for training each other.
- **Word-Embedding and Word-Segmentation**: Word-embedding is another concept for handling low-resource language in unsupervised NMT. It is a special way of illustrating a word in a vector space such that the semantic similarity of each word can be identified. There are several concepts in word-embedding techniques such as Word2Vec, GloVe [5]. The BPE is one of the techniques used for word-segmentation [22]. It aims to break down words into subword units, making it easier to deal with rare and unknown words. It is a data compression method in which the most often occurring a pair of bytes in a sequence are replaced.
- **Transfer learning**: The concept is to transfer some limited parameters from NMT trained model of large resource language pair to a language model of low-resource language pair [27]. The idea behind transfer learning involves two steps: first, training of machine learning models in high-resource settings and then, secondly, by taking advantage of high-resource unlabeled data, utilize a semi-supervised approach on more minor labeled data downstream classification task. A transfer learning also shows improvement for parent models by taking advantage of its lexical resemblance to another low-resource language [18].
- **Zero-Shot NMT**: In zero-shot approach, there is no parallel corpus accessible for a language pair. Conventional methods for zero-shot NMT include pivoting, transfer learning, multi-NMT, and unsupervised NMT. Pivoting methods are applied in the absence of a parallel corpus by using a third language also called pivot language. The translation model breaks up into two models: the first model and the second model. The first model is used to translate a source sentence, and the output is then translated again using the second model to produce the target sentence. However, it has a drawback of error propagation. Combining pivoting with transfer learning is one technique for reducing the issue of error propagation.
- **Multilingual Approach**: NMT systems that handle translation across many language pairs are known as multilingual NMT systems. It can be categorized into three varieties: single-source language to multiple-target languages, multiple-source languages to single-target language, and multiple-source languages to multiple-target languages. The key concept underlying these techniques is to transfer knowledge from high-resources to low-resource translation. According to recent studies, multilingual models produce better output as compared to bilingual models, especially with a limited number of languages in the system and the relatedness of the languages also determines the performance.
- **Multimodal Approach**: Apart from textual low-resource NMT, the researchers have been investigating a multimodal concept where a combination of textual

and visual features helps to improve low-resource NMT. The multimodal NMT approach extracts information from several modalities including textual, picture, and audio. It is done with the expectation that the various modalities will provide valuable perspectives on the supplied data [23].

5 Research Gap for English–Mizo Pair

The above approaches discuss the techniques for improving low-resource languages on NMT. Many researchers have investigated different methods on a variety of low-resource languages. The following Table 1 summarizes recent research papers on several techniques for resolving the issue of limited resources in NMT. The barriers of a low-resource language may be addressed to some extent using these strategies. However, these new strategies have yet to be implemented for low-resource Mizo language. Despite the fact that several limited languages have improved their NMT output by adopting various techniques, there is a significant research deficit in the field of MT for the English–Mizo language pair.

Aside from the various strategies developed by researchers to enhance restricted languages in NMT, one of the most essential metrics is linguistic challenges of a specific language. Consequently, dealing with the linguistic challenges of specific low-resource language is one of the factor that will enhance the performance of the NMT. Therefore, there are research gaps for English–Mizo pair in the aspect of Machine Translation, they are as follows:

- **Language origin**: Each languages has its unique linguistic framework. The similarity of the language pairs has a significant impact on the output. In terms of English–Mizo language pair, both languages are substantially distinct from one another with respect to their originating languages. The English language is a part of the Indo-European family of languages, whereas the Mizo language is a part of the Sino-Tibetan family.
- **Syntactic structure**: Both languages have distinct syntactic structures, resulting in a research gap in the field of machine translation. The English language has a subject-verb-object (SVO) syntactic structure. However the Mizo language employs an object-subject-verb (OSV) syntactic structure.
  - **English**: I am going to school ***(SVO)***
  - **Mizo**: Schoolah ka kal ***(OSV)***
- **Gender distinction**: The proper names in English does not distinguish between gender. Eg: Dave, Robin. But in Mizo, a gender indicator is appended to the suffix of all proper names to identify them.
  - Rami *(indicates female since a letter* **i** *is append at the end)*
  - Mawia *(indicates male since a letter* **a** *is append at the end).*

Table 1 Survey of recent papers for techniques in low-resource NMT

Techniques (Language pair)	Paper	Research gap
Back-Translation (Bribri–Spanish)	Neural Machine Translation Models with Back-Translation for the Extremely Low-Resource Indigenous Language Bribri (2020) [7]	In this paper, a group of models augmented with iterative back-translation have been trained producing a gain in the output and concluded that NMT techniques can have acceptable results. As a result, working with iterative back-translation approaches on the English Mizo language pair is expected to produce improved results
Word-Embedding and Word-Segmentation (English–Tamil and English–Malayalam)	Neural Machine Translation for Low-Resourced Indian Languages (2020) [5]	An NMT model employing self-attention Multihead alongside pre-trained Byte-Pair-Encoded with MultiBPE embeddings is proposed to build an effective translation technique that solves the Out Of Vocabulary barrier for limited-resourced morphologically rich Indian languages. Though the performance output is solely dependent on the linguistic structure, it is anticipated that dealing with the linguistic challenges of the Mizo language would produce in a superior output for the English–Mizo pair
Hierarchical Transfer Learning Architecture (Uygur–Chinese Turkish–English)	Hierarchical Transfer Learning Architecture for Low-Resource Neural Machine Translation (2019) [15]	The method adds an intermediate layer. It not only benefits from the high-volume data on high-resource languages with synthetic similarity on intermediate languages, it also improves in flexibility. Though the hierarchical transfer learning architecture achieves substantial improvements over a number of baselines, it had not yet been implemented on the English–Mizo language pair
Multilingual Neural Machine Translation (Arabic, German, Tagalog, Turkish, and Xhosa into English)	An Analysis of Massively Multilingual Neural Machine Translation for Low-Resource Languages (2020) [17]	Cross-lingual performance of low-resource multilingual NMT is highly variable. The interdependency of multiple languages has an effect on efficiency. Adding more data to a testing corpus result in higher performance if performed correctly. Using multilingual NMT on English Mizo pairs with other closely related languages may enhance the performance
Multimodal Machine Translation System (Japanese–English)	TMU Japanese–English Multimodal Machine Translation System for WAT 2020 [24]	The translation model is trained on augmented noisy data before being fine-tuned with clean data. In the decoder, the probabilistic dropping of the visual or textual context vector is also examined. The experiment indicates the performance of the translation can be enhanced. Therefore, a multimodal approach on a low-resource language like Mizo language is expected to produce better results

Table 2 Vocabulary distinction

Types of clothes	Different vocabulary for 'wear' in Mizo
Pants or shirt	Hà
Skirt	Féng
Shoes or Socks	Bún
Hat	Khúm
Belt	Hréng

- **Tonal language**: Furthermore, the Mizo language has several challenges. It is a tonal language where words with varied tonal inflections convey diverse meanings. There are four tones in the Mizo language (high, low, rising, and falling) while English is not a tonal language.
- **Lexical ambiguity**: There are several ambiguous words called polysemy words that exist in the Mizo language, that is, words with the same intonation yet distinct interpretations. For example, the Mizo word '*là*' with a low tone can signify two distinct English interpretations '*take*' and '*yarn*'.
- **Vocabulary distinction**: Although the vocabulary of English is undeniably rich. However, for a few specific words, Mizo has a richer vocabulary than English. For example, as in Table 2 the word '*Wear*' in English can refer to different vocabulary based on the clothes we wear in Mizo.

Thus, bridging the linguistic gap between the two languages by addressing the issues and implementing recent NMT strategies for low-resource languages would result in a major improvement for the English–Mizo language pair.

6 Conclusion and Future Work

In this paper, we have presented the related works on the English–Mizo pair and the linguistic challenges with research gap. In addition, a systematic survey is explored on the existing traditional and knowledge-transfer approaches of low-resource NMT. In future work, we will develop parallel and monolingual corpora by addressing the linguistic challenges of the English–Mizo pair and investigating NMT by employing suitable techniques, namely, BT, word-embeddings, multilingual settings, to improve translation quality.

References

1. Adouane W, Bernardy J-P (2020) When is multi-task learning beneficial for low-resource noisy code-switched user-generated Algerian texts? In: Proceedings of the 4th workshop on computational approaches to code switching, CodeSwitch@LREC 2020, May, 2020, Marseille, France. European Language Resources Association, pp 17–25
2. Bentham J, Pakray P, Majumder G, Lalbiaknia S, Gelbukh A (2016) Identification of rules for recognition of named entity classes in Mizo language. In: 2016 fifteenth Mexican international conference on artificial intelligence (MICAI), pp 8–13
3. Burlot F, Yvon F (2018) Using monolingual data in neural machine translation: a systematic study. In: Proceedings of the third conference on machine translation: research papers, Brussels, Belgium, October 2018. Association for Computational Linguistics, pp 144–155
4. Cheng Y, Xu W, He Z, He W, Wu H, Sun M, Liu Y (2016) Semi-supervised learning for neural machine translation. In: Proceedings of the 54th Annual meeting of the association for computational linguistics (Volume 1: Long Papers), Berlin, Germany, August 2016. Association for Computational Linguistics, pp 1965–1974
5. Choudhary H, Rao S, Rohilla R (2020) Neural machine translation for low-resourced Indian languages. CoRR arXiv:2004.13819
6. Conneau A, Lample G (2019) Cross-lingual language model pretraining. In: Advances in neural information processing systems 32: annual conference on neural information processing systems 2019, NeurIPS 2019, 8–14 December 2019, Vancouver, BC, Canada, Canada. Advances in Neural Information Processing Systems (NeurIPS), pp 7057–7067
7. Feldman I, Coto-Solano R (2020) Neural machine translation models with back-translation for the extremely low-resource indigenous language Bribri. In: Proceedings of the 28th international conference on computational linguistics, Barcelona, Spain (Online), December 2020. International Committee on Computational Linguistics, pp 3965–3976
8. Hoang VCD, Koehn P, Haffari G, Cohn T (2018) Iterative back-translation for neural machine translation. In: Proceedings of the 2nd workshop on neural machine translation and generation, Melbourne, Australia, July 2018. Association for Computational Linguistics, pp 18–24
9. Imamura K, Sumita E (2018) NICT self-training approach to neural machine translation at NMT-2018. In: Proceedings of the 2nd workshop on neural machine translation and generation, Melbourne, Australia, July 2018. Association for Computational Linguistics, pp 110–115
10. Imankulova A, Sato T, Komachi M (2019) Filtered pseudo-parallel corpus improves low-resource neural machine translation. ACM Trans Asian Low-Resour Lang Inf Process 19(2)
11. Koehn P, Knowles R (2017) Six challenges for neural machine translation. In: Proceedings of the first workshop on neural machine translation, Vancouver, August 2017. Association for Computational Linguistics, pp 28–39
12. Lalrempuii C, Soni B (2020) Attention-based English to Mizo neural machine translation. In: Bhattacharjee A, Borgohain SK, Soni B, Verma G, Gao X-Z (eds) Machine learning, image processing, network security and data sciences, Singapore. Springer Singapore, pp 193–203
13. Lalrempuii C, Soni B, Pakray P (2021) An improved English-to-Mizo neural machine translation. ACM Trans Asian Low-Resour Lang Inform Process 20:1–21
14. Laskar SR, Pakray P, Bandyopadhyay S (2019) Neural machine translation: Hindi–Nepali. In: Proceedings of the fourth conference on machine translation (Volume 3: Shared task papers, Day 2), Florence, Italy, August 2019. Association for Computational Linguistics, pp 202–207
15. Luo G, Yang Y, Yuan Y, Chen Z, Ainiwaer A (2019) Hierarchical transfer learning architecture for low-resource neural machine translation. IEEE Access 7:154157–154166
16. Majumder G, Pakray P, Khiangte Z, Gelbukh AF (2016) Multiword expressions (MWE) for Mizo language: literature survey. In: Gelbukh AF (ed) Computational linguistics and intelligent text processing—17th international conference, CICLing 2016, Konya, Turkey, April 3–9, 2016, Revised Selected Papers, Part I, volume 9623 of Lecture notes in computer science. Springer, pp 623–635

17. Mueller A, Nicolai G, McCarthy AD, Lewis D, Wu W, Yarowsky D (2020) An analysis of massively multilingual neural machine translation for low-resource languages. In: Proceedings of the 12th language resources and evaluation conference, Marseille, France, May 2020. European Language Resources Association, pp 3710–3718
18. Nguyen TQ, Chiang D (2017) Transfer learning across low-resource, related languages for neural machine translation. In: Proceedings of the eighth international joint conference on natural language processing (Volume 2: Short papers), Taipei, Taiwan, November 2017. Asian Federation of Natural Language Processing, pp 296–301
19. Pakray P, Pal A, Majumder G, Gelbukh AF (2015) Resource building and parts-of-speech (POS) tagging for the Mizo language. In: Sidorov G, Galicia-Haro SN, Pichardo-Lagunas O, Herrera-Alcántara O, Arroyo-Figueroa G (eds) Fourteenth Mexican international conference on artificial intelligence, MICAI 2015, Cuernavaca, Mexico, October 25–31, 2015. IEEE Computer Society, pp 3–7
20. Pathak A, Pakray P (2019) English-Mizo machine translation using neural and statistical approaches. Neural Comput Appl 31
21. Sennrich R, Haddow B, Birch A (2016) Improving neural machine translation models with monolingual data. In: Proceedings of the 54th annual meeting of the association for computational linguistics (Volume 1: Long papers), Berlin, Germany, August 2016. Association for Computational Linguistics, pp 86–96
22. Sennrich R, Haddow B, Birch A (2016) Neural machine translation of rare words with subword units. In: Proceedings of the 54th annual meeting of the association for computational linguistics (Volume 1: Long papers), Berlin, Germany, August 2016. Association for Computational Linguistics, pp 1715–1725
23. Shah K, Wang J, Specia L (2016) SHEF-multimodal: grounding machine translation on images. In: Proceedings of the first conference on machine translation: Volume 2, Shared task papers, Berlin, Germany, August 2016. Association for Computational Linguistics, pp 660–665
24. Tamura H, Hirasawa T, Kaneko M, Komachi M (2020) TMU Japanese–English multimodal machine translation system for WAT 2020. In: Proceedings of the 7th workshop on Asian translation, Suzhou, China, December 2020. Association for Computational Linguistics, pp 80–91
25. Thihlum Z, Khenglawt V, Debnath S (2020) Machine translation of English language to Mizo language. In: 2020 IEEE international conference on cloud computing in emerging markets (CCEM), pp 92–97
26. Wang Y, Xia Y, Zhao L, Bian J, Qin T, Chen E, Liu T (2019) Semi-supervised neural machine translation via marginal distribution estimation. IEEE/ACM Trans Audio Speech Lang Process 27(10):1564–1576
27. Zoph B, Yuret D, May J, Knight K (2016) Transfer learning for low-resource neural machine translation

HybridHAR-Net: Recognizing Human Activities Using a Hybrid Deep Learning-Based Model for Mobile Health Applications

Debarshi Bhattacharya, Karam Kumar Sahoo, Pawan Kumar Singh, and Mufti Mahmud

Abstract The classification of human motions into their corresponding activity classes is an integral challenge for Human Activity Recognition (HAR) since the data is available in multiple modalities; hence, a single framework or approach is not enough to handle this. In this paper, we have applied Continuous Wavelet Transformation (CWT) to convert the raw time-series sensor data into 3D matrices of n-channels analogous to multi-channel images. These matrices are fed into our proposed hybrid model named HybridHAR-Net, which comprises a two-dimensional Convolutional Neural Network (CNN) architecture and a multi-layered Gated Recurrent Unit (GRU) network. The CNN architecture of HybridHAR-Net extracts structural features from the n-channel inputs while the layered GRU network processes the sequential features, producing a better HAR evaluation result. The proposed HybridHAR-Net model is evaluated on two publicly available benchmark datasets- WISDM and

D. Bhattacharya
Department of Electronics and Communication Engineering, Techno Main Salt Lake, EM-4/1, Sector V, Bidhannagar, Salt Lake City, Kolkata 700091, West Bengal, India
e-mail: bhattacharyadebarshi4@gmail.com

K. K. Sahoo
Department of Computer Science and Engineering, National Institute of Technology, Mahatma Gandhi Avenue, A-Zone, Durgapur 713209, West Bengal, India
e-mail: karamsahoo@gmail.com

P. K. Singh (✉)
Department of Information Technology, Jadavpur University, Jadavpur University Second Campus, Plot No. 8, Salt Lake Bypass, LB Block, Sector III, Salt Lake City, Kolkata 700106, West Bengal, India
e-mail: pawansingh.ju@gmail.com

School of Science and Technology, Nottingham Trent University, Clifton, Nottingham, UK

M. Mahmud
Department of Computer Science, Nottingham Trent University, Clifton, Nottingham, UK

Medical Technologies Innovation Facility, Nottingham Trent University, Nottingham, UK

Computing and Informatics Research Centre, Nottingham Trent University, Nottingham, UK

S. K. Biswas et al. (eds.), *Intelligent Computing Systems and Applications*, Lecture Notes in Networks and Systems 1307, https://doi.org/10.1007/978-981-96-3860-4_12

UCI-HAR and achieved classification accuracies of 97.43% and 97.15%, respectively. The experiment results show that our proposed method of utilizing CWT on raw sensor data and applying the proposed HybridHAR-Net model for predicting daily-life human activities outperforms recent multiple HAR methods to which our method is compared.

Keywords Human activity recognition · HybridHAR-Net · Continuous wavelet transformation · Hybrid deep learning model · WISDM · UCI-HAR

1 Introduction

The rapid technological advancement creates a vast pool of data through various state-of-the-art IoT devices. The task of HAR (Human Activity Recognition) is becoming increasingly indispensable in our daily lives. From health care to surveillance, HAR has many use cases in this modern era. The task of human activity recognition is mainly of two types: sensor-based [1] and image or video-based [2]. Till now, the HAR models that various researchers have proposed are primarily divided into two categories: one is classical machine learning models, viz., Decision Tree [3], Random Forest [4], Support Vector Machine (SVM) [5] and the other is based on various deep learning architectures [6]. The main drawback of using traditional machine learning algorithms is that we need to create features out of the raw sensor data; hence, it is time-consuming and complex. One needs proper domain knowledge of that data to apply feature engineering. At the same time, deep learning architectures like CNN (Convolutional Neural Network) and LSTM (Long Short Term Memory) networks can derive features on their own from raw sensor data. Cruciani et al. [7] used CNN as a feature extractor, and Zhao et al. [8] used a Deep Bidirectional LSTM network on raw sensor data to classify human activities. Ullah et al. [9] formed a hybrid architecture by combining CNN with layered bidirectional LSTM. Bhattacharya et al. [10] implemented an ensemble of such above-mentioned deep learning architectures to achieve the state of the art.

Owing to the advancements that deep learning and image classification have made in the past few decades, converting raw sensor data into images and implementing HAR methods on those is drawing the attention of many researchers nowadays. Xu et al. [11] used Gramian Angular Field (GAF) conversion on raw sensor data to apply the HAR method for activity classification. Wang et al. [12] decomposed ECG signals using CWT to derive contrasting time–frequency components. Our paper employs image classification techniques using 2D convolutional neural networks (CNN) models in the field of HAR. Hence, the choice of features is 3D matrices of n-channels corresponding to the different sensors used for collecting data. These n-channel matrices are analogous to multi-channel images, for instance, RGB and RGB-D images with 3 and 4 channels, respectively, which can be fed to CNN architectures fine-tuned for the multi-channel input data. The kernel filters in the CNN enable the model to pick up essential feature maps from our input. Additionally, data

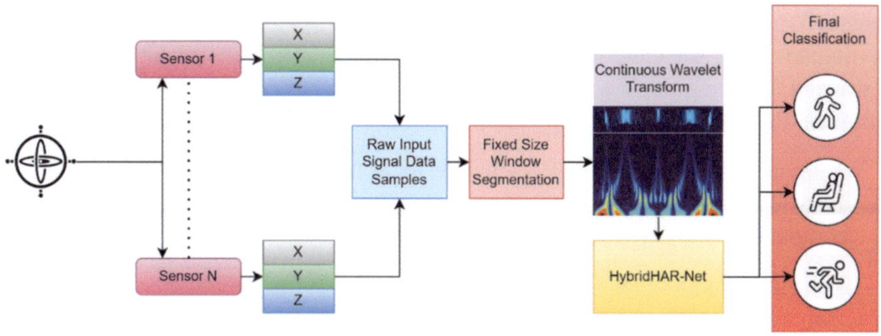

Fig. 1 The flow of work for the proposed experiment in this paper includes the collection, preprocessing, and window segmentation of the raw data, the transformation of raw data using CWT, and finally, the prediction of activities by our proposed HybridHAR-Net model

is present in not one but n-channels which carry information from different sensors, giving our model more information to train on for better performance.

This paper introduced a hybrid framework that takes CNN for structural feature extraction and multi-layered GRU for sequential feature representation. Before feeding the raw sensor data to the model, we applied Continuous Wavelet Transformation (CWT) on the raw sensor data to produce n-channel 3D matrices. The overview of our proposed work is shown in Fig. 1.

2 Proposed Methodology

2.1 *Continuous Wavelet Transformation (CWT) on Raw Sensor Data*

In many sectors of science and engineering, the continuous wavelet transform (CWT) has played an important role in studying time–frequency data. It is based on the short-time Fourier transform but with variable time–frequency resolution. The CWT is the result of multiplying the shift wavelet by the total sum of the time signals. The wavelet coefficients are the result of the continuous wavelet transform. Scale and position are functions of these coefficients. Both simple and discontinuous signals can be processed using this method. A sine wave is followed by a medium sine wave to create a discontinuous wavelet. The CWT is conceptually a sliding cross-correlation between a signal, $s(t)$, and a group of wavelets, as in:

$$W_s(a, b) = \frac{1}{\sqrt{a}} \int s(t)\psi^*\left(\frac{t-b}{a}\right)dt \tag{1}$$

where $W_s(a, b)$ is a representation of the signal's time scale; Ψ^* denotes the complex conjugate of the wavelet; a, b, and c represent the amplitudes and time shifting of a wavelet Ψ. The mother wavelet is another name for the reference wavelet. In this instance, the transformation depends on the ideas of scalability and shifting.

1. A signal is scaled when its size changes over time based on a scaling factor.
2. Transferring wavelets of various scales from the signal's beginning to its completion is known as shifting.

The scaling factor, which shows how much a signal has been scaled over time, is inversely related to frequency. This suggests that the scale is higher for more refined scale discretion. The various wavelets in terms of shift parameters and scaling along with time are transformed across the signal to obtain the physical significance. The resulting coefficients are then multiplied by the sampling interval and wavelet scaling along with shifting variables. For instance, multiplying a signal with 1000 timesteps by 64 scales (ranging from 1 to 65) results in 64,000 coefficients. Characterizing oscillated behavior in CWT signals is facilitated by this.

In our suggested solution, we have to determine an efficient mother wavelet as well as a scaling factor using the continuous wavelet transform. Its most palatable shape led to the selection of the Morlet Mother Wavelet. To select an appropriate range of scales, we need to observe the CWT coefficients that have three distinct ranges—32, 64, and 128—displayed as a scalogram. In our example, 32, a lower scale size makes it possible to concentrate more on quick transitions. As stated earlier, these sudden shifts are often the most noticeable characteristics. Figure 2 illustrates how a wide range of scales (64 or 128 in our case) can increase classification accuracy by providing additional information (relating to gradually shifting situations). Conversely, the second strategy will demand a more in-depth CNN.

Visualization of the scalograms for different human activities makes it humanly possible to distinguish among the various activities. Additionally, any CNN model could pick this information from the scalograms and make a decent classification. This is the theory behind our research in this paper.

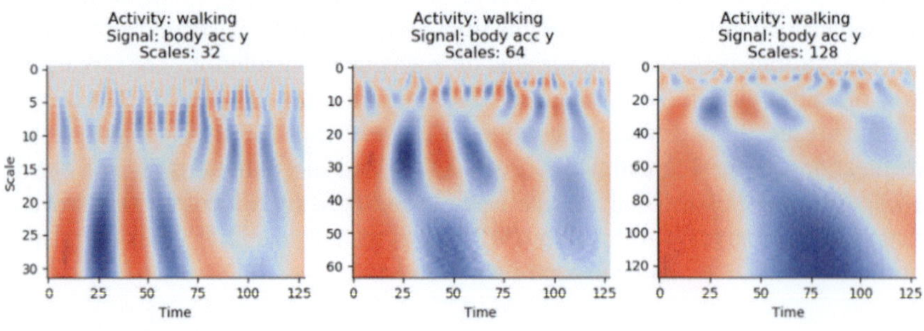

Fig. 2 Visualization of signal's behavior for a different range of scales

2.2 N-Channel CWT Input

Our proposed model accepts input in the shape of 3D-CWT, which is constructed by stacking together the different CWT scalograms for the various sensors from which the data samples have been collected. N number of different sensors have been employed for HAR data collection that includes a gyroscope and accelerometer to name a few. Stacking the various CWT scalograms together enables the model to simultaneously make convolutional operations on all the CWT for all N sensor data.

2.3 Proposed HybridHAR-Net Architecture

CNNs, a type of artificial neural network that has been successful in computer vision, are gaining popularity in various fields. CNN uses several building blocks such as convolution layers, pooling layers, and fully connected layers to learn spatial hierarchies of data automatically and adaptively through backpropagation. The Gated Recurrent Unit (GRU) is a form of recurrent neural network (RNN) that has advantages over long short-term memory (LSTM)—GRU is faster and utilizes less memory than LSTM, and also deals with the problem of vanishing gradient from which standard RNN suffers. The GRU solved this problem by implementing two gates: a reset gate that governs the assimilation of new input with the prior memory and an update gate that controls the maintenance of the valuable memory. The reset and update gates govern how much each hidden unit remembers or forgets when reading or adaptively creating a sequence. Figure 3 shows the overall architectural diagram of our proposed HybridHAR-Net model. Figure 4 shows the overall architectural diagram of our proposed HybridHAR-Net model comprised of deep CNN architecture followed by GRU.

- The proposed model consists of three 2D convolutional layers with different numbers of filters, which are of different kernel sizes ((3, 3), (5, 5), (7, 7)), respectively. These layers are equipped with the ReLU activation function.

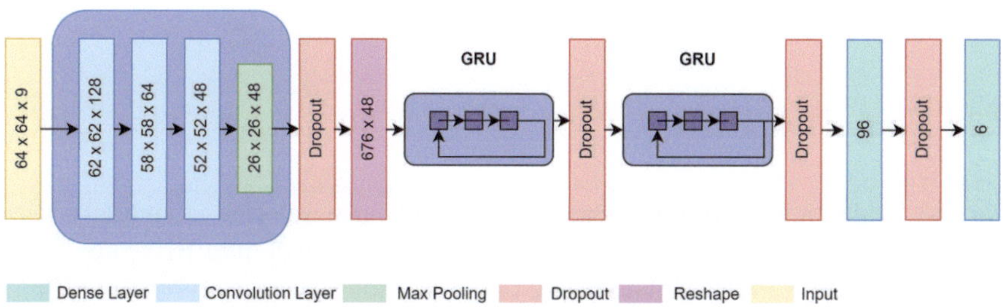

Fig. 3 Architectural overview of the HybridHAR-Net model

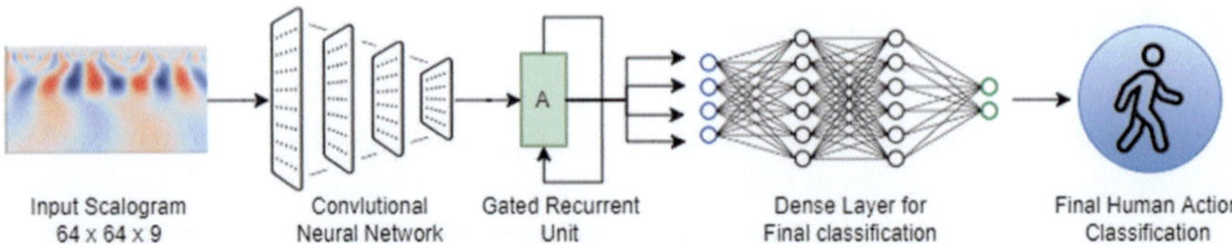

Fig. 4 The end-to-end workflow of the proposed HybridHAR-Net model

- The n-channel 3D matrices are passed through these convolutional layers serially, and the summary of the extracted features is generated by the Maxpool layer of pool size 2, reducing the computation cost.
- The outputs coming from the CNN layers and subsequent MaxPooling layer are of four-dimensional shape, as GRU, a recurrent neural network, requires an input of three-dimensional shape (batch, timestep, and feature). Therefore, there is a need to reshape the output tensor before feeding it to the GRU layer. Hence, a reshaping layer has been added before GRU to make the input shape as desired.
- Two consecutive GRU layers have been added for sequence prediction, each with a subsequent dropout layer to minimize the likelihood of the model being overfitted.
- The output from the above layers is fed to the classification layer, which consists of two dense layers. The first one is equipped with ReLU activation, while the final dense layer uses SoftMax activation on its input tensor to generate the prediction of activity classes.

3 Experimental Results and Analysis

3.1 Datasets Used

We have used two publicly available benchmark datasets for our experiment, WISDM [13] and UCI-HAR [14]. We have done preprocessing—firstly, segmentation into fixed-sized windows (128 readings per window) and secondly, Min–Max ([−1, 1]) scaling.

3.2 Evaluation Metrics

For the evaluation of our proposed model, we have selected commonly used performance measures, namely, Accuracy, Precision, Recall, and F1-score. These parameters are already described in one of our previous works, proposed in [10].

Table 1 Experimental parameters used for training of our proposed HybridHAR-Net model

Experimental parameter	Parameter setting
Loss function	Categorical cross-entropy
Optimizer	Adam
Learning rate	0.001
Batch size	64
Epoch	150

3.3 Experimental Parameters Setting

We have trained the proposed HybridHAR-Net model on both datasets according to the parameter setting mentioned in Table 1. These parameters help the model minimize the prediction error, increasing the classification accuracy. We obtained these parameter values mentioned in Table 1 by experimenting multiple times with a combination of different parameter settings.

3.4 Experimental Results and Analysis on the WISDM Dataset

Before applying CWT, the pre-processed window segmented samples of the WISDM dataset were split into the train (70%) and test (30%) sets, and then we applied the CWT on both train and test sets separately. We trained our proposed HybridHAR-Net model on the training set using the parameters mentioned in Table 1. We obtained the confusion matrix shown in Fig. 5 by evaluating our proposed trained model on the test set. From the confusion matrix, we can see that we have achieved over 97% accuracy in four of six activity classes, and the overall accuracy is 97.43%.

It can be observed that the misclassifications mainly occurred for two activity classes–"Upstairs" and "Downstairs." Around 5% of samples associated with the activity label, "Upstairs" are misclassified into activity "Downstairs." In comparison, around 4.5% of samples belonging to the activity label "Downstairs" are misclassified to "Upstairs" as these activities are of the opposite kind. This can also be observed in Fig. 6; while the other four activity labels have higher scores, the "Upstairs" and "Downstairs" activities have lesser scores in terms of Precision, Recall, and F1-score. The accuracy and loss curves during training and testing of the HybridHAR-Net model are shown in Fig. 7.

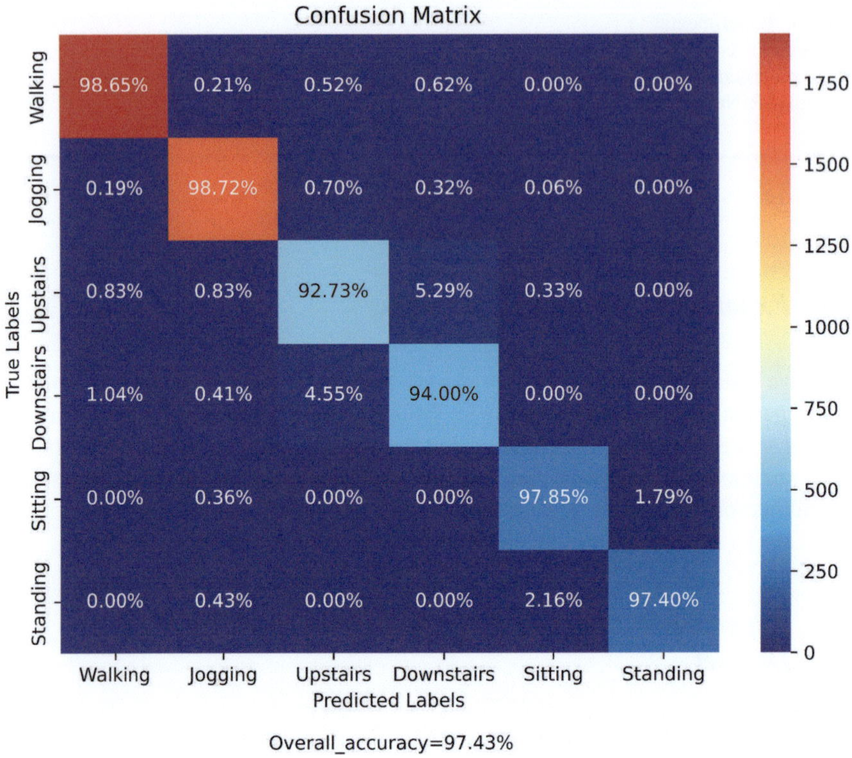

Fig. 5 Confusion matrix obtained by evaluating our trained proposed HybridHAR-Net model on the test set of the WISDM dataset

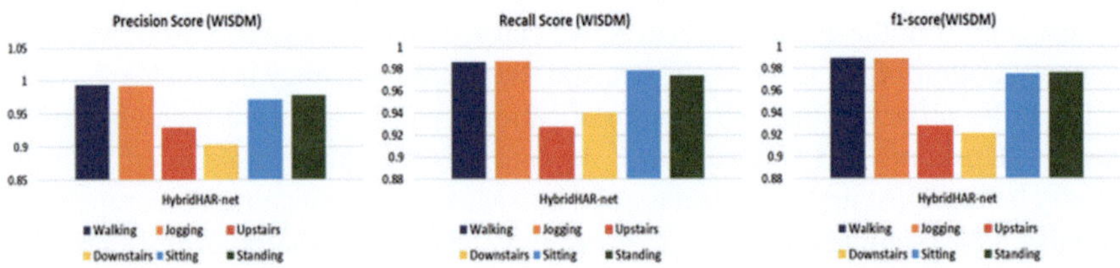

Fig. 6 Precision, Recall, and F1-score of our proposed HybridHAR-Net model for each activity class in the WISDM dataset

3.5 Experimental Results and Analysis on the UCI-HAR Dataset

Among the total samples in the dataset, the train and test sets contain 7352 and 2947 samples, respectively. We have trained the proposed HybridHAR-Net model on the train set using the parameters mentioned in Table 1, evaluated the model performance on the test data, and obtained the confusion matrix shown in Fig. 8.

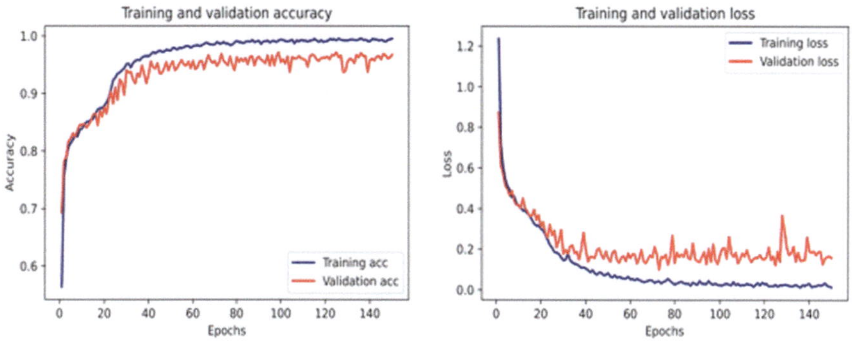

Fig. 7 Visualization of classification accuracies and losses during training and testing on the WISDM dataset

From the confusion matrix, it can be observed that our proposed model has predicted the activity labels with almost 99% accuracy in four out of six activity classes and achieved an overall accuracy of 97.15%.

The model did the misclassification mainly the two activity labels—"Sitting" and "Standing." Around 10% of samples from the test data belonging to the "Sitting" activity label have been misclassified into the activity label "Standing," while approximately 5% of samples from the "Standing" activity label are misclassified as "Sitting." This misclassification arose due to similar linear accelerations related to these two activities. This can also be observed in Fig. 9; while the other four activity labels have higher scores, the "Sitting" and "Standing" activities have lower scores in precision, recall, and F1-score. The accuracy curve and loss curve during training and testing of the HybridHAR-Net model on the UCI-HAR Dataset are shown in Fig. 10.

The proposed HybridHAR-Net model performed well in evaluating the test set of the mentioned datasets. A summary of its performance is shown in Table 2 based on the earlier mentioned evaluation metrics.

3.6 *Performance Comparison with Cutting-Edge Methods*

Table 3 shows the performance comparison of our proposed model with some recently developed HAR models. It can be seen from Table 3 that our proposed model performs significantly better than the state-of-the-art models.

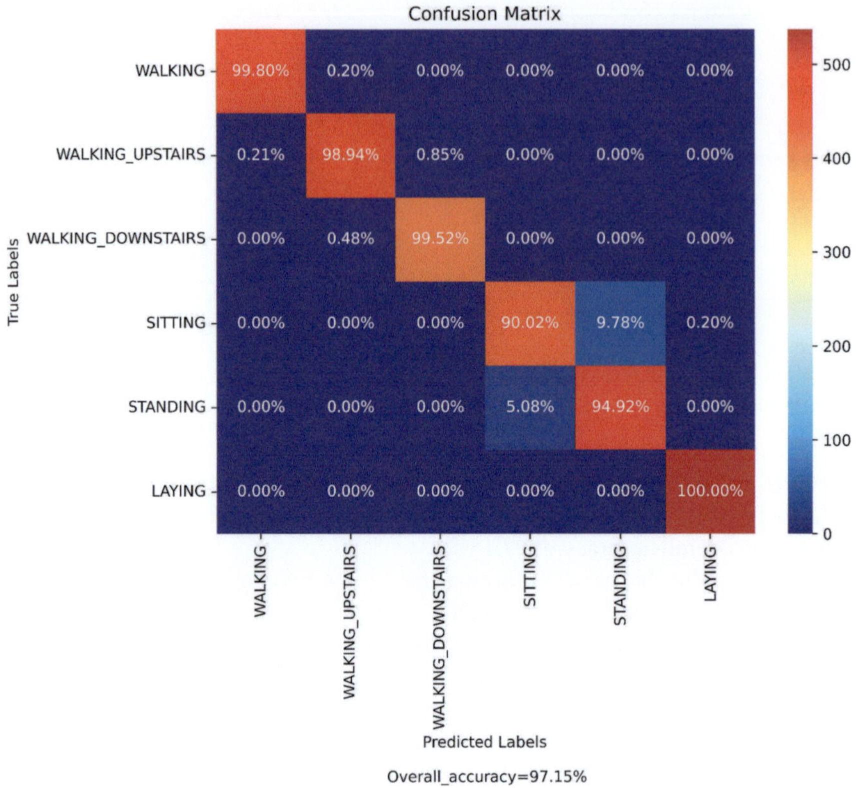

Fig. 8 Confusion matrix obtained by evaluating our trained proposed HybridHAR-Net model on the test set of the UCI-HAR dataset

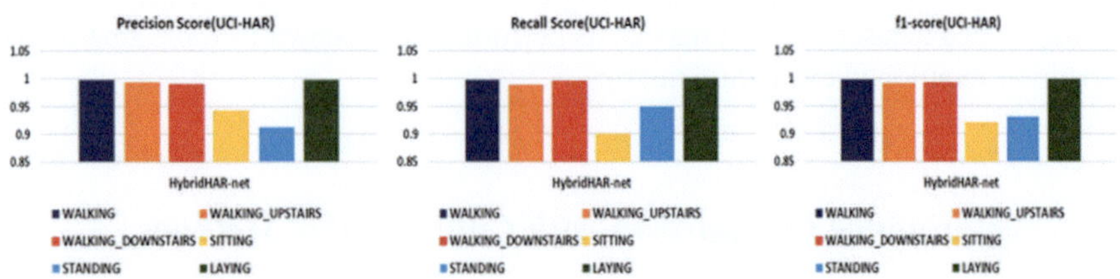

Fig. 9 Precision, Recall, and F1-score of our proposed HybridHAR-Net model for each activity class in the UCI-HAR dataset

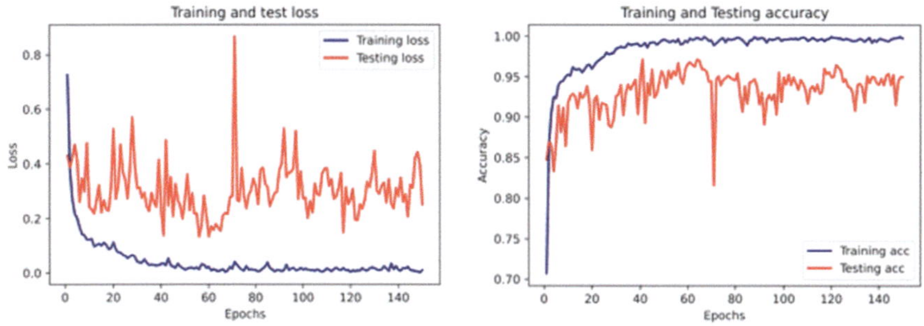

Fig. 10 Visualizing classification accuracies and losses during training and testing on the UCI-HAR dataset

Table 2 Summarization of performance evaluation of the proposed HybridHAR-Net model on both WISDM and UCI-HAR datasets

Dataset	Precision (%)	Recall (%)	f1-score (%)	Accuracy (%)
WISDM	96.10	96.55	96.32	97.43
UCI-HAR	97.26	97.20	97.21	97.15

Table 3 Comparison of accuracy between our proposed HybridHAR-Net model and state-of-the-art HAR methods

Dataset	Author	Year	Method	Accuracy (in %)
WISDM	Yin et al. [15]	2022	ConvBLSTM-PMwA	95.86
	Li et al. [16]	2022	Residual Network and BiLSTM	97.32
	Xu et al. [11]	2020	GAF + Fusion-Mdk-ResNet	97.27
	Proposed	2022	CWT + HybridHAR-Net	97.43
UCI-HAR	Yin et al. [15]	2022	ConvBLSTM-PMwA	96.71
	Shilpa et al. [17]	2021	Multi CNN-ELM	96.86
	Xu et al. [11]	2020	GAF + Fusion-Mdk-ResNet	89.48
	Proposed	2022	CWT + HybridHAR-Net	97.15

4 Conclusion

In this paper, we have proposed a hybrid deep learning architecture made up of CNN and GRU, which processes *n*-channel 3D matrices to predict human activity efficiently. The matrices are produced by performing CWT on raw sensor data. Although a significant association between activities led to misclassification in some cases, the proposed HybridHAR-Net model has shown better performance than the other recent approaches in the previously mentioned datasets. The experiments in this paper are currently being conducted on publicly available activity datasets. The future goal of this paper is the real-time implementation of HAR in wearable devices with built-in sensors such as smart wristbands. Hence, the future work will focus on the model's compression and computation cost reduction to make it more practical to be used in real time.

References

1. Mondal R, Mukhopadhyay D, Barua S, Singh PK, Sarkar R, Bhattacharjee D (2021) A study on smartphone sensor-based Human Activity Recognition using deep learning approaches. In: Handbook of computational intelligence in biomedical engineering and healthcare. Academic Press Publishers, Elsevier, Chapter 14, pp 343–369. (https://doi.org/10.1016/B978-0-12-822260-7.00006-6)
2. Bhattacharya S, Shaw V, Singh PK, Sarkar R, Bhattacharjee D (2019) SV-NET: a deep learning approach to video based human activity recognition. In: Proc. of 11th international conference on soft computing and pattern recognition (SoCPaR 2019). AISC, vol 1182, pp 10–20, Springer, Cham. https://doi.org/10.1007/978-3-030-49345-5_2
3. Pärkkä J, Cluitmans L, Ermes M (2010) Personalization algorithm for real-time activity recognition using PDA, wireless motion bands, and binary decision tree. IEEE Trans Inf Technol Biomed 14(5):1211–1215. https://doi.org/10.1109/TITB.2010.2055060
4. Feng Z, Mo L, Li M (2015) A random forest-based ensemble method for activity recognition. In: 2015 37th Annual international conference of the IEEE engineering in medicine and biology society (EMBC), pp 5074–5077. https://doi.org/10.1109/EMBC.2015.7319532
5. Chen Z, Zhu Q, Soh YC, Zhang L (2017) Robust human activity recognition using smartphone sensors via CT-PCA and online SVM. IEEE Trans Industr Inf 13(6):3070–3080. https://doi.org/10.1109/TII.2017.2712746
6. Mukherjee D, Mondal R, Singh PK, Sarkar R, Bhattacharjee D (2020) EnsemConvNet: a deep learning approach for human activity recognition using smartphone sensors for healthcare applications. In: Multimedia tools and applications. Springer Publishers, vol 79, no 41–42, pp 31663–31690
7. Cruciani F, Vafeiadis A, Nugent C et al (2020) Feature learning for human activity recognition using convolutional neural networks. CCF Trans Pervasive Comp Interact 2:18–32. https://doi.org/10.1007/s42486-020-00026-2
8. Zhao Y, Yang R, Chevalier G, Xu X, Zhang Z (2018) Deep residual bidir-LSTM for human activity recognition using wearable sensors. Math Probl Eng 1–13. https://doi.org/10.1155/2018/7316954
9. Ullah FUM, Ullah A, Haq U, Rho S, Baik SW (2019) Shortterm prediction of residential power energy consumption via CNN and multilayer bi-directional LSTM networks. IEEE Access. https://doi.org/10.1109/ACCESS.2019.2963045
10. Bhattacharya D, Sharma D, Kim W, Ijaz MF, Singh PK (2022) Ensem-HAR: an ensemble deep learning model for smartphone sensor-based human activity recognition for measurement of elderly health monitoring. Biosensors 12:393. https://doi.org/10.3390/bios12060393
11. Xu H et al (2020) Human activity recognition based on gramian angular field and deep convolutional neural network. IEEE Access 8:199393–199405. https://doi.org/10.1109/ACCESS.2020.3032699
12. Wang T, Lu C, Sun Y, Yang M, Liu C, Ou C (2021) Automatic ECG classification using continuous wavelet transform and convolutional neural network. Entropy 23:119. https://doi.org/10.3390/e23010119
13. Kwapisz JR, Weiss G, Moore SA (2011) Activity recognition using cell phone accelerometers. ACM SIGKDD Explor Newsl 12:74–82
14. Anguita D, Ghio A, Oneto L, Parra-Llanas X, Reyes-Ortiz J (2013) A public domain dataset for human activity recognition using smartphones. In Proceedings of the 21th international

european symposium on artificial neural networks, computational intelligence and machine learning. Bruges, Belgium
15. Yin X, Liu Z, Liu D et al (2022) A Novel CNN-based Bi-LSTM parallel model with attention mechanism for human activity recognition with noisy data. Sci Rep 12:7878. https://doi.org/10.1038/s41598-022-11880-8
16. Li Y, Wang L (2022) Human activity recognition based on residual network and BiLSTM. Sensors 22(2):635. https://doi.org/10.3390/s22020635
17. Ankalaki S, Thippeswamy MN (2022) Static and dynamic human activity detection using multi CNN-ELM approach. In: Shetty NR, Patnaik LM, Nagaraj HC, Hamsavath PN, Nalini N (eds) Emerging research in computing, information, communication and applications. Lecture notes in electrical engineering, vol 789. Springer, Singapore. https://doi.org/10.1007/978-981-16-1338-8_18

Hate Speech Detection in the Assamese Language with Cross-Lingual Testing

Koyel Ghosh and Apurbalal Senapati

Abstract Warning: *This paper contains examples of the language that some people may find offensive.* Social media has become an open platform for all its users. In the comment section, anyone can express their opinion, anger, frustration, and taunt toward a person or a group based on their religion, race, sex, or sexual orientation. Detecting this type of hate comment, which provokes people to become fierce unnecessarily, is another challenging target for all social media platforms like Facebook, Twitter, YouTube, etc. A little experiment is done in Indian languages, where European languages took massive advantage of being the most popular language in the world. The Assamese language is one of the languages in which hate speech detection work has not been done yet. This paper tried to explore Assamese hate, its issues, and challenges and finally perform cross-lingual testing on it. Firstly, we trained the state-of-the-art BERT (bidirectional encoder representations from transformers) models on the existing Bengali hate speech dataset and then evaluated the model's performance on both Bengali and Assamese hate data.

Keywords Hate speech detection · Assamese hate speech · Bengali hate speech · BERT · Cross-lingual testing

1 Introduction

On the internet, cyberbullying is becoming more common. People use social media to vent their frustrations and emotions, which can inflict pain on others or lead to the action of violence [1]. Regarding hate speech identification, social media sites have been taking action recently. Although the problem of online abuse is on the rise in India [2, 3], especially for females, it is still prevalent [4]. We proceed into the

K. Ghosh (✉) · A. Senapati
Central Institute of Technology, Kokrajhar, India
e-mail: ghosh.koyel8@gmail.com

A. Senapati
e-mail: a.senapati@cit.ac.in

S. K. Biswas et al. (eds.), *Intelligent Computing Systems and Applications*, Lecture Notes in Networks and Systems 1307, https://doi.org/10.1007/978-981-96-3860-4_13

hate detection research and discover very few works in Indian languages [5]. Hate speech detection research has been straightforward till now. Machine learning and deep learning techniques are used to train models on several datasets and then test the model's performance [6, 7]. Unlike traditional print media, social media users use different linguistic terms. It has a wide range of grammatical features. As a result, it is tough to detect hate speech automatically. Indian researchers have a significant shortage of resources in the hate speech detection field, even though lots of this work have already been done in English. Machine learning, deep learning, and transformer models for real-world problems are challenging because the model's fuel is data. In India, HASOC (hate speech and offensive content identification),[1] a shared task, is organized by FIRE (Forum for Information Retrieval Evaluation),[2] published hate datasets in Indian languages such as Hindi and Marathi. Besides the HASOC, some other datasets are available in Hindi [8], Bengali [9], etc.

However, online abuse is also increasing in Assam [10–12], mainly based on race, religion, sex, and sexual orientation. Designing a system for Assamese hate detection is a little problematic as no datasets and other tools are available in the Assamese language.

In this paper, we fine-tune pre-trained BERT models on Bengali Hate Speech dataset (BenHateSpeech) and test the trained model's performance on our Assamese hate dataset (AssameseHate), which we have collected from different social media platforms. Variation of BERT (bidirectional encoder representations from transformers) [13], RoBERTa (robustly optimized BERT) [14], ALBERT (A Lite BERT) [15], and their pre-trained models such as mBERT [13] MuRILBERT [16], Indic Bert [17], XLM-RoBERTa [18], and BanglaBert [19] are used for this work. Here, our contribution is

- To create a small Assamese hate speech dataset (AssameseHate).
- Train five variants of the BERTs [8] model with Bengali hate speech dataset (BenHateSpeech).
- Perform cross-lingual testing with our AssamHate dataset to calculate the accuracy of our experiment.

This paper is structured as follows. Section 2 is the related work on hate speech detection in Hindi, Bengali, etc. Section 3 describes the experimental setup, like the BenHateSpeech and AssameseHate dataset, preprocessing steps, and a short description of used BERT models. Section 4 summarizes the results and findings from all of the experiments. Finally, it is concluded in Sect. 5.

[1] https://hasocfire.github.io/hasoc/2022/index.htm.

[2] http://fire.irsi.res.in/fire/2022/home.

2 Related Work

There is a very small study on the Indian hate speech dataset as less data is available publicly. Creating labeled datasets of hate speech in the Indian language is tedious and challenging. In Hindi, few datasets are available as a shared task has been organized since 2019 HASOC [7, 20, 21]. Here, some top participants used baseline models to perform the task like word embedding, (wor2vec, fasttext) converting the texts into vectors. Then for classification, they used CNN (convolutional neural network), LSTM (long short-term memory networks), BiLSTM (bidirectional LSTM), etc., deep learning models [22, 23]. Most of the top participants also used transformer-based BERT models [24–28].

In the paper [8], author used the pre-trained multilingual BERT (mBERT) model for computing the input embedding on the Hostility Detection Dataset (Hindi), later SVM, Random Forest, Multilayer Perceptron (MLP), and Logistic Regression models as classifiers. In [29], they used Gated Recurrent Unit (GRU)-based deep neural network model for classifying users' comments on Facebook pages. They have collected 5,126 Bengali comments and classified them into six classes—Hate Speech, Communal Attack, Inciteful, Religious Hatred, Political Comments, and Religious Comments. Karim et al. [30] published a Bengali dataset with 35,000 hate statements (political, personal, geopolitical, and religious) and applied a multichannel CNN- and LSTM-based approach. Later DeepHateExplainer [31] added more than 5,000 labeled examples with it and used an ensemble method of transformer-based neural architectures to classify them into political, personal, geopolitical, and religious hates and achieved F1-scores of 78%, 91%, 89%, and 84%, for political, personal, geopolitical, and religious hates. In the paper [9], they published a Bengali Hate Speech corpus with 30,000 comments labeled with "1" for hate comments; otherwise, "0". In [32], encoder-decoder-based machine learning model has been used on 7,425 Bengali comments of seven distinct categories of hate speeches to predict hate speech categories. [33] collected 7,245 reviews from YouTube and apply classical machine learning (CML) classifiers, stochastic gradient descent (SGD) optimizer, a deep learning (DL)-based architecture, and lexicon-based methods to identify presence of vulgarity. This paper [34] created a political news corpus and then developed a keyword or phrase-based hate speech identifier using a semi-automated approach.

3 Experimental Setup

3.1 Dataset

Table 1 shows the dataset statistics for training and test sets along with hate and non-hate class distribution.

Table 1 Class distribution analysis for BenHateSpecch and AssameseHate datasets

Datasets	Hate		Non-hate		Total
	Train	Test	Train	Test	
BenHateSpeech	8,000	2,000	16,000	4,000	30,000
AssameseHate	300	80	500	120	1,000

sentence	hate	category
আর একটা কথা রাখে আল্লা মারে কে	0	Meme, TikTok and others
নাস্তিক হারমজাদারা	1	religion
বাল ছাল তর সাউয়া😡	1	sports
তাসকিন একটা ফালতু	0	sports
এই গাধার বাচ্চা, কি বলিস তুই	1	crime

Fig. 1 Snap of BenHateSpeech dataset where hate comments are tagged as "1" and otherwise "0"

sentence	hate	Category
তেওঁলোকে ঐক্যভাৱে একত্ৰিত হৈ এটা দ্বেষবিহীন সমাজ	0	social
মই মুছলমান লোকৰ এক্স-ৰে নকৰোঁ। এয়া মোৰ শপ	1	religion
আকৌ এবাৰ মোডী চৰকাৰ' গানটো বজাই দিলেই না	1	political
বিজেপিৰ বিধায়ক থকা সমষ্টিত বানপানী হোৱা নাই	1	political

Fig. 2 Snap of AssameseHate dataset where hate comments are tagged as "1" for hate and otherwise "0"

Bengali Hate Speech Dataset (BenHateSpeech): This section describes the data with instances. The BenHateSpeech dataset is with two classes which contains 30k Bengali comments collected from Facebook and YouTube [9]. It has 10k hate speech and 20k non-hate speech comments. The dataset includes the sentence, hate, and category fields. **sentence** field contains comments collected from Facebook and YouTube. Manually **hate** field are labeled as "1" for hate comments; otherwise, "0". Original sentences are taken from sports, entertainment, crime, religion, politics, celebrity, Meme, TikTok, etc. **Category**. Figure 1 shows the sample of BenHateSpeech dataset.[3]

Assamese Hate Speech Dataset (AssameseHate) We have collected Assamese data mainly from Facebook comments and YouTube comments. We followed a few Assamese Facebook pages such as news, entertainment, celebrity, and politics and select some uploads to manage its related comments. The same we did with some Assamese local news, celebrity, politics, etc., YouTube channels. Until now, we managed to gather very few comments, not more than 1k, in categories of social, religious, political, sports, entertainment, etc. Figure 2 shows the sample of AssameseHate data.

[3] https://www.kaggle.com/naurosromim/bengali-hate-speech-dataset.

3.2 Preprocessing

Text data need to be cleaned and noise free before employing data to the transformers to enrich the performance. The BenHateSpecch data is clean, except for emojis and unwanted characters. We didn't remove punctuations and numbers because they may add some meaning while using the transformer. We have applied a few steps to increase the performance. In most cases, the following steps are used:

1. **Replaced emojis with words:** The new generation often uses emojis and emoticons in social media to express their emotion, humor, etc. So, instead of removing emojis and emoticons, we decided to keep it using a Bangla Emot module [35], which detects the feelings hidden behind these emojis and emoticons for Bangla language. This Python module extracts information from emoticons and emojis in English and then converts it to Bengali. It then becomes easy to understand the feeling of the user who wrote the comment.
2. **Remove unwanted characters:** Some unwanted characters like hyphen (−), underscore (_), hash (#), at the rate (@), etc. are removed.
3. **Stemming:** Stemming removes each word's inflections and converts them to the root word. Verbal and nominal inflections mainly inflect Bengali words [36, 37].
4. **Removal of Bangla stopwords:** We remove the most commonly used words, i.e., stopwords, from the dataset. A stopword list is available, containing around 400 words [38].

3.3 Experiments

After several experiments, we have selected five pre-trained BERT models that perform best on the BenHateSpeech dataset. We apply the same to our AssameseHate data to test its cross-lingual performance. Bengali and Assamese share almost identical scripts,[4] so we tried these experiments. Our entire experiment is based on two parts:

1. **Experiment 1:** We fine-tuned pre-trained BERT models on only the BenHateSpeech dataset (training set—80%) and test on the same dataset (test set—20%).
2. **Experiment 2 :** We fine-tuned pre-trained BERT models on the BenHateSpeech dataset (training set—80% from BenHateSpeech) and test on the entire collected AssameseHate dataset.

Table 2 shows the data statistics maintained for all the experiments.

[4] https://www.raiseupwa.com/miscellaneous/is-assamese-similar-to-bengali/.

Table 2 BenHateSpecch and AssameseHate Dataset distribution for all two experiments

Experiment no	BenHateSpecch		AssameseHate		Total
	Train	Test		Test	
Experiment 1	24,000	6,000	–	–	30,000
Experiment 2	24,000	–	–	1000	25,000

3.4 Transformer Language Models

Google develops BERT, a state-of-the-art transformer-based technique primarily for text-related tasks. Contextualized embeddings can be generated by BERT. Almost similar vectors for synonyms and different vectors are produced if the use of words is other. During the training, the transformer learns the details from both sides of the word's context. So, it is called a bidirectional model. We pre-trained mBERT, MuRILBERT, XLM-RoBERTa, IndicBERT, and BanglaBERT. First, pre-trained on BenHateSpeech dataset only then tested on the same data, later tested the performance only on the AssameseHate data. Figure 3 shows a general architecture for all the experiments, and we took the input from the AssameseHate dataset.

After the above steps of preprocessing, we employ the m number of rows or texts (comments) of the training set (T) indicated as

$$T = \{C_1, C_2, C_3, .., C_i, ...C_m\},$$

where C_i is the ith number of comments and m is equal to the total number of comments in a training set. The comment C_i having n words, i.e., length of the comments, is denoted as

$$C_i = \{w_{i,1}, w_{i,2}, w_{i,3}, ..., w_{i,p}, .., w_{i,n}\},$$

where $w_{i,p}$ denotes the pth word in the ith comment.

1. **mBERT**[5]: A masked language modeling (MLM) approach is used to train on the largest dataset of Wikipedia, over 104 top languages worldwide, including Hindi, Bengali, and Marathi.
2. **MuRILBERT**[6]: Multilingual Representations for Indian Languages (MuRIL) is a BERT model pre-trained on 17 Indian languages and their transliterated counterparts, i.e., monolingual and parallel segments.
3. **XLM-RoBERTa (base-sized model)**[7]: The XLM-RoBERTa model is pre-trained on 2.5TB of filtered CommonCrawl data containing 100 languages, and it is a multilingual version of the RoBERTa model.

[5] https://huggingface.co/bert-base-multilingual-cased.

[6] https://huggingface.co/google/muril-base-cased.

[7] https://huggingface.co/xlm-roberta-base.

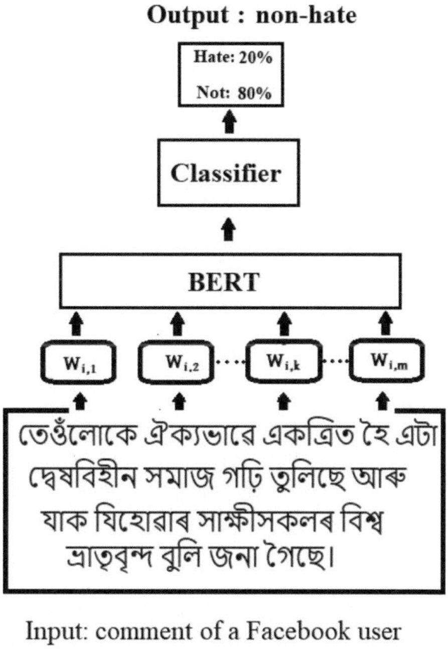

Fig. 3 The whole architecture followed to detect hate from Assamese data

4. **IndicBERT**[8]: A multilingual ALBERT model is released by Ai4Bharat which is trained on large-scale datasets covering 12 major Indian languages (such as Assamese, Bengali, English, Gujarati, Hindi, Kannada, Malayalam, Marathi, Oriya, Punjabi, Tamil, and Telugu).
5. **BanglaBERT**[9]: Using MLM, bangla-Bert-Base was pre-trained on data downloaded from OSCAR and Bengali Wikipedia Dump Dataset.

4 Result

Table 3 shows the precision, recall, and weighted F1 score obtained by training all the variants of BERT on BenHateSpeech and AssameseHate datasets for the two experiments. We intentionally prefer a weighted F1 score over an accuracy score to evaluate models' performance because imbalanced class distribution exists in both datasets; almost half are "1" compared to "0" classes. So, a weighted F1 score (W_{F1}) is a better metric to consider in this scenario. At last, we calculate *Accuracy*.

[8] https://huggingface.co/ai4bharat/indic-bert.

[9] https://huggingface.co/sagorsarker/bangla-bert-base.

Table 3 Precision, Recall, F1 score, and Accuracy of various transformer models for Experiment 1 and Experiment 2, respectively

Experiment 1	Precision			Recall			F1 score			Accuracy
(train and test on BenHateSpeech)	0	1	w.avg.	0	1	w.avg.	0	1	w.avg.	
mBERT	0.9303	0.8630	0.9078	0.9155	0.8362	0.8890	0.9228	0.8493	0.8983	0.8980
MuRILBERT	0.9225	0.8507	0.8985	**0.9406**	**0.8835**	**0.9215**	**0.9314**	**0.8667**	**0.9098**	**0.9095**
XLM-RoBERTa	**0.9463**	**0.8975**	**0.9300**	0.9047	0.8249	0.8781	0.9250	0.8596	0.9032	0.9023
IndicBERT	0.9042	0.8030	0.8704	0.9300	0.8515	0.9038	0.9169	0.8265	0.8867	0.8876
BanglaBERT	0.9333	0.8685	0.9117	0.9207	0.8456	0.8956	0.9269	0.8568	0.9035	0.9033
Experiment 2 (train on BenHateSpeech and test on AssameseHate)										
mBERT	0.9212	0.8623	0.8988	0.8115	0.6521	0.7509	0.8532	0.7591	0.8174	0.8200
MuRILBERT	**0.9435**	**0.8817**	**0.9200**	0.8309	0.6868	0.7761	**0.8836**	**0.7721**	**0.8412**	**0.8460**
XLM-RoBERTa	0.9334	0.8745	0.9112	0.8438	**0.7020**	0.7899	0.8752	0.7622	0.8322	0.8393
IndicBERT	0.8941	0.8303	0.8698	**0.8910**	0.6602	**0.8032**	0.8409	0.7306	0.7989	0.7998
BanglaBERT	0.8845	0.8202	0.8600	0.8819	0.6712	0.8018	0.8301	0.7201	0.7883	0.7891

$$W_{F1} = \frac{F1_0 * T_0 + F1_1 * T_1}{T_0 + T_1} \quad (1)$$

$$Accuracy = \frac{True_0 + True_1}{T_0 + T_1}, \quad (2)$$

where $True_0$ = True-negative (model predicted the texts as "0" and the actual value of the same is also "0"), $True_1$ = True-positive (model predicted the texts as "1" and the actual value of the same is also "1"), $F1_0$ = F1 score of "0" class, $F1_1$ = F1 score of "1" class, T_0 = The total number of "0" class text present in test set, and T_1 = The total number of "1" class text present in test set.

For both Experiments 1 and 2, we notice that MuRILBERT scores top in terms of F1 score (W_{F1}) and $Accuracy$, but recall for the class "1" is significantly less for almost all of the cases. It happens because of the considerably less and imbalanced data. BanglaBRT and IndicBERT perform poorly here, whereas mBERT and XLM-RoBERTa produce a good F1 score.

Figure 4 shows the confusion matrix for Experiments 1 and 2. We intended to check whether a cross-lingual evaluation of Assamese data on Bengali is successful, so we can say the performance depends entirely on the data here.

Hate speech detection is very much tonal and context based. Being a human, how do we decide that someone is hating me or hurting me by saying a curse word or passing a vulgar comment, or they are just extra friendly? We observe their body language, way of talking, previous context about the person, and the tone they are using while talking, then we decide whether to take the saying as hate or not. In a friendly get-together, people frequently say hateful words or slang while talking. A human can understand even if it is written based on previous knowledge. Still, it is difficult to make a machine understand that only a text as a text inherently loses tonal information. The same thing may happen if someone comments abusive things with sophisticated words. Figure 5 shows some examples of the text's machine misclassified.

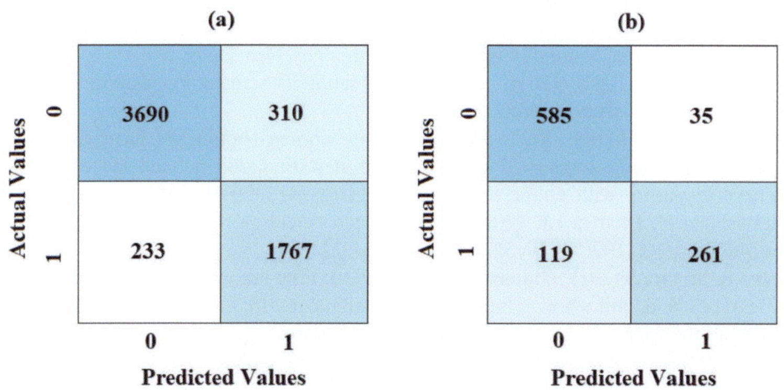

Fig. 4 Confusion matrix of best models such as MuRILBERT for BenHateSpeech **a** and MuRILBERT for AssameseHate **b**

1. যদি তুমি মৰিবা মই উদযাপন কৰিম
2. তুমি অসম গৌৰৱ গড় কেইটাৰ লগতে কাজিৰঙা ৰাষ্ট্ৰীয় উদ্যান খন ধ্বংস কৰি ফটা লেকচাৰ মাৰিব আহিছা।
3. হে বেশ্যৰ চোড, কি জানি কি হবো, আজি বৰষুণ হ'লে চাপ থাকে

Fig. 5 First two examples are hate comments but misclassified as non-hate and last one is non-hate comment but misclassified as hate

5 Conclusion and Future Scope

Hate speech has flared more rapidly through the daily use of social media and, mainly, by conveying users' thoughts or emotions on social media comment sections in a negative aspect. We told earlier that many companies are working on European languages, which includes collecting data, preprocessing the data, and building a proper system to detect hate. Now, what about low-resourced languages? Few works have been done in Indian languages like Hindi, Bengali, Marathi, Tamil, and Malayalam, but the northeast is yet to be explored. So, we intended to collect the Assamese hate data. In this paper, we collect only 1k data; it is significantly less to experiment separately like to train deep learning or transformer models, as we know they need huge data to learn from. Despite that, we tried and got an inferior result, so we did only test to check whether BERTs can relate almost similar data or not. Here, we use BERT not only because it's a state-of-the-art model but also has self-supervised properties and context understanding in it. In the future, we hope to create a 5k dataset, make it available, and do different experiments. I hope the data, experiments, and results can help further research.

References

1. Kemp S (2021) Digital 2021: Global overview report, 27 JANUARY 2021. https://datareportal.com/reports/digital-2021-global-overview-report. Accessed 25 May 2021
2. Crime in India (2019) Statistics, vol. I, National crime records bureau, Ministry of home affairs, Government of India, National highway–8, Mahipalpur, New Delhi
3. Sentinel Digital Desk (2016) https://www.sentinelassam.com/news/online-harassment-remains-a-serious-issue-in-india/
4. MINT (2022) 08 March 2022: 55 indians say women should not have to put up with online abuse survey. https://www.livemint.com/news/india/55-indians-say-women-should-not-have-to-put-up-with-online-abuse-survey-11646741309919.html
5. Balakrishnan K, Dhanya LK (2021). Hate speech detection in Asian languages: A survey. https://doi.org/10.1109/ICCISc52257.2021.9484922
6. Mishra S, Mishra S (2019) 3Idiots at HASOC 2019: Fine-tuning transformer neural networks for hate speech identification in Indo-European languages. In: Proceed- ings of the 11th annual meeting of the forum for information retrieval evaluation
7. Thomas M, Sandip M, Prasenjit M, Daksh P, Mohana D, Chintak M, Aditya P (2019) Overview of the HASOC track at FIRE 2019: Hate speech and offensive content identification in Indo-European languages. In; Proceedings of the 11th forum for information retrieval ealuation. CEUR, pp 14–17. http://cur-ws.org/Vol-2517/T3-1.pdf

8. Bhardwaj M, Md Shad A, Asif E, Amitava D, Tanmoy C (2020) Hostility detection dataset in Hindi. ArXiv preprint arXiv:2011.03588
9. Nauros R, Mosahed A, Hriteshwar T, Saiful I, et al. (2021) Hate speech detection in the bengali language: A dataset and its baseline evaluation. In: Proceedings of international joint conference on advances in computational intelligence Springer, pp 457–468
10. NDTV digital, TOI digital, November 02, 2020. https://www.ndtv.com/india-news/man-arrested-from-assam-for-harassing-delhi-woman-online-cops-2319599
11. Roy S, Shukla AK (2016) Social media in the domain of communal violence: a study of assam riot 2012. Int J Adv Res Innov Ideas Educ
12. Berenyi K (2020) Mapping minorities' vulnerability to hate speech and denationalisation with a focus on east and Southeast Asia. Statelessness & Citizenship Rev 2:5
13. Jacob D, Ming-Wei C, Kenton L, Kristina T (2018) BERT: pre-training of deep bidirectional transformers for language understanding. CoRR arXiv:1810.04805
14. Yinhan L, Myle O, Naman G, Jingfei D, Mandar J, Danqi C, Omer L, Mike L, Luke Z, Veselin S (2019) Roberta: A robustly optimized BERT pretraining approach. CoRR arXiv:1907.11692
15. Zhenzhong L, Mingda C, Sebastian G, Kevin G, Piyush S, Radu S (2019) ALBERT: A lite BERT for self-supervised learning of language representations. CoRR arXiv:1909.11942
16. Simran K, Diksha B, Sarvesh M, Savya K, Atreyee D, Balaji G, Dilip KM, Pooja A, Rajiv TN, Shachi D, Shruti G, Subhash CBG, Vish S, Partha PT (2021) Muril: Multilingual representations for indian languages. CoRR arXiv:2103.10730
17. Divyanshu K, Anoop K, Satish G, Gokul NC, Avik B, Mitesh MK, Pratyush K (2020) Indic-NLPSuite: Monolingual corpora, evaluation benchmarks and pre-trained multilingual language models for Indian languages. In: Findings of the association for computational linguistics: EMNLP 2020, pp 4948–4961. Online. Association for Computational Linguistics
18. Alexis C, Kartikay K, Naman G, Vishrav C, Guillaume W, Francisco GA, Edouard G, Myle O, Luke Z, Veselin S (2019) Unsupervised cross-lingual representation learning at scale. CoRR arXiv:1911.02116
19. Sagor S (2020) Banglabert: Bengali mask language model for bengali language understading
20. Mandl T, Modha S, Shahi GK, Jaiswal AK, Nandini D, Patel D, Majumder P, Sch J (2020) Overview of the HASOC track at FIRE 2020: Hate speech and offensive content identification in Indo-European languages. CEUR, In FIRE (Working Notes)
21. Satapara S, Modha S, Mandl T, Madhu H, Majumder P (2021) Overview of the HASOC Subtrack at FIRE 2021: Conversational hate speech detection in code-mixed language. In: Working notes of FIRE
22. Md. Abul B, Richi N (2020) Qutnocturnal@hasoc'19: CNN for hate speech and offensive content identification in hindi language. CoRR arXiv:2008.12448
23. Roushan R, Shivangi S, Sunil S (2020) Nsit & iiitdwd@hasoc 2020: Deep learning model for hate-speech identification in indo-european languages. In: FIRE
24. Jean-Christophe M, Pierre-Antoine J, Andon T, Sebastien H (2019) Imt mines ales at hasoc 2019: automatic hate speech detection, vol. 2517. In: FIRE 2019-11th forum for information retrieval evaluation, p 279
25. Shubhanshu M, Sudhanshu M (2019) 3idiots at hasoc 2019: Fine-tuning transformer neural networks for hate speech identification in indo-european languages. In: FIRE (Working Notes), pp 208–213
26. Vandan M, Pruthwik M, Dipti MS (2019) Iiit-hyderabad at hasoc 2019: Hate speech detection. In: FIRE (Working Notes), pp 271–278
27. Kumari S (2020) Nohate at hasoc2020: Multilingual hate speech detection. In: Forum for information retrieval evaluation, FIRE
28. Mehar B, Tenzin SB, Akshat A, Prakash R, Shubham G, Kumar S, Felix L, Ayushman D (2021) One to rule them all: Towards joint indic language hate speech detection. CoRR arXiv:2109.13711
29. Ishmam AM, Sadia S (2019) Hateful speech detection in public facebook pages for the bengali language. In: 2019 18th IEEE international conference on machine learning and applications (ICMLA). IEEE, pp 555–560

30. Md Rezaul K, Bharathi RC, John PM, Michael C (2020). Classification benchmarks for under-resourced bengali language based on multichannel convolutional-lstm network. In: 2020 IEEE 7th international conference on data science and advanced analytics (DSAA). IEEE, pp 390–399
31. Md Rezaul K, Sumon KD, Tanhim I, Sagor S, Mehadi HM, Kabir H, Md. Azam H, Stefan D (2021) Deephateexplainer: Explainable hate speech detection in under-resourced bengali language. In: 2021 IEEE 8th international conference on data science and advanced analytics (DSAA), pp 1–10
32. Das AK, Al Asif A, Paul A, Hossain MN (2021) Bangla hate speech detection on social media using attention-based recurrent neural network. J Intell Syst 30(1):578–591. https://doi.org/10.1515/jisys-2020-0060
33. Sazzed S (2021) Identifying vulgarity in Bengali social media textual content. PeerJ Comput Sci 7:e665. https://doi.org/10.7717/peerj-cs.665
34. xx
35. Das AK, Abdullah AA, Anik P, Md Nur H (2021) Bangla hate speech detection on social media using attention-based recurrent neural network. J Intell Syst 30(1):578–591
36. Mahmud MR, Afrin M, Razzaque MA, Miller E, Iwashige J (2014) A rule based bengali stemmer. In: 2014 international conference on advances in computing, communications and informatics (ICACCI). IEEE, pp 2750–6. https://doi.org/10.1109/ICACCI.2014.6968484
37. Biswas E, Das AK (2019) Symptom-based disease detection system in bengali using convolution neural network. In: 2019 7th international conference on smart computing communications (ICSCC), pp 1–5. https://doi.org/10.1109/ICSCC.2019.8843664
38. Haque R, Mehera P, Mridha MF, Hamid A (2019) A complete Bengali stop word detection mechanism. In: 2019 joint 8th international conference on informatics, electronics & vision (ICIEV) and 2019 3rd international conference on imaging, vision & pattern recognition (icIVPR). IEEE, pp 103–7. https://doi.org/10.1109/ICIEV.2019.8858544

Efficiency of Machine Learning in Disease Prediction

Anil Kadu and Swarali Pathak

Abstract Machine learning is a rapidly growing trend in the healthcare industry. The technology is being employed on a large scale for the prediction of various diseases. With the growth of data analysis in biomedical, accurate predictive analysis of biomedical data benefits early disease detection and improved care of the patient. In this paper, we have attempted to perform predictive analysis of various diseases using Machine learning algorithms and tried to draw inferences about their efficiency in accurate predictions. We have employed models for Diabetes, Heart Disease, and Liver disease prediction, which provide considerably more accurate results of 91.37%, 82.67%, and 71.79%, respectively.

Keywords Classification · Disease · Machine learning

1 Introduction

In recent years, Machine learning has taken a leap and proven to be effective in solving real-world problems using various algorithms. It has proven to play a crucial role in biomedical analysis, price predictions, span and risk detection, and advancements in the automobile industry, automation, and analyzing businesses. Similarly, it has proven to be effective in Disease Predictions. Healthcare is one of the most important sectors that creates a huge amount of data. The application of ML to healthcare data can actually be a matter of life and death. This paper attempts to use various machine learning approaches for the analysis of prediction for multiple diseases and tries to draw inferences about their efficiency to perform classifications.

The health industry creates a large amount of data, which makes it much more useful to analyze it which enabling it to be interpreted in predicting the causes

A. Kadu · S. Pathak (✉)
Department of Instrumentation and Control Engineering, Vishwakarma Institute of Technology (VIT), Pune, India
e-mail: swarali.pathak19@vit.edu

A. Kadu
e-mail: anil.kadu@vit.edu

S. K. Biswas et al. (eds.), *Intelligent Computing Systems and Applications*, Lecture Notes in Networks and Systems 1307, https://doi.org/10.1007/978-981-96-3860-4_14

and their possible consequences. The aim of this predictive research analysis is to understand the relationship between various data features and use these values to build an accurate predictive model that can predict the risk of having a disease. It implements a comparison of various classification algorithms to find the best fitting model to predict each disease. We compared the accuracy of each algorithm on the problem statements to identify the most accurate model. Moreover, this paper also tried to create a UI for the doctors and patients to make the most frequent use of the technique. Thus, this analysis would be helpful for both doctors as well as patients to keep track of their health and take necessary treatments and save time as well. The ultimate motive of this research project is to build an ML model that can precisely predict the risk of diseases and thus compute its efficiency.

2 Literature Survey

2.1 Comparing Different Supervised Machine Learning Algorithms for Disease Prediction

In this study, research efforts are made among 48 research articles to identify studies that implemented multiple supervised learning algorithms for predicting a single disease. The conclusion states that most of the studies applied the Support Vector Machine (SVM) algorithm, followed by the Naïve Bayes. However, the Random Forest has given the best accuracy compared to others. Thus, this study provided an overview of the performance of different types of ML algorithms for disease prediction. [Ekramul Hossain, Arif Khan, Mohammad Ali Moni and, Md Shahadat Uddin, BMC Medical Informatics and Decision Making].

2.2 Disease Prediction Using Machine Learning Over Big Data

This paper proposes a system that uses a latent factor model to fill out the missing data and overcome the difficulty of null data. It uses data from hospitals and implements the Decision Tree algorithm. The comparison shows that the proposed algorithm reaches an accuracy of 94.8%. [Vinusha H, Sajini S, Vinitha S, and, Sweetlin S, Computer Science & Engineering: An International Journal (CSEIJ), Vol.8, No.1, February 2018].

2.3 Diabetes Prediction Using Machine Learning Algorithms

In this study, several machine learning algorithms are applied to the dataset, of which Logistic Regression gives the highest accuracy of 96%. Furthermore, the pipeline model has been applied to the AdaBoost classifier as the best model with an accuracy of 98.8%. Comparison of machine learning algorithm accuracies is carried out with two different datasets. This paper added some more factors that affect the risk of having diabetes and created a new dataset. [Dr. Vaidehi V, Aishwarya Mujumdar, International Conference On Recent Trends in Advanced Computing 2019, Icrtac 2019].

2.4 Heart Disease Prediction Using Machine Learning and Data Mining Technique

This paper experimented with various machine learning algorithms, and as a result of that J48 tree technique has proven to be the most accurate for predicting heart disease, and it has given better accuracy and the shortest time to build. It has the highest accuracy of 56.76%, and the build time of the model was 0.04 s. This paper also suggested the need for combinations of multiple algorithms that can increase the accuracy. [Samir Patel, Jaymin Patel, Prof.Tejal Upadhyay, Dr, IJCSC].

2.5 Liver Disease Prediction Using Machine Learning

This paper proposes a model for predicting liver disease through information mining. In liver issue infections, there are 500 informational indexes with 10 factors. This system incorporates the use of classification algorithms for categorization. J48 algorithm resulted to be the best algorithm and hence gave accuracy of 95.04%. [Dinesh Kalthireddy, Vasan Durai, Suyan Ramesh, International Journal of Advance Research Ideas and Innovations in Technology].

3 Methodology

3.1 Exploratory Data Analysis (EDA)

Exploratory Data Analysis is the procedure of performing analysis on data so as to discover patterns in the data, to spot abnormalities, check for outliers, and to check predictions with the help of statistical representations. It is necessary to know the columns and their corresponding data types and find whether they contain null values or not. In Python, we have a function called info() and describe(), which gives us a statistical overview of the data. It is the most important process in Data analysis.

Python has visualization libraries such as Seaborn and Matplotlib. They provide very attractive statistical graphs and plots in order to perform both Univariable and Multivariable analysis. The main purpose of EDA is to look at the distribution of the data precisely before making any conclusions and assumptions. It helps to identify obvious errors, detect outliers, and find interesting relations among the variables. Thus, EDA includes Screening of the data, identifying outliers, checking assumptions, and characterizing differences among groups.

3.2 Algorithms

Machine Learning is further divided into two concepts of learning, which are Supervised and Unsupervised Learning. In Supervised Learning, the data for training the algorithm, which already contains some samples of accurate outcomes, is employed, which helps the machine to learn. Hence, it is the most widely used method as it increases the likelihood of getting more accurate predictions. In this paper, we have implemented a supervised learning approach. Supervised Learning is further subdivided into two types: Regression and Classification. The outcome variable is Numerical and continuous in the case of regression, whereas it is categorical for classification as it contains predictions in the form of categories like either true or false (yes or no). Classification is further divided into Binary and multiclass classification. As we are dealing with categorical data in this system, we will be using Classification Techniques or Algorithms for predicting whether or not a person has a Disease or not. For example, classification will tell us if the temperature is either hot or cold, whereas Regression will tell us the exact numerical value of the temperature.

The algorithms that will be used in this paper are described below.

3.2.1 Logistic Regression

This algorithmic rule will facilitate predicting the chance of an incident happening or an alternative being created. To take an example, we may want to know the chances of a customer choosing an offer made on a website, which is an outcome variable. This analysis will be carried out using some known features of customers, such as resources they redirected from, repeated visits to the site, actions, and surfing on your site, which will be independent variables. It helps to determine the probability of the type or category of customers who are likely to accept the offer. Consequently, better decisions about promoting the offer can be made.

In healthcare, this algorithm is often used to predict the chances of having a disease for a particular patient, which means that early prevention measures can be carried out.

3.2.2 Support Vector Classifier

Support Vector Classifier (SVC) or also known as Support Vector Machine (SVM) is another type of supervised machine learning. It can be used for both classification and regression. The SVM algorithm finds a hyperplane and uniquely classifies the data points. The dimensions of the hyperplane are dependent upon the number of features. When the number of features exceeds three, it becomes hard to visualize. The objective of the SVM algorithm is to identify the best fitted line that can differentiate n-dimensional space into classes. Hence, it is easily possible to add a new data point in the appropriate category later. This best boundary of decision is thus called a hyperplane.

3.2.3 Decision Tree Classifier

Decision Tree is a tree-like structured algorithm. It has three types of nodes, namely: Root, Interior, and Leaf node. The features of a data set are the Interior Nodes, and the branches are the rules for decision.

The topmost node is called the root node. A decision tree selects the best attribute to split the data, makes it a decision node, and divides the dataset into smaller subsets. Then it starts building a tree by repeating this process for each child until all subsets belong to the same attribute value.

3.2.4 Random Forest Classifier

Both classification and regression can be performed using a Random forest classifier. A forest consists of trees. Similarly, In Random forest, decision trees on randomly selected data samples are created. It has a versatile set of applications, including recommendation systems, loan applicants, identifying fraud activity, and predicting diseases.

It is considered one of the highly accurate and robust methods for classification because of the number of decision trees participating in the process. Overfitting of the features is often not the problem in Random Forest because it cancels out biases by taking the average of all the predictions.

3.2.5 K-nearest Neighbor Classifier

K-Nearest Neighbor (KNN) is the simplest, versatile, and one of the most important machine learning algorithms. KNN is widely used in the analysis of various applications in the fields like healthcare, finance, handwritten digit and letter detection, and image recognition. This algorithm is based on a similarity approach and lazy learning. A lazy algorithm means that it does not need any training data points for generating models. All training data is used in the testing process thus makes training faster and testing slower.

In the KNN algorithm, K is the number of nearest neighbors. The distance between two points can be determined using distance measures such as the Euclidean distance.

4 Implementation

For the implementation of the analysis and prediction, we first need a dataset with a good set of entries that can be considered for reliable predictions. In this paper, we have collected datasets from Kaggle, which is a data science company that provides free datasets with a reliable set of features. Further, we have performed data preprocessing in which we will deal with data cleaning and removal of null values. Then we need to perform Exploratory Data Analysis in which outliers and model building will be carried out. Then we find the best fitting model out of the various classification algorithms, and using this cross-validation, we will build the final model for prediction and calculate its accuracy. The flow of implementation goes as follows (Fig. 1).

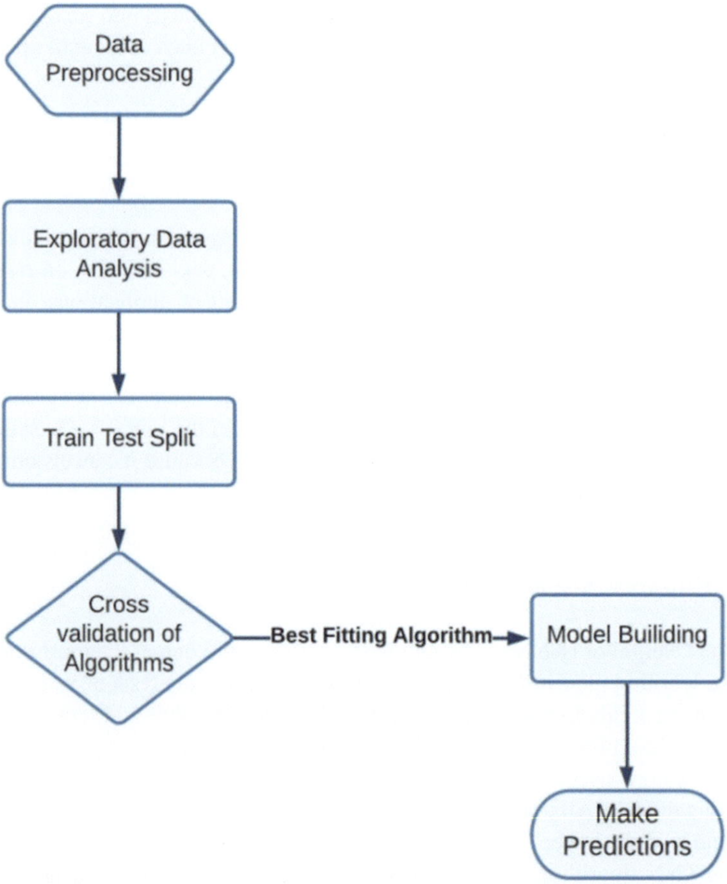

Fig. 1 Implementation flowchart

4.1 Heart Disease Prediction

One of the most common causes of death in human beings is Heart Disease. Heart disease describes a range of factors that affect the heart. Heart disease generally refers to conditions that involve the risk of heart attack, chest pain, heart stroke, or lack of blood supply to the heart. Prediction of heart disease is becoming one of the most important topics in the field of clinical data analysis. By using Machine Learning for Heart Disease prediction, we can identify the major causes and thus provide better treatment and preventive measures.

For this analysis, we used the Kaggle UCI dataset. It has a total of 14 features, out of which 13 are independent variables and 1 is the target column, which is the dependent variable. We will be predicting this target variable, which will be based on the rest of the factors. The data set contains entries of 303 patients. In the data cleaning process, we have removed 2 columns, namely, Old peak and exang, which did not seem to cause much difference to the output prediction. After this, we plotted a histogram of all data columns to see the distribution (Fig. 2).

Once all the data preprocessing and visualization of the distribution are done, model building is carried out. First, we need to split the data into train and test data, for which we will be using the standard 80/20 ratio (80% train data and 20% test data). For finding out the best fitting algorithm, we calculate the accuracy of every algorithm on the train and test data, and the model that gives the most accuracy on test data prediction will be the final best fitting model. In our analysis, Logistic

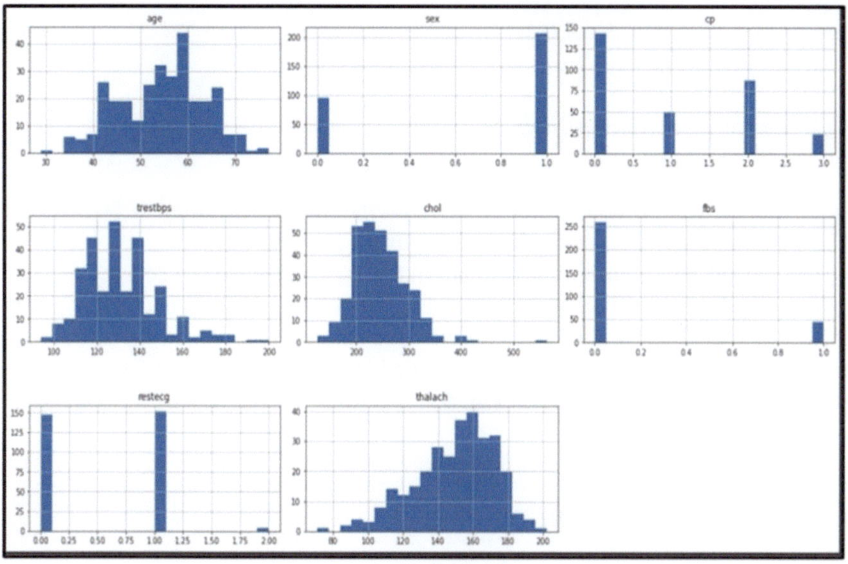

Fig. 2 Histogram of independent data columns

```
Logistic Regresssion Accuracy :  82.6772131147541
Decision Tree Accuracy        :  75.40983606557377
Support Vector Classifier     :  62.295081967213115
Random Forest Accuracy        :  78.68852459016394
K nearest neighbors Accuracy  :  62.295081967213115
```

Fig. 3 Model accuracies for heart disease prediction

```
input_data = (62,0,0,140,268,0,0,160,0,3.6,0)

input_data_as_numpy_array= np.asarray(input_data)

input_data_reshaped = input_data_as_numpy_array.reshape(1,-1)

prediction = model.predict(input_data_reshaped)
print(prediction)

if (prediction[0]== 0):
  print('The Person does not have a Heart Disease\n')
else:
  print('The Person has Heart Disease\n')

[0]
The Person does not have a Heart Disease
```

Fig. 4 Heart disease predictive model

Regression proved to be the best model with the highest accuracy of 86.36% and 82.67% on training and testing data, respectively (Fig. 3).

The second-best algorithm was Random forest, which gave an accuracy of 78.68 followed by Decision Tree, which gave an accuracy of 75.40%. The lowest accuracy of 62.29% was given by the SVC and KNN algorithms. Thus, we can say that the best predictive model for Heart disease prediction is Logistic Regression. The predictive function of the model is shown in Fig. 4.

4.2 *Diabetes Prediction*

Diabetes is one of the most common diseases that many people suffer from. Recent studies have shown some adverse effects of diabetes on the eyes (diabetic Retinopathy) in which a person might lose his/her eyesight. Also, diabetes patients have a higher risk of having heart attacks and heart failure. Hence, keeping track of your health regularly and early diagnosis of diabetes can help patients as well as doctors in order to provide necessary treatment. Therefore, having a reliable prediction system will be a step forward for keeping track of diabetes.

For this analysis, we used the Kaggle PIMA dataset. It has a total of 9 features, out of which 8 are independent variables and 1 is the outcome column, which is the dependent variable. We will be predicting this outcome variable, which will be based on the rest of the factors. The data set contains entries of 768 observations. After this, we plotted a histogram of all data columns to see the distribution (Fig. 5).

During the process of EDA, it was discovered that the data columns Blood pressure, skin thickness, and pregnancies, and DPF contained more than 300 null values, which we tried to fill out using the mean and standard deviation of the present entries. However, these columns seemed to be irrelevant in making predictions and affecting the overall performance of the model due to the missing values. Hence, we decided to drop these columns. So the final predictions are based on the four important features, namely, Insulin level, Glucose level, BMI, and Age. Then model building is carried out. First, we split the data into train and test data. For finding out the best fitting algorithm, we calculate the accuracy of every algorithm on the train and test data, and the model that gives the most accuracy on test data prediction will be the final best fitting model. In our analysis Support Vector Classifier (SVC) resulted to be the best algorithm with the highest accuracy of 91.37% (Fig. 6).

The second-best algorithm was Logistic Regression, which gave an accuracy of 82.07%, followed by the Random Forest classifier, which gave an accuracy of 80.97%. Thus, we can say that the best predictive model for Diabetes prediction, SVC, fits the best. The confusion matrix of the model is shown in Fig. 7.

We deployed the model into a web application using the Flask framework in Python. The predictions of the model are shown in Fig. 8.

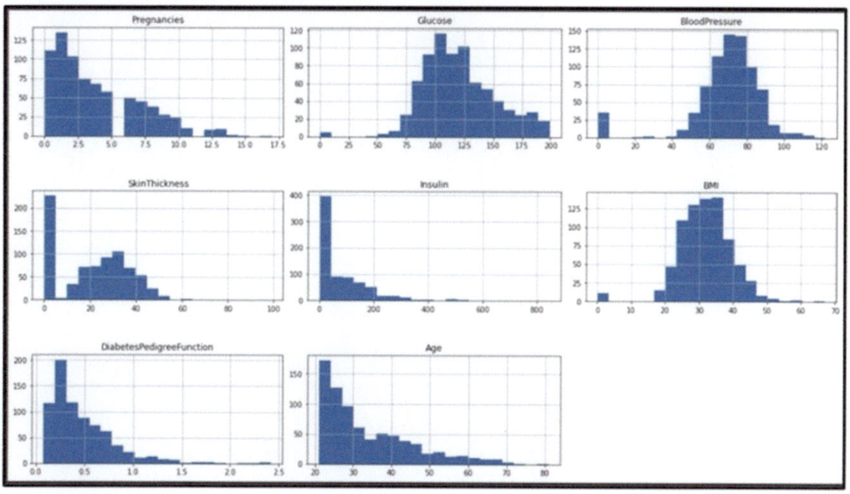

Fig. 5 Histogram of independent data columns

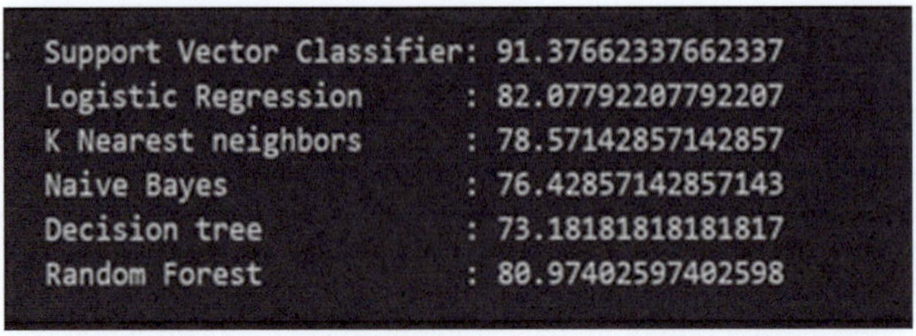

Fig. 6 Model accuracies for diabetes prediction

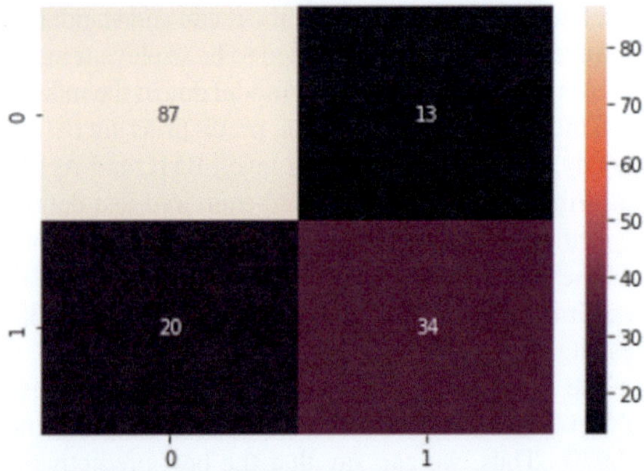

Fig. 7 Confusion matrix of diabetes prediction

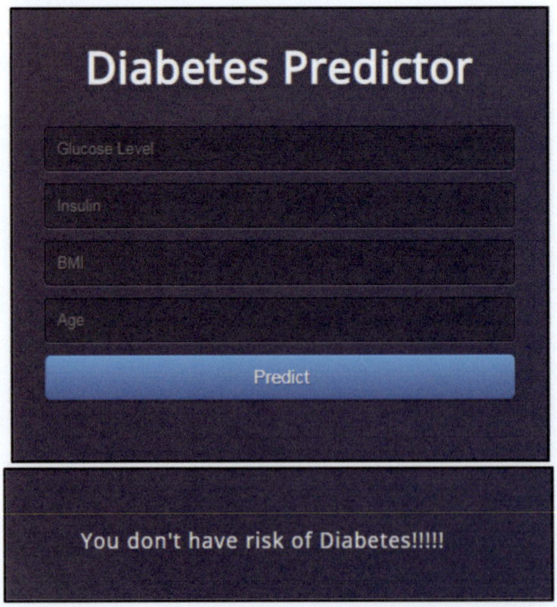

Fig. 8 Deployed predictive model of diabetes prediction

4.3 Liver Disease Prediction

Liver Disease is also becoming a cause of concern in the health industry. Liver problems are not easily discovered at an early stage. By the time the disease is discovered, it has already caused partial damage to the body. Hence, early diagnosis of liver disease will increase the survival rate in patients. Thus, a reliable predictive model will increase the chances of survival and early diagnosis of the disease.

For this analysis, we used the Liver dataset from Kaggle. It has a total of 11 features, out of which 10 are independent variables and 1 is the dependent variable. We will be predicting this variable based on the rest of the factors. The data set contains entries of 583 patients. In the Exploratory Data Analysis, we plotted a correlation plot of all data columns to see the distribution (Fig. 9).

After the data preprocessing and visualization of the distribution are done, model building is performed. We split the data into two parts: train and test data in an 80/20 ratio. For finding out the best fitting algorithm, we cross-validate the accuracy of every algorithm on train and test data, and the model that gives the most accuracy on test data prediction will be the final best fitting model. In our analysis, Random Forest Classifier proved to be the best model for Liver disease prediction with the highest accuracy of 100% and 71.79% on training and testing data, respectively (Fig. 10).

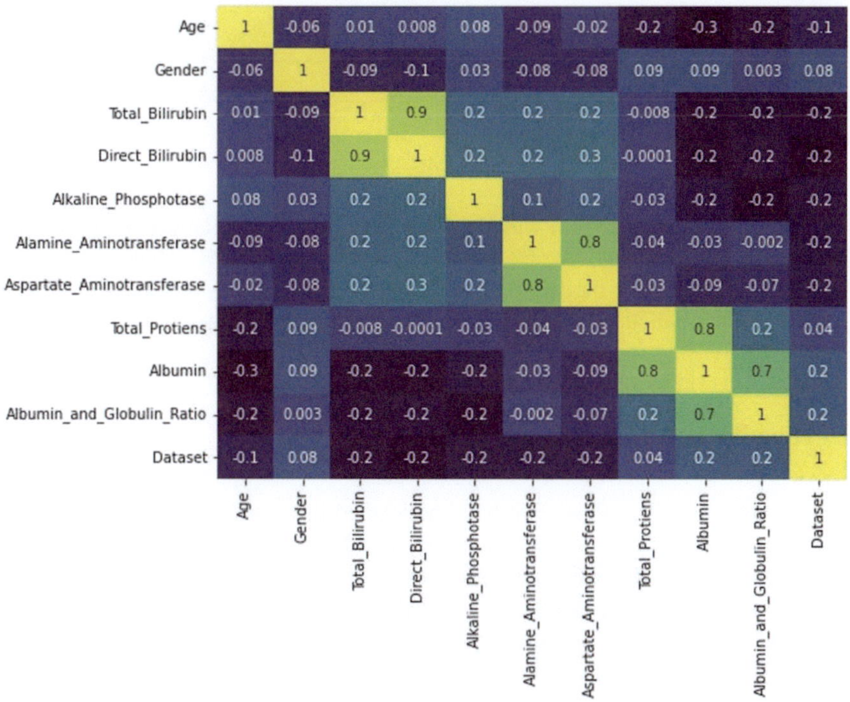

Fig. 9 Correlation plot of liver disease data columns

	Model	Training Accuracy %	Testing Accuracy %
0	Logistic Regression	74.498567	65.384615
1	Random Forest Classifier	100.000000	71.794872
2	Decision Tree Classifier	100.000000	67.521368

Fig. 10 Model accuracies for liver disease prediction

```
input_data = (62,2,1.7,140,268,0,0,160,0,3.6)


input_data_as_numpy_array= np.asarray(input_data)


input_data_reshaped = input_data_as_numpy_array.reshape(1,-1)

prediction = ran.predict(input_data_reshaped)

if (prediction[0]== 2):
  print('According to the given details person does not have a Liver Disease')
else:
  print('According to the given details person has Liver Disease')

According to the given details person has Liver Disease
```

Fig. 11 Liver disease predictive model

The second-best algorithm was Decision Tree, which gave an accuracy of 67%, followed by Logistic, which gave the least accuracy of 65.38%. Thus, we can say that the best predictive model for Liver disease prediction is the Random Forest Classifier. The predictive function of the model is shown in Fig. 11.

5 Results and Discussion

In this experimental analysis, we have performed data analysis on three different diseases, namely, Diabetes, Heart Disease, and Liver Disease. For Heart Disease Prediction, Logistic Regression has given the best accuracy of 82.67%, for Diabetes Prediction, Support Vector Classifier has given the best accuracy of 91.37%, and for Liver Disease Prediction, Random Forest Classifier has given the best accuracy of 71.79%. Thus, Machine Learning Algorithms have proven to be very efficient and reliable in Disease prediction if a Large Dataset is provided for training the model,

Table 1 Comparative chart of accuracies of diseases for various algorithms

Disease →	Diabetes (%)	Heart disease (%)	Liver disease (%)
Algorithm ↓			
Support vector classifier	**91.37**	62.29	64.30
Logistic regression	82.07	**82.67**	65.38
Random forest	80.97	78.68	**71.79**
Decision tree	73.18	75.40	67.52
K-nearest neighbors	78.57	62.29	62.21

and proper Data analysis is done so that more accurate results can be obtained. The comparative results are displayed in the Table 1.

6 Conclusion

Machine Learning has the ability to predict diseases based on existing big data. In this study, we have performed Exploratory Data Analysis (EDA) on 3 different Healthcare Datasets and predicted the causes and conditions of having a disease or not. This paper has applied various Machine Learning classification Algorithms to identify the best fitting model that gives the best accuracy. We found that performing Data preprocessing becomes an important aspect in identifying the predictions as the raw data cannot be in an appropriate form as needed for processing, and thus it may lead to errors and outliers. Removing outliers, plotting the correlations and confusion matrices of the data becomes necessary to analyze the models and predict with more accuracy. The more the model, the more correct predictions the model will be able to make. Thus, this study summarizes that if proper processing, analysis, and use of the most suitable algorithms is made, then Machine Learning can boost the disease prediction accuracies and provide more appropriate results, and thus improved treatment measures for doctors as well as patients.

References

1. Uddin S, Khan A, Md Hossain E, Moni MA Comparing different supervised machine learning algorithms for disease prediction. BMC Med Inf Decis Mak
2. Vinitha S, Sweetlin S, Vinusha, Sajini S (2018) Disease prediction using machine learning over big data. Comput Sci Eng Int J (CSEIJ) 8(1)
3. Mujumdar A, Vaidehi V (2019) Diabetes prediction using machine learning algorithms. International conference on recent trends in advanced computing
4. Patel J, Upadhyay T, Patel S Heart disease prediction using machine learning and data mining technique. IJCSC
5. Durai V, Ramesh S, Kalthireddy D Liver disease prediction using machine learning. Int J Adv Res Ideas Innov Technol

6. Joshi TN, Chawan PM Diabetes prediction using machine learning techniques. S. Dewangan.et.al. Int J Eng Res Appl
7. Mohan S, Thirumalai C, Srivastava G Effective heart disease prediction using hybrid machine learning techniques. IEEE access special section on smart caching, communications, computing and cybersecurity for information-centric internet of things
8. Chen M, Hao Y, Hwang K, Wang L, Wang L Disease prediction by machine learning over big data from healthcare communities. IEEE Access Spec Sect Healthc Big Data
9. Ramalingam VV, Dandapath A, Karthik Raja M Heart disease prediction using machine learning techniques: a survey. Int J Eng Technol
10. Jamgade AC, Zade SD Disease prediction using machine learning. Int Res J Eng Technol (IRJET)
11. Mohan S, Thirumalai C, Srivastava G Effective heart disease prediction using hybrid machine learning techniques. IEEE
12. Banu Priya M, Laura Juliet P, Tamilselvi PR Performance analysis of liver disease prediction using machine learning algorithms. Int Res J Eng Technol (IRJET)

Specialized Lexicon Development for Dermatological Diseases

Somiya Rani and Amita Jain

Abstract Dermatological diseases are among the most common medical conditions, affecting nearly 900 million people worldwide. Despite their prevalence, they remain among the most neglected diseases. Dermatological diseases can largely impact patients' quality of life and pose serious repercussions. Therefore, accurate diagnosis and treatment by healthcare professional, such as a dermatologists, are essential. However, the ambiguous use of terminology in the textual descriptions of dermatological conditions and their associated signs/symptoms, poses a challenge for healthcare professionals in correctly identifying the underlying skin diseases. To address this challenge, we propose a corpus-based approach to develop a lexicon of dermatological diseases. The proposed framework utilizes a seed word list extracted from PubMed abstracts. This lexicon is further enhanced with UMLS knowledge sources (UMLS Metathesaurus and SPECIALIST Lexicon) to incorporate synonymous terms and concept variants, thereby enhancing the quality and coverage of the generated lexicon. Additionally, the lexicon is evaluated on drug reviews from two benchmark datasets (Druglib.com and WebMD.com), by classifying the reviews into dermatological and non-dermatological disease categories. The specialized lexicon enables dermatology researchers and professionals to uncover critical information about dermatological diseases, ultimately improving the delivery of dermatological care.

Keywords Corpus-based approach · Dermatological disease · Lexicon · Natural language processing · PubMed · Skin disorder

---

S. Rani
Department of Computer Science and Engineering, Netaji Subhas University of Technology, East Campus (Formerly AIACTR), Delhi, India
e-mail: somiya1093@gmail.com

A. Jain (✉)
Department of Computer Science and Engineering, Netaji Subhas University of Technology, Delhi, India
e-mail: amita.jain@nsut.ac.in

S. K. Biswas et al. (eds.), *Intelligent Computing Systems and Applications*, Lecture Notes in Networks and Systems 1307, https://doi.org/10.1007/978-981-96-3860-4_15

1 Introduction

Dermatological diseases are one of the most common nonfatal diseases worldwide and impose a considerable burden that affects patients and their families socially, psychologically, and financially [1]. These diseases can be infectious, inflammatory, degenerative, congenital, or cancerous, and if left untreated, they may lead to serious consequences such as long-term disfigurement, disability, and social stigma. It largely impacts patients' quality of life, yet the most neglected in the past years due to the misconception that they pose fewer risks compared to other medical conditions.

However, awareness of skin health has been increasing, and people are becoming more mindful of their well-being. Dermatological diseases include all types of conditions that may irritate, clog, or damage skin, including some of which can cause cancer [2]. These dermatological conditions may be inherited or developed, and can have temporary or permanent effects [3]. In this direction, we aim to develop a dermatological diseases lexicon that can serve as a valuable resource for healthcare professionals, new researchers in this field, and patients alike.

The dermatological diseases lexicon can be extensively used by clinical researchers for natural language processing (NLP) subtasks such as concept tagging, relationship extraction between concepts, and term normalization, but there remains a lack of a domain-specific lexicons in this area. Existing domain-specific lexicons are typically created manually, which is labor-intensive and time-consuming. To overcome this limitation, we propose a semi-automated method for developing a lexicon of dermatological diseases using abstracts from PubMed dermatology articles.

The main motivation behind the development of the dermatological diseases lexicon is to support researchers and healthcare professionals in performing clinical NLP tasks such as concept tagging in the clinical narratives and extracting relationships between symptoms and skin lesions.

The main contributions of this paper are.

- Propose and implement a semi-automatic system for developing a dermatological diseases lexicon.
- Exploit neural embeddings to expand the seed list of dermatological-related terms generated from PubMed abstracts.
- Use UMLS knowledge sources (UMLS Metathesaurus and SPECIALIST Lexicon) for incorporating synonymous terms and concept variants to enhance the generated lexicon.
- Demonstrate the effectiveness of the lexicon on two drug review datasets by classifying the reviews into dermatological disease and non-dermatological disease categories.

2 Literature Review

The clinical domain still lacks the disease-specific lexicons and requires more attention from researchers in this field. While many authors have worked on the lexicon development for various domains such as hate speech, product review sentiment analysis, etc., but very few focused on the clinical domain. For example, the author in [4] presented a lexicon of profane and obscene terms in the Bengali language. A semi-automatic approach is used for developing the obscene lexicon by utilizing an obscene corpus, word embeddings, and POS taggers. In addition, the author in [5] introduced a domain-specific sentiment lexicon using an incremental approach that employed a vector model for IMDb movie reviews and confessions from the Experience Project website. Similarly, the author in [6] proposed an approach for building domain-specific lexicons for IMDb movie reviews and Amazon product reviews. Additionally, [7] proposed a new deep model that incorporates intensity scores into word representations to generate 24 domain-specific sentiment lexicons from 142.8 million English language reviews on Amazon.com.

Recent work has also emerged in the clinical domain. For instance, Parikh [8] developed a lexicon of COVID-19-related concepts using open-source word embedding sources, which are evaluated through both intrinsic and extrinsic measures. The author in [9] developed a sentiment lexicon for healthcare-related topics using a hybrid approach. Luo, Z. created a sentiment lexicon for clinical research eligibility criteria based on UMLS [10]. Velupillai [11] developed a vocabulary for extracting substance abuse information from psychiatric notes. A brief overview of state-of-the-art studies on lexicon generation is presented in Table 1.

3 Proposed Methodology

The proposed methodology for the development of the lexicon is divided into two phases. The first phase is corpus collection and pre-processing, and the second phase is lexicon generation. The first phase involves constructing a corpus of dermatology-related data. For this purpose, we constructed a corpus of abstracts from PubMed. The second phase involves lexicon generation and is further divided into three subphases: seed word induction, lexicon expansion, and analysis, discussed in Sect. 3.2. The framework for lexicon development is shown in Fig. 1. The algorithm for lexicon generation and expansion is presented in Algorithm 1.

Table 1 Comparison of state-of-the-art studies on lexicon generation

Reference	Objective	Approach	Dataset
Sazzed [4]	To propose a semi-automatic approach for building a lexicon of profane and obscene words in the Bengali language	Semi-automatic corpus-based approach	Abusive comment from YouTube
Muhammad et al. [5]	To generate a domain-specific sentiment lexicon through an incremental approach	Vector model	IMDB and confessions from Experience Project website
Han et al. [6]	To propose a new method for domain-specific lexicon for review sentiment analysis	SentiWordNet and mutual information scores	IMDB movie and Amazon product review
Islam et al. [7]	To introduce deep models based on vector representations for generating domain-specific sentiment lexicons	Bag-of-words and neural vector	142.8 million English language reviews from Amazon.com
Parikh et al. [8]	To develop a lexicon for COVID-19 by using open-source word embeddings	Biomedical embeddings	MIMIC III Discharge summaries, EPIC PennChart COVID-19 registry
Asghar et al. [9]	To create a health-related sentiment lexicon utilizing a hybrid system	Bootstrapping and tagging	Druglib, Drugsearch, and annotated adverse drug reaction (ADR) reviews
Luo et al. [10]	To create a semantic lexicon for clinical research eligibility criteria from UMLS with a corpus-based approach	Corpus-based approach	Public clinical trial registry maintained by the NLM, Clinicaltrials.gov
Velupillai et al. [11]	To develop a vocabulary for extracting substance abuse from psychiatry notes	Corpus-based, rule-based, and neural network	100 psychiatric clinical notes from University Hospital, Salt Lake City

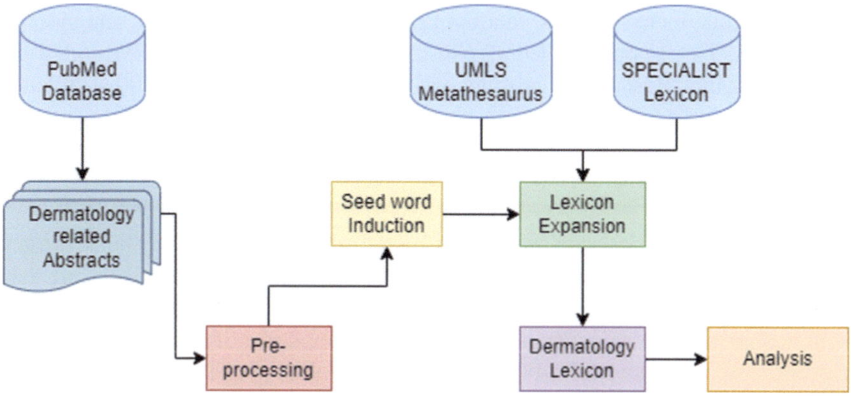

Fig. 1 Framework for lexicon development

Algorithm 1: Lexicon generation and expansion

Input: *SL: Seed List*
Output: *DL: Dermatology Lexicon*
Begin
1. for each w_i *Ɛ SL do*
2. $E_w \leftarrow$ *BuildEmbeddings(*w_i*)*
3. for each A_i *in* D_i *do*
4. Apply E_w *to* A_i *and identify* t_i
5. DL.add(t_i*)*
6. repeat 4 on t_i
7. End for no change
8. End for
9. for each t_i *do*
10. if t_i *exists in UMLS then*
11. DL.add (t_i*, CN)*
12. else
13. no_match_list.add(t_i*)*
14. end if
15. End for
End

3.1 Corpus Collection and Pre-processing

As discussed above, for the development of a corpus related to dermatological diseases, we searched for the term "Dermatology" in article-type such as books and documents, meta-analyses, randomized controlled trials, reviews, and systematic reviews. This search yielded 42,616 results from PubMed [12]. We sorted the results using the "Best-match" display options and selected the first 100 abstracts.

3.1.1 Pre-processing

We performed pre-processing on the constructed corpus to remove any noise present in the data. We first tokenized the text and removed stop words. The tokens are then lemmatized. Each token was tagged with POS taggers, which identify whether a word is a noun, verb, adjective, etc. POS tagging helps identify words with clinical relevance. We used the NLTK POS tagger [13] and the Spacy POS tagger [14].

3.1.2 Word Embedding

Word embedding represents word in a vector space while preserving their contextual usage. Word embedding is leveraged to learn synonyms for building the lexicons [15]. Some common embeddings include Word2Vec [15], FastText [16], global vectors for word representation (GloVe) [17], and recent transformer-based embeddings such as BERT, ELMo, XLNet, etc. Domain-dependent embeddings have attained significant amount of popularity as they can capture richer, domain-relevant information. A few authors have worked on biomedical embeddings. For example, [18] utilized PubMed and PubMed Central articles, along with the clinical notes from MIMIC III, to train embeddings using GloVe, Word2Vec, and FastText. Additionally, BioASQ created a continuous space vectors by applying Word2Vec to a corpus of 10,876,004 English biomedical abstracts from PubMed [19, 20]. Similarly, [21] proposed a clinical embedding corpus by leveraging PubMed articles to train the embedding. Pyysalo et al. [22] trained embeddings using Word2Vec on a combination of PubMed and PubMed Central articles along with Wikipedia to combine open domain and biomedical knowledge (biomedical natural language processing [BioNLP] corpus). A comparison of a few existing medical embeddings is shown in Table 2.

3.2 Lexicon Generation

The lexical resources can be leveraged for finding the presence of dermatology concepts, information retrieval, clinical named entity recognition, etc. The lexicon generation methodology in this paper consists of three steps:

1. Seed word induction
2. Lexicon expansion
3. Analysis

Table 2 Comparison of existing medical embeddings

Reference	Name	Embedding	Trained on	Vocabulary size
Zhang et al. [18]	BioWordVec intrinsic and extrinsic	FastText	PubMed + MeSH	2.3 billion tokens
Tsatsaronis et al. [19, 20]	BioASQ	Word2Vec	PubMed abstract	1.7 billion tokens
Flamholz et al. [21]	Clinical embeddings W2V300	Word2Vec	PubMed, PMC, and MIMIC III	300k token
Pyysalo et al. [22]	BioNLP Lab PubMed + PMC W2V	Word2Vec	PubMed/ Token and PMC articles	4 billion tokens
	BioNLP Lab Wiki + PubMed + PMC W2V	Word2Vec	Wikipedia, PubMed/PMC articles	5.4 billion tokens

3.2.1 Seed Word Induction

The first step of lexicon generation is seed word induction. To initiate the framework, we utilized Lancbox [23] for corpus analysis and exploration. The top 50 words (w_i) that exceeded a frequency threshold of 10 were selected as the seed list (SL).

$$SL = [W_1,\ W_2,\ W_3 \ldots W_n] \tag{1}$$

We then manually annotated the extracted terms as either dermatological and non-dermatological disease classes. Given a dataset D_i of PubMed abstracts $A_i \in x \times C$ where $i = 1,\ldots,n$ and C is the class showing dermatology disorder or non-dermatology disease, i.e., $C = \{dermatology, non{-}dermatology\}$. The mentions of other concepts that were not related to dermatological diseases are also observed and given a non-dermatological disease label. The words with the non-dermatological disease label were not included during lexicon expansion.

A few examples from the seed list are "rosacea," "acne," "blister," "allergy," "scaling," "keratosis," "ulcer," "fungal infection," "crust," "granuloma," "hypertrophy," "eczema," "measles," "warts," "chickenpox," "diaper rash," "hives," "actinic keratosis," "psoriasis," "dermatitis," "itch," "inflammation," "dryness."

3.2.2 Lexicon Expansion

The second step of the lexicon generation methodology is lexicon expansion which involves the word embeddings of the seed words. We applied BioASQ embeddings (E_w) to find similar words (t_i) in the corpus.

$$E_W = BioASQ[w1, w2, w3 \ldots wn] \quad (2)$$

$$DL = E_w \times A_i \quad (3)$$

The entire procedure first involves identifying words (t_i) that are the most similar to the seed words (w_i) and then iteratively finding words similar to dermatology-related words recognized in the first step, and keeps adding them to the dermatology lexicon (*DL*). The iteration stops when no significant expansion of the dermatology-related word is observed.

The duplicate words are removed automatically by checking the occurrence (>1). Furthermore, we utilized UMLS knowledge sources: UMLS metathesaurus and SPECIALIST lexicon tools for incorporating synonymous terms and concept variants to enhance the generated lexicon. We used words from *DL* which have term types: "Skin disease," "Skin disorder," "Dermatosis," "Dermatological disorder," "Dermatological disease," and "Dermatopathy" in UMLS metathesaurus only. If the term is not found, it is added to the no_match_list. Furthermore, all the concepts named (*CN*) having inverse_isa relation (hierarchical links of a semantic relation to its children) [24] were added to the *DL* to expand the lexicon.

3.2.3 Analysis

We investigated the generated lexicon by testing document coverage on drug reviews from Druglib.com [25] and WebMD [26] to find out the effectiveness of the developed lexicon. The reviews from both datasets are evaluated for the dermatology and non-dermatology classification. The case study and evaluation of both datasets are discussed in Sect. 5.

4 Result

The corpus consists of 15,119 tokens and 3,117 lemmas. The pre-processed corpus contains 9,704 tokens and 3,034 lemmas. The developed Dermatology Lexicon (*DL*) contains 2,086 unique terms, including 1981 dermatology-related terms (such as disorders, procedures, signs, substances, etc.) and 103 non-dermatology terms (also including disorders, procedures, signs, substances, etc.).

We considered four dermatology diseases to explore the developed lexicon. According to the American Academy of Dermatology Association, acne is the most common dermatological disease in the US, typically beginning during puberty and affecting adolescents and young adults. Psoriasis affects approximately 7.5 million people in the US. The third dermatological disease, rosacea, affects around 16 million people. The fourth condition, dermatitis, was also chosen for lexicon exploration. Table 3 shows the UMLS codes for the selected dermatological diseases.

Table 3 Sample list of dermatological diseases with UMLS code

Dermatological disease	UMLS CUI
Acne	C0702166
Psoriasis	C0033860
Rosacea	C0035854
Dermatitis	C0011603

Table 4 Count of diseases mentions "Acne," "Psoriasis," "Rosacea," and "Dermatitis" in the PubMed corpus, Druglib.com, and WebMD reviews

Terms	PubMed abstract	Druglib.com	WebMD
Acne	1468	973	21,267
Psoriasis	6306	25	2281
Rosacea	484	88	1083
Dermatitis	2540	33	1174

Table 4 presents the frequency of these diseases mentioned in PubMed corpus, Druglib.com, and WebMD drug reviews. In addition, we utilized Lancbox [23] to identify collocates associated with the terms "Acne," "Psoriasis," "Rosacea," and "Dermatitis." Table 5 presents the top 5 collocates for the above-mentioned dermatological diseases in the PubMed abstract corpus. Based on this exploratory information, we created a visual representation of dermatological diseases and their subtypes in Fig. 2. A sample of words from the dermatology lexicon, along with their UMLS semantic type is shown in Table 6.

5 Case Study and Evaluation

The lexicon generated through the proposed method can be utilized for various tasks, such as clinical named entity recognition (CNER), clinical report identification/retrieval, disease classification, or sentiment analysis. To show the effectiveness of the developed lexicon, we used it to classify drug reviews into dermatology and non-dermatology disease categories by matching lexicon terms with tokens in the reviews. The performance of the lexicon in identifying drug reviews related to dermatological and non-dermatological conditions on the Druglib.com and WebMD datasets is shown in Table 7. The developed lexicon performed outstandingly and demonstrated strong classification performance. It was also observed that some reviews mentioned dermatological conditions as side effect. These reviews were included in both dermatology and non-dermatology classes to minimize bias.

As an evaluation metric, we used document coverage to assess effectiveness. The document coverage is denoted as Dcov and is defined using the formula.

Table 5 Collocates for dermatological diseases from PubMed abstracts corpus with frequency >10

Condition	Collocates for dermatological diseases				
Acne	Acne vulgaris	Pustular acne	Propionibacterium acne	Cutibacterium acne	Cystic acne
Psoriasis	Plaque Psoriasis	Pustular Psoriasis	Palmoplantar Psoriasis	Psoriasis Vulgaris	Chronic Psoriasis
Rosacea	Ocular Rosacea	Papulopustular Rosacea	Acne Rosacea	Erythematotelangiectatic Rosacea	Phymatous Rosacea
Dermatitis	Contact Dermatitis	Atopic Dermatitis	Seborrhoeic Dermatitis	Lichenoid Dermatitis	Vesiculobullous Dermatitis

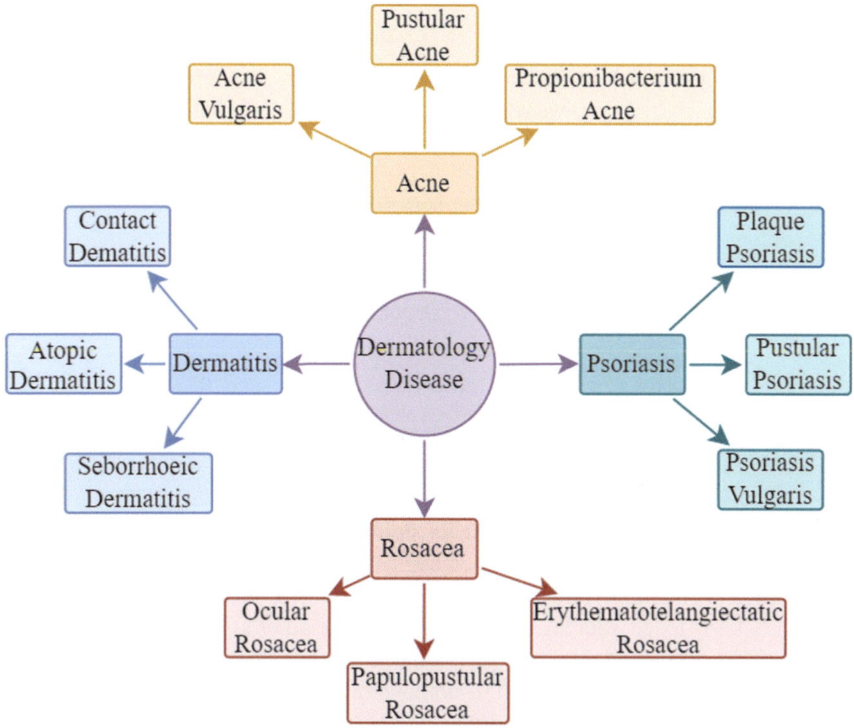

Fig. 2 Visual representation of dermatological diseases and their subtypes

Table 6 Sample words from the dermatology lexicon with UMLS semantic type

Word	UMLS semantic type
Vitiligo	Disease or syndrome
Alopecia areata	Disease or syndrome
Acne vulgaris	Disease or syndrome
Topical corticosteroids	Pharmacologic substance
Melasma	Disease or syndrome
Lichen planus	Disease or syndrome
Wound inflammation	Finding
Comedone	Disease or syndrome
Levonorgestrel	Pharmacologic substance
Acne keloid	Disease or syndrome
Nodular folliculitis	Disease or syndrome
Pyoderma	Disease or syndrome
Trichophytide	Disease or syndrome
Galangin	Pharmacologic substance
Abnormal pigmentation	Finding

Table 7 Performance of proposed lexicon-based method on Druglib.com and WebMD datasets

Datasets	Total reviews	Correctly identified
Druglib.com	4143	3906
WebMD	362,806	345,789

$$D_{cov} = \frac{\begin{array}{c} no.\,of\ reviews\ with\ dermatology\ related \\ word\,(\geq 1)identified \end{array}}{\begin{array}{c} total\ of\ no.\,of\ reviews \\ in\ corpus \end{array}} \tag{4}$$

6 Conclusion

This paper presents a lexical resource for dermatological diseases using a corpus-based approach. The proposed approach begins with a seed list of the most common dermatological conditions and expands it using neural embeddings, further enhanced by UMLS knowledge sources, namely, UMLS Metathesaurus and SPECIALIST lexicon. The developed lexicon is evaluated on two benchmark drug review datasets (Druglib.com and WebMD), demonstrating its effectiveness in detecting dermatology-related terms within drug reviews. This lexicon can be leveraged for various clinical NLP tasks, such as concept tagging, disease identification or classification, and sentiment analysis. In the future, we plan to expand the lexicon using larger corpora. Additionally, we aim to build a specialized lexicon for neglected diseases that require greater attention, such as Tuberculosis (TB) and HIV/AIDS.

References

1. Basra MK, Shahrukh M (2009) Burden of skin diseases. Expert Rev Pharmacoecon Outcomes Res 9(3):271–283
2. https://my.clevelandclinic.org/health/diseases/21573-skin-diseases. Accessed 30 Apr 2022
3. https://www.healthline.com/health/skin-disorders. Accessed 30 Apr 2022
4. Sazzed S (2021) A lexicon for profane and obscene text identification in Bengali. In: Proceedings of the international conference on recent advances in natural language processing (RANLP 2021), pp 1289–1296
5. Muhammad SH, Brazdil P, Jorge A (2020) Incremental approach for automatic generation of domain-specific sentiment lexicon. In European conference on information retrieval, pp 619–623
6. Han H, Zhang J, Yang J, Shen Y, Zhang Y (2018) Generate domain-specific sentiment lexicon for review sentiment analysis. Multimed Tools Appl 77(16):21265–21280
7. Islam SM, Dong XL, de Melo G (2020) Domain-specific sentiment lexicons induced from labeled documents. In: Proceedings of the 28th international conference on computational linguistics, pp 6576–6587

8. Parikh S, Davoudi A, Yu S, Giraldo C, Schriver E, Mowery D (2021) Lexicon development for COVID-19-related concepts using open-source word embedding sources: an intrinsic and extrinsic evaluation. JMIR Med Inform 9(2):e21679
9. Asghar MZ, Ahmad S, Qasim M, Zahra SR, Kundi FM (2016) SentiHealth: creating health-related sentiment lexicon using hybrid approach. Springerplus 5(1):1–23
10. Luo Z, Duffy R, Johnson S, Weng C (2010) Corpus-based approach to creating a semantic lexicon for clinical research eligibility criteria from UMLS. Summit Transl Bioinform 26
11. Velupillai S, Mowery DL, Conway M, Hurdle J, Kious B (2016) Vocabulary development to support information extraction of substance abuse from psychiatry notes. Presented at: The 15th workshop on biomedical natural language processing. Berlin, Germany
12. https://pubmed.ncbi.nlm.nih.gov/. Accessed 30 Apr 2022
13. Loper E, Bird S (2002) Nltk: the natural language toolkit. arXiv preprint cs/0205028
14. Honnibal M, Montani I (2017) SpaCy 2: natural language understanding with bloom embeddings, convolutional neural networks, and incremental parsing. To Appear 7(1):411–420
15. Mikolov T, Sutskever I, Chen K, Corrado GS, Dean J (2013) Distributed representations of words and phrases and their compositionality. Adv Neural Inf Process Syst 26
16. Bojanowski P, Grave E, Joulin A, Mikolov T (2017) Enriching word vectors with subword information. Trans Assoc Comput Linguist 5:135–146
17. Pennington J, Socher R, Manning C (2014) GloVe: global vectors for word representation. In: Proceedings of the 2014 conference on empirical methods in natural language processing, pp 1532–1543
18. Zhang Y, Chen Q, Yang Z, Lin H, Lu Z (2019) BioWordVec, improving biomedical word embeddings with subword information and MeSH. Sci Data 6(1):52
19. Tsatsaronis G, Balikas G, Malakasiotis P et al (2015) An overview of the BIOASQ large-scale biomedical semantic indexing and question answering competition. BMC Bioinform 16:138
20. BioASQ. http://bioasq.org/news/bioasq-releases-continuous-space-word-vectors-obtained-applying-word2vec-pubmed-abstracts. Accessed 2 May 2022
21. Flamholz ZN, Ungar LH, Weissman GE (2019) Word embeddings trained on published case reports are lightweight, effective for clinical tasks, and free of protected health information. medRxiv
22. Pyysalo S, Ginter F, Moen H, Salakoski T, Ananiadou T (2013) Distributional semantics resources for biomedical text processing. In: Proceedings of the 5th international symposium on languages in biology and medicine
23. Brezina V, Weill-Tessier P, McEnery A (2020) LancsBox v. 5. x.[Software]. Lancaster University
24. UMLS® Reference Manual [Internet] (2022) Bethesda (MD): National Library of Medicine (US); 2009 Sep-. 5, Semantic Network. [Updated 2021 Aug 20]. https://www.ncbi.nlm.nih.gov/books/NBK9679/, Accessed 13 May 2022
25. Gräßer F, Kallumadi S, Malberg H, Zaunseder S (2018) Aspect-based sentiment analysis of drug reviews applying cross-domain and cross-data learning. In: Proceedings of international conference on digital health, 121–125
26. Kaggle. www.kaggle.com/datasets/rohanharode07/webmd-drug-reviews-dataset?select=webmd.csv. Accessed 13 May 2022

Deep Learning-Based Detection of Surface Anomalies

Rahul, Ashwani Kharola, Varun Pokhriyal, and Prajwal Dangwal

Abstract The proposed article is based on the intelligent system for the detection of defects in magnetic tiles using a transfer learning approach. The training models used in the study were already trained with various images so that the transfer learning approach could be efficiently utilized. The study uses a convolutional neural network (CNN) in multiple layers and processes data in a grid-like model. One main merit of using CNNs is that there is no need for preprocessing. To utilize the benefits of high speed, large bandwidth, and low latency 5G network is used. The standard dataset used in the article for experimental purposes, and the highest accuracy of 97.8% was obtained using the DeepTree model.

Keywords Convolutional neural network · Transfer learning · 5G technologies · Deep learning

1 Introduction

The anomalies in magnetic tiles are detected mainly with manual inspection, which is inaccurate, imprecise, and requires a lot of hard work. At the same time, omissions and error detection are common. A lot of non-destructive tests (NDTs) have been proposed to solve the above problems. Thus, an inexpensive and highly reliable CNN system is required in the field of magnetic technology [1]. This intelligent system increases accuracy and production latency, makes production more reliable,

Rahul (✉)
Sri Venkateswara College, University of Delhi, Delhi, India
e-mail: rvermaresearch@gmail.com

A. Kharola
Department of Mechanical Engineering, Graphic Era (Deemed to be) University, Dehradun, India

V. Pokhriyal
Scabula Labs Private Limited, Dehradun, India

P. Dangwal
Department of Computer Science and Engineering, Delhi Technological University, Delhi, India

S. K. Biswas et al. (eds.), *Intelligent Computing Systems and Applications*, Lecture Notes in Networks and Systems 1307, https://doi.org/10.1007/978-981-96-3860-4_16

and enhances product quality [2]. The CNNs applications are numerous today and have been opened to all sectors of industry [3]. Magnetic tiles are mainly affected by intermittent light, intense vibrations, and noise in the environment. Therefore, detecting the defect in magnetic tiles is a core task [4]. A study by Essid et al. [5] showed that by combining an auto-encoder and deep neural network (DNN) architecture, it was not only possible to detect the defects, but we can also localize the defects in the surface with relatively high precision. The study further highlighted that compared to conventional methods, DNNs were able to acquire visual features that produced the highest results.

A scalable multi-layer CNN-based transfer learning approach was suggested to identify abnormalities in anomaly datasets by Natarajan et al. [6]. The study considered a majority voting system to solve over-fitting issues by integrating deep-learning features with support vector machine (SVM) classifiers. A two-stage model consisting of the CNN and the SVM was applied to detect the defect. However, the localization of the defect was not focused on. A hierarchical damage identification system focused on Faster R-CNN to identify different forms of surface damage, such as concrete cracking, steel corrosion of various levels, and bolt deterioration, was proposed by Cha et al. [7]. Chen et al. [8] suggested a detection method with three CNN-based detection levels, where two levels were used to identify the main defect, and a classifier was used to identify the type of defects. In the manuscript [9], the method evaluates the degree of abnormality of each sub-region of an image under consideration by computing a CNN-based visual similarity with respect to a dictionary of anomaly free sub-regions belonging to a training set. In [10], the framework reconstructs lace videos and performs defect inspection using reconstruction error and the pre-set threshold. This paper proposes a magnetic tile defect detection system using a pre-trained CNN model. The model has been fine-tuned considering the training samples from the metal surface image dataset.

2 Framework of Proposed System

The system is implemented in a 5G environment to decrease latency for a real-time experience. The overall system framework is shown in Fig. 1. The major contributions of the paper include developing a system using pre-trained CNN models to detect magnetic tile defects and implementing the system in a 5G environment.

The proposed magnetic tile defect detection approach uses pre-trained CNN models to transfer knowledge from one domain to another. Since the number of samples of the magnetic tile surfaces is less therefore we utilized the already existing CNN models, which are pre-trained using millions of natural images. The study investigates the following pre-trained CNN models, which showed excellent performance in various image processing tasks: ResNet50, ResNet101, Inception v3, Xception, SqueezeNet, ShuffleNet, MobileNet, and DeepTree as shown in Fig. 2. All the above models except the DeepTree model were trained using millions of images from the

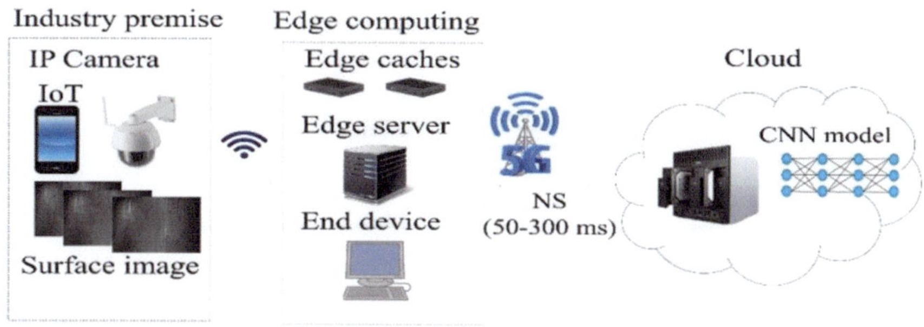

Fig. 1 The framework of the proposed system

ImageNet challenge dataset. The deep-tree model was trained using the Canadian Institute for Advanced Research-10 (CIFAR-10) dataset. Further, Fig. 3 shows the block diagram of the proposed system. An IP camera captures the image of a magnetic tile surface and sends it to a computer or an edge server. The computer or the server does preprocessing, which includes illumination normalization and resizing. The preprocessed image is then sent to the cloud server using the 5G network. The cloud server has the proposed defect detection system. Once the decision is reached, it is fed back to the computer or the edge server. The preprocessing is done for better accuracy and to reduce the size of the image so that it takes less bandwidth to be transferred to the cloud server.

During fine-tuning, the last layer (softmax) of the pre-trained model is replaced by a new softmax layer of six neurons, one each for one of the six classes of the metal surface (described in the next section). The weights of the model were optimized by the Adam optimizer. The Adam optimizer calculates exponential moving averages (x and y) of the gradient (g) of the current mini-batch as given by Eqs. (1) and (2), respectively.

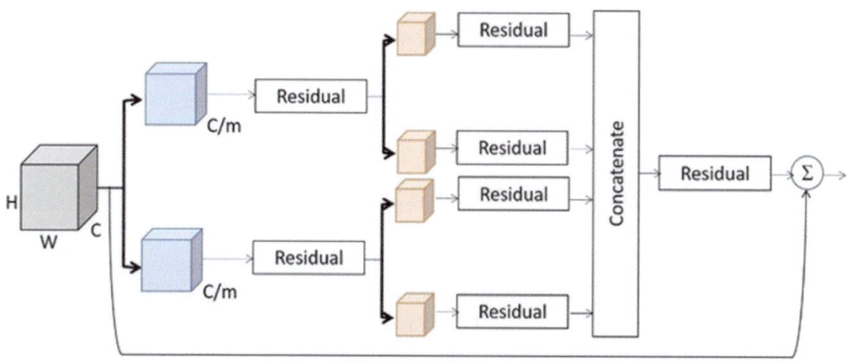

Fig. 2 DeepTree architecture with tree height = 2 and branch = 2

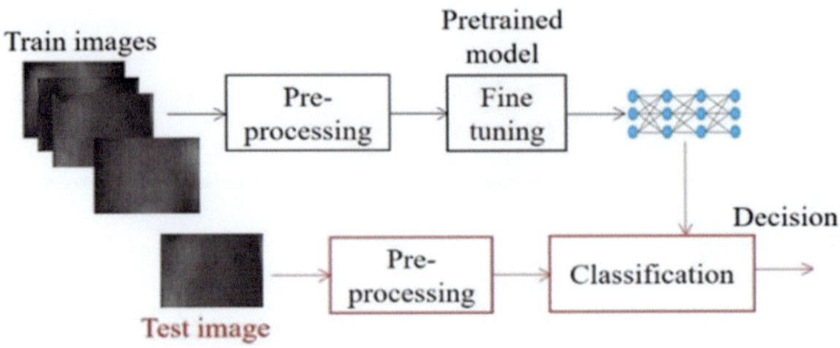

Fig. 3 Block diagram of the proposed magnetic tile defect detection system

$$x_t = \beta_1 x_{t-1} + (1 - \beta_1) g_t \tag{1}$$

$$y_t = \beta_2 y_{t-1} + (1 - \beta_2) g_t^2 \tag{2}$$

where β_1 and β_2 are two parameters of the optimizer. We set these parameters to 0.9 and 0.999, respectively. The final estimates of the moving averages are found using Eqs. (3) and (4) respectively

$$\tilde{x}_t = x_t / (1 - \beta_1^t) \tag{3}$$

$$\tilde{y}_t = y_t / (1 - \beta_2^t) \tag{4}$$

The weights of the model were then updated according to the formula in Eq. (5),

$$w_t = w_{t-1} - \eta \frac{\tilde{x}_t}{\sqrt{\tilde{y}_t} + \varepsilon} \tag{5}$$

where η is the learning rate, and ε is a very small value, which we set to 10E-8.

3 Dataset

In the magnetic tile defect dataset, there were 1344 images of six different types: Blowhole, Crack, Fray, Break, Uneven (induced by the drilling operation), and Free (no defects) [11]. The region of interest (ROI) of magnetic tiles is cropped, and each cropped image has labels at pixel levels. Blowholes and cracks have the biggest effect on the appearance of magnetic tiles. Uneven has the highest possible complexity to detect because of its highly similar color and texture to the background. Fray's color and texture are close to those of dark surface soil. The dirt does not impact the magnetic tile quality. Figure 4 shows one example of each of the five types of defects. The upper row of the figure shows the magnetic tile images, and the bottom row of the figure shows the ground truth of the defects in the images.

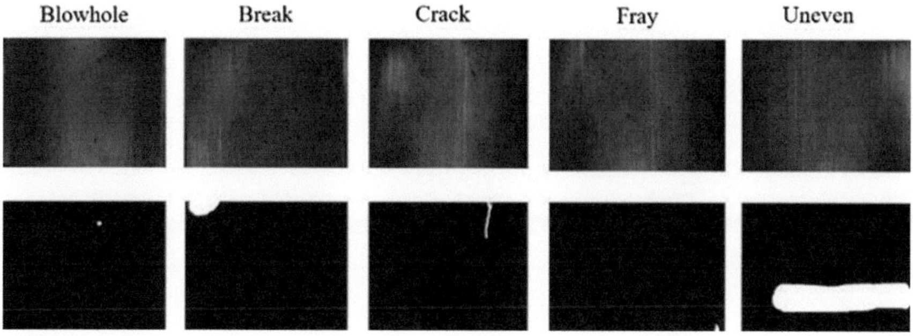

Fig. 4 Sample images (upper row) of magnetic tile defects and the ground truth images (bottom row)

4 Experiments

In the experiments, a five-fold cross-validation approach was adopted. In this approach, the whole dataset was divided randomly into five folds. In each run, four folds were used for pre-training and the other fold for testing. After five runs, the average accuracy over the runs was calculated. For pre-training, the following values of the parameters were set empirically: mini-batch size = 10, maximum number of iterations = 2000, initial learning rate = 3E − 4. Training images were augmented using reflection in both x and y directions, random rotation within the range between −20° and 20°, and random scaling with a factor between 0.8 and 1.2. All the experiments were performed using MATLAB R2020a. A deep-learning toolbox and a 5G toolbox of MATLAB were utilized to develop the codes. The machine configuration was as follows: two CPUs of 2.80 GHz and 6 cores each, 64.0 GB RAM, NVIDIA GPU of GeForce GTX 1080 (8 GB).

5 Results

The figures were drawn using the t-distributed Stochastic Neighbor Embedding (t-SNE) technique as shown in Fig. 5. The t-SNE is a non-linear dimensional reduction technique best adapted for incorporating high-dimensional data for visualization in a low-dimensional space. Figure 5a shows the distribution of the samples of different classes. The distribution was plotted from the output of the first max-pooling layer of the pre-trained model in the first iteration. Figure 5b–i shows the distribution of the samples after the softmax layer of the final iteration. From Fig. 5a–i, we can understand that the DeepTree model is a very good pre-trained model to cluster the samples of different classes. Further, SqueezeNet and Xception models could not cluster the samples properly, which has in turn affected the accuracy of the system. Figure 6 shows the ROC curves of the proposed system using eight different

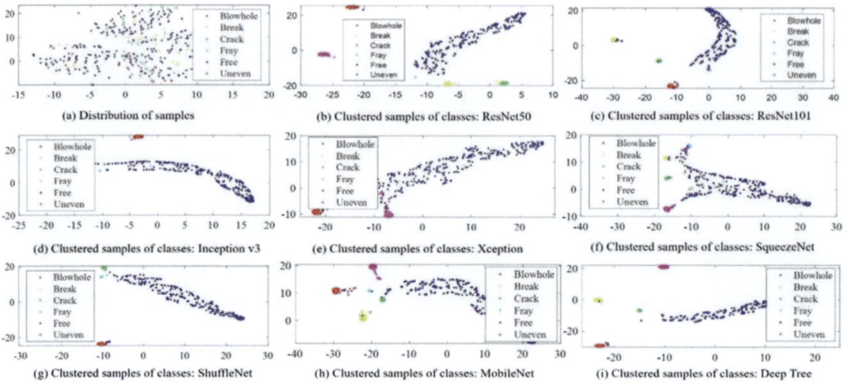

Fig. 5 t-SNE visualization of the samples using different pre-trained models

pre-trained models. From the ROC curve, we find that the DeepTree model had an excellent AUC. Other models also had good AUC however, SqueezeNet and Xception have comparatively lower AUC. Figure 7 shows the learning curve of the system using the DeepTree pre-trained model. Figure 7a shows the accuracy curve, and Fig. 7b shows the loss curve. From the figure, we see that the system converges smoothly.

Table 1 shows the accuracy (%) and the AUC of the system using eight pre-trained models, and of the system developed in the study of Huang et al. [11]. However, the result of a study by Huang et al. [11] does not mention the percentage accuracy. The highest accuracy was obtained by DeepTree, followed by ResNet50 and Inception v3.

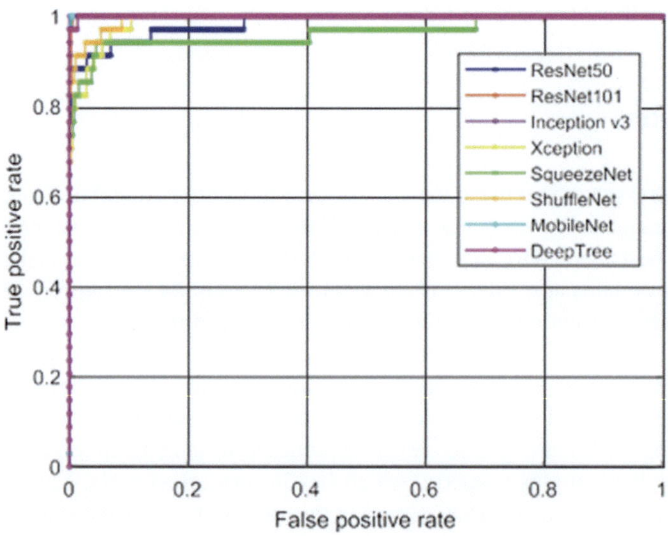

Fig. 6 ROC curves of the system using eight different pre-trained CNN models

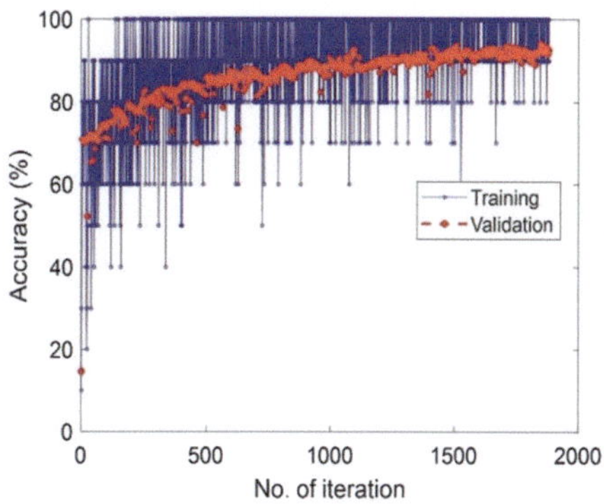

(a) Learning curve: accuracy; for Deep Tree.

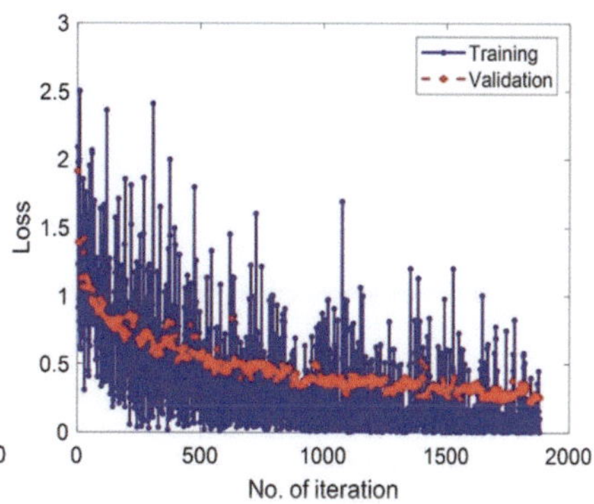

(b) Learning curve: loss; for Deep Tree.

Fig. 7 The learning curve of the system for the DeepTree pre-trained model

Table 1 AUC and accuracy of the systems

Systems	AUC	Accuracy (%)
Huang et al. [11]	0.966	–
ResNet50	0.984	96.3
ResNet101	0.999	91.8
Inception v3	1.000	95.8
Xception	0.989	88.4
SqueezeNet	0.963	84.9
ShuffleNet	0.994	94.6
MobileNet	0.999	90.1
DeepTree	0.999	97.8

6 Conclusion

A magnetic tile defect detection system using the transfer learning of CNN models was proposed. Eight different pre-trained models were investigated in the 5G scenario. The highest accuracy of 97.8 was obtained by the DeepTree model. All the models achieved an accuracy of over 90%, except Xception and SqueezeNet. The high accuracy of the proposed system using pre-trained models suggests that the system is robust and efficient using transfer learning in industrial applications. In future work, other recent lite pre-trained models will be investigated.

References

1. Balodi A, Anand RS, Dewal ML, Rawat A (2020) Computer-aided classification of the mitral regurgitation using multiresolution local binary pattern. Neural Comput Appl 32(7):2205–2215
2. Joshi K, Tripathi V, Bose C, Bhardwaj C (2020) Robust sports image classification using inceptionV3 and neural networks. Proced Comput Sci 167:2374–2381
3. Feng J, Li F, Lu S, Liu J, Ma D (2014) Injurious or non-injurious defect identification from MFL images in pipeline inspection using convolutional neural network. IEEE Trans Instrum Meas 66(7):1883–1892
4. Bhatt J, Gangwar A, Nijhawan R, Gangodkar D (2019) A research on deep learning advance for landslide classification using convolutional neural networks. Int J Innov Technol Explor Eng 8(4):903–906
5. Essid O, Laga H, Samir C (2018) Automatic detection and classification of manufacturing defects in metal boxes using deep neural networks. PLOSONE 13(11):e0203192
6. Natarajan V, Hung TY, Vaikundam S, Chia LT (2017) Convolutional networks for voting based anomaly classification in metal surface inspection. In: Proceedings of IEEE international conference on industrial technology (ICT). Toronto, Canada, pp 986–991. https://doi.org/10.1109/ICIT.2017.7915495
7. Cha YJ, Choi W, Suh G, Mahmoudkhani S, Büyüköztürk O (2018) Autonomous structural visual inspection using region-based deep learning for detecting multiple damage types. Comput Aided Civil Infrastruct Eng 33:731–747
8. Chen J, Liu Z, Wang H, Núñez A, Han Z (2018) Automatic defect detection of fasteners on the catenary support device using deep convolutional neural network. IEEE Trans Instrum Meas 67:257–269
9. Kingma D, Ba J (2015) Adam- a method for stochastic optimization. In: Proceedings of 3rd international conference for learning representations, San Diego. https://doi.org/10.48550/arXiv.1412.6980.
10. Maaten L, Hinton G (2008) Visualizing data using t-SNE. J Mach Learn Res 9:2579–2605
11. Huang Y, Qiu C, Yuan K (2020) Surface defect saliency of magnetic tile. Vis Comput 36:85–96

EV's Lithium-Ion Battery Heating Analysis

Vijaykumar Rayappa Bhanuse, Anita Bapu Dombale, Krishna Kalaskar, Varun Jaybhaye, Swapnil Irkate, and Amogh Dongre

Abstract In the last few months, there have been several reports of fires on electric vehicles, especially electric scooters. This paper determines the cause of those electric vehicles on fire and how much heat the battery gets when charging the electric vehicle. Also, how much heat is generated when EVs are discharged has been analyzed. Essentially, the parameters of the battery, both internal and external, are what cause the heating problems in the battery. As a result, we have observed how the two previously stated parameters' sub-parameters affect battery temperature in the proposed work.

Keywords Battery initial temperature · Battery final temperature · SOC (state of charge) · EV (electric vehicle) ambient temperature · Discharge · Charge · Speed

1 Introduction

Use of the electric vehicle has taken a notable increase worldwide over the last few years. Yet, there is a problem with the solution and many people nowadays are raising questions and waiting for the answers. The battery explosions' cause and method are

V. R. Bhanuse (✉) · A. B. Dombale · K. Kalaskar · V. Jaybhaye · S. Irkate · A. Dongre
Department of Instrumentation Engineering, Vishwakarma Institute of Technology, Pune, India
e-mail: vijaykumar.bhanuse@vit.edu

A. B. Dombale
e-mail: anita.dombale@vit.edu

K. Kalaskar
e-mail: krishna.kalaskar@vit.edu

V. Jaybhaye
e-mail: varunjaybhaya@vit.edu

S. Irkate
e-mail: swapnil.Irkate@vit.edu

A. Dongre
e-mail: Amogh.dongre@vit.edu

S. K. Biswas et al. (eds.), *Intelligent Computing Systems and Applications*, Lecture Notes in Networks and Systems 1307, https://doi.org/10.1007/978-981-96-3860-4_17

the main questions and problems. This is a serious problem as it is directly related to the "life of people" and it must be taken very seriously. Now for solving any problem, initially, we must comprehend it fundamentally. This paper does some analysis for this, which is why this project has this goal.

The temperature of the lithium-ion battery must be properly operated as it resolves the performance of the electric vehicle. During the charging as well as discharging process, the climate of the battery changes due to internal hotness build-up, increasing the need for survey of rate of heat generation in batteries. The heat generated involves reaction heat as well as joule heat and is assumed by two distinctive parts including temperature, battery aging result, SOC, and working current. In some articles, experiment using power-type graphite and lithium manganese oxide batteries has been appealed in different statuses. The specification of the joule temperature as well as the response temperature as well as the overall rate of heat generation is finished and analyzed taking into account the outcome of temperature, aging, and charge consign. Applies a broad battery heat transfer model to confirm the correct of the temperature generation rate, to measure as well as to check the temperature difference, amazingly the calculated differences in temperature have shown a good communication aging and experimental findings under various current circumstances. Thus, the temperature abnormality forecasting method is well suited for real-time applications.

2 Literature Survey

This paper provides an update on welfare issues pertaining to fire and fire insurance strategies, as well as the newest battery technologies EVs. EVs are especially vulnerable to fire hazards and Li-ion (LIB) battery pitfalls because of their high driving and charging demands, inevitable traffic accidents, growing power levels, and large battery packs [1]. Based on heat increase records, thermal data, and pictorial images, this paper reviews and describes the phenomenon of fires in Plug-in Hybrid Electric Vehicles (PHEVs). The analysis utilized two different car types, and the outer short-circuit faults in the battery cell packs cause the fires to spread [2]. This essay has presented a number of risk-reduction strategies that deal with accident [3]. An electrode response and charging transition simulation model was created and specifically used to test power outages in order to forecast battery performance. It was demonstrated that altering the lithium-ion batteries' electrode boundaries allows the development of systems with energy sources appropriate for application in a range of naturally occurring vehicles that are being regarded [4]. This essay examines the problems with managing lithium-ion batteries from a healthcare perspective. Because numerous intricately linked systems damage batteries [5]. Li-ion batteries are excellent energy source. When compared to many other batteries, their power output is high. Li-ion batteries do not need much upkeep. Their battery life is decent. They offer a free monthly rate between 1.5 and 2%. They have hazardous materials like cadmium in them, which can deplete the battery. These batteries cost a lot of money. They can be harmed by high voltage and have severe heat issues. We can use a safety mechanism and place a limit on its voltage to solve these issues, but doing so may also lower battery performance. Li-ion batteries have certain drawbacks, but they will be addressed in the future [6]. In this paper, the relaxation of the electrified chemical procedure takes part in a negative role in talon penetration, emplacement of the cell first is found to be the bulk danger to a battery among the given three cells of the battery. Even if the second and the third cells of the battery, with thicker copper plates, i.e., electrodes, receive lower volume and energy reservation, similarly high thermal expansion in nail penetration is controlled under the extreme heat changing [7]. It concludes that for every approach there is no suitable chemistry; thus, the picking process is very essential as the improper pick of the lithium-ion battery so the chemistry of a particular system has non-positive results on the economic feasibility of the financing [8, 9]. The work in this paper shows a summary of aging procedures and presents models to define them, from electrochemical to experimental. Finally, opportunities for car-to-grid (V2G) services are discussed in conjunction with older battery models [9]. Li-ion battery (LIB) is an essential thing of the present and future technology for electric vehicles, power repositories, conveying, and consumer electronics. Even also, there are many types of LIB that represent the likelihood to catch fire or disclose gases. Even though the probability is uncommon, LIB causes fires with very different hazards to other fire risks in conditions of start-up, spread rate, period, dangerous, and pressure [10].

3 Analysis of EV's Lithium-Ion Battery Heating

Electric vehicle, i.e., an EV is a cut-down composition for an electric vehicle. Electric vehicles are vehicles that are neither partly nor fully electrically powered. Electric vehicles have less unceasing costs as they have lesser displaced things for maintenance and they are most eco-friendly for the environment as they use a smaller number of fossils in the case of hybrid vehicles. Some of the electric vehicles have used various types of batteries like Pb-acid or Ni-metal composite batteries, Li-ion batteries are now supposed the standard of modern batteries in EVs.

4 How Do Lithium-Ion Batteries Operate and What Does It Entail?

Lithium atoms are ionized in the anode during discharge, and subsequent electron separation occurs. Reaching the cathode, where the electrons reassemble and attain electrical neutrality, lithium ions pass first through the anode and subsequently the electrolyte. Owing to their small size, lithium ions can move between the anode and cathode through a micro-permeable separator. With regard to charge storage per unit mass and volume, Li-ion batteries boast exceptionally high voltages. Their electrodes can be made of a variety of materials. Lithium cobalt oxide (cathode) and graphite (anode) are therefore the most commonly used combinations in devices like laptops and mobile phones. Two components, a cathode and an anode, mostly composed of graphite.

5 Charge

The design of a lithium-ion battery is made such that it receives and stores certain amount of data in it. When these LIBs are overcharged and its limits are reached, performance of the cells is affected or sometime even damaged. The term SOC that is state of charge is used to determine the charging level. Most normally 0–100% are the operational limit set which demonstrates that a battery is deemed fully charged when it reaches 100% SOC. Although the battery capacity can go beyond it defines limit. Use of wrong charger or breakdown of charger can cause overcharging. Sometimes the anode apparatus becomes overly lithiated, while overcharging. This stops lithium inter-balance as well as the accumulation of metal lithium on the anode. And these deposits grow in metal fingers called dendrites.

6 Discharge

The discharging of LIBs causes the flow of lithium ions to move from the negative current collector as well as from the anode to the cathode and positive current collector. If discharge levels grow, they can dissolve as well as negative current collectors and those containing copper. Consequently, the electrolyte is exposed to minute conductive copper particles, increasing the possibility of an internal short circuit. Moreover, this may cause cell plating and cathode venting in addition to the evolution of oxygen and hydrogen.

7 Methodology

A technique for examining the factors that raise the temperature of the Li-ion battery in electric vehicles is offered in an effort to identify the primary factors that contribute to battery thermal expansion, which raises the risk of a fire or explosion. After conducting research on the heating issues relating to lithium-ion batteries, we have determined that the main focus of this study is on three external and internal parameters: charging, discharging, and ambient temperature, or outside temperature. This system processes the lithium-ion battery data based on variables such as the battery's initial temperature, ambient temperature, charging, and discharging. Given that the

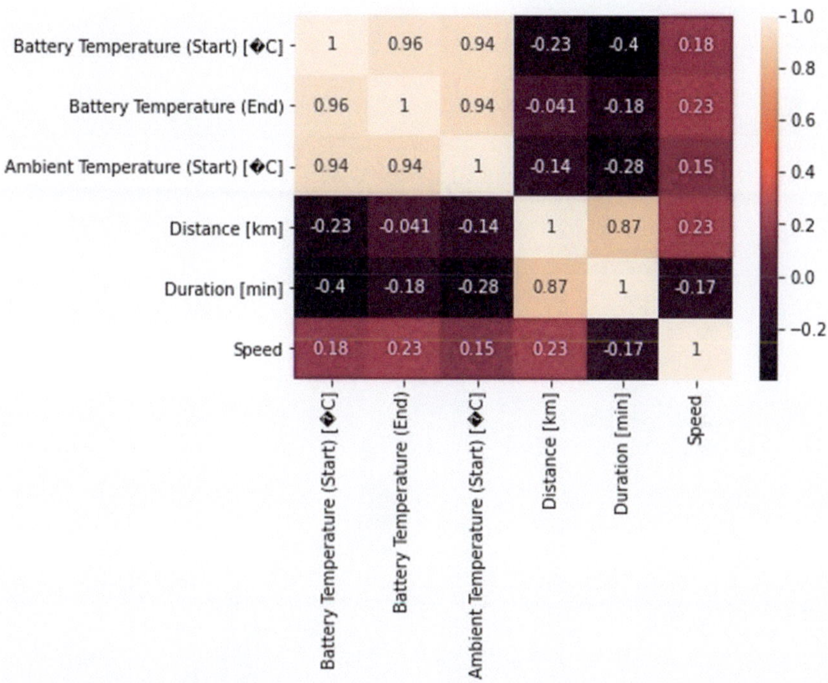

Fig. 1 Relationship graph of variables

moving vehicle dataset was utilized, we also have variables for speed, distance traveled, battery, the temperature at start and stop weather, and other time-related variables. EDA is a machine learning concept that is used to analyze data and identify patterns and trends.

8 Analysis Result

We did the EDA which is a machine learning concept which analyzes the given huge data and from that data it will find out the trends, patterns in the different variables of the given dataset for the statistical summary. So, we analyze each and every variable in the dataset w.r.t. each other to find out the patterns between them (Figs. 1, 2, 3, 4, 5, 6, 7, 8, and 9).

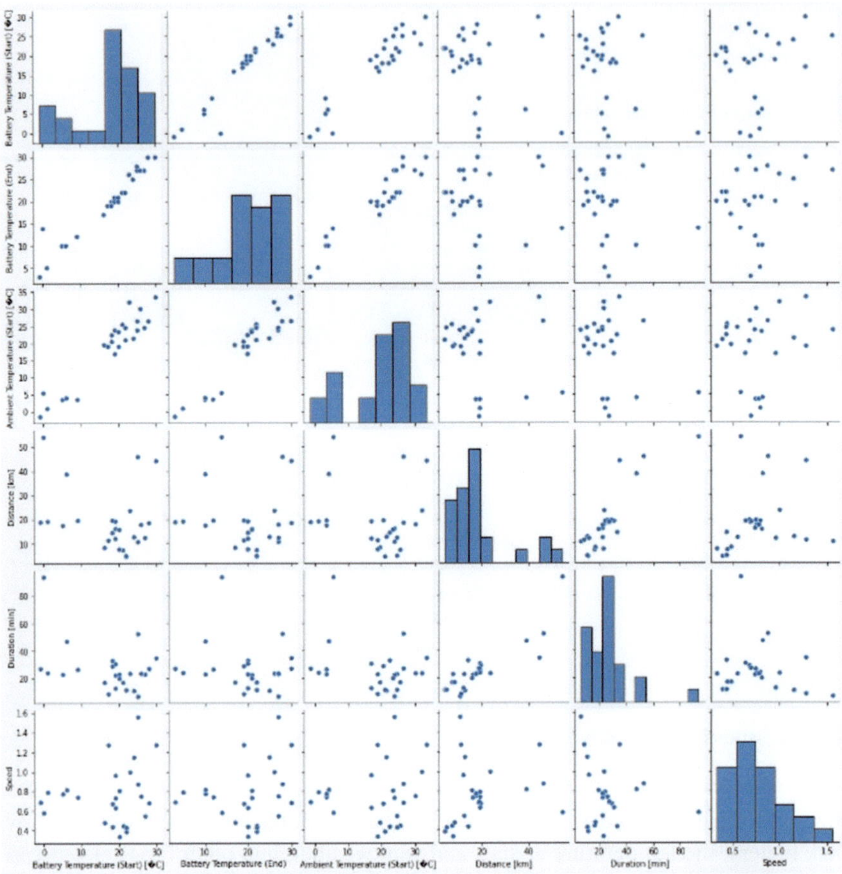

Fig. 2 Matrix correlation for every variable

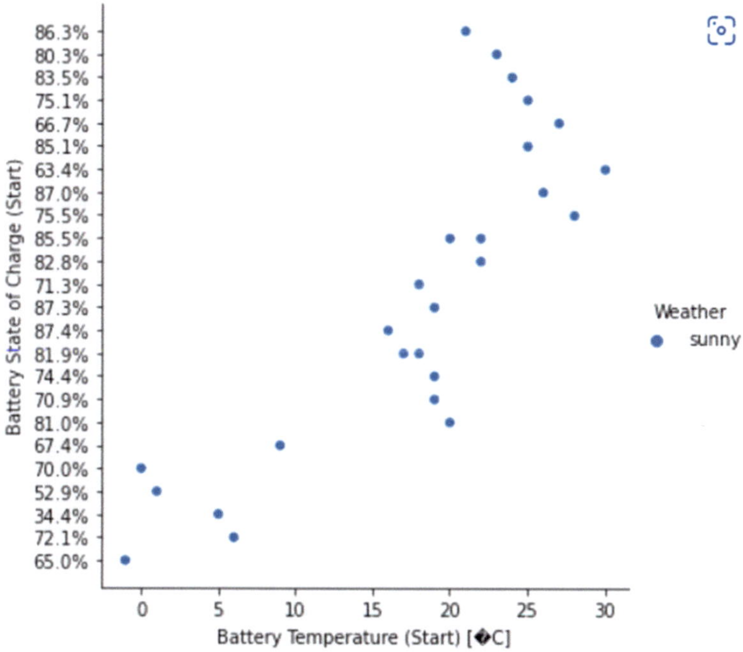

Fig. 3 Start the battery charge according to battery temperature

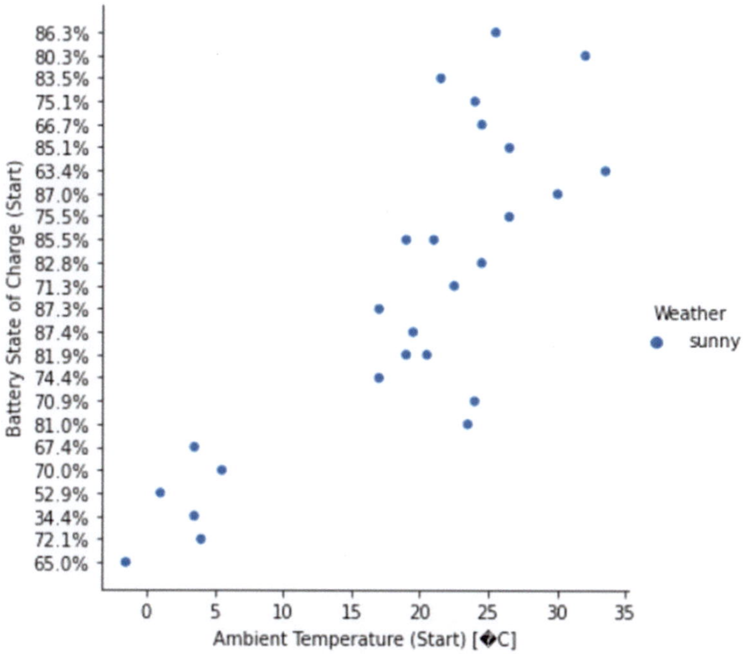

Fig. 4 Battery start charge with regard to ambient temperature

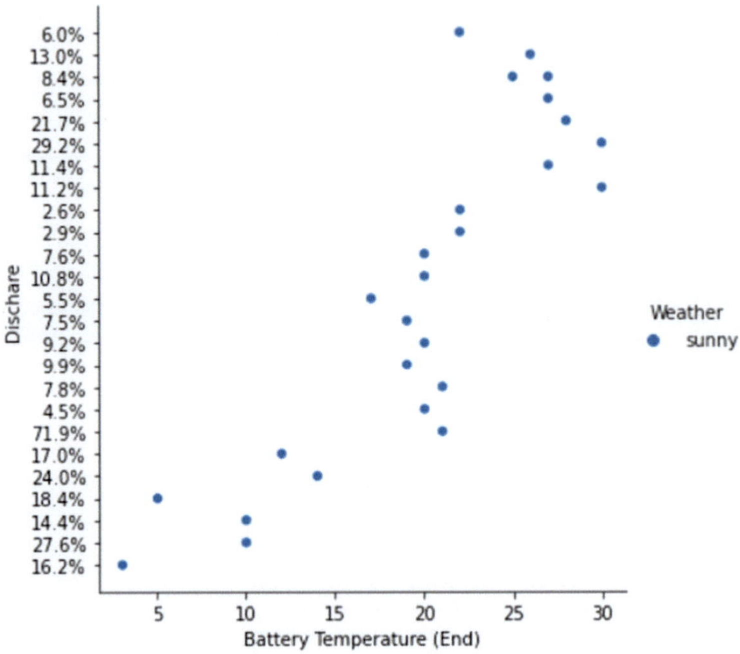

Fig. 5 Discharge in relation to battery temperature

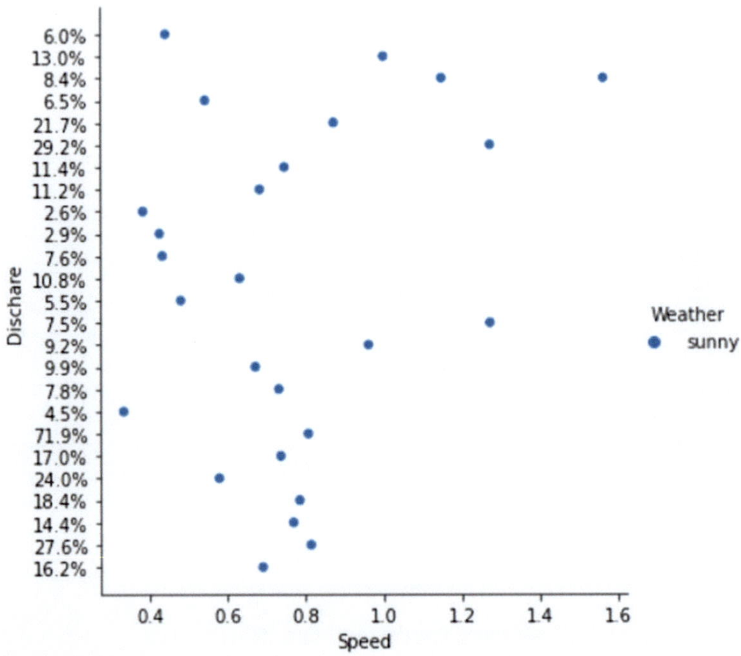

Fig. 6 Discharging according to speed

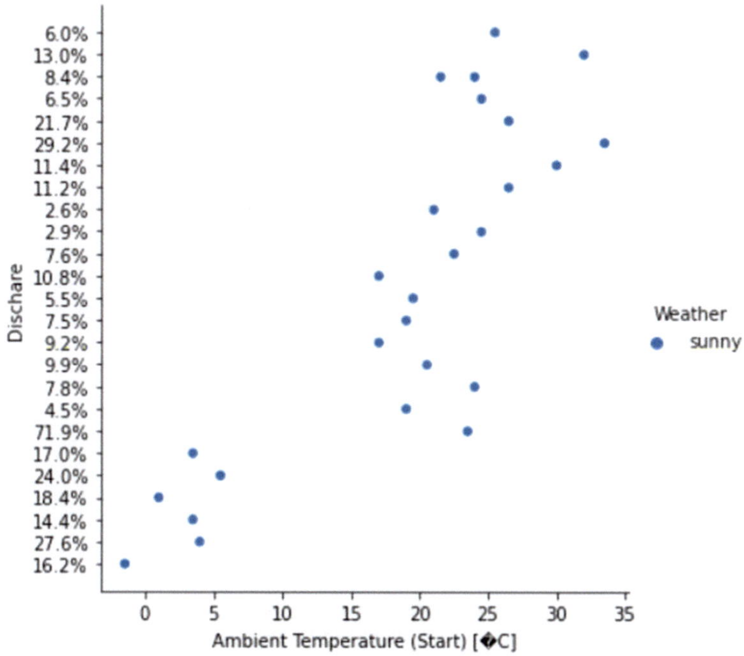

Fig. 7 Discharge in relation to ambient heat

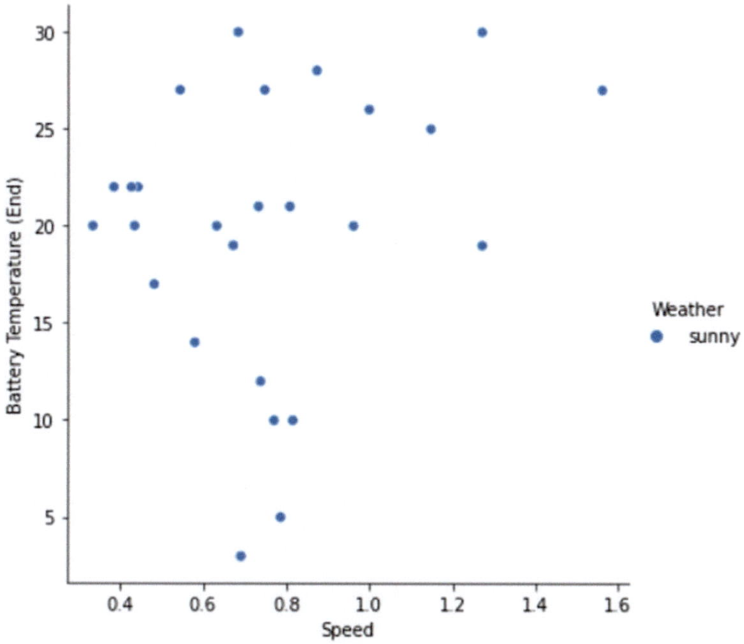

Fig. 8 Battery temperature w.r.t. speed

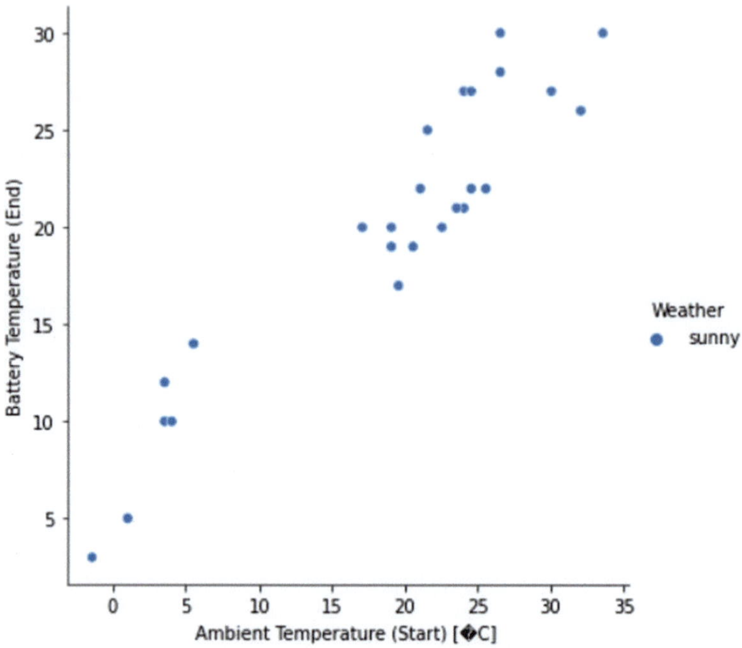

Fig. 9 Battery temperature end in relation to ambient temperature

9 Conclusion

This paper effectively studies the factors that influence the heating problems in Li-ion batteries as well as the analysis of those problems. We were able to effectively examine how heat is produced in the lithium-ion battery while it is being charged and discharged and record the various temperature fluctuations. In order to examine the temperature variation in batteries, we examined exploratory data analysis, or EDA, techniques in this paper. Furthermore, by taking real-time readings of the battery's charging state temperature, ambient temperature, and initial state of charge—and then verifying the battery's state of charge—we were also able to successfully create the charging dataset. We obtain the accurate heat generation analysis output using the reading that we computed in both the real-time and unprocessed datasets. Thus, we deduce that when battery temperature rises, a number of internal factors also contribute to the increase in heat generated by the battery in addition to the ambient temperature outside.

References

1. Tomaszewska A, Chu Z, Feng X, O'Kane S, Liu X, Chen J, Ji C, Endler E, Li R, Liu L, Li Y, Zheng S, Vetterlein S, Gao M, Du J, Parkes M, Ouyang M, Marinescu M, Offer G, Wu B (2019) Lithium-ion battery fast charging: a review. Elsevier B.V.
2. Perrone D, Di Stefano S (1997) Survey of lithium-ion battery performance for potential use in NASA missions. In: Intersociety conference on energy conversion engineering. IEEE
3. Cui Y, Cong B, Liu J, Qiu M, Han X (2022) Characteristics and hazards of plug-in hybrid electric vehicle fires caused by lithium-ion battery packs with thermal runaway. Frontiers
4. Scarfogliero M, Carmeli S, Castelli-Dezza F, Mauri M, Rossi M, Marchegiani G, Rovelli E (2018) Lithium-ion batteries for electric vehicles: a review on aging models for vehicle-to-grid services. In: 2018 International conference of electrical and electronic technologies for automotive. IEEE
5. Zhao R, Zhang S, Gu J, Liu J (2016) Safety study of three types of lithium-ion batteries. In: 2016 IEEE electrical power and energy conference (EPEC)
6. Bisschop R, Willstrand O, Rosengren M (2020) Handling lithium-ion batteries in electric vehicles: preventing and recovering from hazardous events. https://doi.org/10.1007/s10694-020-01038-1
7. Liu G, Ouyang M, Lu L, Li J, Han X (2014) Analysis of the heat generation of lithium-ion battery during charging and discharging considering different influencing factors. J Therm Anal Calorim
8. Hatzell KB, Sharma A, Fathy HK (2012) A survey of long-term health modeling, estimation, and control of lithium-ion batteries: challenges and opportunities. In: 2012 American control conference (ACC). IEEE
9. Sun P, Bisschop R, Niu H, Huang X (2020) A review of battery fires in electric vehicles. Fire Technol
10. Diaz LB, He X, Hu Z, Restuccia F, Marinescu M, Barreras JV, Patel Y, Offer G, Rein G (2020) Review—meta-review of fire safety of lithium-ion batteries: industry challenges and research contributions. J Electrochem Soc

Investigation of Handwritten Cancer Notes Classification Using Transformer-Based Deep Neural Network Pipeline

Tiyasa Chakraborty, Ashok Kumar Bhadra, and Debashis Nandi

Abstract Analysis of clinical notes is a popular as well as challenging field in Natural language processing, especially in the healthcare domain. The handwritten text dataset is an image dataset that further needs to be processed. Hence, primarily we need to extract the textual part from the images properly. Identifying the specific disease condition from clinical notes thus can be elaborated as a junction of image processing and natural language processing. Prediction of cancer disease conditions from clinical notes can prevent doctors from overlooking possible risks. Doctors can advise patients about taking hospital admission. Also, the patient can plan for further treatment accordingly. According to domain experts, after providing the treatment, the cancer condition can be classified into four different categories. Those are complete response, progressive disease, partial response, and stable disease. In our proposed approach, we need to start with an Optical Character Recognition (OCR) engine to extract proper text information from handwritten notes. Further, the NLP process was executed. Finally, a deep neural network is used to classify different cancer disease conditions. Our approach is compared to some recent implementations, and we achieved more promising results compared to them.

Keywords Optical character recognition · Natural language processing · Clinical notes

T. Chakraborty (✉) · D. Nandi
Department of Computer Science and Engineering, National Institute of Technology Durgapur, Durgapur, India
e-mail: tiyasachakraborty@gmail.com

A. K. Bhadra
Medical College and Hospital, Kolkata, India

S. K. Biswas et al. (eds.), *Intelligent Computing Systems and Applications*, Lecture Notes in Networks and Systems 1307, https://doi.org/10.1007/978-981-96-3860-4_18

1 Introduction

Classification of the clinical note is one of the important pieces of work that can be solved through several machine learning approaches [1, 2]. This work comes under the area of Natural Language Processing (NLP) [3]. The main intention is to accumulate an appropriate pipeline so that we can extract the text portion properly, and further those texts can be applied to the model for classification.

A deep neural network is an advanced version of artificial neural networks, and this concept is becoming very popular nowadays [4]. It is useful to populate the features automatically from a given dataset. Thus, after the training, the model becomes capable of predicting as per the requirement [5]. This promising technique is currently being applied in several fields, including speech processing, image/video analysis, or prediction, and also in natural language processing [6, 7]. In conventional neural networks, the model contains several neurons, i.e., computational units, that involve the concept of activation to decide which neuron should be activated next. By using backpropagation, the model learns continuously and finally tries to reach an optimal solution based on the gradient descent technique [8].

Handwritten clinical notes classification study is greatly influenced by deep neural architecture as it requires less knowledge of feature engineering [9]. The deep neural network model involves a multi-layered architecture. The response from previous layers propagates through the next layers sequentially. These models can take raw input samples and automatically create prominent features through the internal computation, and finally map the created feature set to the output label corresponding to that input sample [10]. Thus, the model also automatically learns the pattern matching between input and output values. The accuracy of other conventional machine learning methods is solely dependent on domain expert knowledge to create a proper schema of features [11, 12], however, deep learning only takes the dataset into account. Thus, deep neural architecture models can reduce human effort as well, and they can also give improved results.

In the clinical notes use case, the notes are mainly taken by the nurses or care personnel, or any medical experts. Thus, different people with different kinds of handwriting can confuse the model to understand the text within it [13]. Thus, a relevant OCR technique followed by an appropriate cleaning process and data preparation can support the model to perform more accurately. In this paper, we have started with OCR extraction from the handwritten notes related to cancer conditions, followed by text cleaning and vectorization. Finally, the text vectors are fed into the deep neural net model for training. Our proposed approach is successfully able to classify different cancer conditions.

2 Related Literature

People used various machine learning and statistical approaches specifically for note classification, among them Naïve Bayes (NB), and Support Vector Machines (SVMs) based on the rule-based approach got much popularity [14]. SVM also merged with NB to improve the result [15]. In the last few years, deep learning architectures have been providing promising results in several subdomains of NLP, such as note classification [16]. But choosing the appropriate model among those is always a challenging task. Different models react in different ways based on the dataset. Several such models are available, like a simple deep neural network(DNN), a convolutional neural network (CNN), a recurrent neural network (RNN), etc. Each type of that model has having own complexity. Thus, if the dataset follows some level of similarity, we can strengthen our pre-processing and choose a simpler model.

Recently, CNN embedded with LSTM has shown high-level results in some aspects. On the other hand, the simple deep belief network model can also perform better in some use cases related to text classification. The deep belief network is capable of absorbing the features appropriately using hidden units and understanding the data in a better way. Here, each hidden layer captures the statistical connection among the neurons. The higher layer can produce a more complex feature map. In the case of the RNN model, the proper sequence needs to be chosen for language modeling. It also demands flexibility for the sequential modeling to understand the context of the dataset. Thus, not only in note classification, but RNN is also popular in other NLP tasks as well, like sentiment classification, spam data filtering, website context classification, etc. [17]. But at the same time, RNN includes exploding and vanishing gradient problems as it sometimes becomes unstable while incorporating the rectified linear unit (ReLU). LSTM sometimes overcomes the above issue, but limits the performance from one dataset to another. The batch size and amount of hidden layers also make big differences in outputs. Various deep learning models have been applied in the different areas of Natural Language Processing, like text classification and language modeling.

In the healthcare domain, NLP also has a deep impact nowadays. It can be used to extract information from prescriptions or clinical notes [18]. It is also useful for text summarizing and extracting meaningful information from text. Morbidity analysis from the clinical notes is another use case in this domain where both classical machine learning as well as deep learning models were applied. Unstructured clinical notes analysis is another challenging task where research is still ongoing [18]. Thus, it can be observed that machine learning and deep learning have been supporting and providing solutions for NLP-related challenges for many years.

In our approach, we focused on strengthening the initial data preparation, as it is very important in this study. In clinical notes analysis limited dataset is a major issue, especially to applying a deep neural network model. Thus, initial processing of the dataset is important, which involves enhancement as well as augmentation of the note images. It helps to balance the class in a multi-class classification problem. We can observe that proper pre-processing and then applying the modified transformer-based deep neural network model gives more promising results compared to others.

3 Proposed Work

In this study, we have worked on a full pipeline for NLP. We have started with some image processing work, like increasing brightness, adjusting contrast, cropping, and enlarging the text area from images. It is described in Algorithm 1.

Algorithm 1: Image data preparation

Initialize $contrast_{low} = -127$
Initialize $contrast_{high} = 127$
Initialize $brightness_{low} = -255$
Initialize $brightness_{high} = 255$
for *all images in input directory* **do**
    Show the image
    Take user input for contrast and apply to the image
    Take user input for brightness and apply to the image
    Take user input for top-left and bottom-right corner to crop the image
    Resize all image in same size
    Save images in output directory maintaining naming convention for proper mapping with classification level
end

Then images are fed into the Enhanced- TrOCR engine to extract text from handwritten images. TrOCR (Transformer-based Optical Character Recognition) is a complete solution for text identification. It was proposed by Li et. al. in 2021, and this engine is publicly available for use [19]. This solution was built based on pre-trained Computer vision and Natural language processing models. Thus, this architecture is useful for both image understanding and word piece-level text generation. It starts with resizing the input text images into 384 x 384, and then it breaks into a sequence of 16 patches that are the input to image Transformers. But, in our case, we modified the structure by resizing the image to 512 x 512 after cropping as discussed in the previous algorithm. Thus, instead of passing 16 patches, we are passing 32 patch sequences, thus, we are enhancing the OCR engine by providing models with more granular information.

Conventional transformer architecture embedded with a self-attention technique consists of both encoder and decoder units, and finally, word pieces are identified from the input images.

The extracted text the cleaned and vectorized. After that, vectorized text is propagated up to the model to give a classification result. The workflow for the same is shown in Fig. 1.

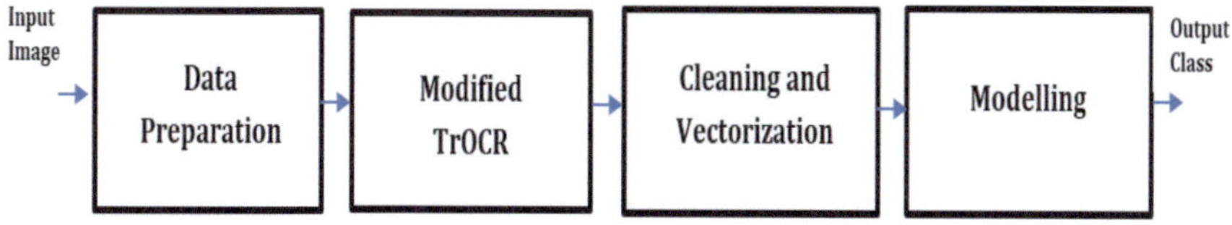

Fig. 1 Block diagram of working methodology

4 Experiments and Analysis of Performance

4.1 Description of Notes-Image Dataset

In this study, we have incorporated clinical notes of cancer patients that have been taken by medical experts of Medical College, Kolkata, India. Some samples are presented in Fig. 2.

4.2 Experimental Results

We started our study with 121 cancer clinical notes. Those include both lung and abdominal cancer notes and images, respectively. We consolidated 72 lung cancer notes and 49 liver cancer note images. After data augmentation, we achieved 726 images. The result of our OCR approach is given as follows in Fig. 3. As we can observe in Fig. 3, OCR followed by initial cleanup can identify the words correctly in

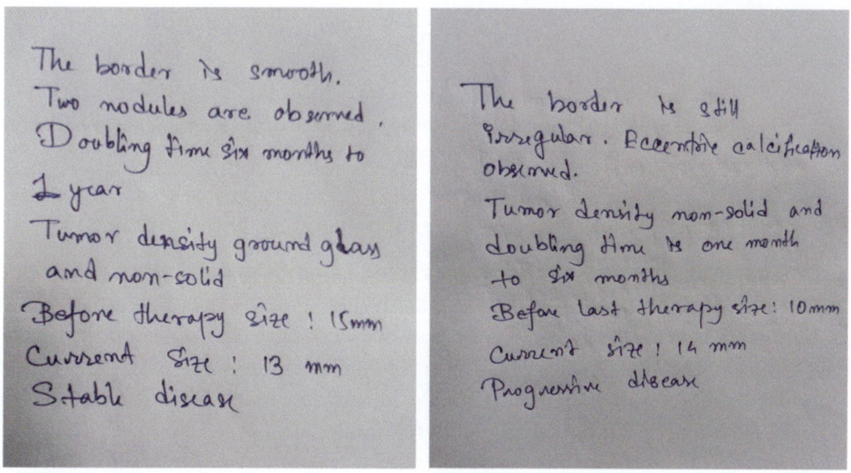

Fig. 2 Example cancer note images

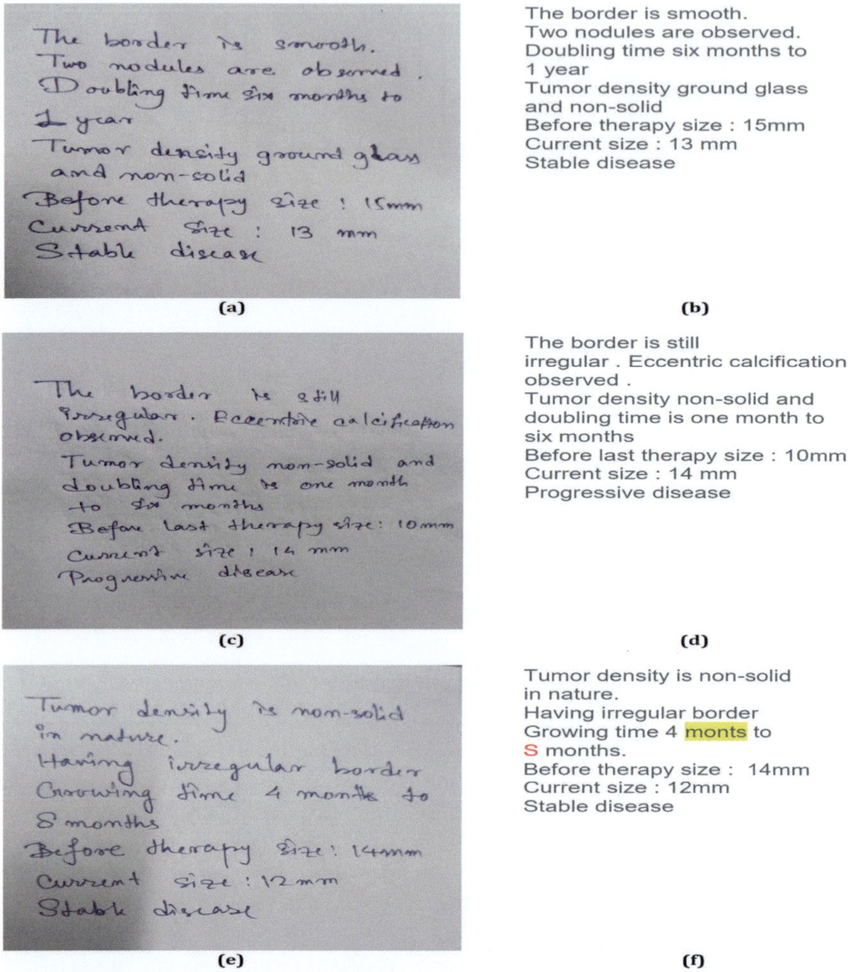

Fig. 3 Notes-images and respective extracted cleaned OCR text outputs. **a** Sample note 1 **b** Cleaned OCR text of sample note 1 **c** Sample note 2 **d** Cleaned OCR text of sample note 2 **e** Sample note 3 **f** Cleaned OCR text of sample note 3

most of the cases, as we can see, in samples note 1 and 2 under Fig. 3a, b, respectively. But in sample note 3, under Fig. 3c, it is identifying "months" word incorrectly, and the number 8 is recognized as "S". Also, it does not have an impact on the grammar. The cleaning process involves the medical context as well. After getting the output from OCR, it is executed through the TensorFlow tokenizer and vectorization engines. Then finally fed into a deep network model.

Deep Neural Network Model Architecture: Our model has 8 fully connected layers. Each layer has 16 hidden neurons, and the last one has 4 neurons. All layers except the last one include the ReLU activation function, and the last layer has a Sigmoid activation function.

Model Parameters: In model compilation, the optimizer we used is Mini-batch gradient descent with batch size 512. The loss function is sparse categorical cross-entropy.

At last, our method is compared with other recent implementations as well. In the next section, the comparison result is presented.

4.3 *Performance Analysis*

Several performance metrics are available for evaluating classification results. In our study, we have used Dice Coefficient and Accuracy. We compared our proposed method with three other recent methods, in which it is visible that our approach is providing very good performance compared to others. The comparison results are shown in Table 1 as well as in Fig. 4. As we know, higher values for Accuracy and

Table 1 Performance analysis for Comparison of disease status

Category	Metrics mean normalized value	Different methods			
		Deep learning by fine-tuning BERT	LSTM with dense layer	TrOCR with dense layer	Proposed approach
		Zhang et al. [20] **2019**	**Kumar et al.** [21] **2020**	**Li et al.** [19] **2021**	**2022**
Complete response	Dice coefficient	0.9332	0.9323	0.9512	**0.9821**
	Normalized elapsed time	0.78	0.51	0.38	**0.26**
	Accuracy	0.9504	0.9245	0.9437	**0.9815**
Partial response	Dice coefficient	0.9376	0.9298	0.9387	**0.9834**
	Normalized elapsed time	0.62	0.57	0.41	**0.21**
	Accuracy	0.9518	0.9176	0.9291	**0.9811**
Progressive disease	Dice coefficient	0.9159	0.9225	0.9476	**0.9817**
	Normalized elapsed time	0.58	0.61	0.39	**0.27**
	Accuracy	0.9051	0.9181	0.9451	**0.9871**
Stable disease	Dice coefficient	0.9478	0.9416	0.9566	**0.9789**
	Normalized elapsed time	0.62	0.59	0.46	**0.23**
	Accuracy	0.9369	0.9356	0.9378	**0.9863**

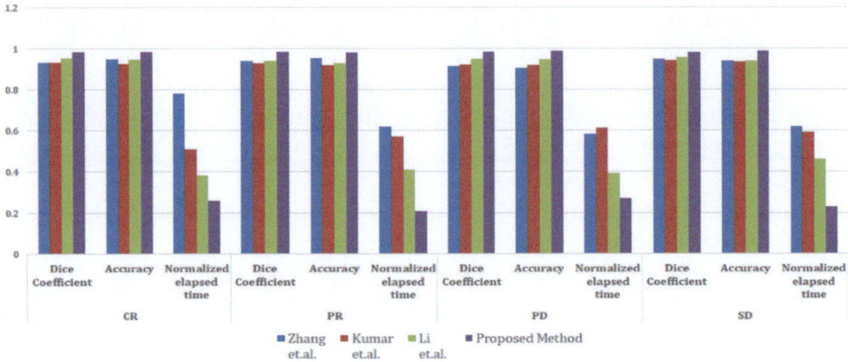

Fig. 4 Analysis of performance of disease status comparison

Dice Coefficient are preferable. It can be observed that our approach is running in less time, and the lower values for Normalized elapsed time denote better performance.

5 Conclusion

In our study, a new investigation strategy is presented for handwritten cancer note classification corresponding to different conditions of disease after giving certain treatment. The approach starts with initial image data preparation, followed by the OCR model. Then OCR output was cleaned and further vectorized to feed into our deep neural net architecture. Here we used both Lung and abdominal cancer handwritten notes and images. We are able to classify different cancer disease conditions successfully. Further, we have analyzed the performance by comparing it with recent implementations. According to the observation, our proposed algorithm outperforms others by giving better performance.

Acknowledgements The authors are grateful to Medical College and Hospital, Kolkata, for providing the major dataset along with some valuable feedback. This study is supported by the Ministry of Electronics and Information Technology, Government of India.

References

1. Dong X, Qian L, Guan Y, Huang L, Yu Q, Yang J (2016) A multiclass classification method based on deep learning for named entity recognition in electronic medical records. In: 2016 New York scientific data summit (NYSDS). IEEE, pp 1–10
2. Sheikhalishahi S, Miotto R, Dudley JT, Lavelli A, Rinaldi F, Osmani V et al (2019) Natural language processing of clinical notes on chronic diseases: systematic review. JMIR Med Inf 7(2):e12239

3. Weng WH, Wagholikar KB, McCray AT, Szolovits P, Chueh HC (2017) Medical subdomain classification of clinical notes using a machine learning-based natural language processing approach. BMC Med Inf Decis Making 17(1):1–13
4. Pouyanfar S, Sadiq S, Yan Y, Tian H, Tao Y, Reyes MP, Shyu ML, Chen SC, Iyengar SS (2018) A survey on deep learning: Algorithms, techniques, and applications. ACM Comput Surv (CSUR) 51(5):1–36
5. Torfi A, Shirvani RA, Keneshloo Y, Tavaf N, Fox EA (2020) Natural language processing advancements by deep learning: A survey. arXiv preprint arXiv:2003.01200
6. Dong S, Wang P, Abbas K (2021) A survey on deep learning and its applications. Comput Sci Rev 40:100379
7. Dargan S, Kumar M, Ayyagari MR, Kumar G (2020) A survey of deep learning and its applications: a new paradigm to machine learning. Arch Comput Methods Eng 27(4):1071–1092
8. Abiodun OI, Jantan A, Omolara AE, Dada KV, Mohamed NA, Arshad H (2018) State-of-the-art in artificial neural network applications: A survey. Heliyon 4(11):e00938
9. Vaidya R, Trivedi D, Satra S, Pimpale M (2018) Handwritten character recognition using deep-learning. In: 2018 second international conference on inventive communication and computational technologies (ICICCT). IEEE, pp 772–775
10. Liu X, Song L, Liu S, Zhang Y (2021) A review of deep-learning-based medical image segmentation methods. Sustainability 13(3):1224
11. Li Y, Lipsky Gorman S, Elhadad N (2010) Section classification in clinical notes using supervised hidden markov model. In: Proceedings of the 1st ACM international health informatics symposium. pp 744–750
12. Chakraborty T, Banik SK, Bhadra AK, Nandi D (2021) Dynamically learned pso based neighborhood influenced fuzzy c-means for pre-treatment and post-treatment organ segmentation from ct images. Comput Methods Programs Biomed 202:105971
13. Luo Y (2017) Recurrent neural networks for classifying relations in clinical notes. J Biomed Inf 72:85–95
14. Silva J, Coheur L, Mendes AC, Wichert A (2011) From symbolic to sub-symbolic information in question classification. Artif Intell Rev 35(2):137–154
15. Zulqarnain M, Ghazali R, Hassim YMM, Rehan M (2020) A comparative review on deep learning models for text classification. Indones J Electr Eng Comput Sci 19(1):325–335
16. Zhang T, Schoene AM, Ananiadou S (2021) Automatic identification of suicide notes with a transformer-based deep learning model. Inter Intervent 25:100422
17. Vu NT, Adel H, Gupta P, Schütze H (2016) Combining recurrent and convolutional neural networks for relation classification. arXiv preprint arXiv:1605.07333
18. Mahbub M, Srinivasan S, Danciu I, Peluso A, Begoli E, Tamang S, Peterson GD (2022) Unstructured clinical notes within the 24 hours since admission predict short, mid & long-term mortality in adult icu patients. Plos One 17(1):e0262182
19. Li M, Lv T, Cui L, Lu Y, Florencio D, Zhang C, Li Z, Wei F (2021) Trocr: Transformer-based optical character recognition with pre-trained models. arXiv preprint arXiv:2109.10282
20. Zhang X, Zhang Y, Zhang Q, Ren Y, Qiu T, Ma J, Sun Q (2019) Extracting comprehensive clinical information for breast cancer using deep learning methods. Int J Med Inf 132:103985
21. Kumar V, Recupero DR, Riboni D, Helaoui R (2020) Ensembling classical machine learning and deep learning approaches for morbidity identification from clinical notes. IEEE Access 9:7107–7126

An Expert System for Student Career Prediction (ESSCP) Using Machine Learning Classifiers

Arijit Bhattacharya, Dipankar Roy, Joydip Roy, Saroj Kr. Biswas, Ardhendu Mandal, Akhil Kumar Das, Debasmita Saha, and Ekram Alam

Abstract Career selection is a key concern for all students to decide a suitable field of study after completion of high school. Students sometimes feel hesitant to make the right choice when it comes to their specialization because it directly impacts their future careers. Knowing the student's interests and academic background will help in selecting their job role properly. In this research, we suggested an Expert System for Student Career Prediction (ESSCP) model based on machine learning techniques to suggest the job roles they should pursue in their career. In this study, five algorithms—two standalone classifiers and three ensemble classifiers were used. The dataset utilized for this work was collected from "Github," which predominantly contains null values. Data imputation has been done using K-nearest neighbor and Multivariate Imputation using Chained Equations. According to the findings, XGBoost performed superior than all the other classifiers in terms of accuracy and kappa

A. Bhattacharya · D. Roy (✉) · J. Roy · A. K. Das · E. Alam
Gour Mahavidyalaya, Malda, India
e-mail: dipankarroy7662@gmail.com

A. Bhattacharya
e-mail: barijit@hotmail.com

J. Roy
e-mail: royjoydeep820@gmail.com

A. K. Das
e-mail: dasakhi@gmail.com

E. Alam
e-mail: ealam4u@gmail.com

S. Kr. Biswas
NIT, Silchar, India
e-mail: saroj@cse.nits.ac.in

A. Mandal
University of North Bengal, Darjeeling, India
e-mail: am.csa.nbu@gmail.com

D. Saha
University of Gour Banga, Malda, India
e-mail: debasmita.saha@hotmail.com

S. K. Biswas et al. (eds.), *Intelligent Computing Systems and Applications*, Lecture Notes in Networks and Systems 1307, https://doi.org/10.1007/978-981-96-3860-4_19

score. Tenfold cross-validation was engaged in conjunction with the 70:30 train-test split to validate the performance. The greatest accuracy for MICE with a 70:30 split was 92.31%, while it reached 92.07% for tenfold cross-validation. Although the findings are extremely encouraging, testing with more real-world datasets will correctly evaluate the model's performance.

Keywords Student career prediction · Null value imputation · Ensemble learning · Bagging · Boosting

1 Introduction

Population increase is a significant challenge for countries such as India and China. The growth in population leads to a significant surge in competitiveness within the market. In today's technologically advanced society, it is especially challenging. Students must be organized and systematic from the outset of their education to effectively compete and accomplish their goals. It is essential to regularly analyze their performance, identify their interests, evaluate their progress toward their goals, and ensure they are heading in the right direction. Any structured method helps people evaluate themselves before they achieve their career peak and self-improve. It also helps them self-motivate for a better professional route if their talents fall short of their goals.

Every year, thousands of students graduate from college and enter the industry, but many struggle to determine their career paths. Many students choose their stream after graduation due to a lack of information or self-interest, making their life challenging. Because there are so many career options and job competitions today, selecting the right occupation has become a difficult task. Researchers have even advocated using machine learning categorization technologies to help people find their ideal career choice. Without competent professional assistance, students struggle to select a suitable job. They frequently mismatch their job direction based on their personality, skills, and interests. Because their families want them to make a good living, students are even forced to pick a job in a different field that they don't really want. In certain circumstances, students who have graduated and begun working for multinational companies become dissatisfied with their jobs over time due to a lack of interest or expertise. Because of this, the next generation has started to choose the streams that appeal to them [1].

Machine learning is a technique in which machines are precisely designed to respond to a certain input or scenario using knowledge they have previously learnt. Giving computers the ability to learn using statistical approaches is all that is required. Machine learning has enabled computers to behave spontaneously without the need for explicit training [2]. This tries to reduce human involvement in situations and concerns that machines can reliably tackle. It facilitates the resolution of complex tasks and problems with minimal human intervention. Today's real-world applications for machine learning include natural language processing, picture recognition, medical diagnosis, self-driving vehicles, and many others.

The dataset is the most important component for any supervised learning task. The dataset serves as the basis for training and testing. Data pre-processing and feature selection are a couple of procedures that must be carefully considered before training. Another critical aspect of every dataset is the presence of missing values. The learning model may have a large negative impact on missing values, which can be severe. As a result, the model's performance may decline, and its validity may be challenged. To address this issue, rows with missing values are frequently eliminated from the dataset. Univariate analysis, such as imputation with the mean value, is also common in this sort of research. However, others may raise concerns about its appropriateness. In the current study, we used a dataset that contained null values in the majority of cases. For data imputation, we used two methodologies to compare acceptability. In the first scenario, we used a machine learning approach called k-nearest neighbor for data imputation. For the second trial, we've employed Multivariate Imputation by Chained Equations (MICE) [3].

The principles of machine learning are utilized in this study to determine the optimum career route for an engineering student. This forecast is critical for educational institutions and recruiters across the board. Companies that offer jobs also spend a lot of money to identify the appropriate candidate for every available position. The model is created via machine learning approaches such as Decision Tree Classifier (DTC) [4], Support Vector Machine (SVM) [5], XGBoost [6, 7], AdaBoost [8], and Random Forest Classifier (RFC) [9]. The paper has been laid out as follows. In Sect. 2, a literature review with the most recent study was shown. The several approaches used in this study have been summarized in Sect. 3. In Sect. 4, the experimental setup and dataset description have been included. Section 5 covers the findings and remarks. Following that, the conclusion was laid out in Sect. 6.

2 Related Works

In the life of every student, the choice of a profession is a significant decision-making aspect. Traditionally, expert practitioners used interviews to discover the key characteristics influencing career choices. However, predicting an occupation seems challenging due to the complexity of each student's goals and desires. A recent study suggests using student data based on their behavioral characteristics to estimate their career route. ACCBOX, which stands for Approach Cluster Centers Based on XGBOOST, is a novel model that was suggested by Nie et al. [10] to facilitate the prediction of students' profession choices. The final result revealed that the current strategy outperformed other predictive methods. This model was built using 13 behavioral data points obtained from 4000 pupils. For the purpose of developing a model for predicting student careers, Roy et al. [1] utilized various machine learning techniques, including the Support Vector Machine, Decision Tree, and XGBoost. With an accuracy rate of 90.3%, the Support Vector Machine method showed superior performance compared to the other algorithms.

An artificial neural network-based approach was suggested by Baashar et al. [11] in order to perform academic performance prediction on students. Using the student's demographics, academic history, and residence location, the researchers in this study developed a model to predict the student's grade point average. In a similar vein, Pushpa et al. [12] have forecasted students' grades using several ML models. They contrasted various machine learning techniques such as Gradient Boosting, Random Forest, Support Vector Machine, and Naïve Bayes. Random Forest was more accurate than other Machine Learning Techniques, according to their study, at 84.3%. In the field of education, Kaur et al. [13] also used machine learning methods to forecast students who would be slow learners. In order to achieve this goal, approaches such as Multilayer Perceptron and Naïve Bayes were used. It was noted that the former produced more precise results than the latter.

3 Different Machine Learning Classifiers

3.1 *Decision Tree (DT)*

A DT is a representation of a structure that is generated with the help of a dataset and consists of nodes and edges. Every node serves a dual purpose: as a decision-making tool ("Decision Node") or as a result-reflecting node ("Leaf Node"). The decision-making process is symbolized by the root and intermediate nodes, while the outcomes are depicted by the leaf nodes. In order to construct a decision tree, ID3 employs a top-down greedy technique that was initially developed by Ross Quinlan [4]. The most well-known attribute is sought after by means of Information Gain (IG), or just Gain. IG calculates entropy decrease and evaluates how well a particular characteristic separates or categorizes target classes. Feature with the highest IG is considered the best and is chosen to serve as the root. S is the name of the dataset, and the following is how the entropy is determined (Fig. 1):

$$\text{Entropy (S)} = -\sum \text{P}_\text{i} * \log_2(\text{P}_\text{i});\ \text{[for i ranges from 1 to n]}.$$

In this case, P_i represents the likelihood of class "I" and "n" represents the overall count of categories in the result column.

3.2 *Support Vector Machine (SVM)*

SVM [5, 14] is commonly employed for multiclass classification and is among the most popular supervised learning techniques. The goal of the SVM algorithm is

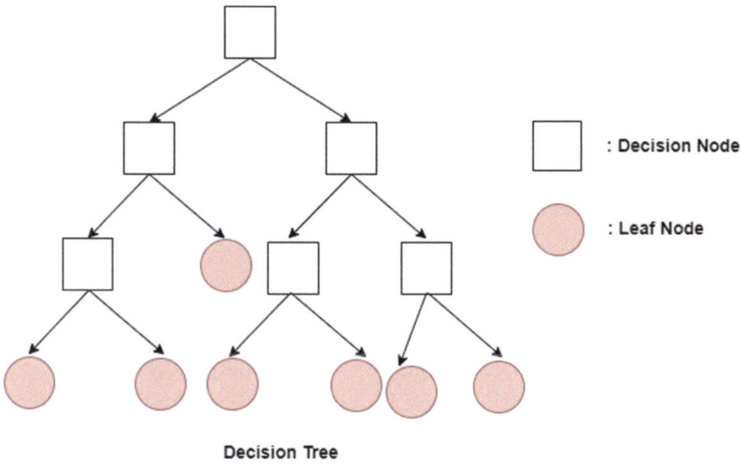

Fig. 1 DT

to find the best line or decision boundary for dividing n-dimensional space into classes, so that we may quickly categorize fresh data points in the future. This ideal decision-making boundary is called a hyperplane. During the process of creating the hyperplane, the SVM will select the extreme vectors and points. These severe cases are called support vectors, and the resulting method is called a Support Vector Machine. A decision boundary, often called a hyperplane, is used to divide two discrete groups, as shown in Fig. 2.

Fig. 2 SVM

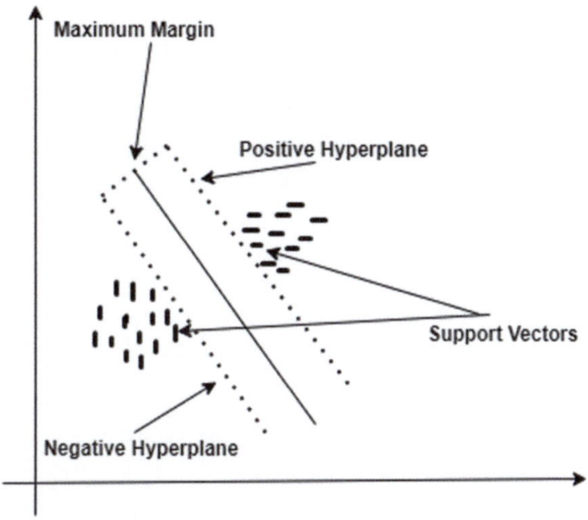

3.3 Ensemble Learning

With the help of the weaker learners in the ensemble, the goal of any classifier is to produce a strong learner. Instead of relying on a single estimator's predictions alone, as in LR, SVM, or DT methods, Ensemble Learning augments generalizability and robustness by combining predictions from multiple base estimators trained with the same learning algorithm. When various models' predictions are combined, the two most popular approaches are:

a. **Bagging**: Bagging, often known as Bootstrap Aggregation, is a technique used to lower DT variance. The objective is to generate distinct data subsets via replacement-based random sampling of the training samples. For each data collection, separate decision trees are constructed and trained simultaneously. The decision trees produce more trustworthy results than a single DT could on its own because they either use majority voting or an average of all potential outcomes. When it comes to bagging techniques, RF is an example.

   - **Random Forest**: An ensemble supervised learning algorithm called Random Forest (RF) [15] builds a group of decision trees using the bagging method. The variance of the method tends to be reduced with the use of bagging. Once all of the Decision Trees have been executed, RF will forecast the most frequently occurring class. This process is done for randomly selected data sets. It aids in overcoming the over-fitting problem in Decision Tree classification, handles a large number of input variables with ease, and gives excellent accuracy. It can also assess the importance of variables for classification.

b. **Boosting**: It all starts with the idea of correcting the mistakes made by one learner by passing them on to the next. For boosting, weak learners are used, who do just a little better than random chance. It works by progressively layering these weak learners and removing the observations that a learner receives correctly at each segment. As the efficacy of the initial model is evaluated, the then sample is modified. A greater amount of attention is directed toward the incorrect sample, as is predicted by the basic model. Training the subsequent base model is done with the updated base sample. Following the continuation of the method, the base models were finally weighted and combined in order to give the final outcome. When it comes to boosting algorithms, AdaBoost was one of the earliest one's ever developed. To provide a more robust and effective solution, gradient boosting enhanced some of AdaBoost's features.

   - **AdaBoost**: One such ensemble learning algorithm that makes use of boosting is AdaBoost. Increasing the accuracy of any learning mechanism can be accomplished through the use of "boosting." By using a linear aggregation of simple weak classifiers, AdaBoost uses a universal strategy to generate a strong classifier. It can be trained quickly (depending on the learner's weakness) and evaluated quickly (linear-additive). It is versatile, and integration can be done with any of the underachieving learners. However, the model's performance is dependent on the available data that is provided, as well as the weak learner. If weak classifiers are either overly complex and under-fit or over-fit, the system may fail to function.

- **XGBoost**: One method, called eXtreme Gradient Boosting (XGBoost) [6], involves building new models to predict the residuals or errors of previous models, which are then combined to get a final prediction. It incorporates new models with minimal loss by means of a gradient descent technique. In an effort to improve performance and speed, XGBoost implements Gradient Boosted Decision Trees (GBDT).

4 Methodology

To evaluate the model, the experiments are conducted using the Python 3.7.12 version in a Windows 11 system using a 70:30 train-test arrangement. The Career Compute dataset, which contains the data of 7525 students, is used in this paper from github [16]. The dataset has 26 attributes {'sslc', 'hsc', 'cgpa', 'school_type', 'no_of_miniprojects', 'no_of _projects', 'coresub_skill', 'aptitude_skill', 'problemsolving_skill', 'programming_skill', 'abstractthink_skill', 'design_skill', 'first_computer', 'first_program', 'lab_programs', 'ds_coding', 'technology_used', 'sympos_attend', 'sympos_won', 'extracurricular', 'learning_style', 'college_bench', 'clg_teachers_know', 'college_performence', 'college_skills', 'role'} among them two are categorical. Among these 7525 records, only 49 records have null values, whereas the rest 7476 records have null or missing values. To overcome this problem, the proposed model—ESSCP shown in Fig. 3, utilized two null value imputation approaches—k-NN [17] and MICE [3]. Five neighbors were employed while implementing k-nearest neighbors. Distribution of data before data imputation and after imputation using k-NN and MICE has been shown in Figs. 4, 5, and 6, respectively.

After outlier detection and removal, the dataset consists of 6092 examples with eight job roles—Technical Writer: 1418, Web Developer: 1002, Software Developer: 892, Technical Support: 871, Software Tester: 776, Business Analyst: 532, Data Analyst: 331, UI/UX Designer: 270. To rectify the imbalance in the dataset, the Synthetic Minority Over-Sampling Technique—SMOTE [18]—was employed. Following this, the dataset was normalized and made ready for training. Training and testing used 70% of the dataset each, and tenfold cross-validation was applied to both sets of data during training and testing to ensure accurate results. It took six distinct machine learning classifiers to solve this multiclass classification task. Individual classifiers, such as Decision Trees, Support Vector Machines, and Artificial Neural Networks, made up three of them. The other three classifiers were ensemble classifiers like XGBoost, AdaBoost, and Random Forest. The models' performance is assessed by looking at their accuracy and kappa score. The confusion matrix is used to determine the machine learning algorithm's accuracy. To see how well the model's predictions matched up with the actual numbers, a confusion matrix is useful. True Positives (TP), False Negatives (FN), True Negatives (TN), and False Positives

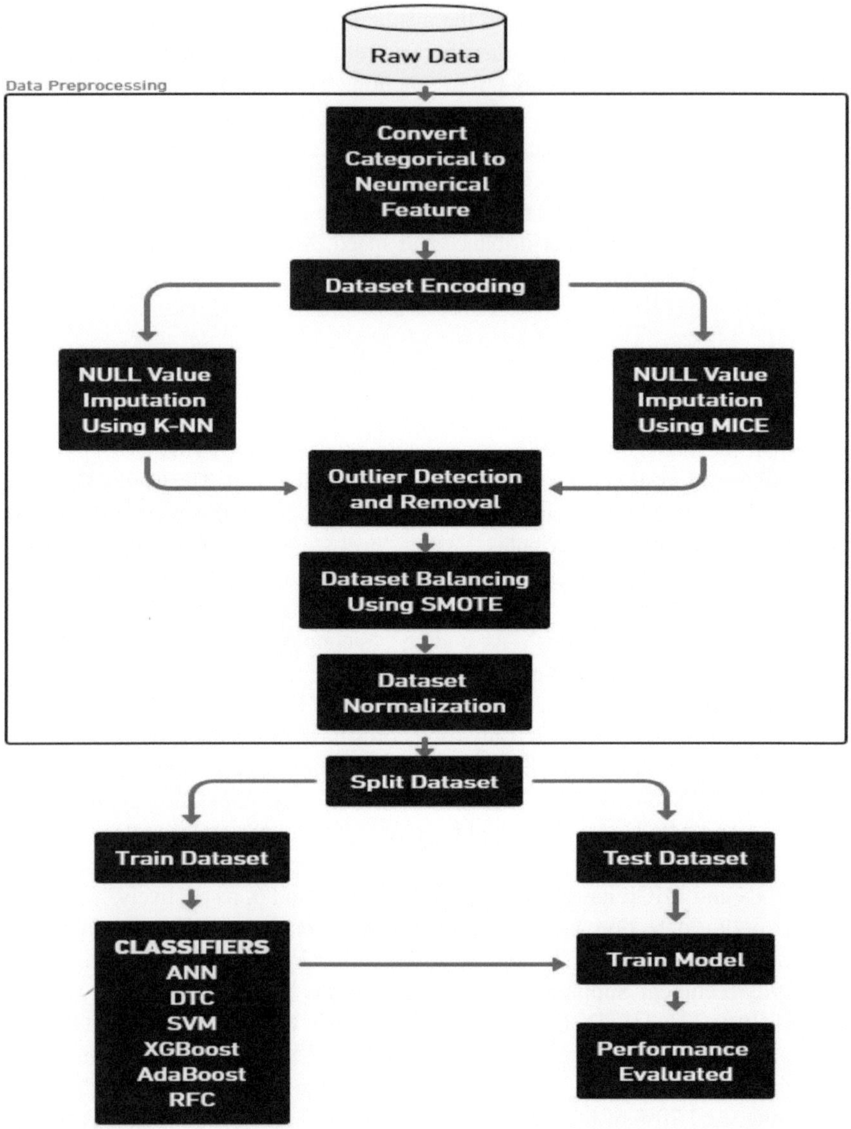

Fig. 3 Flowchart

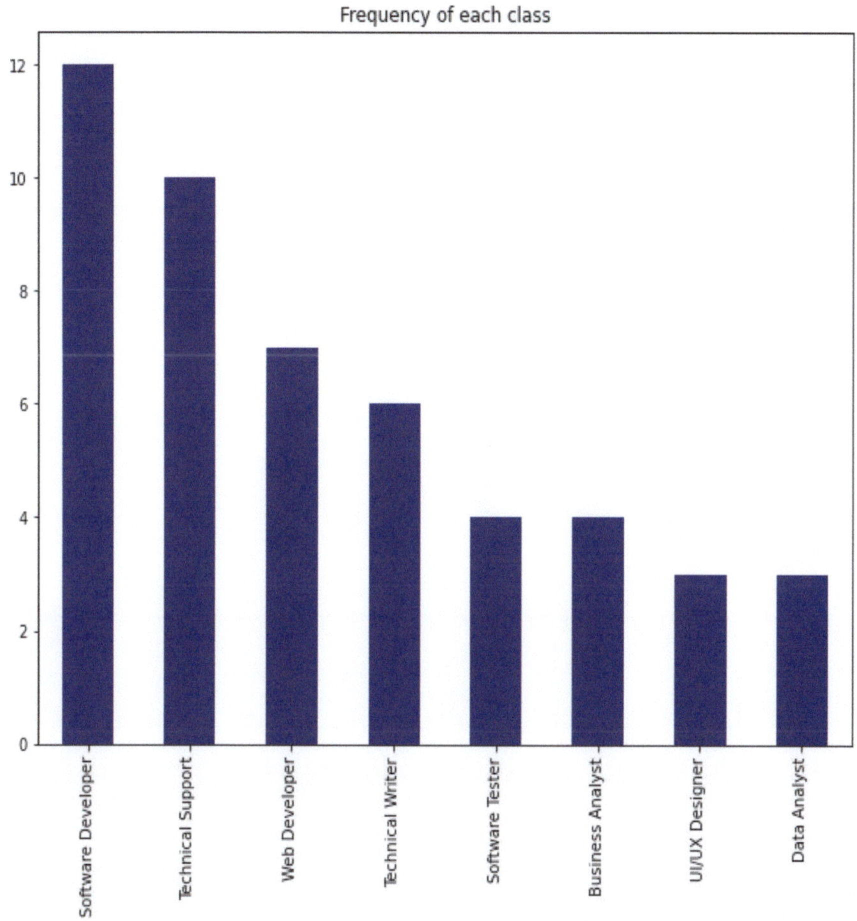

Fig. 4 Before performing imputation

(FP) are the variables shown in the confusion matrix in Table 1. Accuracy is defined as the percentage of correct predictions relative to the total number of forecasts. It makes obvious sense to use it as the model's defining parameter, but the kappa score is frequently preferred as well. On many occasions, the kappa score might be low despite high accuracy. The best-case scenario would be to exclude the possibility of the model incorrectly categorizing anything that actually fits into a different group. For this kind of multiclass classification problem, accuracy isn't enough to validate the hypothesis; high accuracy will not always ensure the correct classification, and so, the kappa score has been utilized to validate the classification.

Mathematical formulations are as follows.

$$\text{Accuracy} = \frac{\text{TP} + \text{TN}}{\text{TP} + \text{TN} + \text{FN} + \text{FP}}; \text{ Kappa Score } (K) = \frac{c \times s - \sum_k^K p_k \times t_k}{s^2 - \sum_k^K p_k \times t_k}$$

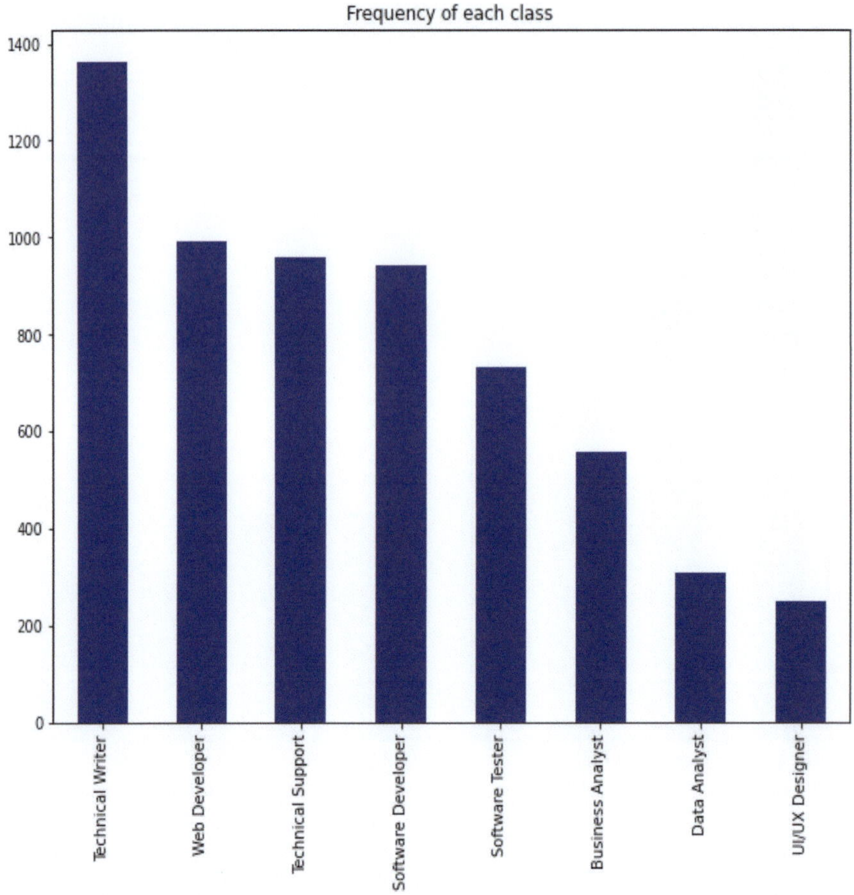

Fig. 5 After k-NN imputation

$c = \sum_{k}^{K} C_{kk}$ the total number of elements correctly predicted

$s = \sum_{i}^{K} \sum_{j}^{K} C_{ij}$ the total number of elements

$P_k = \sum_{i}^{K} C_{ki}$ the number of times that class k was predicted (column total)

$t_k = \sum_{i}^{K} C_{ik}$ the number of times that class k truly occurs (row total).

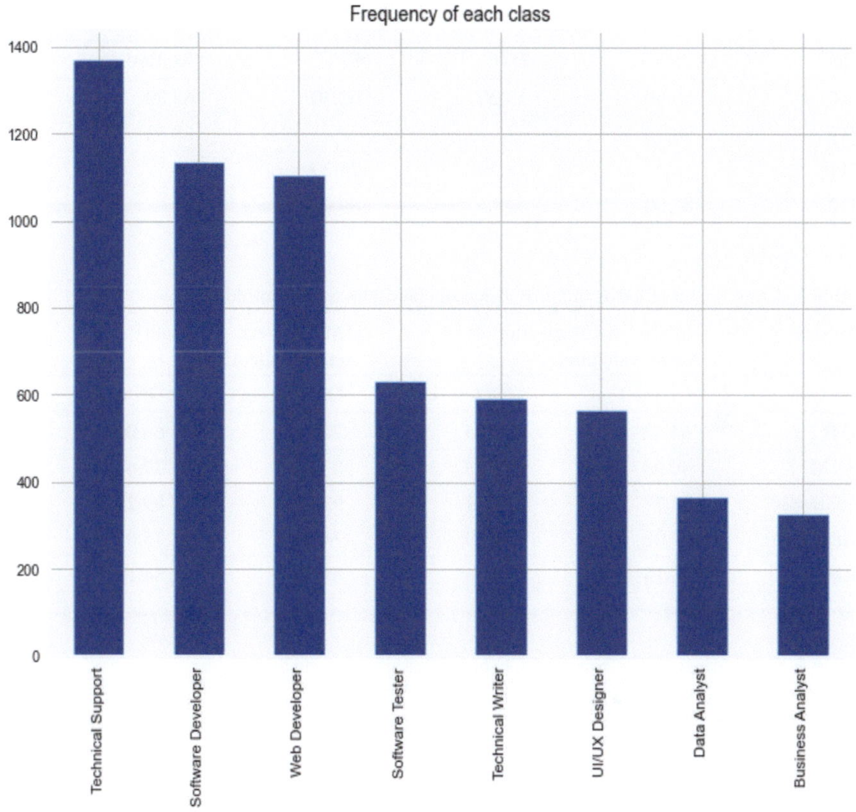

Fig. 6 Using MICE

Table 1 Confusion matrix

	Predicted good (1)	Predicted bad (0)
Actual good (1)	True positive (TP)	False negative (FN)
Actual bad (0)	False positive (FP)	True negative (TN)

5 Experimental Results and Discussion

The first half of the experiment tested k-nearest neighbor as an imputer for missing values, while the second half looked at MICE as a potential data imputation method. Table 2 shows the accuracy level of four different models—SVM, DT, RF, and XGBoost achieved by earlier works on different datasets. Five classifiers were used in the initial k-NN experiment, with two methods of performance evaluation (a 70:30 train-test split and a tenfold cross-validation). The accuracy of all five classifiers was compared in Table 3 using tenfold cross-validation using k-NN and MICE data

Table 2 Accuracy of previous works

Refs.	SVM	DTC	RFC	XGBoost
[10]	60.60	53.20	62.30	63.80
[1]	90.30	–	–	88.33
[11]	–	78.20	78.20	–
[12]	87.50	–	89.06	–

Table 3 Comparison of k-NN and MICE-based classifiers using tenfold CV

Model	k-NN—accuracy with cross-validation		MICE—accuracy with cross-validation	
	On train set	On test set	On train set	On test set
DTC	64.42	56.05	72.85	66.09
SVM	80.94	74.23	83.86	77.64
XGBoost	**82.77**	**75.41**	**92.07**	**89.24**
AdaBoost	59.08	56.69	74.65	75.05
RFC	80.15	73.56	88.03	84.14

Table 4 Comparison of k-NN and MICE-based classifiers using 70:30 split

Model	k-NN with 70:30 split		MICE with 70:30 split	
	Accuracy score	Kappa score	Accuracy score	Kappa score
DTC	63.80	0.58	72.04	0.68
SVM	79.85	0.76	85.87	0.83
XGBoost	**81.41**	**0.78**	**92.31**	**0.91**
AdaBoost	56.75	0.50	73.68	0.69
RFC	80.26	0.77	88.81	0.87

imputation approaches. Ensemble classifiers based on decision trees perform better than standalone classifiers like support vector machines in terms of accuracy and kappa score, according to the experiments. The results show that XgBoost works quite effectively with tenfold cross-validation and 70:30 train-test splits.

Table 4 compares k-NN and MICE-based results in terms of accuracy and kappa score. Figures 6 and 7 compare the classifiers for the two models. In terms of accuracy and kappa score for the two models, it is observed that XGBoost outperforms all other classifiers. Support vector machine comes second for the k-NN-based data imputation model with tenfold cross-validation, whereas for all other cases, random forest achieves the second slot (Fig. 8).

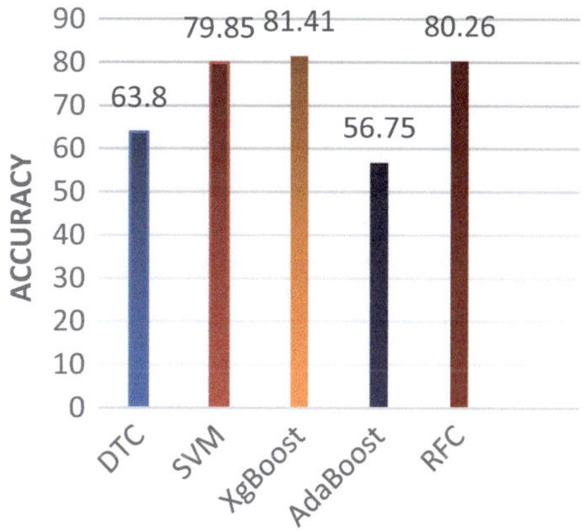

Fig. 7 Performance evaluation using k-NN

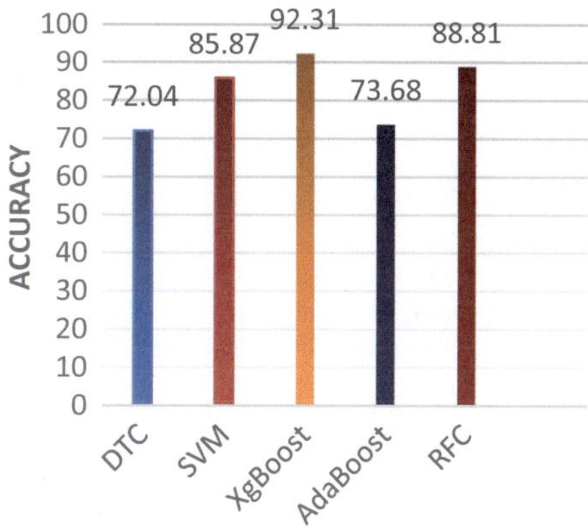

Fig. 8 Performance evaluation using MICE

6 Conclusion

In this contemporary digital age, one of the crucial study fields is student career prediction. Traditionally, a student's career has been predicted using a variety of survey-based methodologies. However, it takes a longer time to anticipate, and the outcome using such approaches isn't always stood with reality. Implementing an automated expert system for students' career prediction using machine learning principles has been projected in this article. It takes less time and produces more accurate outcomes than the other conventional procedures. This study makes career predictions for students utilizing DTC, SVM, XGBoost, AdaBoost, and RFC methods. When compared to the other classifiers, XgBoost has the best accuracy and kappa score. Using the results of this research, educational institutions may better care for slower or underperformed students, and recruiters can choose qualified applicants for their organizations without experiencing numerous setbacks.

References

1. Roy KS, Roopkanth K, Teja VU, Bhavana V, Priyanka J (2018) Student career prediction using advanced machine learning techniques
2. YLCYBGH (2015) Deep learning. Macmillan Publishers Limited. https://doi.org/10.1038/nature14.539
3. Azur MJ, Stuart EA, Frangakis C, Leaf PJ (2011) Multiple imputation by chained equations: what is it and how does it work? Int J Methods Psychiatr Res 20(1):40–49
4. Quinlan JR (1986) Induction of decision trees. Mach Learn 1:81–106. https://doi.org/10.1007/BF00116251
5. Cortes C, Vapnik V (1995) Support-vector networks. Mach Learn 20(3):273–297
6. Chen T, Guestrin C (2016) XGBoost: a scalable tree boosting system. ACM ISBN
7. Chen T, Guestrin C (2016, August) XGBoost: a scalable tree boosting system. In: Proceedings of the 22nd ACM SIGKDD international conference on knowledge discovery and data mining, pp 785–794
8. Freund Y, Schapire R (1999) A short introduction to boosting. J Jpn Soc Artif Intell. https://cseweb.ucsd.edu/~yfreund/papers/IntroToBoosting.pdf
9. Breiman L (2001) Random forests. Mach Learn 45(1):5–32
10. Nie M, Xiong Z, Zhong R, Deng W, Yang G (2020) Career choice prediction based on campus big data—mining the potential behavior of college students. Appl Sci 10:2841. https://doi.org/10.3390/app10082841
11. Baashar Y, Alkawsi G, Mustafa A, Alkahtani AA, Alsariera YA, Ali AQ et al (2022) Toward predicting student's academic performance using artificial neural networks (ANNs). Appl Sci 12(3):1289
12. Pushpa SK, Manjunath TN, Mrunal TV, Singh A, Suhas C (2017, August) Class result prediction using machine learning. In: 2017 International conference on smart technologies for smart nation (SmartTechCon). IEEE, pp 1208–1212
13. Kaur P, Singh M, Josan GS (2015) Classification and prediction based data mining algorithms to predict slow learners in education sector. Procedia Comput Sci 57:500–508
14. Hearst MA, Dumais ST, Osuna E, Platt J, Scholkopf B (1998) Support vector machines. IEEE. https://doi.org/10.1109/5254.708428
15. Breiman L (2001) Random forests. Mach Learn 45:5–32. https://doi.org/10.1023/A:1010933404324
16. https://github.com/Nantha-1998/404-CODER/blob/master/career_compute_dataset.csv
17. Jerez JM, Molina I, García-Laencina PJ, Alba E, Ribelles N, Martín M, Franco L (2010) Missing data imputation using statistical and machine learning methods in a real breast cancer problem. Artif Intell Med 50(2):105–115
18. Chawla NV, Bowyer KW, Hall LO, Kegelmeyer WP (2002) SMOTE: synthetic minority over-sampling technique. J Artif Intell Res 16:321–357

Co-reference Resolution in the Assamese Language: A Comprehensive Study

Mridusmita Das and Apurbalal Senapati

Abstract The co-references are all the terms that define the same thing in a text. For example, consider the sentence "The ***James Webb Space Telescope*** is a ***space telescope*** and ***it*** is situated at the L2 point". In this example, ***"James Webb Space Telescope"***, ***"space telescope"*** and ***"it"*** are the co-references. To find out these co-references is known as co-reference resolution. Co-reference resolution is an important task for linguistic as well as computational aspects. It is used in various applications of Natural Language Processing (NLP) tasks like Information Retrieval (IR), Machine translation (MT), Text summarization, Question answering systems, etc. No significant work has been made in co-reference resolution in the Assamese language. This paper attempted to investigate all major existing approaches to co-reference resolution and identify the issues and challenges specific to the Assamese language.

Keywords Co-reference resolution · Natural language processing · Linguistic analysis · Assamese language

1 Introduction

In certain ways, co-references function like natural language hyperlinks. On the one hand, it provides the human writer with a feeling of style and consistency, but it also adds another layer of ambiguity to the computer interpretation of language. Co-reference resolution has been a key focus of NLP research. It has numerous applications, including automatic summarization, named entity extraction, question answering, and machine translation. Literature on the subject is abundant and expanding. Since the 1970s, algorithms for pronoun-resolving problems have been developed [3].

M. Das (✉) · A. Senapati
Central Institute of Technology Kokrajhar, Assam, India
e-mail: ph21cse1003@cit.ac.in

A. Senapati
e-mail: a.senapati@cit.ac.in

S. K. Biswas et al. (eds.), *Intelligent Computing Systems and Applications*, Lecture Notes in Networks and Systems 1307, https://doi.org/10.1007/978-981-96-3860-4_20

More recent approaches have shown a tendency for knowledge-lean strategies, which were made possible by the creation of more powerful automatic parsers and taggers. Early systems heavily relied on domain and linguistic expertise. A survey of some of the more important articles in co-reference resolution is presented in this paper, and then we find the issues and challenges for the Assamese language.

The predominant language spoken in the northeastern Indian state of Assam is Assamese (Oxomiya), an Indo-Aryan language. Assamese, the official language of Assam, is one of the 22 languages officially recognized by the Republic of India. It serves as the common language of the region, with over 23 million. The largest number of speakers of any language is Indo-Iranian. However, Assamese is a language with few resources and lacks the required language-specific tools, according to linguistic studies.

This paper is organized as follows: (1) An in-depth look into co-reference resolution, (2) Co-reference Resolution in the Assamese language and its linguistic features, (3) Technical and linguistic challenges for the Assamese language, and (4) Need for an Assamese dataset or corpus.

2 Existing Approaches in Co-reference Resolution

The issue of co-reference resolution has existed for a very long time. It is a classic NLP linguistic difficulty. Fifty years ago, people attempted to solve it [3]. There has previously been a lot of work done in this area and the associated field of anaphora resolution. Rule-based methods are available initially. By the time after the availability of resources for computing power, people tried Deep learning and machine learning techniques. In this paper, we present a Literature survey of the existing approaches that can be divided broadly into three categories: (1) Rule-based Co-reference Resolution approaches, (2) Machine learning-based Co-reference Resolution approaches, and (3) Hybrid Co-reference Resolution approaches (Fig. 1).

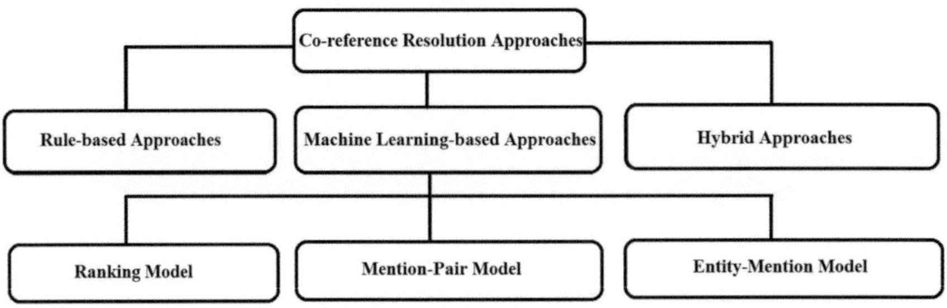

Fig. 1 Co-reference resolution approaches

2.1 Rule-Based Approaches

A pioneering attempt was made by Hobbs [3] on English pronominal co-reference. It was a rule-based approach applied to parse trees based on some syntactic feature. Some researchers described the usefulness of this issue in the pipeline of other tasks like machine translation [4, 5], dialog systems [6], and text summarization [7]. An end-to-end co-reference resolution model was accomplished by Lee et al. [8] on the CoNLL-2012 shared task [9]. The task first identifies mentions from the document and then predicts the co-reference relations among these mentions. Grosz [10] proposed the centering theory, based on the focus of attention of sentences and its referent is on the focused expression within a discourse segment. Lu et al. [26] examine the state of the art of co-reference in the biomedical sector in their study. Sukthanker et al. [29] conducted a thorough study of the data sets, evaluation criteria, and research methodologies on the NLP issues of anaphora and co-reference.

Some scholars apply the existing approaches in Indic languages. Dutta et al. [2] used the Hobbs algorithm [3] for Hindi. Senapati et al. [12] used Guitar system done by Massimo [11]. Senapati et al. [13, 14] also used a rule-based approach where rules are hand-crafted based on grammatical features. Sobha et al. [31] developed VASISTH, a multilingual system for anaphora resolution. Initially, it was supported by the languages Hindi and Malayalam. They claim that the system is easily adopted for other Indian languages too. The system used features like morphological markers to identify subjects, objects, phrases, etc.

2.2 Machine Learning-Based Approaches

Charniak et al. [15] introduced a statistical approach to anaphoric co-reference resolution. They have used several syntactic and semantic features related to pronoun resolution. Distance (between the pronoun and the nearest possible antecedent), syntactic structure based on the Hobbs algorithm, Gender, number, and animacy agreements, etc. are used. The decision-tree-based approach for co-reference resolution by Soon et al. [16]. They have treated the co-reference resolution problems as a classification problem and tried to identify whether two markable co-referring? In the pre-processing phases, all the markables are identified and learned by the decision-tree approach based on various features of the pairs of markables. Here, also syntactic features, distance features, and agreements are used. Ng et al. [17, 18] extend the Soons methods with some modifications, with some additional features and heuristics. They also did another task to check whether a noun phrase was anaphoric and then used this result in the subsequent steps. Stoyanov et al. [28] adopted the Reconcile system, which is a development environment for learning-based noun phrase (NP) co-reference resolution systems. Chai [1] presented Transformer-based co-reference resolution methods that have made significant advancements in the

CoNLL data set in recent years. Their approach outperforms SOTA baselines, particularly when it comes to pronoun resolution in lengthy texts, formal well-structured material, and clusters with sporadic references.

Ng [20] categorized co-reference based on supervised learning into three categories. These are (i) Mention-pair models, which imply a binary classifier to determine whether a mention pair is co-referent. (ii) An entity-mention models, which determine whether a mention is co-referent to a prior cluster. A cluster has collective features compared with respect to a mention in mention-pair models. and (iii) Ranking-based models, which rank all candidate antecedents to identify the most potential antecedent. It is a ranking-based methodology that takes mentions and clusters into account. As a result, the model learns about potential antecedents for a mention and takes into account all potential mentions. While Ma et al. [23] enhance the model with better attention mechanisms using loss functions, Joshi et al. [22] improve Lee et al.'s model with the BERT-based word embeddings [21].

A generic machine learning approach by Sobha et al. [19] especially for the Indian languages, which are mostly resource-limited languages. They have analyzed the similarities and dissimilarities between pronouns and their agreement with antecedents. The system is tested with the Indo-Aryan and Dravidian languages, i.e., Hindi, Bengali, and Tamil. They have reported that the results are encouraging. A shared task [24] was conducted on the anaphora resolution of the micro-blog texts from Twitter for Indian languages such as Tamil, Hindi, and Malayalam. A deep-learning-based framework [25] is in the literature that attempted anaphora resolution from social media text.

2.3 Hybrid Approaches

Some significant approaches to solving co-reference resolution are covered in this section. These approaches are machine learning-based (which relies on data-driven approaches) and linguistics-based (which heavily relies on linguistic and domain knowledge), or hybrid approaches.

Labeled data for the co-reference resolution process is a limited resource that requires significant human labor. While A Stolfo et al. [27] provide an approach that uses an end-to-end neural model in cases where labeled data is not available, modern co-reference models rely on such data.

O Uryupina [30] in his research addresses the problem of statistical co-reference resolution. The goal of this research was to bridge the gap between theoretical studies on co-reference resolution and cutting-edge statistical techniques for the job. The author looked into the feasibility of integrating various linguistic knowledge types into a machine learning framework as recommended by theoretical research. The research yielded several conclusions, some of which were related to the theoretical claims under investigation's empirical applicability and others to their applicability to a data-driven resolution process. An empirical multi-dimensional investigation of co-reference resolution is offered by L Marquez et al. [32] in their work. They examined multilingualism with Spanish, Catalan, and English corpora. They employed mention-pair models RELAXCOR and RECONCILE, as well as the entity-mention system CISTELL, in their co-reference resolution approaches.

3 Co-reference Resolution in the Assamese Language

There has been a lot of work done on natural language processing (NLP) for Western languages, especially English, over the past few decades. Due to the scarcity of language processing studies on their equivalents in the East, notably those of the Indian subcontinent, these languages need special consideration. Indian languages are behind in this field, as compared to Western languages, which have access to a variety of dictionaries, WordNet, and related technologies. The Assamese language has seen little development due to a lack of resources, complicated linguistic details, and the integration of widely used neighboring languages. This study aims to give readers an understanding of the numerous linguistic resources, tools, and state-of-the-art methods used to process the Assamese language. Before discussing the Assamese language's characteristics, morphological descriptions of the Assamese language are first given. Following that, an overview of the tools, methodologies, and corpora that can be utilized to create NLP tasks for the Assamese language is given.

3.1 Linguistic Features in the Assamese Language

3.1.1 Assamese Morphology and Grammar

The Assamese script is used in the modern Assamese language. It strongly resembles both the Bengali script and the Mithilakshar script used for the Maithili language. There are twenty-three consonants, ten diphthongs, and eight vowels in the Assamese phonemic inventory (including two semivowels). As a result, Assamese resembles the non-Indic languages of Northeast India. The only other language with fronted retroflex stops into alveolars is Bengali, which has a similarly related group of eastern dialects. The Assamese language contains the following distinctive morphological characteristics:

(1) The gender and number are not marked grammatically.

(2) The third-person pronoun has a lexical distinction for gender.

(3) Intransitive and transitive verbs are different.

(4) The agentive case is clearly distinguished from the accusative.

(5) Kinship nouns are inflected for possessive personal pronominals.

(6) From the verb roots, adverbs can be derived. It's acceptable to use a passive construction idiomatically.

4 Challenges for the Assamese Language

4.1 Technical Challenges

There aren't many materials available in Assamese. No dedicated data (corpus) is available to work on co-reference resolution [36]. There is no complete NER (Named Entity Recognition) system, and Assamese does not have a specific parser. There is a lack of a comprehensive framework for studying co-reference resolution in Assamese. As a result, it is difficult to establish a co-reference resolution mechanism for the Assamese language. Several technologies for text processing jobs have lately been developed [37]. Many of these resources are freely available to academics and can be used to do complicated NLP tasks.

4.2 Linguistic Challenges

In comparison to other languages, Assamese is distinctive. The Assamese language is composed of numerous scripts, phonemes, and dialects. Assamese words begin with the letter "Barna". The Assamese language relies on "Sandhi" and "Samaas," which combine existing words to form new words. Assamese text is challenging to process from an IR standpoint due to its unique characteristics. The inherent features of the Assamese language are as follows: (1) There is no capitalization. (2) The use of several words, (3) Different writing techniques, (4) Word Order, and (5) Punctuation alter the interpretation. The machine may struggle to understand the meaning of the text, resulting in negative impacts and altered sentences.

4.3 Assamese Data Set/Corpus

The most important aspect of studying and experimenting with languages is having the appropriate data set. There is no suitable data set for co-reference resolution in Assamese [36]. An annotated data set will improve the study in this area. When creating the Assamese data set, two main elements are considered: (a) Unicode encoding and (b) XML storage type. The quantity of data sets and corpora accessible for fundamental Assamese language NLP tasks varies depending on the application. 1.5 million Assamese words were constructed using the Standard Assamese Raw Text Corpus [33–35]. The text was extracted from books, magazines, and newspapers that were either keyed in or crawled off the internet. Words that are hard to locate in books can often be found in newspapers and periodicals. However, a unique corpus with mention detection and reference detection, both of which are currently lacking in the Assamese language, is required for the co-reference perspective.

5 Conclusion

This paper offers a survey of the co-reference resolution research in chronological order and the latest state of the art. This gives a clear idea about the pre-requisite and essential requirements in technical as well as linguistic resources. The paper also reveals that the trends in resolution algorithms are shifting to the variant model of deep learning. It then focuses on the present status of the Assamese language, its linguistic agreements and features, and its work on a co-reference resolution. This paper also draws attention to the resource scarcity of the Assamese language. That will help and motivate the research on the Indian language, especially in the Assamese research community.

References

1. Chai H, Strube M (2022) Incorporating centering theory into neural coreference resolution. In: Proceedings of the 2022 conference of the North American chapter of the association for computational linguistics: Human language technologies, pp 2996–3002
2. Kamlesh D, Nupur P, Saroj K (2008) Resolving pronominal anaphora in hindi using hobbs algorithm. Web J Formal Comput Cogn Linguist 1(10):5607–11
3. Jerry RH (1978) Resolving pronoun references. Lingua 44(4):311–338
4. Ruslan M, Sung-Kwon C (1995) Anaphora resolution in machine translation. In: TMMT
5. Lapshinova-Koltunski E, Hardmeier C, Krielke P (2018) ParCorFull: a parallel corpus annotated with full coreference. In: Proceedings of LREC
6. Michael S, Christoph M (2003) A machine learning approach to pronoun resolution in spoken dialogue. In: Proceedings of ACL, pp 168–175
7. Josef S, Massimo P, Mijail AK, Karel J (2007) . Two uses of anaphora resolution in summarization. In: Information processing and management
8. Kenton L, Luheng H, Mike L, Luke Z (2017) End-to-end neural coreference resolution. Proceed EMNLP 2017:188–197
9. Pradhan S, Moschitti A, Xue N, Uryupina O, Zhang Y (2012) Conll-2012 shared task: Modeling multilingual unrestricted coreference in ontonotes. In; Proceedings of CoNLL, pp 1–40
10. Grosz BJ, Weinstein S, Joshi AK (1995) Centering: a framework for modeling the local coherence of discourse. Comput Linguist 21(2):203–225
11. Massimo P, Mijail KA (2004) A General-purpose, Off-the-shelf anaphora resolution module: Implementation and preliminary evaluation. In: Proceedings of the fourth international conference on language resources and evaluation
12. Senapati A, Garain U (2013) Guitar-based pronominal anaphora resolution in bengali. In: Proceedings of the 51st annual meeting of the association for computational linguistics (vol. 2: Short Papers), pp 126–130
13. Senapati A, Garain U (2012) Anaphora resolution in Bangla using global discourse knowledge. In: 2012 international conference on Asian language processing, pp 49–52. https://doi.org/10.1109/IALP.2012.25
14. Senapati A, Garain U (2011) Anaphora resolution system for bengali by pronoun emitting approach. In: 2011 Proceedings of the ICON-2011, NLP Tool Contest
15. Hale NGJ, Charniak E (1998) A statistical approach to anaphora resolution. In: Proceedings of the sixth workshop on very large corpora
16. Soon WM, Ng HT, Lim DCY (2001) A machine learning approach to coreference resolution of noun phrases. Comput Linguist 27(4):521–544

17. Ng V, Cardie C (2001) Improving machine learning approaches to coreference resolution. In ACL '02: Proceedings of the 40th annual meeting on association for computational linguistics. Morristown, NJ, USA. Association for Computational Linguistics, pp 104–111
18. Ng V, Cardie C (2002) Identifying anaphoric and non-anaphoric noun phrases to improve coreference resolution. In: Proceedings of the 19th international conference on computational linguistics. Morristown, NJ, USA, Association for Computational Linguistics, pp 1–7
19. Sobha LD, Vijay SR, Pattabhi RR (2014) A Generic Anaphora resolution engine for Indian languages. In: Proceedings of COLING 2014, the 25th international conference on computational linguistics: Technical papers. Dublin, Ireland. Dublin City University and Association for Computational Linguistics, pp 1824–1833
20. Ng V (2017) Machine learning for entity coreference resolution: A retrospective look at two decades of research. In: AAAI. AAAI Press, pp 4877–4884
21. Devlin J, Chang M, Lee K, Toutanova K (2019) BERT: pre-training of deep bidirectional transformers for language understanding. In: NAACL-HLT (1). Association for Computational Linguistics, pp 4171–4186
22. Joshi M, Levy O, Zettlemoyer L, Weld DS (2019) BERT for coreference resolution: Baselines and analysis. In: EMNLP/IJCNLP (1). Association for Computational Linguistics, pp 5802–5807
23. Ma J, Liu J, Li Y, Hu X, Pan Y, Sun S, Lin Q (2020) Jointly optimized neural coreference resolution with mutual attention. In: WSDM. ACM, pp 402–410
24. Sobha LD (2020) Anaphora Resolution from Social Media Text in Indian Languages (SocAnaRes-IL)- Overview, FIRE 2020: Forum for Information Retrieval EvaluationDecember, pp 9–13
25. Saha BN, Senapati A, Garain U (2022) A deep learning framework for Anaphora resolution from social media text. In: Advanced machine intelligence and signal processing. Lecture Notes in Electrical Engineering, vol 858. Springer, Singapore. https://doi.org/10.1007/978-981-19-0840-8-53
26. Pengcheng L, Massimo P (2021) Coreference resolution for the biomedical domain: A survey. arXiv preprint arXiv:2109.12424
27. Alessandro S, Chris T, Vikram G, Mrinmaya S (2022) A simple unsupervised approach for coreference resolution using rule-based weak supervision. In: Proceedings of the 11th joint conference on lexical and computational semantics, pp 79–88
28. Veselin S, Claire C, Nathan G, Ellen R, David B, David H (2010) Coreference resolution with reconcile. Technical report, Cornell univ ithaca ny dept of computer science
29. Sukthanker R, Poria S, Cambria E, Thirunavukarasu R (2020) Anaphora and coreference resolution: A review. Inf Fus 59:139–162
30. Olga U (2007) Knowledge acquisition for coreference resolution
31. Sobha L, Patnaik BN (1999) Vasisth: An Anaphora resolution system. Unpublished doctoral dissertation. Mahatma Gandhi University, Kottayam, Kerala
32. Llu ıs M, Marta R, Emili S, (2013) Coreference resolution: an empirical study based on semeval-2010 shared task 1. Lang Resour Eval 47(3):661–694
33. Priyanjana C, Swarnav DB, Catherina B, Sandeep DM, Sanghamitra N, Utpal S (2023) An acous-tic/prosodic feature-based audio dataset for assamese speech summarization. In: Proceedings of international conference on frontiers in computing and systems. Springer, pp 579–588
34. Ringki D, Thoudam DS (2022) A multi-stage multimodal framework for sentiment analysis of assamese in low resource setting. Expert Syst Appl 117575
35. Shikhar KS, Himadri B, Ambeswar G, Ratul D, Anup B (2012) A structured approach for building assamese corpus: insights, applications, and challenges. In: Proceedings of the 10th workshop on Asian language resources, pp 21–28

36. Mridusmita D, Apurbalal S (2023) Development of the Co-reference resolution tagged data set in assamese@ a semi-automated approach. In: 2023 IEEE Guwahati subsection conference (GCON), vol 23. IEEE, pp 1–4
37. Mridusmita D, Apurbalal S (2023) Coreference resolution in the assamese language: A pioneering attempt. International conference on data science and communication, vol 23. Singapore, Springer Nature Singapore, pp 577–588

Object Aware Visual Question Answering Model for Malayalam

Abhishek Gopinath Kovath and O. K. Sikha

Abstract Visual Question Answering (VQA) is a challenging task in natural language processing and computer vision that requires sophisticated reasoning over the visual content to provide an accurate answer to a question. This paper proposes an efficient deep learning framework for Malayalam Visual Question Answering (MVQA) that can answer a specific natural language question in Malayalam about an image. A Malayalam Visual Question Answering (MVQA) dataset was created by translating English question answer pairs from the Visual Genome dataset. By extracting the object-level spatial information from contextual parts of the image using FR-CNN (Faster R-CNN), the proposed model predicts a vernacular answer from multiple answers. The proposed MVQA framework comprises (1) a region-based convolutional neural network (CNN) for visual feature extraction, (2) Long Short-Term Memory cells (LSTMs) for language modeling, and (3) multimodal fusion for integrating visual and language representations. We evaluate the proposal for visual question answering using the proposed framework, and experimental results on the self-created Malayalam VQA dataset show the effectiveness of this framework for Malayalam VQA.

Keywords VQA · Cross-lingual · Multimodal · Computer vision · Natural language processing · Malayalam visual question answering

A. G. Kovath (✉)
Department of Computer Science and Engineering, Amrita School of Computing, Coimbatore, India
e-mail: cb.en.p2aid20002@cb.students.amrita.edu

O. K. Sikha
Amrita Vishwa Vidyapeetham, Vengal, India
e-mail: ok_sikha@cb.amrita.edu

S. K. Biswas et al. (eds.), *Intelligent Computing Systems and Applications*, Lecture Notes in Networks and Systems 1307, https://doi.org/10.1007/978-981-96-3860-4_21

1 Introduction

VQA is a computer-assisted technique for responding to inquiries regarding a particular image [1, 5]. This is a challenging problem since the computer system must understand both the query and the image (which serves as the item's identification) (computer vision). A sophisticated VQA system could create an Artificial Intelligence (AI) entity that receives natural language questions and anticipates the correct response based on an analysis of the problematic situation (s). Verbal understanding, fine-grained visual processing, and numerous thinking processes are necessary to provide an appropriate response on VQA. There has been considerable research on VQA; however, much of this research has been limited to English, limiting the system's application breadth. Because each language is unique in terms of complexity, features, and thought patterns, there is an urgent need to develop VQA systems for various languages [3]. Due to a paucity of linguistic resources, the development of such systems is hampered. Individuals fluent in more than one language frequently communicate in their primary and secondary languages. A language-neutral and multilingual VQA system should be developed to facilitate the creation of an AI framework capable of serving situations in which end users communicate in several languages. In these instances, it is prudent to investigate the VQA technique, which would benefit most people who speak or write in more than one language. The final focus of the work resides in developing a VQA system in Malayalam.

We provide the first-ever solution for Malayalam VQA, allowing us to impose on English-language questions to boost the performance of our Malayalam VQA system. Two contributions were made: first, a Malayalam VQA dataset was generated, which is believed to be the first of its kind. Second, the Malayalam VQA system for natural language question answering in Malayalam using image-level and object-level features is being developed [3, 16–18].

The rest of the paper is as follows: Sect. 2 describes the related works. Details of the Malayalam Visual Question Answering Dataset are given in Sect. 3. The proposed Baseline MVQA model on the Malayalam VQA dataset is described in Sect. 4. Section 5 shows the experimental setup, and the results are discussed in Sect. 6. Finally, the paper concludes with Sect. 7.

2 Related Works

As deep learning spreads and evolves into a variety of fields, an increasing number of activities can be accomplished as a combination of computer vision and natural language processing. Visual question answering [1, 4, 7] is one example of a task-specific optimization technique that has recently gained popularity. Because of this VQA difficulty, academics have made great strides toward increasing the usefulness of the multimodal framework.

VQA has garnered more interest from academics, increasing the number of datasets [1, 2, 5, 15]. The datasets for this multimodal approach to learning linguistic processes for associating language with images are being enhanced. Most photographs are derived from large image libraries, such as MS-COCO and YFCC100M, and from computer-generated photography (dataset reference). Malayalam is not one of the few multilingual languages (Bilingual Image Understanding, Multilingual and Code-Mixed).

Multilingual Recent advances in natural language processing have enabled improved neural network translation of languages. Since the advent of transformers, language modeling has made significant progress in improving the efficiency of machine translation through its attention mechanisms and parallel design.

Improved feature modeling for vision and language and cross-modal integration are critical for thinking about images and correlating them to natural language concepts. The iBOWIMG baseline model [4] was used. With sufficient setup and training, this simple baseline model outperforms numerous recurrent network-based visual quality assurance techniques (Question Answer). This is the simplest method for the VQA assignment. For this study, which focused on natural language processing, an adequate level of accuracy was achieved by utilizing the VQA dataset (NLP). Fukui et al. [20] proposed a method for combining the image and question feature maps, dubbed "Multimodal Compact Bilinear Pooling," to improve on this (MCB).

In contrast to the point-wise product, the outer product of two vectors using bilinear pooling requires that all components of the two vectors be multiplicatively related. It was important to decompose and reduce high-dimensionality parameters to categorize the data using this strategy. By canonizing the existing state-of-the-art model, the accuracy of reaching it was increased.

The current models perform well on the English dataset, although they are the same model. When fed a Malayalam query, even strong VQA models cannot reliably predict a response in Malayalam. Malayalam must be processed to construct a VQA system capable of performing this work. This article proposes a three-phase procedure. To begin, the Malayalam question is converted to a feature vector. This phase focuses on the multilingualism of the Malayalam language. Image characteristics are required to accurately represent the image at the pixel and object levels. The last model is multimodal fusion, which is responsible for one of the most critical components of the VQA framework: the representations obtained from both the encoded question and the learnt image are tightly coupled.

3 Dataset

Most of the VQA models reported in the literature are limited to natural language in English. The scarcity of language resources corresponding to regional languages such as Malayalam, Tamil, Telugu, etc., obstructs the development of such systems. Hence, it is necessary to begin with the creation of the regional language VQA

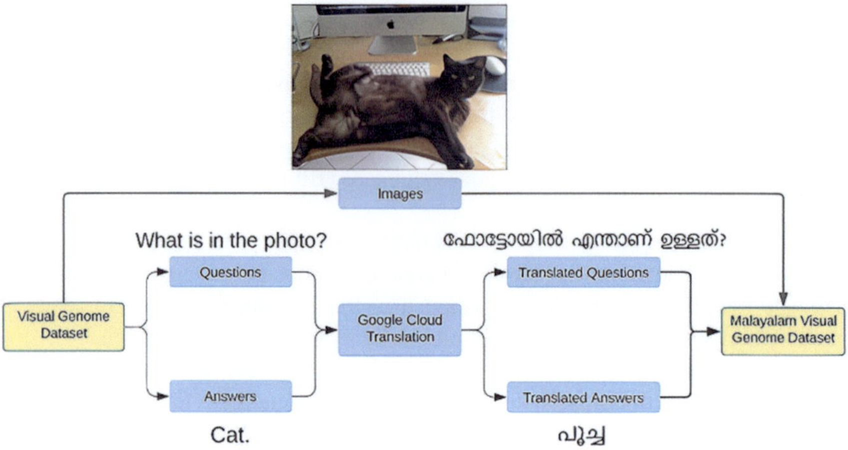

Fig. 1 MVQA dataset generation pipeline

dataset. Following the Visual Genome dataset by [2], which includes 108,077 images and their associated 1.7 million questions and specific answers, a Malayalam VQA dataset was developed using the Google translation engine. Figure 1 illustrates the pipeline followed for the generation of the MVQA dataset.

The dataset's whole 1,445,322 question-answer pairs were translated to Malayalam, of which 1,011,725 and 216,798 question-answer pairs are used, respectively, for training and validation, spread over 97,685 images of the visual genome dataset, and the remaining 216,799 questions are used for testing. The dataset was verified by conducting a short survey on a few random samples manually verified by the language experts for errors.

4 Baseline MVQA

This section describes the baseline MVQA architecture for Malayalam VQA. Given a digital image and a Malayalam question, the primary objective of the baseline Malayalam VQA is to offer an accurate natural language answer to the question by incorporating all the possible alternatives. Mathematically,

$$\widehat{A} = \underset{\widehat{A} \in \mathcal{A}\_M}{\text{argmaxp}}(\widehat{\mathcal{A}} \mid Q, img)$$

where $\widehat{A}$ is the predicted answer, $\widehat{\mathcal{A}}$ is the expected answer, Q is the Malayalam question, and *img* is the image associated with the question. The proposed baseline MVQA model explores the effectiveness of both image-level (obtained from generic

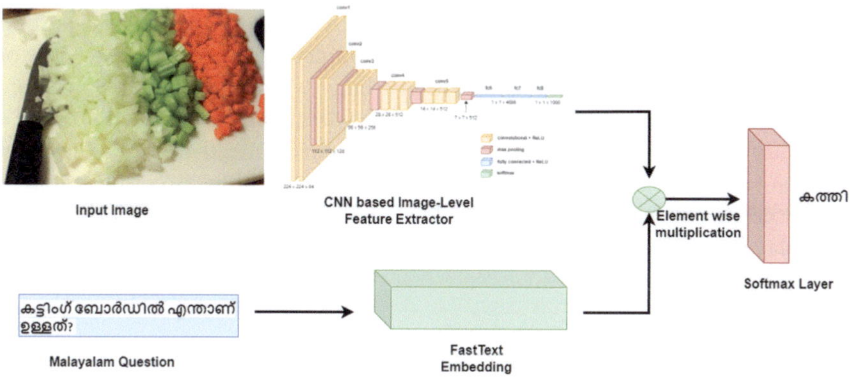

Fig. 2 Image-level feature-based MVQA architecture

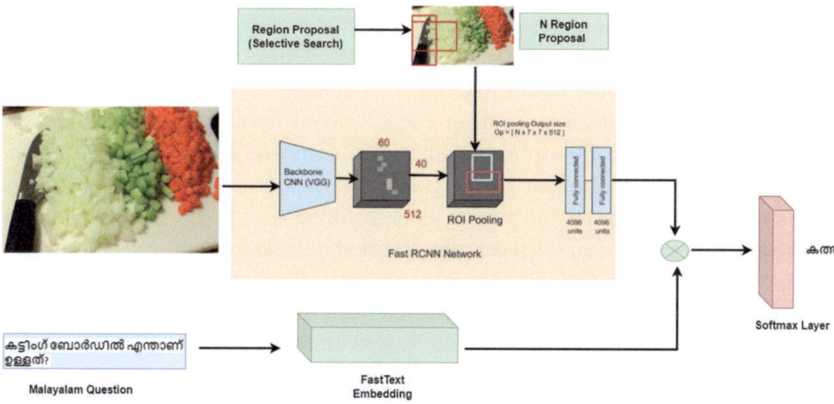

Fig. 3 Object-level feature-based MVQA architecture

CNN models) and object-level features (obtained from FR-CNN) [6, 19] integrated with FastText language embeddings, as shown in Figs. 2 and 3, respectively.

4.1 Malayalam Question Embedding

The Malayalam questions are tokenized and turned into a vector embedding using FastText. FastText, proposed by Joulin et al. [13], is an improvement over Word2vec and GloVe language representations. FastText representations can handle words not in the dictionary. FastText uses a bag of character n-grams notation to represent each word in a sentence/question. For example, for the word "matter", with n = 3, the FastText representation includes < ma, mat, att, tte, ter, er > . The boundary symbols an < , > are used to distinguish the n-gram of a word from the word itself;

for instance, if the word "mat" itself is a part of the vocabulary, then it is symbolized as mat >. This enhances the language representation by preserving the meaning of shorter words that may appear as n-grams of other words, enabling us to capture the meaning of prefixes/suffixes. This work considers Malayalam questions with a maximum of 14 words to make the model computationally effective. Each Malayalam question is mapped into a 1×300-dimensional representation as per the proposal from Facebook research [14], resulting in a 14×300-dimensional embedding for the Malayalam VQA dataset.

4.2 Image Representation

This paper compared the effectiveness of both image-level and object-level features for visual representation. Furthermore, the image-level features are global features extracted from the entire image using generic CNN architectures such as VGG16, ResNet152, EfficientNet_b1, and Wide_Resnet_50. Object-level features are local features extracted using the Faster R-CNN architecture.

VGG16 [8] is a sixteen-layer deep convolutional network that fared well in classifying object categories from the ImageNet database and was trained on the same. We utilize the final linear layer right before the classifier, which generates a 4086-dimensional vector for the picture level features.

ResNet152 [9] is another deep neural network that makes use of skip connections to circumvent the problem of disappearing gradients. Similarly, the final linear layer output has a dimension of 2048 and is derived from the input image.

Wide_Resnet_50 [10] by making the ResNet network shallower but wider, the network performed with the same or improved accuracy. The output dimension is identical to Resnet's 2048-dimensional output.

EfficientNet_b1 [11] is an efficient network widely utilized in mobile devices at the time of this paper. The network is scalable under the available computational resources. We utilized the balanced b1 form of the model. Here, the image provided the output dimension of 1280.

Faster R-CNN [12] is widely used for object detection, which uses a method called Region Proposal Network. This makes it feasible to achieve high performance with the specified task. The object-level feature extracted has a dimension of 2048 for each object detected in the image.

These extracted features are then passed on to the final phase, the multimodal fusion, after passing them through a linear layer to scale the vector to a standard dimension of 2048, matching the vector dimension from the language model.

Table 1 Sample Malayalam questions and answers from the MVQA dataset

English Question	Answer	Malayalam Question	Answer
How many players are shown?	Two.	എത്ര കളിക്കാരെ കാണിക്കുന്നു?	രണ്ട്.
What sport is being played?	Baseball.	ഏത് കായിക ഇനമാണ് കളിക്കുന്നത്?	ബേസ്ബോൾ.
What is the animal?	Dog.	മൃഗം എന്താണ്	നായ.
How many carrots are there?	One.	എത്ര കാരറ്റ് ഉണ്ട്?	ഒന്ന്.
Who is in the photo?	A guy.	ഫോട്ടോയിൽ ആരാണ്?	ഒരാള്.

4.3 *Multimodal Feature Fusion*

We conduct element-wise multiplication on the image-level features and language-level features retrieved from the CNN and LSTM architectures, respectively. Multiplication on an element-by-element basis multiplies each element from the image features by the equivalent element from the language features. To achieve this matrix multiplication, each feature must have the same dimension.

5 Experimental Setup

5.1 *Malayalam Visual Question Answering Dataset*

The Malayalam Visual Genome dataset is used for the questions and answers in the experiments. No prior VQA framework was set up for the Malayalam language, and the primary objective was to create an effective VQA system for the Malayalam dataset. The dataset has 1,445,322 question answer pairs, of which 216,799 questions were set aside for testing the resulting VQA framework. The frameworks are trained and evaluated for the Malayalam dataset. A few samples from the MVQA dataset are shown in Table 1.

5.2 *Hyperparameters*

For the Malayalam question encoding, the words are tokenized and passed through the FastText models to get a weighted embedding vector for each word. The resulting

300-dimensional embedding for each word is the input for the LSTM framework. The model is trained by Facebook Research [14] and proved to be more effective. These were evaluated manually by comparing the nearest neighbors of a given word. The given word is translated from the word 'photo' and was used as an example for the model outputs tabulated in Table 1. The answer class of the model was filtered to include only those classes that would occur at least 30 times, resulting in a total target class of 3,344.

The language model for all the experiments followed the optimal hyperparameter of having a question word sequence length of a maximum of 14 words, resulting in an embedding matrix of 14×300 dimensions after the sequence of words n passed through a FastText model. The embedded output is fed to an LSTM model having 1 layer and a hidden dimension of 1024.

The image feature extraction is carried out by each model's last fully connected layer, which is 4086 dimensions for VGG16, 1280 dimensions for EfficientNet-b1, 2048 dimensions for ResNet152, and Wide-ResNet-50-2. At the same time, the object-level feature extraction was taken from the top 36 objects with the highest probability classification, from which a spatial dimension of 2048 features was extracted, giving the final output dimension of 1×4086. The image models follow an initial learning rate $= 0.001$ and are iterated through 100 epochs. Of these models, the use of the feature representation of the images, more precisely the objects in the images, of the dataset from the FR-CNN framework was pre-extracted by Anderson et al. 2018 [6]. The optimizer used was the Adam optimizer, which helps in the optimization of the weights of the model.

6 Results and Analysis

This section evaluates the performance of the proposed MVQA models using accuracy as the evaluation metric. Table 2 summarizes the performance of the experimental models. Table 3 shows the predictions from each model.

Following the same evaluation metrics from [2], where out of m predictions one matches precisely with the ground truth. The table shows that the FR-CNN-based model produced a better accuracy of 50.93% compared to other generic CNN-based models. FR-CNN extracts object-level spatial features of the input image, whereas

Table 2 Summarizes the performance of the experimental models

Model	Train accuracy	Val accuracy	Test accuracy
VGG16 + FastText	49.75	49.15	49.06
ResNet152 + FastText	49.00	49.47	50.03
EfficientNet-b-1 + FastText	47.47	47.48	48.51
WideResnet50 + FastText	48.72	48.98	49.16
FR-CNN + FastText	51.90	47.78	50.93

Table 3 Image-MVQA results

Image					
Malayalam Question	എത്ര ബസുകൾ പാർക്ക് ചെയ്തി ക്കുന്നു?	ഫ ാഫടാ യിചെ പ്രധാന നിറം എന്താണ് ?	ട്രെയി നിന് എത്ര അറക ളുണ്ട്?'	പുല്ലിൽ ഏര് നിറത്തി ലുള്ള പൂക്കളാ ണ് ഉള്ളത്?	ഫ ാഫടാ യിൽ എത്ര കുരിരക ളുണ്ട്?
answer	4	വെള്ള	3	വെള്ള	3
Predicted VGG	വെള്ള	വെള്ള	ട്രെയിൻ	പച്ച	നാഗ്
Predicted ResNet152	ഗ്രേ	ഗ്രേ	ഫറാഡി ന്ഛ ശെം	ഗ്രേ	വെള്ള
Predicted EfficientNet_b	ഗ്രേ	ഗ്രേ	ഫറാഡി ന്ഛ ശെം	ഫറാഡി ന്ഛ ശെം	ഫറാഡി ന്ഛ ശെം
Predicted WideResnet_50	ഗ്രേ	കൗണ്ടർ	നടത്തം	ഫറാഡി ന്ഛ ശെം	ഗ്രേ
Predicted FasterRCNN	നാഗ്	വെള്ള	ഫറാഡി ന്ഛ ശെം	ഗ്രേ	ഇഫപാൾ

the generic CNN-based models produce a global image representation. The results prove that object-level features can effectively capture the visual content, contributing more to the MVQA model's performance.

7 Conclusion

This work aims to propose a Malayalam VQA framework for visual question answering in Malayalam and to generate a Malayalam version of the Visual Genome dataset. This dataset can be used to perform additional research on visual linguistics subjects relevant to Malayalam, such as image captioning. Based on the appearance of an image, the proposed object-based approach can predict the answer to a Malayalam question. This can aid in developing Malayalam language artificial intelligence systems for translation, image captioning, and dialog systems. There are plans to conduct future research on transformer-based frameworks for VQA in Malayalam.

References

1. Agrawal A, Lu J, Antol S, Mitchell M, Zitnick CL, Parikh D, Batra D (2016) VQA: visual question answering. Int J Comput Vision 123(1):4–31. https://doi.org/10.1007/s11263-016-0966-6
2. Krishna R, Zhu Y, Groth O, Johnson J, Hata K, Kravitz J et al (2017) Visual genome: Connecting language and vision using crowdsourced dense image annotations. Int J Comput Vis 123(1):32–73
3. Gupta D, Lenka P, Ekbal A, Bhattacharyya P (2020) A unified framework for multilingual and code-mixed visual question answering. In: Proceedings of the 1st conference of the Asia-Pacific chapter of the association for computational linguistics and the 10th international joint conference on natural language processing, pp 900–913
4. Zhou B, Tian Y, Sukhbaatar S, Szlam A, Fergus R (2015) Simple baseline for visual question answering. arXiv:1512.02167
5. Goyal Y, Khot T, Summers-Stay D, Batra D, Parikh D (2017) Making the v in vqa matter: Elevating the role of image understanding in visual question answering. In: Proceedings of the IEEE conference on computer vision and pattern recognition, pp 6904–6913
6. Anderson P, He X, Buehler C, Teney D, Johnson M, Gould S, Zhang L (2018) Bottom-up and top-down attention for image captioning and visual question answering. In: Proceedings of the IEEE conference on computer vision and pattern recognition, pp 6077–6086
7. Wu Q, Teney D, Wang P, Shen C, Dick A, Van Den Hengel A (2017) Visual question answering: a survey of methods and datasets. Comput Vis Image Underst 163:21–40
8. Simonyan K, Zisserman A (2014) Very deep convolutional networks for large-scale image recognition. arXiv:1409.1556
9. He K, Zhang X, Ren S, Sun J (2016) Deep residual learning for image recognition. In: Proceedings of the IEEE conference on computer vision and pattern recognition, pp 770–778
10. Zagoruyko S, Komodakis N (2016) Wide residual networks. arXiv:1605.07146
11. Tan M, Le Q (2019) Efficientnet: Rethinking model scaling for convolutional neural networks. In: International conference on machine learning. PMLR, pp 6105–6114
12. Ren S, He K, Girshick R, Sun J (2015) Faster r-cnn: Towards real-time object detection with region proposal networks. In: Advances in neural information processing systems, p 28
13. Bojanowski P, Grave E, Joulin A, Mikolov T (2017) Enriching word vectors with subword information. Trans Assoc Comput Linguist 5:135–146
14. Grave E, Bojanowski P, Gupta P, Joulin A, Mikolov T (2018) Learning word vectors for 157 languages. arXiv:1802.06893
15. Zhang P, Goyal Y, Summers-Stay D, Batra D, Parikh D (2016) Yin and yang: balancing and answering binary visual questions. In: Proceedings of the IEEE conference on computer vision and pattern recognition, pp 5014–5022

16. Khan HR, Gupta D, Ekbal A (2021) Towards developing a multilingual and code-mixed visual question answering system by knowledge distillation. arXiv:2109.04653
17. Shimizu N, Rong N, Miyazaki T (2018) Visual question answering dataset for bilingual image understanding: A study of cross-lingual transfer using attention maps. In: Proceedings of the 27th international conference on computational linguistics, pp 1918–1928
18. Tran KQ, Nguyen AT, Le ATH, Van Nguyen K (2021) ViVQA: Vietnamese visual question answering. In: Proceedings of the 35th Pacific Asia conference on language, information and computation, pp 546–554
19. Teney D, Anderson P, He X, Van Den Hengel A (2018) Tips and tricks for visual question answering: learnings from the 2017 challenge. In: Proceedings of the IEEE conference on computer vision and pattern recognition, pp 4223–4232
20. Fukui A, Park DH, Yang D, Rohrbach A, Darrell T, Rohrbach M (2016) Multimodal compact bilinear pooling for visual question answering and visual grounding. arXiv:1606.01847

Signal and Image Processing

Machine Learning Expert System for Breast Cancer Detection

Rajdeep Paul, Saroj Kr. Biswas, Biswajit Purkayastha, Arpita Nath Boruah, Manomita Chakraborty, and Akhil Kr. Das

Abstract Among all types of cancer, one of the most serious types of cancer that causes the unpredictability of mortality in the majority of women across the world is breast cancer (BC). Numerous researches have been done for the BC treatment and also are going on at present at utmost rate which leads to the better understanding of the disease and hence provides better results. Thus accurate prediction of BC plays a vital role in efficient treatment and also for its early stage detection. Since a few years ago, many Machine Learning (ML) algorithms have been suggested for precise BC prediction. In order to improve BC prediction using machine learning analytics, this article also suggests the MLESBCD (Machine Learning Expert System for Breast Cancer Detection). Prior to using the classification algorithm, the MLESBCD the BC data is preprocessed to handle outliers and the missing values. On the BC dataset from the UCI machine learning repository, the suggested MLESBCD generates satisfactory accuracy.

Keywords Breast cancer · Machine learning · MICE imputation · Random forest · AdaBoost · XGBoost

R. Paul (✉) · S. Kr. Biswas · B. Purkayastha · A. N. Boruah
Computer Science and Engineering, National Institute of Technology, Silchar, India
e-mail: rajdeep_pg_21@cse.nits.ac.in

B. Purkayastha
e-mail: biswajit@nits.ac.in

M. Chakraborty
School of Computer Science and Engineering, VIT-AP University, Amaravati, India

A. Kr. Das
Computer Science, Gour College, Malda, India

S. K. Biswas et al. (eds.), *Intelligent Computing Systems and Applications*, Lecture Notes in Networks and Systems 1307, https://doi.org/10.1007/978-981-96-3860-4_22

1 Introduction

In recent years, the rate of breast cancer among women has increased alarmingly. BC is a life-threatening cause of cancer death among women that develops from breast tissues. In biological studies mammograms are generally used for early detection of breast cancer. A number of signs, such as a new breast or armpit lump, breast swelling or thickening, or breast skin dimpling, aid in the diagnosis of BC [1]. For better treatment decisions and faster detection, the BC is divided into 5 general stages from 0 to 4 depending upon size of the lymph, location of the lymph, spread-ness to nearby places, presence of certain protein or hormone receptors and grade of the tumour.

In stage 0, there is no evidence of tumour cells that have spread to the body's other organs. In stage 1, the cancer cells start affecting the nearby cells or tissues. This stage is further divided into two subcategories i.e. Stage 1A and Stage 1B. In stage 1A the tumour extends upto 2 cm and lies within the breast without involving any lymph node whereas in stage 1B a small group of the lymph nodes contain cancer cells that range in size from 0.2 to 2 mm [2]. Stage 2 is divided into stage 2A and stage 2B. Stage 2B has a tumour that is between 2 and 5 cm in size, whereas stage 2A has no detectable tumour in the breast, but one to three detected malignancies in axillary lymph nodes. Stage 3 is divided into stage 3A, stage 3B, and stage 3C. In stage 3A the tumour has reached one to three axillary lymph nodes and is greater than 5 cm. The tumour may be of any size at stage 3B, and it has reached the chest wall and up to nine axillary lymph nodes. In stage 3C the tumour is under the collarbone spreading upto 10 or more lymph nodes but has not spread to further bodily parts. Stage 4 cancer is when the disease has progressed to the body's other organs, including the lungs, skin, bone, liver, or brain.

Medical institutions are gathering enormous volumes of data every day as a result of the introduction of digital technology in the healthcare industry. Because such institutes contain a wealth of data for various conditions, there is a potential or chance that the analysis of data will reveal some fascinating patterns among the data using specific data analysis techniques or procedures [3]. Due to its quick and easy capacity to identify patterns in internal data to perform some specified tasks during the past ten years, machine learning has seen significant adoption in a variety of applications. Given the availability of many types of breast cancer datasets, it is simple to classify and forecast breast cancer using machine learning [4]. As a result, numerous studies have been conducted to propose various types of models utilising ML for BC prediction [5].

The BC dataset comprises missing values that leads to the imputation on missing values for better model preparation. MICE (Multiple Imputation by Chained Equations) is one of the well known data imputation techniques for handling missing values. Several imputation incorporates uncertainty by using the iterative method and to estimate missing data by using multiple regression models. Chained equations make several imputations using the broad distributions of the observed data to fill in the missing data values. MICE has several advantages over other missing data approaches like it fills the missing values number of times based on the previous observations to make a complete dataset. This is a robust method for data imputation [6].

The large datasets containing numerous data values are supposed to consist of numerous biased data which creates unusual deviations from the rest of the data points which are known as outliers. Outliers can cause a model's fit to be poor, resulting in poor predictive modeling performance. To overcome such problems, proper analysis of the outliers and its removal must be done with a specific outlier detection technique. Nowadays, almost every ML model uses outlier detection in its pre-processing stage so that prediction can be made more accurate. Outlier removal not only improves the performance of the model but also causes the result to be statistically significant.

In most of the ML tasks, prediction from a single model does not produce satisfactory results. So, a predictive model is used which combines predictions from two or more models generally known as Ensemble Learning (EL). By creating a large number of hypotheses for a classifier, ensemble learning improves its performance and produces the best possible hypothesis for the job at hand. Alternatively, an Ensemble is a collection of classifiers whose individual predictions are combined in some manner (usually via weighted or unweighted voting). This is why Ensemble Learning methods need higher processing power. Ensemble models are under the topic of supervised learning models because they may be taught and used to make predictions. One prerequisite must be met for a group of classifiers to perform better than any one of them alone: the classifiers must be reliable and diverse. Both when there is a surplus of data and when there is not enough, ensemble-based systems can be used.

A BC prediction model called Machine Learning Expert System for Breast Cancer Detection (MLESBCD) is proposed in this study in order to improve BC prediction using machine learning techniques using MICE data imputation technique. The MLESBCD eliminates the outliers after doing preprocessing on using MICE to impute the missing values from the BC data using different outlier detection techniques to make the dataset more perfect for training MLESBCD.

MLESBCD uses Isolation Forest, Local Outlier Factor, One Class Support Vector Machine and Minimum Covariance Determinant for outlier detection. The preprocessed data is then fitted into ensemble learning techniques for classifications [7]: Extreme Gradient Boosting, Adaptive Boosting with Random Forest. To improve the efficacy and precision of the breast cancer categorization [8], missing value treatment and outlier detection are done.

The paper is divided into 5 sections—Sect. 2 gives the literature survey, Sect. 3 depicts the methodology followed by Sect. 4 which discusses the results and finally Sect. 5 draws the conclusion.

2 Literature Review

Many researchers have given their thoughts to develop suitable models for breast cancer detection [9]. The majority of machine learning approaches, some of which are described below, have been utilised to obtain high accuracy.

For the categorization and diagnosis of breast cancer, a hybrid approach was provided by Bibhu et al. [10] using a multivariate statistical technique based on artificial intelligence methods. To reduce dimensionality, they used PCA (Principal Component Analysis). For the classification stage, Artificial Neural Networks (ANN) were utilised, extracting relevant data, and PCA as a preprocessing step. Their model achieves 97% categorization accuracy.

In order to predict breast cancer, an ensemble model created by Gupta and Gupta [11] incorporates four distinct classifiers: K-Nearest Neighbor, Decision Tree, Support Vector Machine (SVM), Logistic Regression, and They assigned weights to each of the classification models using the Using the soft-voting methodology, the SLSQP (Sequential Least Squares Programming) method combines the predictions of different classifiers. Their collective model had a 97.88% accuracy.

A technique called GA-MOO-NN was proposed by Ahmad et al. [12] for the identification of breast cancer (Genetic Algorithm-based Multi-Objective Optimization of an Artificial Neural Network). It carries out a simultaneous exploration to identify the feature subsets that are most significant. The hidden nodes with optimal number and the initial network weights are also included in the exploration's output. The best classification accuracy achieved by the system was 98.85%.

Support Vector Machine (SVM), Random Forest (RF), and Bayesian Networks (BN) were three different machine learning techniques that Bazazeh and Shubair [13] examined for the identification and breast cancer diagnosis. Their findings demonstrate that SVM outperforms the other two approaches in terms of accuracy, precision, and specificity. They managed to achieve a 97% accuracy rate.

In order to predict breast cancer, Alwidian et al. [14] investigated the Weighted Classification Based on the Association Rules algorithm, which is based on association rules. It produced class association rules by addressing the flaws in numerous AC (Association Classification) methodologies. Their findings demonstrate that WCBA offers accurate guidelines that include the most helpful traits for predicting breast cancer.

A system for diagnosing cancer using both methods was proposed by Karthik et al. [15], using a deep neural network (DNN) along with the feature selection and the classification model method known as recursive feature elimination (RFE). Fidelity of the DNN model was very high at 98.62%.

For the purpose of predicting breast cancer, Islam et al. [16] used the supervised machine learning techniques like Support Vector Machine (SVM) and K-Nearest Neighbors (KNN). The SVM they suggest, and KNN models yield accuracy 98.57% and 97.14%, respectively.

According to Bektas et al. [17], early identification of breast cancer is essential for preserving lives and taking the necessary actions to treat the condition. Breast cancer mostly affects females and is one of the leading causes of death among females. The 139 patient traits were mostly utilised. The following paper used approaches from random forest, k-stars, and selective sensor neural networks. A 61.85% accuracy rate was discovered. Earlier studies either employed a single classifier or only ensemble learning, whereas the result was not compared after implementing different outlier detection techniques. Therefore, by correcting the shortcomings of the aforementioned research, our work demonstrates improved model performance. Therefore, with this model's greater performance is achieved by selecting the best outlier detection technique that results in more accuracy which increases the overall efficiency.

3 The Proposed Methodology

This study can be divided into following three sections

- Data Collection
- Data Preprocessing
- Classification.

The steps are described in full below. The MLESBDC model's process diagram is depicted in Fig. 1.

3.1 Data Collection

3.1.1 Breast Cancer Dataset

The dataset, often known as the Wisconsin Breast Cancer dataset, is a typical dataset for employing bioinformatics and machine learning techniques [18]. The dataset was chosen since it has been referenced in a number of articles, and it is from there that the Ensemble Learning model was built. The dataset contains a total number of 9 features labelled in either benign or malignant classes. The dataset has a target column, 9 characteristics, and 698 occurrences. The characteristics are listed in Table 1 and the dataset just includes numbers.

3.2 Data Preprocessing

Each feature in the BC dataset is checked for missing values, and any corrupted data that is inappropriate, out of range, or otherwise flawed is then eliminated. The procedure of extracting relevant and important information from the dataset will have the biggest impact on the outcome of the subsequent forecast.

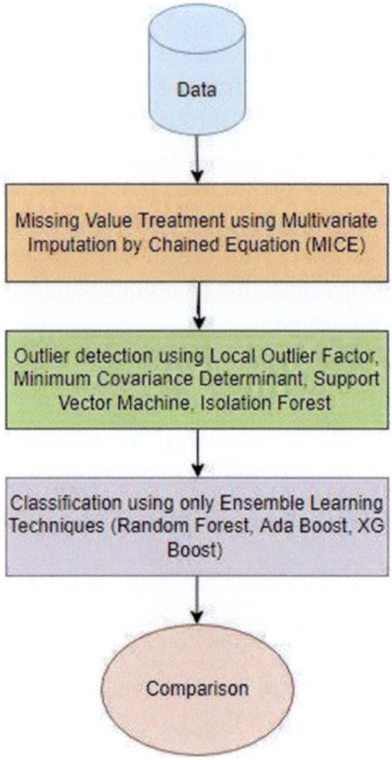

Fig. 1 Work-flow of MLESBCD

Table 1 List of features

Dataset attributes	Domain
Thickness of clump	1: 10
Uniformity of cell size	1: 10
Cell shape uniformity	1: 10
Marginal adhesion	1: 10
Single epithelial cell size	1: 10
Bare nuclei	1: 10
Bland chromatin	1: 10
Normal nucleoli	1: 10
Mitoses	1: 10
Class	2 for benign 4 for malignant

3.2.1 Missing Value Treatment

16 instances of the feature Bare Nuclei in Wisconsin Breast Cancer contain a single unavailable (missing) attribute value, denoted by "?" This attribute consists of

Table 2 Missing values

Attributes	Quantity of missing values
Thickness of clump	0
Uniformity of cell size	0
Cell shape uniformity	0
Marginal adhesion	0
Single epithelial cell size	0
Bare nuclei	16
Bland chromatin	0
Normal nucleoli	0
Mitoses	0
Class	0

improper data and hence the missing values are imputed with advanced imputation technique MICE.

In this research the iterative imputer uses the features other than the Bare Nuclei to predict the missing value of Bare Nuclei using Linear Regression in an iterative fashion. This process is repeated for max iteration of 10 rounds till the difference amongst the consecutive imputed values gets minimized. Table 2 shows the missing values of the dataset.

3.2.2 Outlier Detection

To determine whether or not there are outliers, box plot analysis is utilized. As can be seen in Fig. 2, the homogeneity of cell shape has two outliers, the marginal adhesion has three, the naked nuclei have one, the normal nucleoli have two, and the mitoses have several.

The proposed MLESBCD employs four different outlier detection methods in order to identify outliers and compare them. Here, the One Class Support Vector Machine, Isolation Forest, Local Outlier Factor, and Minimum Covariance Determinant are utilized to identify outliers. The box plot representation of the dataset is shown in Fig. 2. Therefore, it may be concluded that there are a lot of outliers, but as box plots are only useful for single-dimensional data, they cannot be used to identify outliers.

Local Outlier Factor

Local Outlier Factor (LOF) is a density based unsupervised anomaly detection method where each point is assigned a scoring of how isolated or how likely it is to be outliers based on the size of its neighborhood. It is measured with the help of

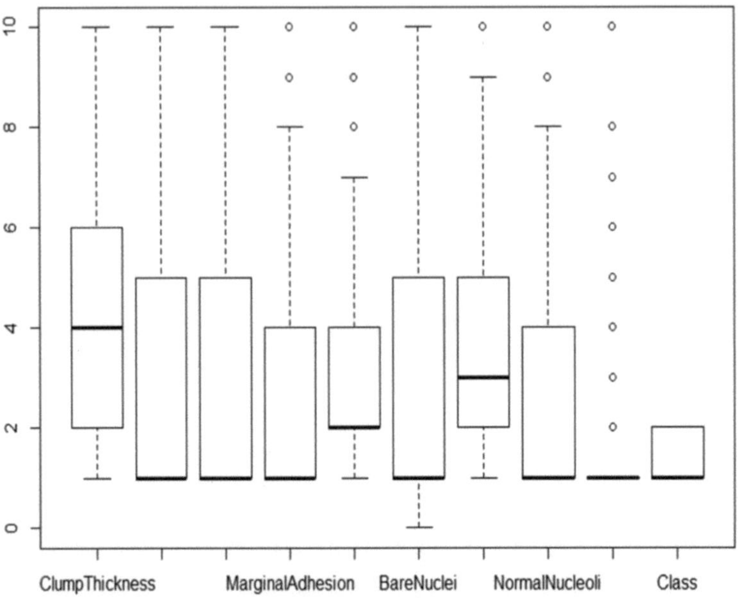

Fig. 2 Box plot representation of the dataset

statistical means of a data point like k-distance, k-nearest neighbors and reachability distance.

One Class Support Vector Machine

One Class Support Vector Machine (One-Class SVM) is an unsupervised outlier detection method which can be used for both regression and classification tasks. It is more likely to be used in imbalanced datasets. The underlying quality of this method is that it creates nonlinear decision boundaries by projecting the data belonging to a similar class using nonlinear functions like a hyperplane. This hyperplane plays a major role in differentiating one class from another i.e. for detecting the presence of outliers.

Isolation Forest

Isolation Forest (iForest), which employs the decision-tree method, is comparable to Random Forest. By selecting a feature at random, followed by a split value between that feature's maximum and minimum values, Isolation Forest isolates observations. When a forest with a number of random trees produces a shorter path, it is highly likely to be an anomaly. Outliers have attribute values that are drastically different from the normal instances.

Minimum Covariance Determinant

This basic statistical approach may be used to find outliers if the input variables have a Gaussian distribution. The Minimum Covariance Determinant (MCD) is a robust covariance estimator which is relevant on data drawn from a unimodal, symmetric distribution.

3.3 *Classification*

Classification is one of the problems with supervised learning in machine learning. It's used to divide large amounts of data into different values or classes. The task's outcome is to determine which of the aforementioned class labels a specific data point belongs to.

3.3.1 Random Forest

Random Forest (RF) or in the training phase as an ensemble learning approach for classification, regression, and other issues, random decision forests create a lot of decision trees [19]. The result of the random forest for classification issues is the class that most of the trees choose. For regression tasks, the mean or average forecast of each individual tree is returned.

3.3.2 AdaBoost (Adaptive Boosting)

A machine learning technique called AdaBoost can be used to enhance other machine learning methods' performance. These models achieve accuracy significantly above chance on a classification task. Decision trees with one level are the best fit and, hence, the most frequently employed technique with AdaBoost. These trees are commonly known as decision trees since they are so brief and contain only one classification option stumps.

3.3.3 Extreme Gradient Boosting (XGBoost)

The subgroup of boosting strategies in ensemble learning to which XGBoost belongs. A collection of predictors from many models is used in ensemble learning to increase prediction accuracy [20]. The Boosting technique tries to fix errors from prior models by giving the models specific weights. The XGBoost method is a decision tree-based ensemble machine learning technique that works inside the gradient boosting framework. In its most basic versions, decision trees are simple to see and comprehend, but it may be difficult to develop intuition for the next generation of tree-based algorithms.

4 Result and Analysis

Three phases of analysis are used to look at the output of the suggested model MLESBCD. In stage 1, the evaluation metrics were analysed by implementing the classification models in the original dataset. In stage 2, the original dataset was pre-processed to handle missing values using MICE imputation technique and then the evaluation metrics were recorded accordingly. In stage 3, outlier detection with data imputation is used which results in further improvement in the evaluation metrics. The result of the model is analyzed with Confusion Matrix and its different statistical metrics [21] which are shown below. The Confusion Matrix for binary classification issues is shown in Table 3.

From Table 3, accuracy is defined as:

$$\text{Accuracy} = (\text{TP} + \text{TN})/(\text{TP} + \text{TN} + \text{FN} + \text{FP})$$

Table 4 shows the accuracy comparison before and after data imputation using MICE with RF, AdaBoost and XGBoost.

Table 5 compares the proposed MLESBCD's accuracy utilizing RF, AdaBoost, and XGBoost with proposed outlier detection techniques i.e. LOF, One-Class SVM, iForest, MCD.

Table 5 displays the proposed MLESBCD with RF, AdaBoost, and XGBoost classification accuracy against the outlier detection algorithms i.e. LOF, One-Class SVM, iForest, MCD. Also, in the proposed MLESBCD, all the missing values are handled and imputed with MICE imputation technique which enhances the performance of the classifiers. Further analyzing, it is observed that after removing outliers with LOF, MLESBCD with XGBoost provides the greatest accuracy (98.57%) of all. By adding weights to the models and utilizing the suggested MLESBCD, which

Table 3 Confusion matrix

	Positive (1)	Negative (0)
Positive (1)	True positive (TP)	False positive (FP)
Negative (0)	False negative (FN)	True negative (TN)

Table 4 Accuracy comparison (in %)

Algorithm used	Accuracy (in %)
Random forest	93.28
XGBoost	96.54
AdaBoost	93.01
RF with MICE	95.50
XGBoost with MICE	97.01
AdaBoost with MICE	94.12

Table 5 Accuracy comparison (in %)

	MLESBCD with RF	MLESBCD with AdaBoost	MLESBCD with XGBoost
LOF	98.56	96.42	98.57
One-class SVM	98.43	96.87	96.88
iForest	92.85	93.75	96.42
MCD	96.26	94.02	96.26

eliminates and manages missing data and outliers, XGBoost alone corrects the faults created by prior models in the succeeding models, leading to better performance.

Precision which is defined as the ratio of the total of the TP and FP divided by the TP.

$$\text{Precision} = (\text{TP})/(\text{TP} + \text{FP})$$

Table 6 shows the precision comparison before and after data imputation using MICE with RF, AdaBoost and XGBoost.

Table 7 shows the precision comparison of the different models with their corresponding outlier detection methods.

Recall is defined as the ratio of TP divided by the total of TP and FN.

$$\text{Recall} = (\text{TP})/(\text{TP} + \text{FN})$$

Table 6 Precision comparison (in %)

Algorithm used	Precision (in %)
Random forest	93.62
XGBoost	96.02
AdaBoost	92.98
RF with MICE	95.58
XGBoost with MICE	97.01
AdaBoost with MICE	94.03

Table 7 Precision comparison (in %)

	MLESBCD with RF	MLESBCD with AdaBoost	MLESBCD with XGBoost
LOF	98.56	96.41	98.62
One-class SVM	98.51	96.86	96.86
iForest	93.24	95.25	96.42
MCD	96.30	93.97	96.27

Table 8 shows the recall comparison before and after data imputation using MICE with RF, AdaBoost and XGBoost.

Table 9 shows the recall comparison of the different models with their corresponding outlier detection methods.

Better performance in terms of recall and accuracy shows that the forecast is almost accurate since there are relatively fewer false positives and false negatives. From Table 7 it is observed that the proposed model MLESBCD with XGBoost gives the highest precision of 98.62% after removing outliers with the LOF algorithm. Better performance in terms of precision illustrates that the forecast is nearly accurate and there are quite few false positives. Also from Table 9, it is observed that the proposed model MLESBCD with XGBoost after LOF outlier detection gives the highest recall of 98.57%. Better performance in terms of recall illustrates that comparatively few false negatives occur, and the forecast is nearly correct. High accuracy and recall values are crucial for business from an application standpoint, healthcare and medical purposes.

F-measure or F1 score is a measure of a test's accuracy.

$$\text{F-measure} = 2 \times (\text{Precision} \times \text{Recall})/(\text{Precision} + \text{Recall})$$

Table 10 shows the F-measure comparison before and after data imputation using MICE with RF, AdaBoost and XGBoost.

Table 8 Recall comparison (in %)

Methods	Recall (in %)
Random forest	93.57
AdaBoost	93.01
XGBoost	96.01
RF with MICE	95.52
XGBoost with MICE	97.01
AdaBoost with MICE	94.01

Table 9 Recall comparison (in %)

	MLESBCD with RF	MLESBCD with AdaBoost	MLESBCD with XGBoost
LOF	98.56	96.42	98.57
One-class SVM	98.43	96.87	96.87
iForest	92.85	93.75	96.42
MCD	96.26	94.02	96.26

Table 10 F-measure comparison (in %)

Algorithm used	Accuracy (in %)
Random forest	93.58
XGBoost	96.02
AdaBoost	93.01
RF with MICE	95.52
XGBoost with MICE	97.01
AdaBoost with MICE	94.01

Table 11 F-measure comparison (in %)

	MLESBCD with RF	MLESBCD with AdaBoost	MLESBCD with XGBoost
LOF	98.57	96.41	98.58
One-class SVM	98.44	96.87	96.88
iForest	93.00	94.05	96.42
MCD	96.27	93.97	96.27

The F-measure comparison of the various models is shown in Table 11 along with the related outlier identification techniques.

5 Conclusion

BC, the most common type of cancer discovered and the main reason why women die from cancer as being life-threatening in female patients. On the BC, the suggested model, MLESBCD, fared a better dataset with proposed techniques. The techniques were applied on the classification based models for the purpose of enhancement in the accuracy of the MLESBCD. Although it raises the system's computing cost, this research, the exceptionally high detection rate is prioritised over complexity. Ensemble voting improves the system's reliability and efficiency. The suggested approach is therefore more resistant to any unforeseen circumstances that might affect a certain classifier model. MLESBCD is more useful since it will give breast cancer patients better care and a more accurate diagnosis. The model MLESBCD with XGBoost paired with LOF outlier detection generates the highest accuracy, precision, recall, and F-measure among the three ensemble classifiers and four outlier detection techniques. Deep learning techniques may be used in the future to create this system, which may result in great accuracy.

References

1. https://www.ncbi.nlm.nih.gov/pmc/articles/PMC3255438. Accessed 12 Feb 2022
2. https://www.webmd.com/breast-cancer/stages-grades-breast-cancer. Accessed 12 Feb 2022
3. Some basic aspects of statistical methods and sample size determination in health science research. https://www.ncbi.nlm.nih.gov/pmc/articles/PMC4279315/. Accessed 14 Feb 2022
4. Comparative analysis of breast cancer detection using machine learning and biosensors. https://www.sciencedirect.com/science/article/pii/S2667102621000887. Accessed 15 Feb 2022
5. When and how should multiple imputation be used for handling missing data in randomized clinical trials—a practical guide with flowcharts. https://doi.org/10.1186/s12874-017-0442-1. Accessed 15 Feb 2022
6. Breast cancer prediction using machine learning. https://www.researchgate.net/publication/341508593_BREAST_CANCER_PREDICTION_USING_MACHINE_LEARNING. Accessed 16 Feb 2022
7. Types of classification tasks in machine learning. https://machinelearningmastery.com/types-of-classification-in-machine-learning/. Accessed 17 Feb 2022
8. Deep learning to improve breast cancer detection on screening mammography. https://www.nature.com/articles/s41598-019-48995-4. Accessed 17 Feb 2022
9. Predicting breast cancer 5-year survival using machine learning: a systematic review
10. Sahu B, Mohanty SN, Rout SK (2019) A hybrid approach for breast cancer classification and diagnosis. EAI Endorsed Trans Scalable Inf Syst 6(20)
11. Gupta M, Gupta B (2018) An ensemble model for breast cancer prediction using sequential least squares programming method (SLSQP). In: 2018 eleventh international conference on contemporary computing (IC3). IEEE, pp 1–3
12. Ahmad F, Isa NAM, Hussain Z, Sulaiman SN (2013) A genetic algorithm-based multi-objective optimization of an artificial neural network classifier for breast cancer diagnosis. Neural Comput Appl 23(5):1427–1435
13. Bazazeh D, Shubair R (2016) Comparative study of machine learning algorithms for breast cancer detection and diagnosis. In: 2016 5th international conference on electronic devices, systems and applications (ICEDSA). IEEE, pp 1–4
14. Alwidian J, Hammo BH, Obeid N (2018) WCBA: weighted classification based on association rules algorithm for breast cancer disease. Appl Soft Comput 62:536–549
15. Karthik S, Perumal RS, Mouli PC (2018) Breast cancer classification using deep neural networks. In: Knowledge computing and its applications. Springer, pp 227–241
16. Islam MM, Iqbal H, Haque MR, Hasan MK (2017) Prediction of breast cancer using support vector machines and k-nearest neighbors. In: 2017 IEEE region 10 humanitarian technology conference (R10-HTC). IEEE, pp 226–229
17. Majeed B, Iqbal HT, Khan U, Bin Altaf MA (2018) A portable thermogram based non-contact non-invasive early breast-cancer screening device. In: 2018 IEEE biomedical circuits and systems conference (BioCAS), Cleveland, OH, pp 1–4
18. Bharat A, Pooja N, Reddy RA (2018) Using machine learning algorithms for breast cancer risk prediction and diagnosis. In: 2018 3rd international conference on circuits, control, communication and computing (I4C), Bangalore, India, pp 1–4
19. How random forest algorithm works in machine learning. Medium (2020). https://medium.com/@Synced/how-randomforest-algorithm-works-in-machine-learning-3c0fe15b6674. Accessed 20 Feb 2022
20. XGBoost algorithm: long may she reign! Medium (2020). https://towardsdatascience.com/https-medium-comvishalmorde-xgboost-algorithm-long-she-may-rein-edd9f99be63d. Accessed 23 Feb 2022
21. Accuracy, recall, precision, F-score and specificity, which to optimize on? Medium (2020). https://towardsdatascience.com/accuracy-recall-precision-f-scorespecificity-which-to-optimize-on-867d3f11124. Accessed 23 Feb 2022

Adaptive Rank-Based Switching Vector Median Filter for Removing Impulse Noise from Color Medical Image

Thiyam Romita Chanu, Th Rupachandra Singh, and Kh Manglem Singh

Abstract In this paper, a switching filter based on adaptive rank for salt and pepper noise removal from color medical images is presented. Color medical images are widely used to detect and diagnose diseases. But image quality is frequently degraded by noise, and such images are not suitable for the purpose. So, eliminating noise is a crucial stage in the processing of images. In this paper, for detecting impulse noise, an adaptive rank is utilized to classify a pixel as a noise candidate or not. The center pixel is probably noisy in the first step if its rank is more than the calculated rank according to the contamination level. This probably noisy pixel is further checked in the second stage to determine if it is an edge or impulse. The vector median filter's output is used to substitute pixels that are found to be noisy. The simulation results demonstrate that the proposed filter outdoes other state-of-the-art filters in removing and retaining image details.

Keywords Impulse noise · Color medical image · Vector median filter · Salt and pepper noise

1 Introduction

Color image processing is becoming an increasingly significant area in the field of health care. Cervicography, gastrointestinal endoscopy, wound pathology, dermoscopy, etc. are some of the images where color information has been introduced [1]. These images are widely used in the automated computer-aided diagnosis system to detect and diagnose various diseases such as skin cancer, breast cancer, glaucoma, etc. This work mainly consists of two steps—image preprocessing and segmentation. The preprocessing step often consists of image resizing, contrast enhancement,

T. R. Chanu (✉) · T. R. Singh
Department of Computer Science, Manipur University, Imphal, India
e-mail: thromita@manipuruniv.ac.in

K. M. Singh
Department of Computer Science and Engineering, NIT Manipur, Imphal, India

S. K. Biswas et al. (eds.), *Intelligent Computing Systems and Applications*, Lecture Notes in Networks and Systems 1307, https://doi.org/10.1007/978-981-96-3860-4_23

noise reduction, and hair removal. Impulse noise typically degrades medical images because of issues with camera sensors during recording, transmission, communication, processing, etc. This noise degrades the image quality and damages the structure and other essential information required for an accurate diagnosis. So, efficient algorithms are required to detect and remove noise from the images as much as possible for the precise diagnosis and detection of disease and further steps for various purposes.

Impulse noise is one of the most common types of noise found in an image. In a color image, the presence of such type of noise causes total loss of pixel information because a different value substitutes the original channel information. If the impulse noise in the image is salt and pepper, then the pixel's channel value is replaced by either 0 or 255. So, noise filtering should be applied at the beginning of image processing to get accurate results. Few works for eliminating noise from images in the preprocessing stage are found in [1, 2].

Color images are multichannel images. So, in order to preserve the inherent correlation between the color channels, vector ordering techniques are preferred. Vector median filter (VMF) [3], basic vector directional filter (BVDF) [8], and directional-distance filter (DDF) [9] are the classical vector filters for eliminating impulse noise from color images. The median filter of the grayscale image is extended by these filters. They are based on reduced ordering principles, and the output is the sample that gives the minimum aggregated color distances. But these filters have the disadvantage of too much smoothing. It also damages the fine information as they perform filtering operations on every pixel, regardless of whether the pixel is an impulse or not. Various switching filters are proposed in the literature to overcome the drawbacks of these filters. Switching filters consist of two steps: first is detecting noisy pixels, and second is restoring noisy pixels. Various techniques are found in the literature to detect noisy pixels. In AVMF [10], the difference between the average of the low-ranked closest pixel and the center pixel is used as the detection criterion. Modified Switching Median Filter (MSMF) [11] used a combination of AVMF and four one-dimensional Laplacian operators in the detection step. Adaptive rank-weighted switching filter (RWASF) [12] used the weighted sum of distances, but when the noise density is present medium or high, its performance is low. Switching filters that use the rank of the pixels for detecting noisy pixels are found in [13, 14]. A switching filter based on Laplacian convolution and pixel grouping was proposed in [15].

In an image corrupted by salt and pepper noise, if the center pixel in a kernel is corrupted, then the sum of the aggregated distance of the center pixel to all other pixels in the kernel would be very large and would result in the extreme end. The center pixel's rank is used as a noise detection criterion in many papers [13, 14]. Also, in most noise detection techniques, a single rank is considered for different noise contamination ranges, mostly 10–70%. But the number of noise and noise-free pixels in the kernel will vary depending on the noise contamination level. An optimum filtering would be achieved if the rank that partitions the pixels into noisy and noise-free pixels could be determined for different noise contamination levels.

In medical images, the loss of image information is not suited as it may contain important information about the diagnosis. Therefore, a switching filter based on the adaptive rank and Laplacian convolution is formulated to reduce impulse noise from the medical image. If the center pixel's rank is greater than the determined rank for the contamination level, it might be noisy. Then, such a pixel is checked using Laplacian convolution to confirm whether it is an edge or an impulse. Finally, the noisy pixel is restored by the output of VMF.

The presentation of the manuscript is as follows: In Sect. 2, we give some related works. The following section is the proposed work, which is followed by experimental results and analysis. Lastly, a conclusion is drawn in the last paragraph.

2 Related Works

A brief description of vector median filter (VMF) [3], rank conditioned vector median filter (RCVMF), rank conditioned and threshold vector median filter (RCTVMF) [13], and quaternion switching vector median filter (QSVMF) [14] is presented in this section.

2.1 *Vector Median Filter (VMF)*

As stated above, VMF is the most widely used filter for eliminating impulse noise from color images. This filter treats a pixel as a vector signal having multiple correlated channels. A vector pixel consists of n components, that is $\boldsymbol{x}_i = \left(x_{i,1}, x_{i,2}, \ldots, x_{i,n}\right)$. A 3 × 3 kernel, shown in Fig. 1, is used to perform the filtering operation on the center vector pixel. For calculating the vector median of the vectors in the kernel, the sum of distances of every vector to all the other vector pixels in the kernel is calculated as follows:

$$d_l(i,j) = \left(\sum_{k=1}^{n} \left|x_{i,k} - x_{j,k}\right|^l\right)^{\frac{1}{l}} \tag{1}$$

where $l = 1$ for city block distance and $l = 2$ for Euclidean distance.

Fig. 1 A 3 × 3 kernel

x_2	x_5	x_7
x_3	x_1	x_8
x_4	x_6	x_9

The sum of distances δ_i of the vector pixels $\boldsymbol{x}_i$ is calculated using

$$\delta_i = \sum_{j=1}^{N} d_l(i, j) \tag{2}$$

The sum of distances δ_i is arranged in increasing order as follows:

$$\delta_{(1)} \leq \delta_{(2)} \leq \ldots, \leq d_{(N)} \tag{3}$$

which is same as the ordering of the vector pixels

$$\boldsymbol{x}_{(1)} \leq \boldsymbol{x}_{(2)} \leq \ldots, \leq \boldsymbol{x}_{(N)} \tag{4}$$

And the output is $\boldsymbol{x}_{(1)}$, which occupies the lowest rank in the ordering.

2.2 *Rank Conditioned Vector Median Filter (RCVMF), and Rank Conditioned and Threshold Vector Median Filter (RCTVMF) [14]*

RCVMF is a switching filter that applies filtering operations only to corrupted pixels. In this filter, the center vector pixel's rank is used as the main noise detection criterion. It is calculated by assigning a rank to the ordered vector pixels of Eq. (4) of VMF. If the center vector pixel's rank is more than a healthy vector pixel's predetermined rank in the kernel, then the output of VMF replaces it. This filter is expressed as

$$\boldsymbol{x}_{\text{RCVMF}} = \begin{cases} \boldsymbol{x}_{\text{VMF}}, & \text{if } r_N > r_k \\ x, & \text{otherwise} \end{cases} \tag{5}$$

RCTVMF incorporates a distance D to the noise detection method of RCVMF, which is calculated by using

$$D = \Delta\left(\boldsymbol{x}, \boldsymbol{x}_{(k)}\right) \tag{6}$$

where the center vector pixel is $\boldsymbol{x}$ and $\boldsymbol{x}_{(k)}(1 < k < N)$ is a healthy rank-ordered vector pixel inside the kernel.

If the center vector pixel's rank is more than a predefined healthy vector pixel's rank and D is greater than a predetermined value, then it is noisy and the output of VMF replaces it; otherwise, it remains unchanged.

Fig. 2 A 3 × 3 kernel used in QSVMF

q_1	q_2	q_3
q_4	q_5	q_6
q_7	q_8	q_9

2.3 *Quaternion Switching Vector Median Filter (QSVMF)*

In this filter, a quaternion is used to represent differences between two pixels. A 3 × 3 kernel shown in Fig. 2 is used for performing the filtering operation. The following equations are used for computing the difference between two pixels.

$$d(q_1, q_2) = d_1(q_1, q_2) + d_2(q_1, q_2) \tag{7}$$

$$d_1(q_1, q_2) = \frac{1}{\sqrt{3}}\left[r_1 + g_1 + b_1 - r_2 - g_2 - b_2\right] \tag{8}$$

$$d_2(q_1, q_2) = \frac{1}{\sqrt{3}}\begin{bmatrix} ((b_1 - g_1) - (b_2 - g_2))i + ((r_1 + b_1) - (r_2 - b_2))j \\ + ((g_1 - r_1) - (g_2 - r_2))k \end{bmatrix} \tag{9}$$

$$\delta_m = \sum_{n=1}^{N} d(q_m, q_n), \quad (1 \le n \le N) \tag{10}$$

Using Eq. (10), the aggregated color distance of each pixel to other pixels in the kernel is calculated.

Ranks are assigned to the sorted aggregated distances $\delta_1, \delta_2, \delta_3 \ldots \delta_N$ as shown below

$$\delta_{(1)} \le \delta_{(2)} \le \delta_{(3)} \ldots \delta_{(N)} \tag{11}$$

which is the same as the input set ordering, i.e.,

$$q_{(1)} \le q_{(2)} \le q_{(3)} \cdots q_{(N)} \tag{12}$$

If the center vector pixel's rank exceeds a predetermined rank, it is regarded as an impulse, and is further checked by calculating the minimum of the four-edge directional method, i.e., $D = \min(D_1, D_2, D_3, D_4)$ and

$$D_1 = \frac{1}{2}(d(q_5, q_4) + (q_5, q_6)) \quad (13)$$

$$D_2 = \frac{1}{2}(d(q_5, q_3) + (q_5, q_7)) \quad (14)$$

$$D_3 = \frac{1}{2}(d(q_5, q_2) + (q_5, q_8)) \quad (15)$$

$$D_4 = \frac{1}{2}(d(q_5, q_1) + (q_5, q_9)) \quad (16)$$

If D is more than a threshold, the center vector pixel is an impulse and is substituted by the output of VMF. This filter is represented as follows:

$$q_{\mathrm{QSVMF}} = \begin{cases} q_{\mathrm{VMF}}, & \text{if } r_c > r_k \text{ and } D > \mathrm{Tol} \\ q_5, & \text{otherwise} \end{cases} \quad (17)$$

3 Proposed Work

In this filter, we propose a switching filter using a 3×3 kernel, as shown in Fig. 1, for performing the filtering process. The proposed work employs a two-stage noise detection step and utilizes VMF output for those that are found noisy. In the first step, the rank of the center vector pixel is used to classify whether it is noisy or not. For determining the rank of the healthy pixel at different noise densities, the following rank estimation equation [15] is used:

$$\mathrm{Td} = \mathrm{round}(7 - \mathrm{P} \times 6.66) \quad (18)$$

The vector ordering technique used in VMF is performed to find the center pixel's rank. If the center pixel's rank is more than Td, it is considered as a noise candidate. In the second step, this noise candidate pixel is checked using four-dimensional Laplacian convolution to determine whether it is an edge pixel.

In the second step, the pixel distances between the center vector pixel $\boldsymbol{x}_1$ and the other two vectors in directions of 0, 45, 90, and 135 are calculated as follows:

$$V_1 = \frac{1}{2}(d(\boldsymbol{x}_1, \boldsymbol{x}_3) + d(\boldsymbol{x}_1, \boldsymbol{x}_8)) \quad (19)$$

$$V_2 = \frac{1}{2}(d(\boldsymbol{x}_1, \boldsymbol{x}_4) + d(\boldsymbol{x}_1, \boldsymbol{x}_7)) \quad (20)$$

$$V_3 = \frac{1}{2}(d(\boldsymbol{x}_1, \boldsymbol{x}_5) + d(\boldsymbol{x}_1, \boldsymbol{x}_6)) \quad (21)$$

$$V_4 = \frac{1}{2}(d(\boldsymbol{x}_1, \boldsymbol{x}_2) + d(\boldsymbol{x}_1, \boldsymbol{x}_9)) \tag{22}$$

When the center vector pixel is on a smooth area, the pixel differences in four directions are all equal to approximately zero, and when it is on an edge, at least one value among the values will be extremely small. The center vector pixel is considered corrupted if $V_{\min} = \min(V_i\ (i = 1, 2, 3, 4))$ it exceeds a threshold value Tol and will be restored by the output of VMF. Otherwise, the center pixel is considered as noise-free and kept unchanged.

The proposed filter (PF) is represented as

$$\text{PF} = \begin{cases} \text{VMF}, & \text{if } r_c > Td \text{ and } V_{\min} > \text{Tol} \\ x_1, & \text{unchanged} \end{cases} \tag{23}$$

4 Experimental Results and Analysis

The experiment is conducted by using the following impulse noise model [2].

$$n(x) = \begin{cases} o & \text{with probability } 1 - \emptyset \\ (d, o_2, o_3)^T & \text{with probability } \emptyset_1 \cdot \emptyset \\ (o_1, d, o_3)^T & \text{with probability } \emptyset_2 \cdot \emptyset \\ (o_1, o_2, d)^T & \text{with probability } \emptyset_3 \emptyset \\ (d, d, d)^T & \text{with probability } \emptyset \cdot \emptyset\Sigma \end{cases} \tag{24}$$

where $n(x)$ is the noisy signal, $\text{o} = (o_1, o_2, o_3)^T$ represents the uncorrupted pixel, $\emptyset$ is the sample corruption probability, $\emptyset_1$, $\emptyset_2$ and $\emptyset_3$ are the corruption probability of the channel and $\emptyset\Sigma = 1 - \emptyset_1 - \emptyset_2 - \emptyset_3$, d is the impulse value which can be 0 or 255 for a fixed value impulse noise having equal probability. Three objective measures [3] are used for measuring the capabilities of the proposed filter. They are Normalized Color Distance (NCD), Peak Signal to Noise Ratio (PSNR), and Mean Absolute Error (MAE). NCD measures the perceptual error in the CIELab color space, PSNR measures noise suppression capability, and detail preservation capability by MAE. An efficient filter should have a low MAE and NCD while having a high PSNR.

Four different standard color images of size 512×512 (Fig. 3) [4–7] are used for our experiment. State-of-the-art filters meant for impulse noise removal are selected for evaluating the efficiency of the proposed method. VMF [3], AVMF [10], MSMF [11], RWASF [12], and SFLCPG [15] are chosen. Experiments are performed using MATLAB version R2015a on a system with an i5 processor at 1.80 GHz. To maximize the efficacy of the proposed filter, we have to find the optimal value of Tol. Figure 4 shows the PF performance for different threshold values for the noise density ranging from 10 to 70%. It is found that the PF performs best at Tol = 15. The parameters of the other filters used for comparison are shown in Table 1. The parameters are set according to the appropriate recommendation given in the references. Simulation results of the PF on Lena, duodenal_ulcer, and brain images compared with the selected filters VMF, AVMF, MSMF, RWASF, and SFLCPG are shown in Tables 2, 3, and 4, respectively.

From the experimental results shown in Table 2 of Lena image, it is found that the RWASF works well at low noise density (10 and 20%) than other filters, and the results are very close to SFLCPG and PF. The PF performs better than other filters at noise density ranging from 30 to 70%. For the duodenal_ulcer image, the AVMF has the highest PSNR at 20% noise density while the PF has the highest PSNR from 30 to 70%, and the lowest MAE and NCD for all noise density ranges, which shows that the proposed filter can remove noise and can retain image details better than other filters. For the brain image shown in Table 4, the RWASF works better than the other filter at low noise (10 and 20%) density, and our PF gives better results at other

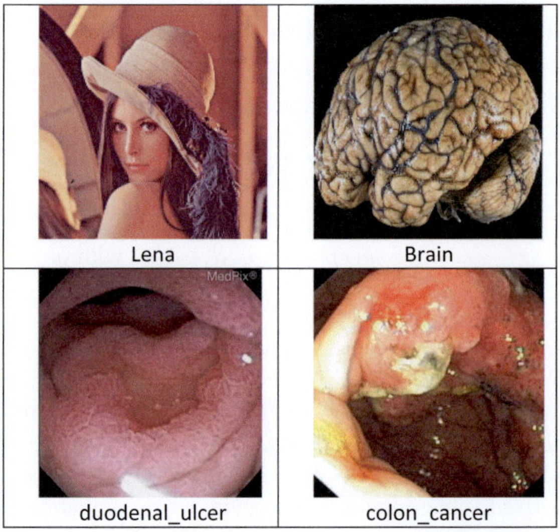

Fig. 3 Images used in this experiment

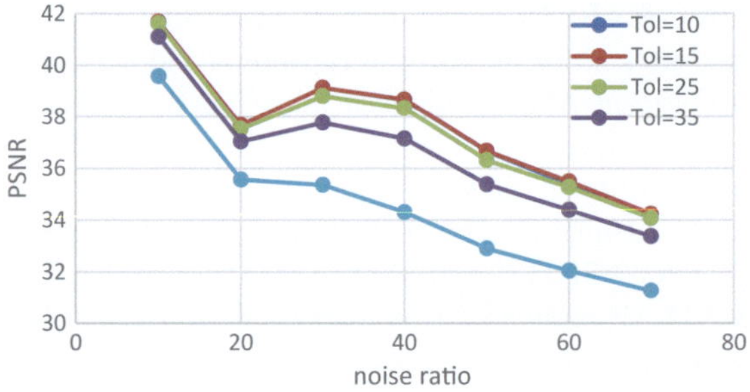

Fig. 4 PSNR values of duodenal_ulcer image at Tol = 15

Table 1 Parameters of the different filters

Filters	Parameters
AVMF [10]	Tol = 35
MSMF [11]	Tol = 30
RWASF [12]	Tol = 65
SFLCPG [15]	LT = 75

Table 2 Comparison of PF in Lena image

Noise (%)	Parameter	VMF	AVMF	MSMF	RWASF	SFLCPG	PF
10	NCD	0.0213	0.0060	0.0065	0.0043	**0.0030**	0.0041
	PSNR	31.76	34.5481	34.5459	**38.6144**	37.5982	37.6112
	MAE	3.72	1.1225	1.1370	**0.5137**	0.6516	0.5901
20	NCD	0.0199	**0.0071**	0.0073	0.0156	0.0158	0.0157
	PSNR	31.31	33.7415	33.5821	**34.6803**	34.1587	34.4279
	MAE	3.9390	1.4881	1.5466	**1.0246**	1.1337	1.0561
30	NCD	0.0168	**0.0076**	0.0080	0.0166	0.0164	0.0163
	PSNR	30.76	32.5868	32.7317	31.6931	32.9593	**33.7486**
	MAE	4.1923	2.0187	1.9457	1.6025	1.6144	**1.5005**
40	NCD	0.0198	0.0100	**0.0088**	0.0181	0.0172	0.0170
	PSNR	30.14	31.4450	31.9360	28.5323	31.5278	**32.5371**
	MAE	4.4895	2.5892	2.3394	2.4019	2.1622	**1.9886**
50	NCD	0.0220	0.0188	**0.0130**	0.0205	0.0183	0.0176
	PSNR	29.34	30.2596	31.0874	25.9293	29.1122	**31.4175**
	MAE	4.8354	3.2416	2.7553	3.4435	2.8268	**2.4026**
60	NCD	0.0212	0.0131	**0.0101**	0.0237	0.0198	0.0184
	PSNR	28.34	29.0210	30.2444	23.7452	27.1659	**30.3663**
	MAE	5.2953	3.9806	3.1704	4.7812	3.6436	**2.9261**
70	NCD	0.0225	0.0214	0.0195	0.0280	0.0246	**0.0192**
	PSNR	27.11	27.6775	29.4524	21.3832	23.0957	**29.4679**
	MAE	5.8971	4.8263	3.5758	6.4549	5.4581	**3.4282**

noise densities. It is also noticed from the experimental results that the performance of the PF is very near to RWASF at low noise density.

Figure 5 shows the performance of the filters for the colon_cancer image. The proposed filter has the highest PSNR value at low, medium, and high noise density. The visual performance of different filters for duodenal_ulcer image is shown in Fig. 6. From the figure, it is seen that the VMF output is blurred because of the

Table 3 Comparison of PF in duodenal_ulcer

Noise (%)	Parameter	VMF	AVMF	MSMF	RWASF	SFLCPG	PF
10	NCD	0.0087	0.0015	0.0022	0.0014	0.0020	**0.0012**
	PSNR	39.4251	40.3114	36.0748	40.5533	36.5626	**41.0927**
	MAE	0.8368	0.2895	0.4264	0.2605	0.3694	**0.2345**
20	NCD	0.0091	0.0022	0.0026	0.0024	0.0028	**0.0020**
	PSNR	38.6535	**38.9878**	35.7685	36.9986	34.3210	37.6787
	MAE	0.9649	0.4429	0.5443	0.4829	0.5856	**0.4168**
30	NCD	0.0104	0.0028	0.0031	0.0036	0.0035	**0.0022**
	PSNR	37.9931	38.0731	35.5174	33.5781	33.8553	**39.1310**
	MAE	0.9649	0.6202	0.6652	0.8027	0.7510	**0.4668**
40	NCD	0.0104	0.0037	0.0032	0.0058	0.0042	**0.0031**
	PSNR	36.7980	36.5620	35.1629	29.7962	33.0379	**37.1516**
	MAE	1.3134	0.8415	0.8054	1.3712	0.9488	**0.6823**
50	NCD	0.0097	0.0047	0.0042	0.0088	0.0057	**0.0038**
	PSNR	35.0863	35.0845	34.6638	26.9240	30.3827	**35.3838**
	MAE	1.5758	1.1238	0.9602	2.1959	1.3690	**0.8605**
60	NCD	0.0070	0.0060	0.0048	0.0132	0.0076	**0.0046**
	PSNR	33.6133	33.6841	33.9621	24.4725	28.0456	**34.3911**
	MAE	1.9287	1.4740	1.1339	3.3915	1.8655	**1.0656**
70	NCD	0.0084	0.0077	0.0055	0.0186	0.0136	**0.0053**
	PSNR	31.9561	31.7308	33.2150	22.4891	23.1483	**34.2377**
	MAE	2.3578	1.9468	1.3164	4.9022	3.5360	**1.2143**

replacement of every pixel. The performance of the other filters, like AVMF, MSMF, RWASF, and SFLCPG, is better than the VMF but lower than the PF. Since the proposed filter used an adaptive rank for different noise densities in the noise detection step and used Laplacian convolution to identify whether that noisy candidate pixel is either an edge or thin lines make the proposed filter outperforms other filters. The performance comparison in terms of MAE for duodenal_ulcer and colon_cancer is shown in Figs. 7 and 8, respectively, and in terms of NCD for the same images are shown in Figs. 9 and 10 respectively.

Table 4 Comparison of PF in brain image

Noise (%)	Parameter	VMF	AVMF	MSMF	RWASF	SFLCPG	PF
10	NCD	0.0408	0.0225	0.0273	**0.0087**	0.0173	0.0092
	PSNR	26.9538	27.8702	27.4906	**30.3929**	29.5016	30.2090
	MAE	4.9976	2.9141	3.1813	**1.2408**	1.5520	1.4610
20	NCD	0.0396	0.0273	0.0277	0.0128	0.0134	**0.0126**
	PSNR	26.5501	27.3712	27.0587	**29.0861**	28.4020	28.8167
	MAE	5.3958	3.5210	3.6854	**1.8998**	2.1473	2.0368
30	NCD	0.0322	0.0277	0.0289	0.0173	0.0216	**0.0168**
	PSNR	26.0290	26.7745	26.6405	27.6556	27.3840	**27.9768**
	MAE	5.8621	4.1811	4.2005	**2.6152**	2.9649	2.8752
40	NCD	0.0422	0.0311	0.0313	0.0225	0.0227	**0.0211**
	PSNR	25.4598	25.9968	26.2062	25.9559	26.4305	**27.0352**
	MAE	6.4388	5.0144	4.7455	**3.4866**	3.8375	3.7871
50	NCD	0.0360	0.0347	0.0333	0.0289	0.0274	**0.0244**
	PSNR	24.7603	25.2952	25.6647	24.1877	25.1996	**26.2980**
	MAE	7.1846	5.9074	5.3448	4.5661	4.6078	**4.4038**
60	NCD	0.0318	0.0358	0.0358	0.0365	0.0332	**0.0287**
	PSNR	23.9237	24.3868	25.1534	22.4737	23.9702	**25.4525**
	MAE	8.0648	7.0116	5.9448	5.8987	5.7165	**5.3655**
70	NCD	0.0364	0.0417	**0.0328**	0.0453	0.0425	**0.0328**
	PSNR	22.9002	23.3209	24.6617	20.9627	21.5324	**24.7440**
	MAE	9.2180	8.3683	6.5572	7.4813	7.6258	**6.2844**

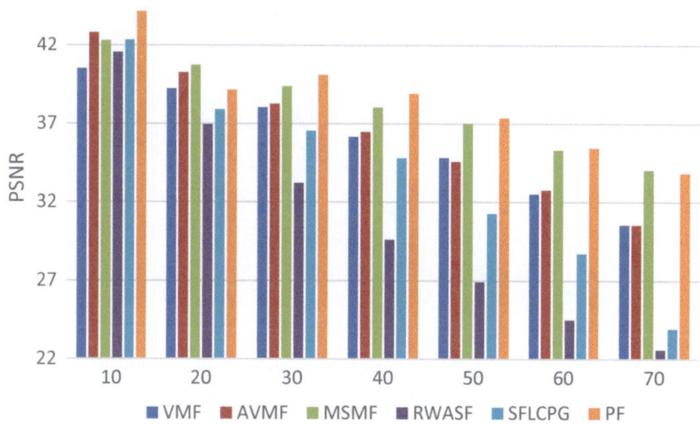

Fig. 5 PSNR values for colon_cancer image

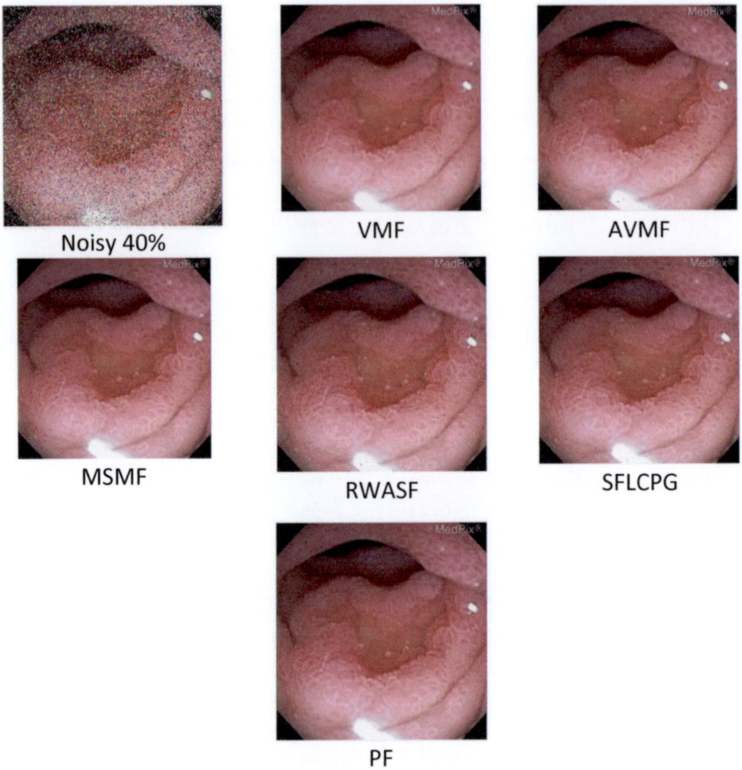

Fig. 6 Performance of various filters in duodenal ulcer image

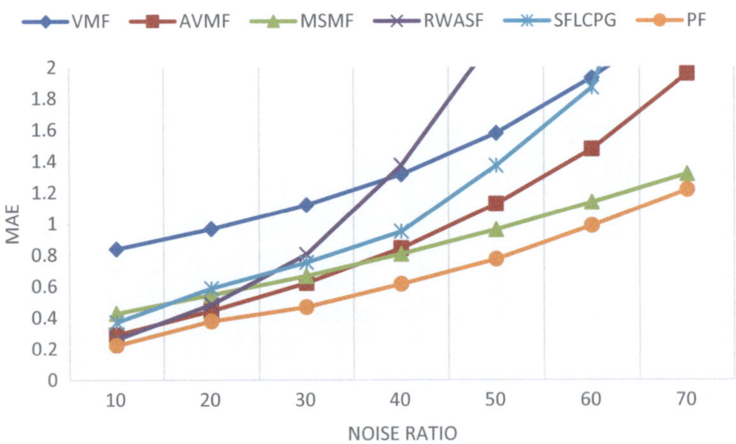

Fig. 7 MAE of the different filters for duodenal_ulcer image

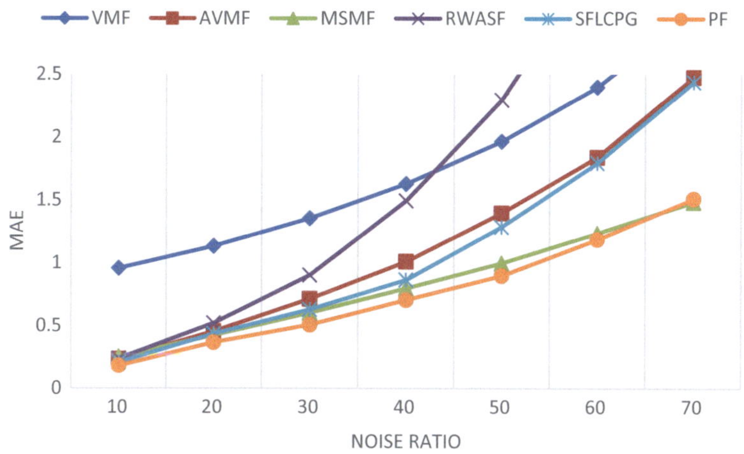

Fig. 8 MAE of the different filters for colon_cancer image

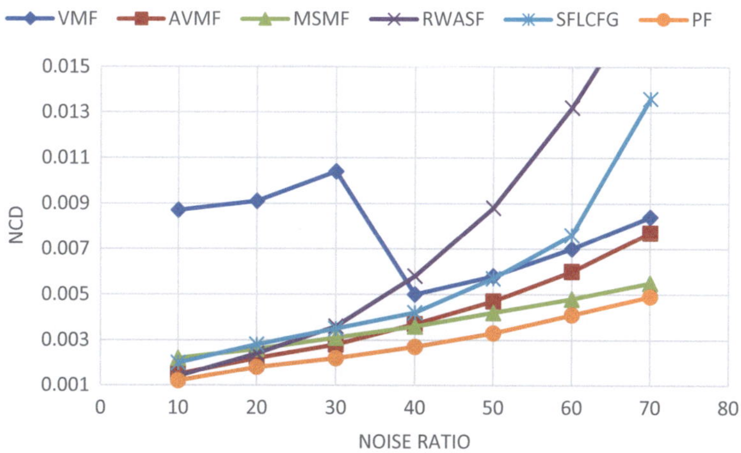

Fig. 9 NCD of the various filters in duodenal_ulcer image

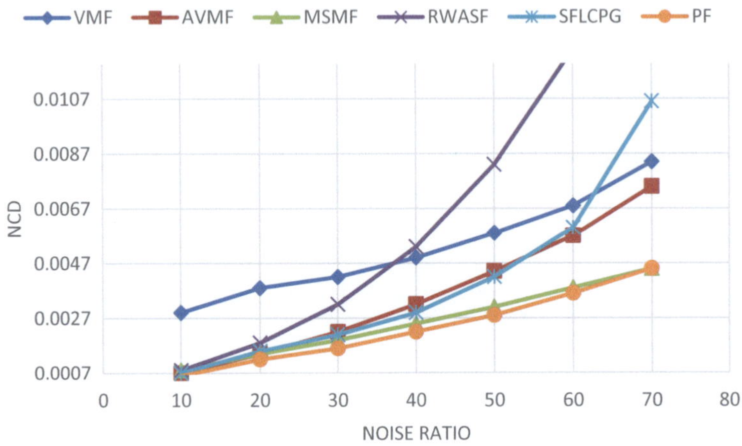

Fig. 10 NCD of the various filters in colon_cancer image

5 Conclusion

An adaptive rank-based impulse noise detection algorithm is formulated for eliminating salt and pepper noise from both standard test images and color medical images. On the basis of adaptive rank for different noise densities, a new method is used to classify whether a pixel is noisy or not. Laplacian convolution in four directions is used to check whether that probably noisy pixel is noise or an edge. Uncorrupted pixels are left unchanged, and the result of the VMF is utilized to replace the corrupted pixel. Experimental results show that the PF outperforms state-of-the-art filters to remove impulse noise from standard tests and color medical images.

References

1. Celebi ME, Gerald S (2013) Color medical image analysis. Springer, Netherlands
2. Plataniotis KN, Venetsanopoulos AN (2000) Color image processing and applications. Springer, Berlin
3. Astola J, Haavisto P, Neuvo Y (1990) Vector median filters. Proc IEEE 78(4):678–689
4. PEIR Digital Library. https://peir.path.uab.edu/library/index.php?/category/244. Accessed 17 Feb 2018
5. Weber AG (2006) The USG-SIPI image database: version 5. http://sipi.usc.edu/database/. Accessed 01 June 2022
6. Chanu PR, Singh KM (2019) A two-stage switching vector median filter based on quaternion for removing impulse noise in color images. Multimed Tools Appl 78(11):15375–15401
7. iGASTROCARE, Red. No. UNHCR/NO/150/DC/216, Ref no. 1088, dated 30.05.2020
8. Trahanias PE, Venetsanopoulos AN (1993) Vector directional filters—a new class of multichannel image processing filters. IEEE Trans Image Process 2(4):528–534. https://doi.org/10.1109/83.242362

9. Karakos DG, Trahanias PE (1995) Combining vector median and vector directional filters: the directional-distance filters. In: International conference on image processing, Washington, DC, USA, vol 1, pp 171–174
10. Lukac R (2003) Adaptive vector median filtering. Pattern Recogn Lett 24(12):1889–1899
11. Wang G, Li D, Pan W, Zang Z (2010) Modified switching median filter for impulse noise removal. Signal Process 90(12):3213–3218
12. Smolka B, Malik K, Malik D (2015) Adaptive rank weighted switching filter for impulsive noise removal in color images. J Real-Time Image Proc 10(2):289–311
13. Singh KM, Bora PK (2004) Adaptive vector median filter for removal impulses from color images. IU-J Electr Electron Eng 4(1):1063–1072
14. Chanu PR, Singh K (2018) Impulse noise removal from medical images by two stage quaternion vector median filter. J Med Syst 42(10):1–10
15. Gao J, Du Z, Shi Z, Xu Z, Cao Q, Tang R (2018) Switching impulse noise filter based on Laplacian convolution and pixels grouping for color images. Signal, Image Video Process 12(8):1523–1529

Waste Classification Using Machine Learning and Image Processing Techniques

P. Maheswari and M. Sivakumar

Abstract The increase in population is giving rise to an exponential increase in the amount of waste that is being generated in day-to-day life. The inappropriate dumping and disposal of these wastes result in major damage to the air, land, and soil by affecting the quality, mineral content, productivity, and nature. In order to carry out efficient waste management, the recyclable wastes should be treated properly. The first step toward recycling is the segregation of wastes into their respective class. The project mainly focuses on the detection and classification of waste materials using the You Only Learn One Representation (YOLOR) object detection algorithm and image processing techniques. A custom dataset has been prepared, and the number of classes of wastes has been increased to 10. The custom dataset has the following classes of waste: cardboard, electronic waste, glass, kitchen waste, medical waste, metal, paper, plastic waste, textile waste, and trash. The classification is performed by the YOLOR algorithm over the custom dataset prepared with a number of preprocessing steps. The detection performance is compared with the previous versions of the You Only Look Once (YOLO) algorithm, such as YOLOV4, YOLOV5, both with and without the preprocessing steps, to prove the effect of preprocessing. The model is analysed in terms of Mean Average Precision (mAP), box loss, class loss, and object loss. The proposed model achieves a Precision of 94.2%, a Recall of 93.3%, and an mAP of 95.9%. The proposed model shows a box loss of 0.025, a class loss of 0.0001, and an object loss of 0.069.

Keywords Deep learning · Feature extraction · Object detection · Waste management

P. Maheswari (✉) · M. Sivakumar
Thiagarajar College of Engineering, Thiruparankundram, India
e-mail: maheswarip@student.tce.edu

M. Sivakumar
e-mail: mskcse@tce.edu

S. K. Biswas et al. (eds.), *Intelligent Computing Systems and Applications*, Lecture Notes in Networks and Systems 1307, https://doi.org/10.1007/978-981-96-3860-4_24

1 Introduction

Rapid growth in industrialization, urbanization, and global population growth has a vital impact on the natural environment. With this higher growth rate, the production rate of the wastes has also been increased, and this resulted in rapid degradation of the environment. Waste classification and segregation have become important areas of interest in the study. In India, most of the traditional waste classification methods involve manual labor, which is highly expensive and time-consuming. Therefore, there is an immediate need to automate the process of waste classification. According to the "Swachhata Sandesh Newsletter" by the Ministry of Housing and Urban Affairs (MoHUA), as of January 2020, 147,613 metric tons (MT) of solid waste are generated per day, from 84,475 wards. It is expected that the volume of waste generated will be increased by 70% to 3.40 billion tons in 2050 [1]. This increase and oscillation in the waste generation are mainly due to various reasons, such as population increase and changing consumption patterns. The main problem in managing solid wastes involves the accumulation of the wastes, segregation into the respective categories of waste, transportation, and lack of public awareness.

Waste recycling is the key to a proper waste management system. In India, the volume of waste generated has been increasing over the past few years. The overall waste management techniques that are adopted in India are actually inadequate. In India, where 71% of the population resides in small-scale towns and villages, proper waste management has to be performed by automating the classification of wastes generated [2]. Leaving trash as it is and not cleaning and disposing of it appropriately will have an adverse effect on the residents in terms of hygiene, appeal, and environmental protection. Automation is essential not only as it improves public health but also decreases the amount of money that is being spent by the government to collect and separate the trash.

As reported, waste management is a costlier operation to be carried out. This is a major process that has a high inclusion of resources and labor power. Many steps have been taken to create awareness among the public with concern toward a proper waste management habitat by advising them to dump waste properly according to the waste's category and also the 3R's (Recycle, Reuse, and Reduce) campaign. CPCB, Central Pollution Control Board, is India's apex authority on pollution control, and it estimated that the country, India, produced about 13 lakhs in the year 2018–2019 and reported the same at year-end.

CPCB also included that India has nearly 6000 units of plastics manufacturing or recycling units, out of which about 5000 units are registered manufacturers, and the rest of the 1000 units are unregistered. Material Recycling Association India (MRAI) reported that the recycling rate in India is 30%. This rate is so slow that it is obvious that the existing system is not effectively recycling in proportion with the generated amount of waste, and also indicates that immediate actions are to be taken toward a healthier recycling environment.

Developments in the field of deep learning have paved the way for great improvements in the automation of trash. Convolutional Neural Network (CNN) is a class of deep learning neural networks that has major applications in image recognition and classification. Deep Learning with state-of-the-art methods provides an obvious solution for understanding human thinking and behaviors. With improvements in the field of deep learning and image processing techniques, the process of waste classification becomes efficient with higher accuracy as well as within a short time interval. Challenges faced in solid waste management include source segregation, collection, treatment, disposal, policy issues, technical issues, and financial issues.

In India, the future challenges for the management of municipal solid waste include increasing quantities and changing composition. Increasing severity of adverse impacts, increasing cost of waste management, limited policy framework, and lack of political priority [3]. The tremendous generation of waste made the traditional methods, such as landfills, fail as they couldn't cope with the amount of waste generated. Infrastructure and incentives are also the factors that make the process of waste management costlier, whereas waste minimization is also an expensive process [4].

2 Related Works

Sultana et al. [5] proposed a system to detect trash in real-time using deep learning and image processing techniques. This system mainly focuses on separating the recyclable items from the landfill's trash. Two convolutional neural networks based on the AlexNet network architecture were used to identify the trash in an image. They trained the system on both indoor and outdoor image datasets. De Carolis et al. [6] presented a real-time waste detection system using YOLO to monitor the abandoned garbage bags with a vision of heading toward a smart city. They focused on detecting and reporting abandoned waste and garbage bags. They achieved decent accuracy and loss, but there was an imbalance between the number of images in the dataset and the number of real-time objects. Deepa et al. [7] developed software to compare the performance of different algorithms and to figure out which of the algorithms is more efficient and takes less computational power to run. They implemented and compared three algorithms, YOLO, SSD MobileNet, and Faster CNN, with only limited parameters. Altikat et al. [8] presented a four-layer and five-layer deep convolutional neural network algorithm to classify the wastes and achieved an accuracy rate of 61.67% and 70%. This research found that, as the number of layers in the DNN decreased, the system performance values decreased as well. The organic waste classification accuracy rate was 83% and 76.7% in the four-layer and five-layer DCNN architectures, respectively. The system reported a lower accuracy rate in the classification of plastic waste, which was 37% and 56.7% respectively, in the four-layer and five-layer DCNN architectures. Baheti et al. [9] proposed an approach for generic object detection with the help of the Local Self-Similarity descriptor. They evaluated the performance of the proposed method on the UIUC car dataset and the ETHZ shape classes. The system was successful in localizing the objects despite variations in wide intraclass, cluttering, and deformation. The proposed method made use of only a single image as a query.

Ahmad et al. [10] used a novel approach named double fusion by combining several deep learning models using both feature fusion and score-level fusion methods, then followed by classification using Support Vector Machines (SVMs). Six fusion methods were implemented along with two feature-level fusion schemes and showed an increase of 3.58% in performance. Later, both early and late fusion methods are combined with a double fusion method and have shown better results. Trashes for which the scores of all the models are significantly lower show relatively less accuracy. Sidharth et al. [11] categorized based on their properties with the help of a self-learning neural network. The designed classifier uses the method of supervised learning where the algorithm learns from a well-defined and labeled data set and is a successful and inaccurate classification and also able to recognize different categories of objects. The system showed a training accuracy of 99.12% and a testing accuracy of 76.19% for 100 epochs. In order to improve the efficiency and accuracy of waste classification processing,

Zhang et al. [12] used a DenseNet169 waste image classification model based on transfer learning. The authors have collected a new set of trash image dataset NWNU-TRASH to specifically address concerns with the existing waste image dataset, such as unbalanced data distribution, single background, obvious features, and a limited sample size of the waste images. This dataset has images with balanced distribution, high diversity, and rich background. The model showed that DenseNet169 with transfer learning provided an accuracy of 82%, which is higher than the other models, such as DenseNet169, AlexNet, GoogleNetV2, and VGG-16. Adedeji et al. [13] simplified the process of classification by using the 50-layer residual net pre-trained (ResNet-50) Convolutional Neural Network model as the extractor, and Support Vector Machine (SVM) to classify the waste. The suggested approach was tested on the TrashImage dataset and achieved an accuracy of 87% in classification. The separation process of the waste proved to be faster and intelligent using the proposed system.

He et al. [14] proposed 2 CNN integrated migration learning neural network models, VGG-16 CNN and LeNet-5 CNN. They showed that by using a labeled dataset of a larger size in a pre-trained CNN model along with suitable network fine-tuning, the image recognition ability and the migration adaptation effect are optimized. Rutqvist et al. [15] mainly focused on the issues during detection for emptying the filled container and proposed a method that included both the manually engineered model, along with its modification, as well as conventional machine learning algorithms. This method uses a Random Forest Classifier on a set of features. The experimental results showed that this method improved the classification accuracy and recall to 99.1% and 98.2%, respectively. A robust model using deep neural networks to classify trash automatically to be applied in smart waste sorter machines was proposed. The VN-trash dataset consists of 5904 images, including Organic, Inorganic, and Medical wastes.

Shi [16] proposed a network enhancement model using channel expansion to classify the garbage images. Using M-b Exception, they improved the predictive ability. The drawback is that it works at an additional computational cost. Vo et al. [17] used an improved ResNeXt model, a deep neural network model for trash classification named DNN-TC, to improve the predictive performance. Experiments were conducted to compare the performance of the proposed model and the state-of-the-art models over two datasets, namely, the TrashNet [18] dataset as well as VN-trash dataset, with an accuracy of 94% and 98%, respectively, compared to the performance of the state-of-the-art methods.

Yang et al. [19] initially constructed a complex data augmentation for model training, followed by the design of a novel, lightweight neural network garbage classification system called WasNet. The accuracy of the system is 64.5%, 82.5%, and 96.10% on the ImageNet data set, the Garbage Classification dataset, and the TrashNet dataset, respectively. This model has been implemented as hardware into an Intelli trash can and also a visualization and decision support platform to assist managers in monitoring traffic.

3 Experimental Dataset

3.1 Contents of the Dataset

In the experiment, the custom dataset with the images of various classes of wastes has been used [20]. These images were collected by clicking photos through mobile devices such as Samsung M21, Apple iPhone 5S, and Apple iPhone SE. It contains 5247 actual waste objects' images in 10 categories, such as cardboard, electronic waste, glass, kitchen waste, paper, plastic, medical waste, metal, textile waste, trash, etc. At present, there is no standard, publicly recognized large-scale image dataset in the field of garbage image research. References show that the custom dataset is the largest multi-class hybrid waste image dataset used by academia. The images are all from real cases. Figure 1 shows sample images in the custom waste images dataset (Table 1).

The dataset has sample images of paper in the higher count, which helps us to train the model with higher accuracy for detecting recyclable objects. Also, some sample images in each of the classes are given in Fig. 1.

3.2 Annotating the Dataset

In order to train the YOLOR algorithm, the images in the custom dataset must be annotated since the network can be trained only using annotated images. Annotations over the images are done using the Roboflow annotating tool. In the proposed system, to improve the classification efficiency of the model, the number of images in the dataset has been increased through several preprocessing steps (Figs. 2, 3, and 4).

Fig. 1 Sample images from the dataset

Table 1 Class-wise distribution of the dataset

Sl. No.	Name of the class	Number of images
1	Paper	594
2	Cardboard	503
3	Glass	501
4	Plastic	582
5	Metal	510
6	Textile	325
7	Electronic	340
8	Medical	680
9	Organic	675
10	Trash	537

4 Proposed Work

Figure 5 describes the architecture of the proposed system. The proposed method mainly consists of two phases, namely, (i) Feature Extraction and (ii) Classification. In the first phase of feature extraction, deep features are extracted from the image. In the second phase of the proposed system, classification is performed using the YOLOR object detection algorithm. Data augmentation and other preprocessing steps are performed over all the images in the dataset, which improves the classification capability of the system and eventually aids in minimizing the time taken by the model for classification. Detailed descriptions of the Feature Extraction and Classification techniques used in the proposed system are provided in the sessions up next.

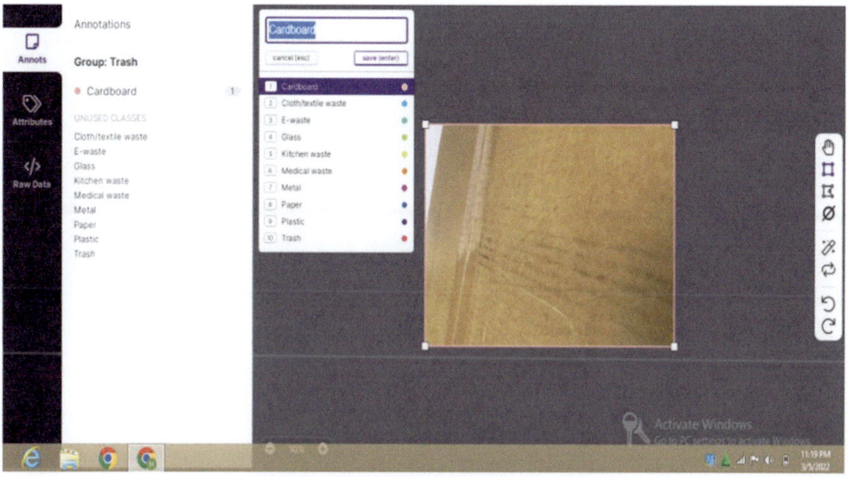

Fig. 2 Annotating a waste image from the public dataset

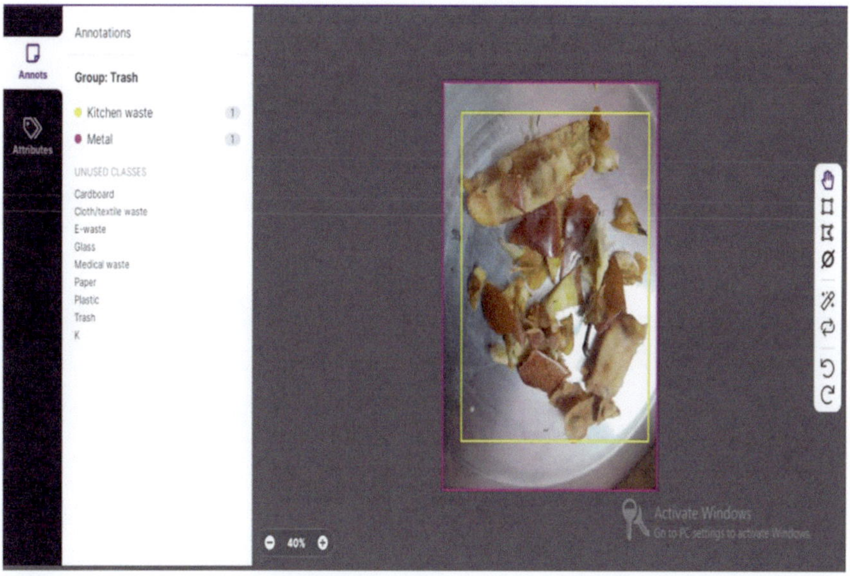

Fig. 3 Annotating a waste image clicked manually

4.1 Transfer Learning

In order to train deep neural networks, there is a need for higher computational power. As the number of images in the used dataset is limited, transfer learning is used. With the help of transfer learning, the proposed method shows better performance over the

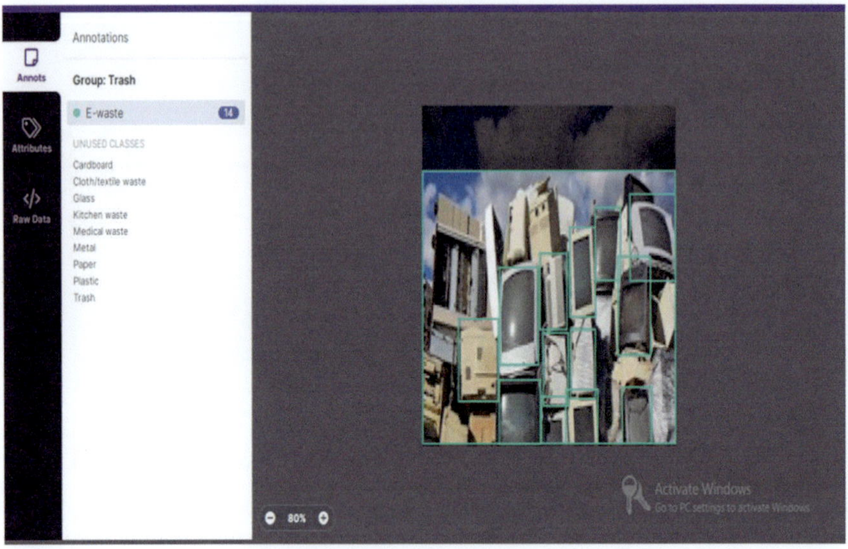

Fig. 4 Annotating a waste image collected from the Internet

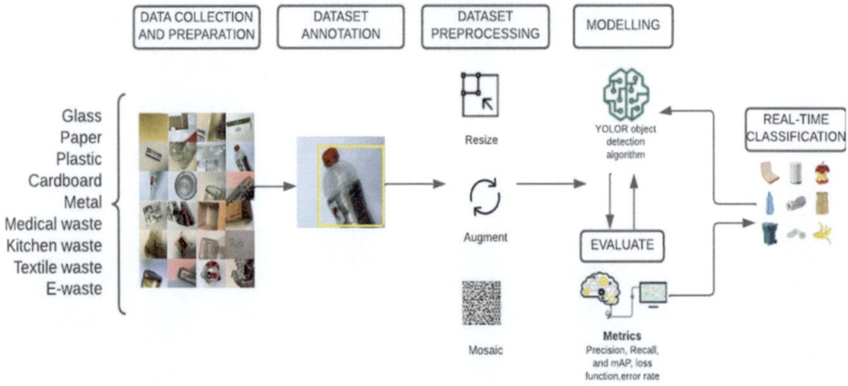

Fig. 5 The architecture of the proposed system

validation set. Various YOLO models, such as YOLOV3, YOLOV4, and YOLOV5, have been used for the comparative analysis. For comparison, the parameters such as mAP and loss are analysed. One of the most important applications of Convolutional Neural Network (CNN) is Image classification. For the purpose of waste classification, an object detection algorithm is to be used. The best-performing object detection models are the YOLO family. Among all the versions of YOLO, the latest version of YOLO is YOLOR.

Deep Learning, in general, requires a lot of training images in order to make the network understand the image features better. With the aim of addressing the problem of insufficient training data, the concept of transfer learning has been applied. In transfer learning, the model parameters are transferred from a similar network that was pre-trained with another large dataset rather than training the model parameters from scratch. In the proposed system, YOLOR has been used. The model parameters of the proposed system are transferred to the YOLOR network pre-trained with the COCO dataset. The proposed system with the transferred parameters is refined with the custom dataset that has been prepared. The transfer learning reduces the training time and the amount of training data that is required to train the model efficiently. Transfer learning helps in increasing the accuracy of the network model.

4.2 Data Preparation

The images in the dataset are taken using mobile phones. All images in the custom dataset prepared are of different pixel sizes. To ensure that all the images are of the same aspect ratio and size, all the images are resized to 512×512 pixels. In addition, this also reduces the training time. It aids in avoiding unnecessary parameters or features to be learnt for waste classification. The mean values for every pixel for all the images in the training set are visualized. To ensure that each input parameter has a similar data distribution, data normalization is used. This aids faster convergence while training the network. To normalize the data, the mean value is subtracted from each pixel, and the resultant is divided by the standard deviation. This data distribution resembles a zero-centered Gaussian curve. A scale of 0–255 was used to scale the data that was normalized. Table 2 shows the normalized mean and standard deviation for each channel. This is because, for some of the input images, the individual pixel value needs to be positive. To increase the number of images to train the neural network, various image augmentation techniques are performed. Horizontal flip, random crop, and zoom are performed to produce variants of the images, so that the system classifies the unseen data precisely. Figure 6 illustrates the various augmentation techniques performed over a sample image. Overfitting is an issue that arises when the algorithm performs well on the training images with a good accuracy rate but fails to perform well and almost generalizes badly with less accuracy for unseen and test images.

Table 2 Normalized mean and standard deviation for each channel

Channel	Normalized mean	Normalized standard deviation
Red	0.6067545935423009	0.1829832537916729
Green	0.5852166367838588	0.1823482159344223
Blue	0.565204377558798	0.1926306225491462

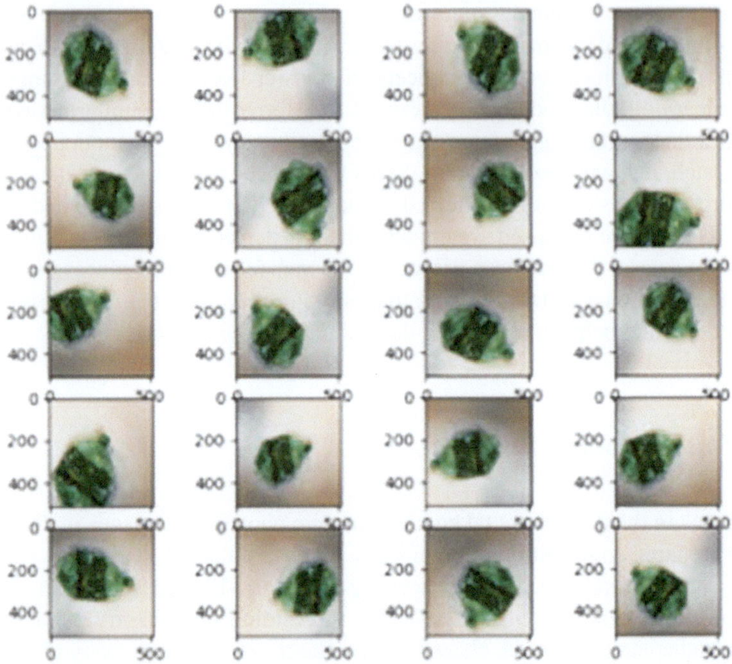

Fig. 6 Data augmentation

It is a common issue caused when the model picks up the noise, fluctuations, and other details and learns these to be the concepts and traits of the object. It leads to the extent where it negatively impacts the result and classifies a waste object as a different class than it doesn't belong to.

4.3 Structure of the New YOLOR Model

Figure 7 represents the structure of the YOLOR model [21].

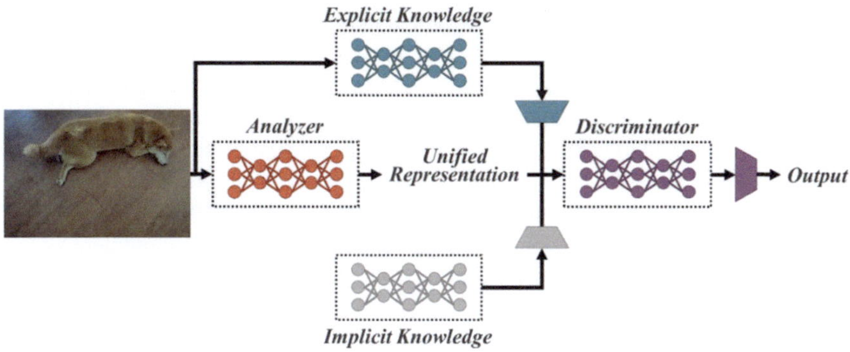

Fig. 7 Structure of the YOLOR model

Algorithm: YOLOR model
Input: Custom Waste Image dataset
Initialization parameters: Epoch, Learning rate, Batch size, Network model configuration yaml file, yolor_p6.pt pre-training weights
Image preprocessing: Data augmentation, Image illumination, Image contrast, Image normalization, Mosaic

(1) Import dataset, perform cross-validation to decide the training and validation set
(2) Load the data configuration along with the initialization parameters and input data and preprocess them
(3) Use the proposed network model to load the input image and conduct feature extraction, object location, and classification
(4) Optimise each set of parameters in the network and if the current epoch is not the last round, the mAP of the current model is calculated in the validation set
(5) Train the set with a number of epochs and obtain the trained optimal performance model

Output: The best-performing detection model in this training

5 Experiments

5.1 Experimental Setup

The implementation of the new model based on the improved YOLOR algorithm can be done in Tensorflow or Pytorch, which possesses the GPU version preferred for deep learning. The NumPy library is used for the numerical computations carried out. Python Pillow is built upon the PIL (Python Image Library) and is used to add, manipulate, new formats of images in the library by creating new file decoders. This library has functions to open, display, resize, flip, rotate, get information and size, enhance the image, adjust the brightness and sharpness, save, blur, merge, and other functions on the images.

The system is evaluated in terms of all the objectives, including classification in a captured image as well as on a live video stream, also in an uploaded image [22]. The individual classification models are evaluated on how well they perform waste classification. The changes to the performance when classification is performed by the proposed method are also analysed. In order to achieve these objectives, several experiments are done. Initially, the individual classification algorithms are analysed, followed by the evaluation of the new model based on the YOLOR algorithm. All the other algorithms compared with the proposed method are also implemented with the same specifications and descriptions.

6 Evaluation and Results

6.1 Evaluation Index

For the verification of the performance improvement of the proposed system, evaluation indices are set. The evaluation metrics used are Precision, Recall, mAP, Loss, and IOU. The average number of detected waste images in the second is given by the FPS value. The proposed system obtains an FPS between the range of 45–50. This project aims to find the best classification algorithm using the following parameters.

6.1.1 Precision

Precision is defined in terms of the ratio of the number of objects (waste items) detected correctly to the number of total objects detected. The mathematical representation of precision is given by the equation,

$$\text{Precision} = \frac{N_{TP}}{N_{TP} + N_{FP}} \tag{1}$$

6.1.2 Recall

Recall is calculated by the ratio between the number of Positive samples correctly classified as Positive to the total number of Positive samples. The mathematical representation of recall is given by the equation,

$$\text{Recall} = \frac{N_{TP}}{N_{TP} + N_{FN}} \tag{2}$$

6.1.3 Intersection Over Union (IoU)

IoU is a well-known evaluation metric in object detection tasks, which is mathematically represented by

$$\mathrm{IoU} = \frac{\mathrm{A} \cup \mathrm{B}}{\mathrm{A} \cap \mathrm{B}} = \frac{\text{Area of Intersection}}{\text{Area of Union}} \tag{3}$$

Algorithm: Generalised Intersection over Union (GIoU)
Input: Two arbitrary convex shapes: A, B $\subseteq$ S $\in$ R Sn
Output: GIoU

1. For A and B, find the smallest enclosing convex object C, where C $\subseteq$ S $\in$ Rn
2. $\mathrm{IoU} = \frac{\mathrm{A} \cup \mathrm{B}}{\mathrm{A} \cap \mathrm{B}}$
3. $\mathrm{GIoU} = \mathrm{IoU} - \frac{|\mathrm{C}(\mathrm{A} \cap \mathrm{B})|}{|\mathrm{C}|}$

6.1.4 Average Precision (AP)

The area under the precision-recall curve is taken, and the average of the average precisions is calculated by

$$AP = \int_0^1 p(r)dr \tag{4}$$

6.1.5 Mean Average Precision (mAP)

Mean Average Precision (mAP) signifies the mean of the average precisions of all classes defined in the test model and is expressed by

$$\mathrm{mAP} = \frac{\sum_{i=1}^{N} AP_i}{N} \tag{5}$$

6.1.6 Loss Function

In addition, the loss function is calculated to evaluate the performance of the new YOLOR model using ALReLU. The loss function for the proposed model can be computed by

Table 3 Comparison of faster R-CNN, YOLOV3, YOLOV4, YOLOV5, and YOLOR

Algorithm used	Precision	Recall	mAP	Box loss	Class loss	Object loss
YOLOV4 without preprocessing	80.5	83.2	81.7	0.04	0.006	0.325
YOLOV4 with preprocessing	81.9	87.6	84.8	0.052	0.003	0.247
YOLOV5 without preprocessing	84.8	88.2	86.4	0.056	0.005	0.265
YOLOV5 with preprocessing	86.4	86.6	87.9	0.047	0.004	0.117
YOLOR without preprocessing	91.7	90.6	92.4	0.03	0.0003	0.109
YOLOR with preprocessing	94.2	93.3	95.9	0.025	0.0001	0.069

$$\text{Loss} = \text{Errorcoord} + \text{ErrorIoU} + \text{Errorcls} \tag{6}$$

Errorcoord represents the coordinate prediction error.

ErrorIoU represents IoU error.

Errorcls represents the classification error.

6.2 *Experimental Results*

Comparison between different versions of the YOLO family, such as YOLOV4, YOLOV5, and the proposed method, is given in Table 3. To prove the effect of the preprocessing of the dataset, all three above-mentioned algorithms are trained 2 times. Firstly, the YOLOR algorithm is trained with a preprocessed and augmented dataset. Secondly, the YOLOR algorithm is trained with the custom dataset without any preprocessing steps or with the bare minimum preprocessing steps applied. Table 3 clearly shows that the proposed system is efficient than the other algorithms in terms of all the performance metrics.

6.3 *Real-Time Classification*

The real-time classification of waste is performed in three ways, namely detecting and classifying the wastes in an uploaded image from files, in a live captured image, and in a live stream. The proposed system adopts different ways for detection and classification. These various detection techniques assist laborers in reducing the manual workload.

Fig. 8 Detection and classification of wastes in an uploaded image

6.3.1 Detection and Classification of Wastes in an Uploaded Image

An image may be uploaded with a number of wastes within it via the given option from files. The system detects the different types of wastes that are present in the image and classifies them into their respective classes among paper waste, cardboard waste, glass waste, plastic waste, metal waste, textile waste, electronic waste, medical waste, organic waste, and trash. The uploaded image is given as the output with the bounding boxes. The accuracy of the object to belong to a particular class of waste is also provided. This helps to classify the wastes in any uploaded image or a set of gathered images. The resultant is given Fig. 8.

6.3.2 Detection and Classification of Wastes in a Captured Image Using Webcam

An image may be captured with a number of wastes within it, using a webcam. The system detects the different types of wastes that are present in the image and classifies them into their respective classes among paper waste, cardboard waste, glass waste, plastic waste, metal waste, textile waste, electronic waste, medical waste, organic waste, and trash. The captured image is given as the output with the bounding boxes. The accuracy of the object to belong to a particular class of waste is also provided. This helps to classify the wastes in the captured image in remote locations. The resultant is given in Fig. 9.

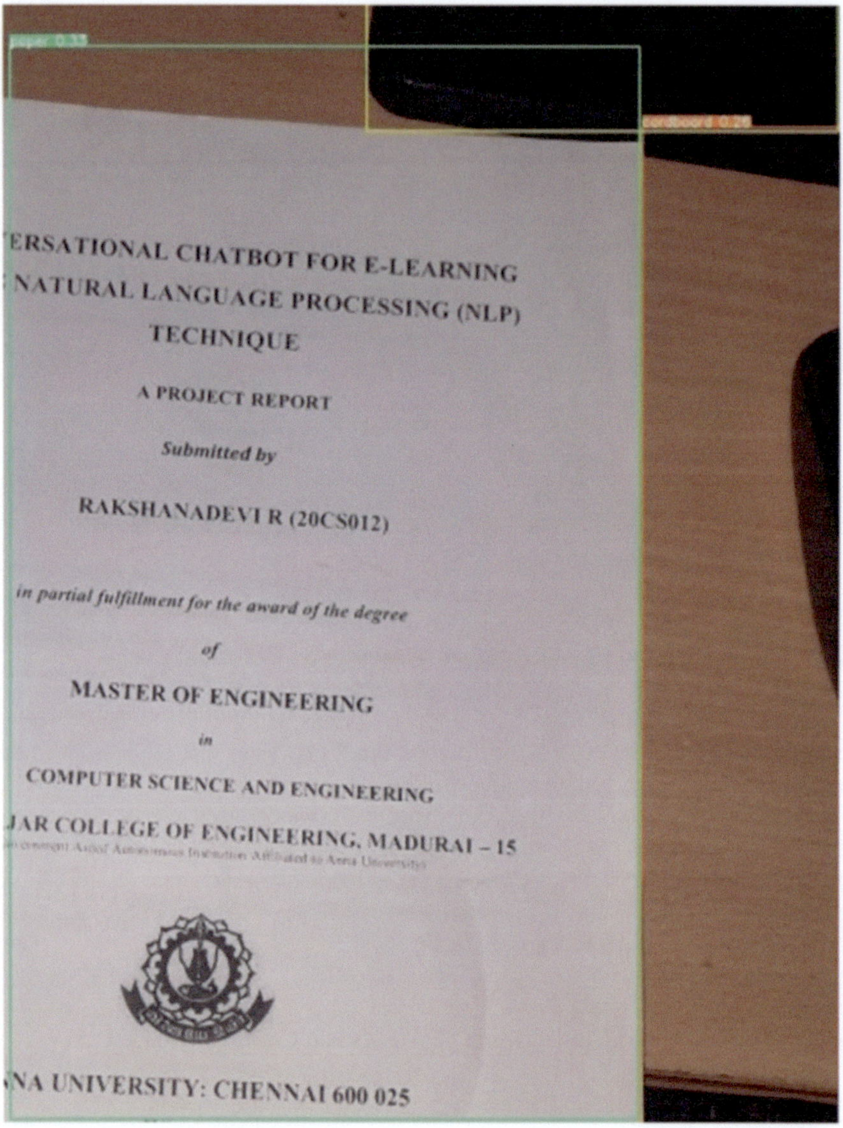

Fig. 9 Detection and classification of wastes in an uploaded image

6.3.3 Detection and Classification of Wastes in Live Using Webcam

The webcam is opened over the area with piles of waste or a single waste via the given option. The system detects the different types of wastes in the image and classifies them into their respective classes among paper waste, cardboard waste, glass waste, plastic waste, metal waste, textile waste, electronic waste, medical waste, organic

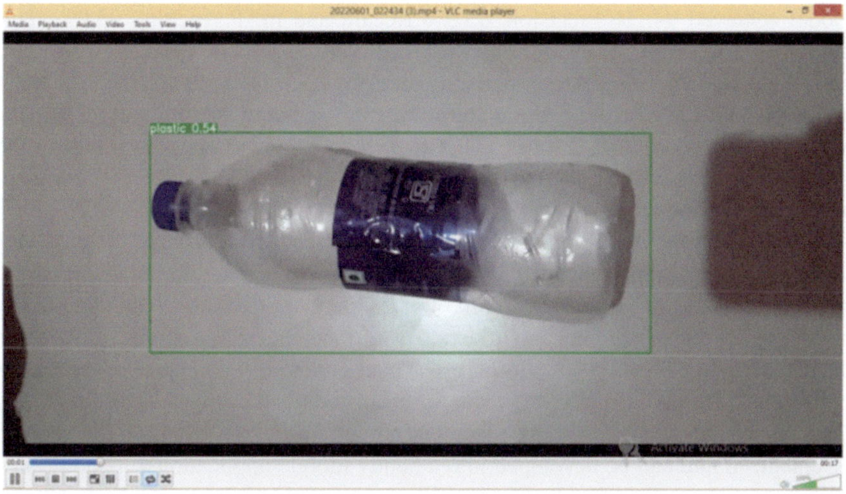

Fig. 10 Detection and classification of wastes in live video using WebCam

waste, and trash. The classification is performed in the live stream along with the bounding boxes. The accuracy of the object to belong to a particular class of waste is also provided. This helps to classify the wastes in a live video stream precisely. It is useful in detecting waste objects in motion. The resultant is given in Fig. 10.

7 Conclusion and Future Work

The proposed system mainly focused on aiding the process of recycling and introduced an automated waste detection and classification model using the YOLOR object detection algorithm and image processing techniques. Therefore, for the implementation, the proposed system prepared a dataset with a large number of images, and the custom dataset was used to train the YOLOR object detection algorithm to detect and classify the waste's type quickly and precisely. In this proposed system, it has been demonstrated how the wastes are classified into 7 major categories as Glass, Paper, Cardboard, Plastic, Metal, Trash, and Kitchen waste. The earlier works that were related to this study were done for only 3–4 categories of wastes and implemented machine learning and deep learning algorithms. In this paper, to overcome the defect in almost most of all existing systems, a custom dataset has been prepared by collecting waste images from various sources. The dataset is preprocessed through several steps such as data augmentation, data normalization, flipping, cropping, gray scaling etc., The detection algorithm not only predicts the class labels but also detects the waste object's location as well that the model can also detect several waste materials within a single image. The classification and localization errors were reduced, and the model showed only 1.003 and 0.989 error rates, respectively. The proposed model achieved 94.2% Precision, 93.3% Recall, and mAP reaching 95.9%. This model used is helpful in classifying the trash with higher class accuracy and also with a really quick time. The comparison is done by applying Faster CNN, YOLOV3, YOLOV4, YOLOV5 classification algorithms on the same dataset, and the experimental results concluded that the YOLOR method provided better results over the existing methods with all the parameters chosen and is highly efficient in terms of both accuracy and loss.

Sometimes, the limited number of classes used leads to misclassifications. This misclassification is because the classes of wastes produced are high, and therefore, when there are only 7 categories, some wastes are classified into closely related or inappropriate classes. For this modification to be done, the dataset should also include the respective class of images within it. The background and ambiance of the wastes in the dataset are a little different from the real-time trash background. Other bulky trash object images should also be included. In the future, the proposed method can be enhanced and can be developed into a handy mobile application with additional classes of wastes. This development will be even more helpful as it'll be handy for the workers and reduce their workloads.

References

1. Kaza S, Yao L, Bhada-Tata P, Woerden FV (2018) What a waste 2.0: a global snapshot of solid waste management to 2050. World Bank Publications
2. Pamnani A, Srinivasarao M (2014) Municipal solid waste management in India: a review and some new results. Int J Civ Eng Technol (IJCIET) 01–08
3. Siddiqqui MA (2018) Municipal solid waste management in India-status and challenges: an overview. Int J Biol Res 3(1):126–133
4. Bisht RS, Ranjan KP (2017) Waste management: prospects and challenges in India. Rai Manag J XIV(I)
5. Sultana R, Adams RD, Yan Y, Yanik PM, Tanaka ML (2020) Trash and recycled material identification using convolutional neural networks (CNN). In: 2020 SoutheastCon. IEEE, pp 1–8
6. Carolis BD, Ladogana F, Macchiarulo N (2020) YOLO TrashNet: garbage detection in video streams. In: 2020 IEEE conference on evolving and adaptive intelligent systems (EAIS), pp 1–7
7. Deepa R, Tamilselvan E, Abrar ES, Sampath S (2019) Comparison of YOLO, SSD, faster RCNN for real-time tennis ball tracking for action decision networks. In: 2019 International conference on advances in computing and communication engineering (ICACCE), pp 1–4
8. Altikat A, Gulbe A, Altikat S (2021) Intelligent solid waste classification using deep convolutional neural networks. Int J Environ Sci Technol
9. Baheti BV, Talbar SN, Gajre SS (2019)"A training-free approach for generic object detection. IETE J Res
10. Ahmad K, Khan K, Al-Fuqaha A (2020) Intelligent fusion of deep features for improved waste classification. IEEE Access 8:96495–96504
11. Sidharth R, Rohit P, Vishagan S, Karthika R, Ganesan M (2020) Deep learning based smart garbage classifier for effective waste management. In: 2020 5th international conference on communication and electronics systems (ICCES), pp 1086–1089
12. Zhang Q, Yang Q, Zhang X, Bao Q, Su J, Liu X (2021) Waste image classification based on transfer learning and convolutional neural network. Waste Manag 135:150–157

13. Adedeji O, Wang Z (2019) Intelligent waste classification system using deep learning convolutional neural network. Procedia Manuf 35:607–612
14. He JY, Yang BL, Su Y (2021) Integrated transfer learning method for image recognition based on neural network. IETE J Res
15. Rutqvist D, Kleyko D, Blomstedt F (2020) An automated machine learning approach for smart waste management systems. IEEE Trans Ind Inf 16(1):384–392
16. Shi C, Xia R, Wang L (2020) A novel multi-branch channel expansion network for garbage image classification. IEEE Access 8:154436–154452
17. Vo AH, Hoang Son L, Vo MT, Le T (2019) A novel framework for trash classification using deep transfer learning. IEEE Access 7:178631–178639
18. Yang M, Thung G (2016) Classification of trash for recyclability status. CS229 project report
19. Yang Z, Li D (2020) WasNet: a neural network-based garbage collection management system. IEEE Access 8:103984–103993
20. Kumar S, Yadav D, Gupta H, Verma OP, Ansari IA, Ahn CW (2020) A novel YOLOv3 algorithm-based deep learning approach for waste segregation: towards smart waste management. Electronics 10(1):14
21. Hua D, Gao J, Mayo R, Smedley A, Puranik P, Zhan J (2020) Segregating hazardous waste using deep neural networks in real-time video
22. Sheng TJ et al (2020) An internet of things based smart waste management system using LoRa and Tensorflow deep learning model. IEEE Access 8:148793–148811

Identification of Human Skin Cancers Using Wavelet Decomposition Technique

Ksh Robert Singh, Subir Datta, Subhasish Deb, Sanjana Chhetri, Malin Enghipi, Zorinpuia, and Lalmalsawma

Abstract This work describes identification of human skin diseases/cancer, namely, Actinic Keratosis, Eczema, Melanoma, Psoriasis, and Vascular Lesion using image processing technique. Colour models, namely, hue–saturation–intensity (HSI) and red–green–blue (RGB) are used for extracting eighteen colour features. The texture features are calculated using co-occurrence matrix, run-length matrix and wavelet decomposition techniques. A total of 34 features pertaining to texture are extracted. The identification of skin diseases is performed using classifier based on Linear Discriminant Analysis (LDA) and its performance is also compared to that of other statistical classifiers. The performance based on the above colour and texture feature extraction techniques towards identification of skin cancers are also compared. It is found that colour features yields better classification accuracy as compared to texture features based on run-length and co-occurrence matrices using LDA classifier. However, it is observed that wavelet features based on wavelet decomposed technique achieves overall best accuracy of 79% unlike other feature evaluation methods presented in this study. This study shows that identification of skin cancer based on RGB colour model achieves better average accuracy of 74% as compare to HSI model. This work also compares the performance of combined features (colour plus texture) and it is found that combination of wavelet and RGB colour features achieves classification accuracy up to 92% as compared to other colour and texture combination. Results also show that the performance of LDA classifier is comparatively better as compare to other classifier considered in this work.

Keywords Image · Colour · Texture · Cancer · Classification

K. R. Singh (✉) · S. Datta · S. Deb · S. Chhetri · M. Enghipi · Zorinpuia · Lalmalsawma
Electrical Engineering Department, Mizoram University, Aizawal, Mizoram, India
e-mail: robert_kits@yahoo.co.in

S. K. Biswas et al. (eds.), *Intelligent Computing Systems and Applications*, Lecture Notes in Networks and Systems 1307, https://doi.org/10.1007/978-981-96-3860-4_25

1 Introduction

Skin exposed to the sun may cause abnormal growth of skin cells and often leads to skin cancer. However, there are also some common forms of skin cancer which are not just due to exposure of skin to the sun. This type of diseases not only damages our skin but also gives a psychological impact to the patient. Thus may disturbed the confidence of the patient and finally put the patient into depression. Many researchers have been carried out to detect various types of skin cancer using different techniques. In [1, 2], the input image is pre-processed and features are extracted and classified using convolution neural network (CNN). Feature extraction using CNN are based on convolution process and thus the physical meaning of the features are not well understood as in the case of other feature extraction techniques, namely, colour and texture. Moreover, it might also be difficult to select the most dominating features without compromising the physical meaning of features. Classification of skin diseases with the help of Support vector machine (SVM) are also presented in [3–6]. However, the performance of the SVM classifier towards successful identification of skin cancer is not compared with other machine learning techniques. The classification task is not that challenging as the types of skin cancer considered are less. Wavelet decomposition technique is considered for the recognition of various skin diseases and the same are discussed in [7, 8]. It is learnt that wavelet features based on wavelet decomposed image are quite helpful for the pattern classification and identification task as it has the potential to extract image information at different resolutions. In [9], different sampling techniques are used where oversampling and data augmentation technique gives the best accuracy for classification of seven cancer images.

It is learnt from the literature that co-occurrence matrix-based texture feature extraction is widely used for the classification task which are presented in [4–6, 8, 10, 11]. Classification of skin disease using CNN is carried out in the literature. However, the physical meaning of the image features are not well defined as in the case of CNN, because the features are extracted based on convolution process. It is also understood that every classification task are specific to its own application. That is a set of features and a particular classification algorithm may not be always suitable for all types of classification task. The physical meaning of the feature is not well defined as the processes involve convolution of the input image by a filter mask. In [12], three machine learning and deep learning algorithms are used and are compared with each other where K-nearest neighbour (KNN) and Bagged Tree Ensemble (BTE) gives best accuracy among all. Classification of skin cancers based on different feature extraction models using artificial neural network are presented in [13–15]. Similar work towards identification and classification of skin cancers using different classifier models are presented in [16–18]. It is learnt from literature that out of all types of skin cancer, melanoma-type skin cancer is the quite dangerous. Works related to identification of melanoma skin cancer using different techniques is discussed in [19, 20]. In rural areas, it is difficult for local people visit dermatologist for skin treatment; so, it is better if they could check whether they are suffering from skin cancer or not using smartphone.

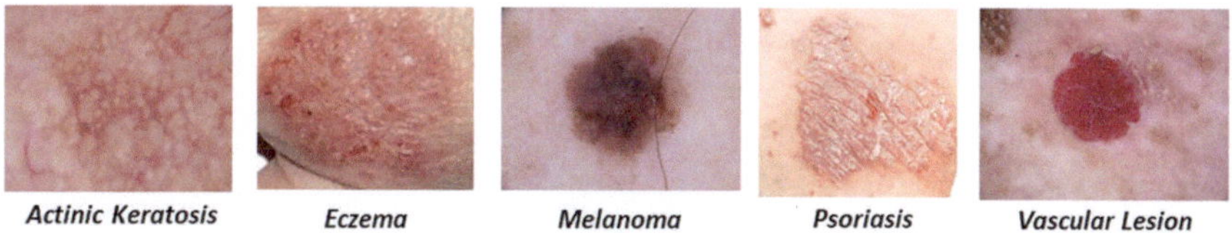

Fig. 1 Sample images

This work presents classification of skin cancers, namely, Actinic Keratosis, Eczema, Melanoma, Psoriasis and Vascular Lesion using different colour and texture feature extraction models. The works aims towards identifying the most suitable feature extraction models and also the best combination of features for this classification task.

2 Description of the Dataset

We collect five different types of skin cancer images from the available dataset. The database has 100 images (20 Actinic Keratosis images, 20 Eczema images, 20 Melanoma images, 20 Psoriasis images and 20 Vascular Lesion images). Five sample images are presented in Fig. 1.

3 Materials and Methods

This section highlights the various steps that are involved for the executing the classification task which is also presented in Fig. 2. The first step includes feature extraction using different techniques followed by classification using those extracted features. The contribution made by different image features for the classification of skin cancer were also analysed and compared.

3.1 Colour Feature Extraction

Colour is nothing but the sensation of light in the visible range. It is also one of the most important attribute of an image which helps in understanding and discriminating colour images. Thus it can be considered as one of the striking features of an image that carries the scenic beauty of an image. Colours are specified in a standard way by defining them in a 3D coordinate system called colour model. Any colour can thus be represented with the help of a suitable colour model. Two different colour

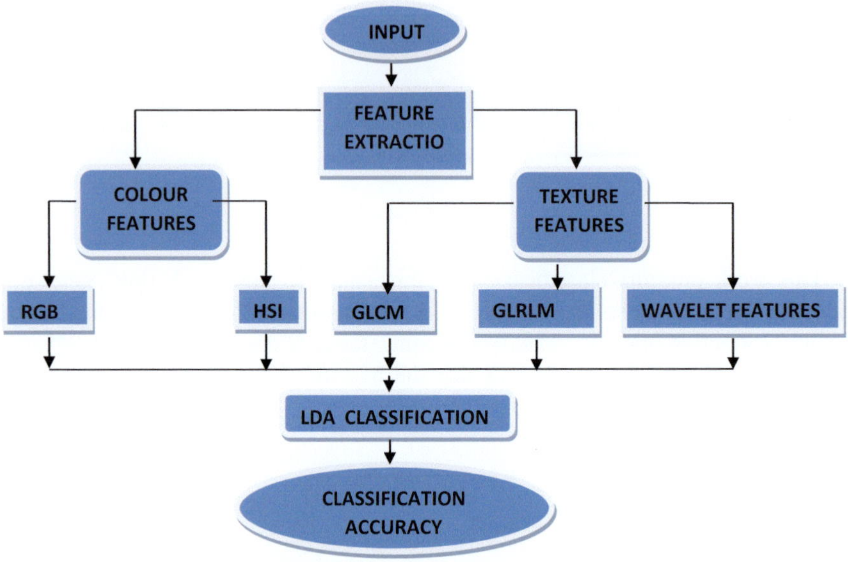

Fig. 2 Block diagram of the proposed system

models, namely, Red–Green–Blue (RGB) and Hue–Saturation–Intensity (HSI) are considered for this work.

RGB Colour Model. In this model colours are represented and are described in the form of a cube. The origin is represented by black, whereas the opposite corner in represented by white. Three statistical measures namely, mean, variance and range are evaluated from this RGB colour model.

HSI Colour Model. The human perception of colour closely resembles the HSI model of colour. In this model hue indicates dominant colour as perceived by human eye. Saturation represents how much is the perceived colour is diluted by white light. Finally, the intensity in this model represents the brightness. Three statistical features namely, mean, variance and range are extracted using Eqs. (1–3) of HSI colour model are calculated.

$$\mathrm{Mean} = \frac{\sum_{x,y} L(x, y)}{N} \tag{1}$$

$$\mathrm{Variance} = \frac{\sum_{x,\,y} [I(x, y) - \mathrm{Mean}]^2}{N} \tag{2}$$

$$\mathrm{Range} = \max(I) - \min(I) \tag{3}$$

where N is the total number of pixels.

3.2 Texture Feature Extraction

Textures are another important attribute that carries information pertaining to the variation in the intensity of an image. That is it reflects how repeatedly a particular pixel or a group of pixels are arranged in a given image. Thus, the presence of repeated patterns in an image provides good information pertaining to texture. This section presents extracting the texture features using different texture models.

Grey level co-occurrence matrix. Texture feature evaluation based on co-occurrence matrix has become one among the most popular and widely used techniques in image processing and pattern recognition task for the past few years. Generation of grey level co-occurrence matrix (GLCM) for a 5 × 5 image with maximum grey level equals 5 (considering the pixels pair occurrence in horizontal direction) is shown in Fig. 3. This technique of extracting the information of how pairs of pixels are arranged in a given image is quite helpful in identifying the texture of that image. However, the process of extracting such information from a given image with high pixel intensity (grey level value) is quite tedious and time consuming, because one has to look for the occurrence of all possible pair of pixels in the entire image. Above all this method is rotation-dependent. That is the texture information will not remain same when the image is rotated. Therefore, one needs to produce a minimum of four such co-occurrence matrixes to overcome its rotation-dependent property, which again increases the computational burden. In order to counter the computational burden, in this work quantizes the input images to an intensity level = 64, which results to a co-occurrence matrix of size 64 × 64. This work also considered four different directions as in [21]. From the resulting matrix, nine statistical measures are evaluated as described in [21, 22].

Grey level run-length matrix. Texture features based on run-length matrix work quite similar to that of co-occurrence matrix discussed above. This technique provides information pertaining to the length of connected pixels (with same pixel value). This might be useful to extract texture information of images having repeated patterns of uniform grey scale. It is also clear that a single run-length matrix might not be able to convey the texture information more precisely (as the matrix itself is

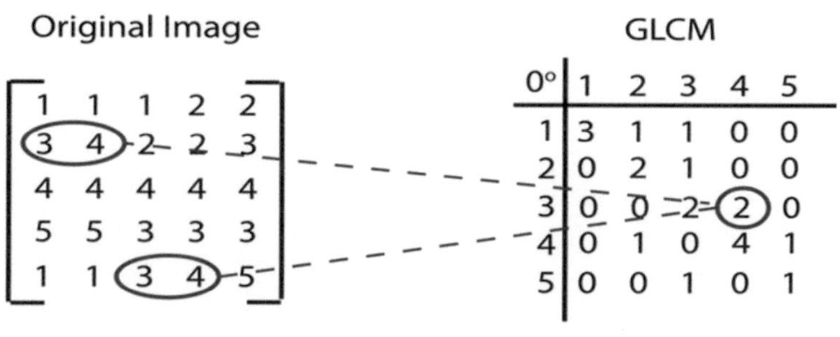

Fig. 3 Grey level co-occurrence matrix at 0° for a given image

rotation-dependent) and therefore, it is always required to generate four run-length matrixes for a minimum of four directions. Generation of run-length matrix (along Zero degree) for a 5×5 input image with maximum pixel intensity = 5 is shown in Fig. 4 (the figure itself is self-explanatory). In view of reducing the computational complexity, the input skin image is quantized as discussed in the above section. Four run-length matrices are generated from the resulting input image (the average of these four matrices is considered for this work). Finally, 11 run-length texture features were evaluated as described in [23, 24].

Wavelet Features. Wavelet decomposition is becoming a very promising technique for image processing and pattern recognition task. Unlike the two texture feature extraction techniques discussed earlier, feature extraction using wavelet decomposition is quite easy and carries valuable information about the image. The ability to analyse image in multiple resolution is the added advantage of wavelet decomposition technique. Thus, this can extract both the detail and coarse information present in an image. This work decomposed the input skin image at two level of wavelet decomposition (in MATLAB platform). Energy and entropy are evaluated from the resulting image as shown in Fig. 5. A total of 14 features were extracted.

The classification was carried out so as to minimize the expected classification cost given in Eq. (4).

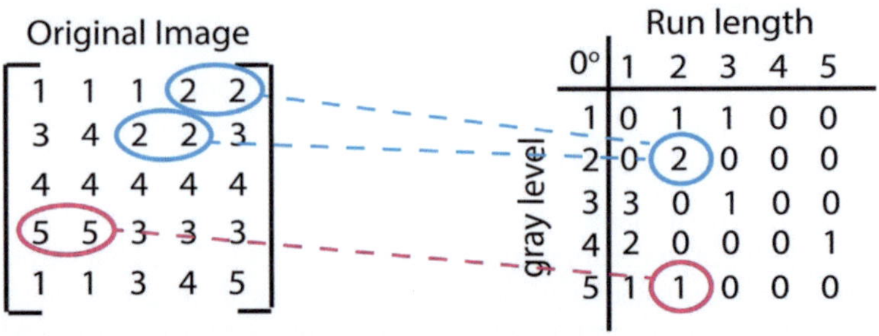

Fig. 4 Grey level run-length matrix at 0° for a given image

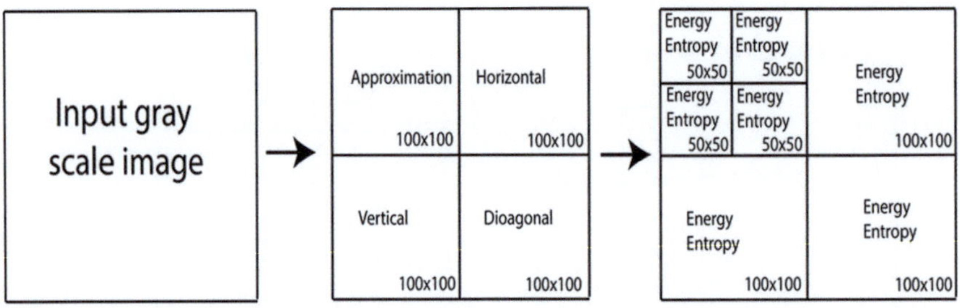

Fig. 5 Fourteen wavelet features

$$\hat{y} = \arg\min_{y=1...K} \sum_{k=1}^{K} \hat{p}(k/x)C(y/k) \quad (4)$$

where $\hat{y}$ is the predicted classification
K is the number of classes
$\hat{p}(k/x)$ is the posterior probability of class k for observation x
$C(y/k)$ is the cost of classifying an observation as y when its true class is k

4 Results and Discussion

Identification of five different human skin cancers is carried out using different colour and texture feature extraction techniques. The classification accuracy for each of the skin cancers using RGB and HSI colour models is presented in Table 1. It is observed that two of the skin cancer types namely, melanoma and Actinic Keratosis are identified with higher accuracy as compared to other three types of skin cancer. It is also observed from Fig. 6, that LDA performs better as compared with other three classifiers. Figure also shows that RGB colour model based colour features yields better results as compared to his-based colour features except for the case of SVM classifier. The performance in terms of classification accuracy using different texture features are presented in Table 2. It can be seen that Actinic Keratosis type skin cancer can be identified easily as compared to the remaining four types of skin cancer. It is also observed that KNN and NB classifier performs quite weak as compared to SVM and LDA. The average classification accuracies based on different texture models using all four classifiers is shown in Fig. 7. It is observed that support vector machine (SVM) classifier is able to achieve better classification accuracy while considering GLCM- and GLRLM-based texture features. However, it is found that LDA classifier is able to achieve overall best classification accuracy with wavelet based features. Thus, results show that wavelet feature is able to provide highest classification accuracy of 79% as compared to all the other remaining features including colour features.

This work also compares the performance of combined colour and texture features towards classification of skin cancers. The classification accuracy of each of the skin cancer types based on combined features is presented in Table 3. Results show that wavelet combined with RGB colour features are able to identify the two types of skin cancer namely, melanoma and vascular lesion with 100% accuracy. Figure 8 shows that combined wavelet and RGB features is able to achieve the highest average classification accuracy of 92% as compared to the other remaining five combination using LDA classifier. Thus, RGB and wavelet features are the most dominating features as compared to other features discussed in this work.

Table 1 Classification accuracy of RGB and HIS features

Base on red green blue (R.G.B)					
Classifier	Accuracy (%)				
	Actinic keratosis (%)	Eczema (%)	Melanoma (%)	Psoriasis (%)	Vascular lesion (%)
Linear discriminant analysis (LDA)	80	60	95	55	80
Support vector machine (SVM)	75	50	90	80	50
K-nearest neighbour (KNN)	95	45	95	25	55
Naive Bayes (NB)	90	40	85	0	60
Base on Hue Saturation Intensity (H.S.I)					
Classifier	Accuracy (%)				
	Actinic keratosis (%)	Eczema (%)	Melanoma (%)	Psoriasis (%)	Vascular lesion (%)
Linear discriminant analysis (LDA)	75	55	85	75	70
Support vector machine (SVM)	85	70	50	80	65
K-nearest neighbour (KNN)	95	65	70	25	55
Naive Bayes (NB)	100	30	75	0	50

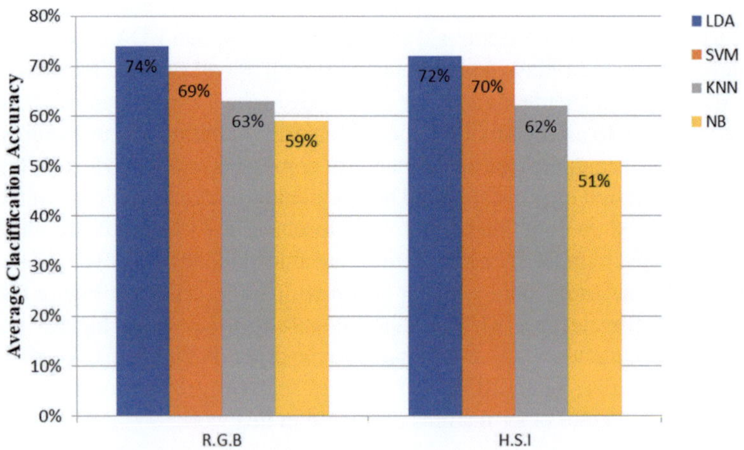

Fig. 6 Average classification accuracy for colour features

Table 2 Classification accuracy of different feature models

Base on grey level co-occurrence matrix (GLCM)					
Classifier	Accuracy (%)				
	Actinic keratosis (%)	Eczema (%)	Melanoma (%)	Psoriasis (%)	Vascular lesion (%)
Linear discriminant analysis (LDA)	75	50	100	60	75
Support vector machine (SVM)	90	55	90	55	70
K-nearest neighbour (KNN)	80	15	90	60	70
Naive Bayes (NB)	70	15	85	60	70
Base on grey level run-length matrix (GLRLM)					
Classifier	Accuracy (%)				
	Actinic keratosis (%)	Eczema (%)	Melanoma (%)	Psoriasis (%)	Vascular lesion (%)
Linear discriminant analysis (LDA)	80	60	90	60	70
Support vector machine (SVM)	95	65	90	60	70
K-nearest neighbour (KNN)	95	55	80	25	55
Naive Bayes (NB)	90	55	50	35	60
Base on wavelet					
Classifier	Accuracy (%)				
	Actinic keratosis (%)	Eczema (%)	Melanoma (%)	Psoriasis (%)	Vascular lesion (%)
Linear discriminant analysis (LDA)	95	80	60	80	80
Support vector machine (SVM)	90	70	75	60	75
K-nearest neighbour (KNN)	85	25	50	60	55
Naive Bayes (NB)	65	20	45	55	75

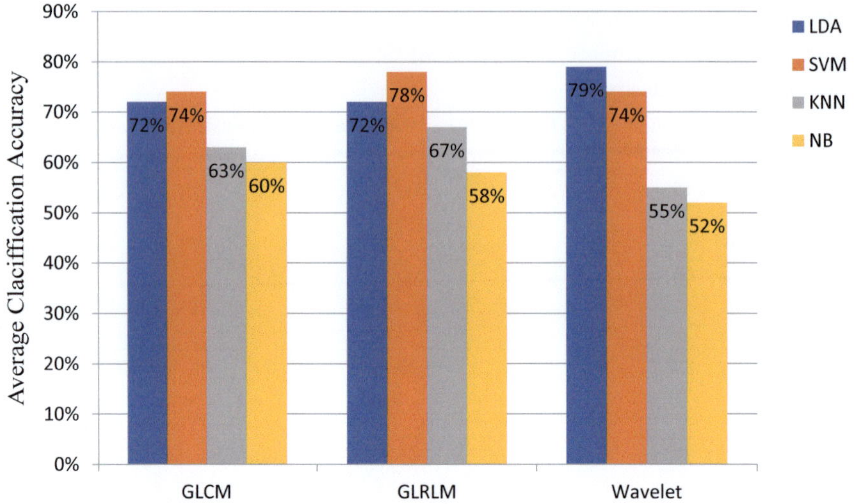

Fig. 7 Average classification accuracy for texture feature

5 Conclusion

Identification of five different types of human skin cancers is carried out using different feature extraction models. Two colour features based on RGB and HSI colour models are consider for colour feature extraction task. Texture features based on GLCM, GLRLM and wavelet decomposition are also considered. It is found that RGB-based colour features are able to achieve better classification accuracy as compared to HSI-based colour features. It is also learnt that features based on wavelet decomposition technique not only out performs GLCM and GLRLM based texture features but also other colour features based on RGB and HSI colour models. Results also show that classification performance based on combine wavelet and RGB colour features yields better classification accuracy as compared to that of other combinations. It is also learnt that LDA classifier is able to perform better as compared to other classifiers presented in this work while considering combined features. It is also found that SVM classifier is also able to perform the classification task little less than LDA but better than KNN and NB. Thus, it can be learnt that wavelet-based features because of its multi-resolution characteristic is found more suitable for identification of skin cancers using LDA classifier. This work can be further extended with more number of cancer data. The performance of the classifier using engineered featured can also be compared with that of the CNN. The performance of the classifier may also be analysed using a suitable feature selection processes and the most dominating feature for this specific skin disease classification task can be identified.

Table 3 Classification accuracy of combined features

Classifier	Accuracy (%)				
Base on wavelet + red green blue (R.G.B)					
Classifier	Actinic keratosis (%)	Eczema (%)	Melanoma (%)	Psoriasis (%)	Vascular lesion (%)
Linear discriminant analysis (LDA)	95	85	100	80	100
Support vector machine (SVM)	70	65	70	100	95
K-nearest neighbour (KNN)	95	45	100	45	75
Naive Bayes (NB)	70	35	65	40	80
Base on wavelet + Hue Saturation Intensity (H.S.I)					
Classifier	Accuracy (%)				
	Actinic keratosis (%)	Eczema (%)	Melanoma (%)	Psoriasis (%)	Vascular lesion (%)
Linear discriminant analysis (LDA)	90	80	90	80	95
Support vector machine (SVM)	65	65	65	100	95
K-nearest neighbour (KNN)	95	30	85	50	90
Naive Bayes (NB)	85	35	65	55	85
Base on grey level co-occurrence matrix (GLCM) +red green blue (R.G.B)					
Classifier	Accuracy (%)				
	Actinic keratosis (%)	Eczema (%)	Melanoma (%)	Psoriasis (%)	Vascular lesion (%)
Linear discriminant analysis (LDA)	90	75	100	75	90
Support vector machine (SVM)	75	65	65	100	95
K-nearest neighbour (KNN)	90	50	95	35	65
Naive Bayes (NB)	100	30	100	65	75

(continued)

Table 3 (continued)

Base on wavelet + red green blue (R.G.B)					
Classifier	Accuracy (%)				
	Actinic keratosis (%)	Eczema (%)	Melanoma (%)	Psoriasis (%)	Vascular lesion (%)
Base on grey level co-occurrence matrix (GLRLM) + hue saturation intensity (H.S.I)					
Classifier	Accuracy (%)				
	Actinic keratosis (%)	Eczema (%)	Melanoma (%)	Psoriasis (%)	Vascular lesion (%)
Linear discriminant analysis (LDA)	100	75	95	60	90
Support vector machine (SVM)	75	65	70	90	100
K-nearest neighbour (KNN)	100	50	85	70	70
Naive Bayes (NB)	100	45	85	60	80
Base on grey level run-length matrix (GLRLM) +red green blue (R.G.B)					
Classifier	Accuracy (%)				
	Actinic keratosis (%)	Eczema (%)	Melanoma (%)	Psoriasis (%)	Vascular lesion (%)
Linear discriminant analysis (LDA)	100	85	95	70	95
Support vector machine (SVM)	75	60	60	80	95
K-nearest neighbour (KNN)	100	45	90	20	80
Naive Bayes (NB)	100	60	85	35	65
Base on grey level run-length matrix (GLRCM) + hue saturation intensity (H.S.I)					
Classifier	Accuracy				
	Actinic keratosis (%)	Eczema (%)	Melanoma (%)	Psoriasis (%)	Vascular lesion (%)
Linear discriminant analysis (LDA)	100	80	90	70	85

(continued)

Table 3 (continued)

Base on wavelet + red green blue (R.G.B)					
Classifier	Accuracy (%)				
	Actinic keratosis (%)	Eczema (%)	Melanoma (%)	Psoriasis (%)	Vascular lesion (%)
Support vector machine (SVM)	70	65	65	70	100
K-nearest neighbour (KNN)	100	25	85	40	65
Naive Bayes (NB)	100	50	80	45	75

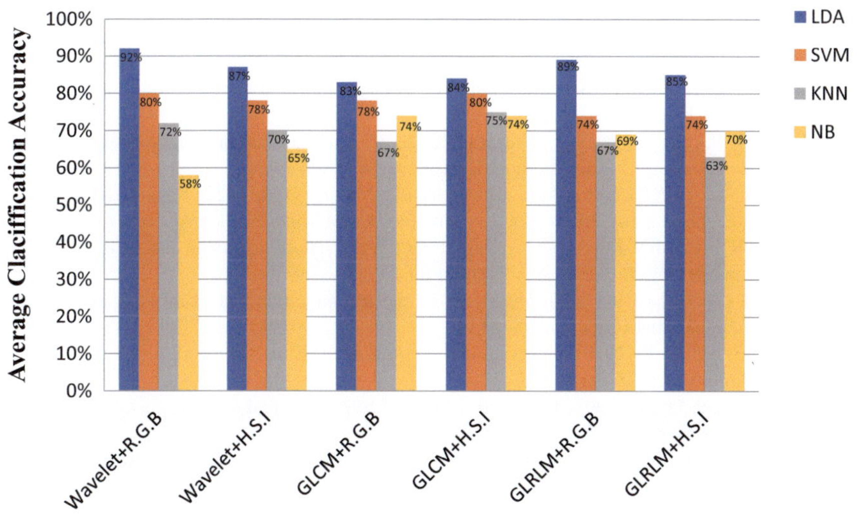

Fig. 8 Average classification accuracy for combined features

References

1. ALEnezi NSALK (2019) A method of skin disease detection using image processing and machine learning. Procedia Comput Sci 162:85–92
2. Leelavathy S, Jaychandra R, Shobana R, Vasudevan SS, Prasand N (2020) Skin disease detection using computer vision and machine learning technique. Eur J Mol Clin Med 7(4):2999–3003
3. Al Shahibi ISS, Koottala S (2020) Detection of skin diseases using Matlab. J Stud Res 1–10
4. Wei LS, Gan Q, Ji T (2018) skin disease recognition method based on image colour and texture features. Comput Math Methods Med 1–10
5. Thakur RC, Panwar V (2020) Skin cancer recognition using SVM image processing technique. BOHR Int J Biocomput Nano Technol 1(1):11–15

6. Divya SDV, Goggi P, Prathibha L, Sushmetha GR, Mary M (2020) Skin disease detection using image processing and neural networks. Int J ProgIve Res Sci Eng 1(4):8–12
7. Shashi RG, Srinivasan MH, Sudarson J (2018) Digital dermatology—skin disease detection model using image processing. Int J Innov Res Sci Eng Technol 7(7):8201–8211
8. Hemavathi S, Jayasakthi K (2021) Skin disease prediction and provision of medical advice using deep learning. J Phys 1724:1–5
9. Jessica V, Cherry P, Jean WA, Jonathan A, John SC, Mark AG, Benjamin JM, Lyndon T, August T, Romeo JJ (2019) A smartphone-based skin disease classification using MobileNet CNN. Int J Adv Trends Comput Sci Eng 8(5):2632–2637
10. Kirubai NK, Mangai V (2019) Skin disease detection using Image processing with CNN. Sci Technol Dev 8(9):190–203
11. Ansari UB, Sarode T (2017) Skin cancer detection using image processing. Int Res J Eng Technol (IRJET) 4(4):2875–2881
12. Payal B, Samir KB, Amiya B, Sandeep P (2021) Skin disease detection: machine learning vs deep learning. Preprints, pp 1–7
13. Rahul L, Vinay K, Ganesh H, Suvaran P (2020) Skin disease detection and classification using image processing algorithm. SAMRIDDHI: J Phys Sci Eng Technol. 12(1):169–171
14. Chaudhuri S, Seema B (2014) Artificial neural network for skin cancer detection. Int J Emerg Trends Technol Comput Sci 3(5):147–153
15. Damilola AO, Oludayo OO, Salomom AO (2013) Automating Skin disease diagnosis using Image classification. In: Proceeding of the world congress on engineering and computer science. WCECS, San Faancisco, USA, pp 1–5
16. Shuchi B, Sachin S, Piyush J, Chitransh K (2019) Machine learning algorithms based skin disease detection. Int J Innov Technol Explor Eng 9(2):4044–4049
17. Mukta J, Sandra PGG, Maria AGL, Thony ADLC, Maleny ECR (2022) Advance study of skin diseases detection using image processing methods. Nat Volatiles Essent Oils 9(1):991–1007
18. Bhavani R, Prakash V, Kumaresh RV, Sundra SR (2019) Vision-based skin disease identification using deep learning. Int J Eng Adv Technol 8(6):3784–3788
19. Prathima G, Shoba RR (2020) Discovery of melanoma carcinoma using image processing approach. Sci Technol Dev 9(2):48–57
20. Amruta S, Sunita K (2019) Detection and classification of melanoma skin cancer. Int Res J Eng Technol 6(4):1038–1041
21. Majumdarand, Jayas (1999) Classification of bulk sample of cereal grains using machine vision. J Agric Eng Res 73:35–47
22. Adi K, Widodo CE, Widodo AP, Gernowo R, Pamungkas A, Syifa RA (2018) Detection of lungs cancer using gray level co-occurrence matrix and back propagation neural network classification. J Eng Sci Technol Rev 11(2):8–12
23. Hamman SA, Purboyo TW, Saputra RE (2017) Cotton texture segmentation based on image texture analysis using gray level run length and Euclidean distance. J Theor Appl Int Technol 95(24):6915–6923
24. Tang X (1998) Texture information in run length matrix. IEEE Trans Image Process 7(11):1602–1609

A Study on Syllabification for Preparation of Pronunciation Dictionary in Manipuri Speech Processing

Hoomexsun Pangsatabam, Yambem Jina Chanu, and Naorem Karline Singh

Abstract The Pronunciation Dictionary is a crucial component of speech processing systems that are used to create language models. Every word in a language is broken down into its constituent phonemes in a pronunciation dictionary. To construct the text corpus, both Automatic Speech Recognition and Text to Speech Synthesis systems require a pronunciation dictionary. The work's main contribution is an examination of several ways in which syllabification plays a significant role in the development of the dictionary. The research will aid in the understanding of the function of features of Natural Language Processing in the development of diverse speech processing applications.

Keywords Syllabification · Speech processing · Manipuri ASR · Manipuri TTS

1 Introduction

In the world of speech processing, a dataset is required to build an accurate and precise system. The quality of the data has a significant impact on the models developed. Data in the form of audio, as well as the matching text, is required in automatic speech recognition [1] and text to speech synthesis [2]. End-to-end models with a large amount of data have recently been implemented. The amount of data is not a problem because of the computing power of large computers. When building a corpus, the typical strategy is to divide it into appropriate pieces, with the matching text and audio needed to complete the system.

H. Pangsatabam (✉) · Y. J. Chanu · N. K. Singh
National Institute of Technology Manipur, Imphal 795004, India
e-mail: humex.pangz@gmail.com

Y. J. Chanu
e-mail: jina.yambem@gmail.com

N. K. Singh
e-mail: naoremkarline@gmail.com

S. K. Biswas et al. (eds.), *Intelligent Computing Systems and Applications*, Lecture Notes in Networks and Systems 1307, https://doi.org/10.1007/978-981-96-3860-4_26

The lack of quality resources has a significant impact on speech tool development. An important resource in the area of Speech Processing is the Grapheme to Phoneme conversion which is also referred to as a pronunciation dictionary. The pronunciation dictionary for highly resourced languages like English or German; already exists. The Carnegie Mellon University Pronouncing Dictionary is an open-source machine-readable North American English pronunciation dictionary with approximately 134,000 terms and their pronunciations [3]. CMUdict is constantly updated and improved.

Syllabification, the division of syllables in a word has played a huge role in Natural Language Processing (NLP) [4]. It is discussed throughout this article using Meiteilon or Manipuri, an under-resourced language spoken by people in Manipur, which is a state in the North-Eastern part of India.

In Manipur, two scripts are widely used. Both these scripts have their own challenges in terms of social factors as well as phoneme imbalance [5]. However, the syllabic nature of the language [6] has provided a way to design and implement speech processing systems and transliteration systems. The corpora building adopted by different authors and researchers is discussed heavily in Sect. 2, while the approaches taken by various researchers to address subsets of the problem are discussed in Sect. 3. Section 4 includes the application of these technologies. In Sect. 5, the findings are summarized.

2 Building the Corpora

2.1 Text Collection

One cannot just use the data available through media as text corpora for research in Speech Processing or Natural Language Processing. This is because the text is not Unicode compatible. To resolve this, Singh et al. used a Bengali OCR from the PDF format but has only 52.6% accuracy and therefore had to be manually corrected [7]. While building the text corpora, the text selected should contain all existing syllables and come from a large vocabulary. H.A. Patil et al. collected the text from a children's storybook "Phunga Wari Singbul" [2], while others collected the text data from internet sources or through media. Nongmeikapam et al. collected data from *Huiyen Lanpao*, Singh, et al. from *poknapham,* etc. [2, 8].

2.2 Meitei Mayek Script and Characters

Mapum Mayek is a set of 21 independent characters in the *Meitei Mayek script*. There are 18 consonants and three vowels in the *Mapum Mayek* alphabet. With the exception of *Ee lonsum* and *cheitap mayek*, which are vowels, *Lonsum Iyek* consists of eight

Table 1 Meitei Mayek characters

Mapum			Lonsum	Cheitap
ꯀ	ꯁ	ꯂ	ꯛ	◌ꯣ
ꯃ	ꯄ	ꯅ	ꯜ	◌ꯤ
ꯆ	ꯇ	ꯈ	ꯝ	◌ꯥ
ꯉ	ꯊ	ꯋ	ꯞ	◌ꯦ
ꯌ	ꯍ	ꯎ	ꯟ	◌ꯧ
ꯏ	ꯐ	ꯑ	ꯠ	◌ꯨ
ꯒ	ꯓ	ꯔ	ꯡ	◌ꯩ
ꯕ	ꯖ	ꯗ	ꯢ	◌ꯪ
ꯘ	ꯙ	ꯚ		

stop consonants in addition to *mapum mayek*. Both of these groups of characters rely on *Mapum mayek* to generate their Manipuri vocabularies. The numeral characters are *Cheising Mayek*, while the punctuations are *Khudam Mayek*. Table 1 lists the Meitei Mayek characters and their pronunciations.

Unicode contains the Meitei Mayek characters [9]. Fonts with UTF-8 encoded characters are accessible under the name Eeyek [10] and Nirmala UI, however alternative fonts such as epaomayek [11] are also available.

2.3 *Syllabic Structure and Patterns*

A syllable structurally has three parts. The nucleus is compulsory and represents high energy while the onset and coda exist before and after the nucleus, respectively. A syllable is generally of the form C*VC*, where C represents consonants, V represents vowels, and C* represents a cluster of consonants or none. The formation of these parts affects the pronunciation of the language. Table 2 contains an example for each type of phoneme or grapheme to form the syllables [2–4, 7, 12].

Table 2 Meitei Mayek phoneme and grapheme type

Type	Single Character	Character group
Independent vowel	ꯑ	ꯑꯦ
Dependent vowel	◌ꯥ	◌ꯥꯏ
Independent consonant	ꯀ	-
Stop consonant	ꯛ	-
Consonant clusters	-	ꯀꯆ

Table 3 Meitei Mayek syllables

Syllable Type	Monosyllabic example
V	ꯑ
VC	ꯑꯡ
CV	ꯃꯤ
CVC	ꯈꯣꯛ
CCV	ꯈ꯭ꯔꯦ
CCVC	ꯈ꯭ꯔꯦꯛ

The Manipuri vocabulary consists of six normal syllables following the C*VC* pattern. It has been studied and observed that CV is the mostly occurring syllable, followed by CVC. The position of the syllable in the word also affects its occurrences [2–4, 7, 12].

A Manipuri syllable can be built in a variety of ways. To create a specific sound unit, two or more characters must also be combined. Table 3 provides a quick summary of how characters are used to produce syllables, as well as an example for each scenario.

2.4 Speech Database

A total of 36 phones are used in the speech corpus of work contributed by Salam et al. [12]. Some phones are merged and later the number of phonetic units gets reduced to 29. The speech has to be recorded in a controlled environment so as to reduce noise. To achieve this, a studio is preferred as well as a noise-canceling microphone. While recording, the speech is sampled between 16 KHz and 44 KHz depending on the work, and the speech files are in wave format.

2.5 Challenges

Various issues are there while building both the text and speech corpora. It is also important to remember that Manipuri is written in two scripts. The borrowed script is prevalent in the present world and is used in general media such as newspapers and old textbooks. However, there has been increasing usage of the original Meitei Mayek Script, and students these days are not legible to the old borrowed script. The Bengali Script has 55 symbols to represent only 38 phonemes in Manipuri Language. This creates an imbalance in phonemes [5].

While building the text corpus, the script is prone to human errors. It has been observed that for every work, there are transcribers involved in order to provide the particular word to form meaningful sentences [1].

Sometimes the spoken sentence, called utterances, might differ from the original text. Each speaker has their own accent and native fashion while pronouncing certain words.

These issues pose certain challenges which are sometimes crucial to the system and if not fixed might develop it into an entirely different product.

3 Related Works

3.1 Automatic Syllabification Rules

There are two general approaches for automatic syllabification, which are rule-based and data-driven. Devi et al. have implemented syllabification automatically using a hash table. In their work, the CV clusters are detected from the words and looked up in a dictionary. The dictionary consists of key-value pairs where the observed sequence is the key and the segmentation rule is the value. It is tested on 1,24,087 distinct words and achieves an accuracy of 99.8% [4].

Singh et al. proposed and implemented three corpus-based automatic syllabification methods, i.e., rule-based approach, sequence labeling approach, and a hybrid of both. In the end, the entropy-based phonotactic segmentation (EPS) is found to achieve 96% accuracy which increases to 98% when combined with Conditional Random Field (CRF) [7].

Salam et al. have used separate syllabification rules manually. It heavily relies on the occurrence of a separate nucleus and a number of consonants which are then divided into coda and onset [12].

3.2 Morpheme Identification

From a linguistic view, the patterns found are of the six types given in 2.3. However additional patterns can be found from a computational view such as VVV, CVV, and CCVC. Nongmeikapam et al. addressed this in their work on morpheme identification [8]. An alternate approach to this is to club the vowels together as proposed by Devi et al. [4]. This works in syllabification but will raise other issues if phonemes are needed in creating a new system.

3.3 Entropy-Based Phonotactic Segmentation (EPS)

Singh et al. tackled three major problems with the help of EPS in Syllabification. The Schwa deletion problem was handled by deploying an entropy estimate of the phone. The EPS algorithm is designed as a recursive algorithm where the cluster entropy of a phone is calculated by considering the preceding and the succeeding phones. Then the syllable is split from the word. In the next step, sequence labeling is implemented by the authors [7].

3.4 Sequence Labeling Approach

Saikia et al. investigated three labeling methods which are CRF, MEMM, and HMM. However, they have implemented this for Manipuri English. Manipuri English is plain English but phonemes differ from the English spoken by American countries because of the accent. However, they have implemented their work using a pronunciation dictionary generated from the CMU dictionary [3].

Singh et al. also used sequence labeling but for Manipuri written using Bengali script. To implement sequence modeling, each letter in the word is tagged by either S for the splitting character or C for the continuous character. Then an algorithm to split the word is applied. They have implemented CRF as well as MEMM in their work [7].

3.5 Semi-automatic Syllabification

A semi-automatic syllabification system was developed by Salam et al. [12]. In the first step, the phonetics segments to be syllabified are segmented and aligned with the speech. Then, automatic syllabification is done by providing the system with syllabification rules developed by the author. In the third step, syllable labeling is done. It was concluded that there is an average deviation of 25 ms is acceptable (Table 4).

3.6 Challenges and Solutions

There are limitations and tradeoffs to achieving good accuracy in the rules developed by various researchers [7].

1. *Schwa* deletion
2. *Heteronyms* are words that have different pronunciations although they are spelled similarly.
3. Loan words are borrowed vocabularies from another language and are unavoidable.

For improving iterative scaling and overfitting, L1 regularizers are deployed by Singh et al. in their sequence labeling implementation [7].

Table 4 Various approaches to syllabification and its applications

Author	Area	Technology used
Devi et al. [4]	Automatic syllabification	Segmentation rules
Singh et al. [7]	Automatic syllabification	Rule-based Approach (Baseline and EPS) and Sequence Labeling
Salam et al. [12]	Semi-automatic syllable labeling	HMM and Syllabification rules
Patil et al. [2]	Unit selection synthesis	Syllable-based synthesis system
Saikia et al. [3]	Pronunciation dictionary	Sequence Labeling
Laitonjam et al. [13]	Transliteration	Dictionary-based and Machine Learning
Devi et al. [14]	Machine transliteration	Mapping and Syllabification/ Segmentation rules
Nongmeikapam et al. [8]	Morpheme Identification	Syllabic Patterns and Morpheme identification system

4 Applications

4.1 Transliteration

The transliteration of Manipuri from English is contributed by various authors. Laitonjam et al. use two approaches, one of which is dictionary-based, while the other is machine learning-based [13]. The authors addressed various issues and in the end, used graphemes as the medium at which the transliteration is implemented. Graphemes are smaller than syllables but bigger than phonemes and are suitable for speech involving tasks. Devi et al. implemented transliteration through various mapping. First, characters of English are mapped to their corresponding Manipuri counterparts. The vowels and consonants are mapped differently. All possible combinations or patterns are studied by the authors and the rules are implemented in their transliteration system [14]. Syllables play important roles in these systems indirectly.

4.2 Automatic Speech Recognition (ASR)

Patel et al. developed an automatic speech recognition system with a large vocabulary for Manipuri. In their work, they used Bengali script and worked on 80 h of data. A new pronunciation dictionary is built by referring to TBT toolkit [1]. Pronunciation plays an important role in building the lexicon to construct the language model. ASR systems use Kaldi for most of their experiment [15].

4.3 *Text to Speech Synthesis (TTS)*

Patil et al. use a syllable-based framework in their TTS system. The texts are collected in UTF-8. In their work, they have mentioned that an under-resourced language hardly has any independent speech recognition system to generate the vocabulary for use in their TTS. The nature of phone sets and syllable structures are studied by the authors for 13 languages [2].

Saikia et al. compared four TTS models in their work. Out of which it can be concluded that CRF performed way better than both MEMM and HMM. The research was conducted on Manipuri English however [3].

5 Summary and Conclusion

According to the findings, the role of syllabification in the construction of the pronunciation dictionary is significant. Various authors have come up with different strategies to build their own dictionaries using phonemes, graphemes, and even syllables. The dictionary so developed is used in a variety of speech and natural language processing applications. The main concerns are the lack of an open-source text dataset in Meitei Mayek and the fact that even if the text is available in Bengali Script, it is not Unicode compliant, making it unusable. To circumvent this, the authors have utilized OCR or even employ a team of linguists to transcribe the text from scratch. There is currently no technology that can convert all words in the Meitei Mayek lexicon to their phonemes without errors.

References

1. Patel T, Krishna N, Fathima N, Shah N, Mahima C, Kumar D, Iyengar A (2018) Development of large vocabulary speech recognition system with keyword search for Manipuri, pp 1031–1035. https://doi.org/10.21437/Interspeech.2018-2133
2. Patil HA et al (2013) A syllable-based framework for unit selection synthesis in 13 Indian languages. In: 2013 international conference oriental COCOSDA held jointly with 2013 conference on Asian spoken language research and evaluation (O-COCOSDA/CASLRE), pp 1–8. https://doi.org/10.1109/ICSDA.2013.6709851
3. Saikia R, Singh SR (2016) Generating Manipuri English pronunciation dictionary using sequence labelling problem. In: 2016 international conference on Asian language processing (IALP), pp 67–70. https://doi.org/10.1109/IALP.2016.7875936
4. Devi MP, Singh I, Devi H (2017) Automatic syllabification rules for Manipuri language 8:57–59. https://doi.org/10.26483/ijarcs.v8i1.2852
5. Singh LS, Tharoijam K, Das PK (2007) Writtten Manipuri (meiteiron) from phoneme to grapheme. Lang India 7(6)
6. Kishore SP, Black AW (2003) Unit size in unit selection speech synthesis. In: Eurospeech, pp 1317–1320
7. Singh LG, Laitonjam L, Singh SR (2016) Automatic syllabification for Manipuri language. In: COLING

8. Nongmeikapam K, Vidya Raj RK, Nirmal Y, Bandyopadhyay S (2012) Manipuri morpheme identification
9. Meitei Mayek, Unicode character database, Unicode Consortium. https://www.unicode.org/charts/PDF/UABCO.pdf. Accessed on 25 Jan 2022
10. Eeyek Unicode Font for Meitei Mayek. http://tabish.freeshell.org/eeyek/download.html. Accessed on 25 Jan 2022
11. Epao Mayek Font for Meitei Mayek. http://e-pao.net/epPageExtractor.asp?src=MeiteiMayek.Install_Meitei_Mayek_Font.html. Accessed on 25 Jan 2022
12. Nandakishor S, Dutta SK, Singh LJ (2015) An HMM based semi-automatic syllable labeling system for Manipuri language. In: International conference on computing, communication and automation, pp 1044–1047. https://doi.org/10.1109/CCAA.2015.7148548
13. Laitonjam L, Singh LG, Singh SR (2018) Transliteration of English loanwords and named-entities to Manipuri: phoneme vs grapheme representation. In: 2018 International conference on Asian language processing (IALP), pp 255–260. https://doi.org/10.1109/IALP.2018.8629141
14. Devi MP (2017) English to Manipuri machine transliteration system based on syllabification. J Glob Res Comput Sci 8:01–05
15. Povey D, Ghoshal A, Boulianne G, Burget L, Glembek O, Goel N, Hannemann M, Motlíček P, Qian Y, Schwarz P, Silovský J, Stemmer G, Vesel K (2011) The Kaldi speech recognition toolkit. In: IEEE 2011 workshop on automatic speech recognition and understanding

DFE-AVD: Deep Feature Ensemble for Automatic Vehicle Detection

Debarshi Bhattacharya, Avirup Bhattacharyya, Maroi Agrebi, Arup Roy, and Pawan Kumar Singh

Abstract Over the past ten years, a lot of research has been done on automatic vehicle recognition utilizing machine learning approaches. The majority of earlier research on vehicle detection was done on datasets like as GTI, which mostly comprise well-organized road scenarios. These datasets do not accurately depict road scenes in areas with a lesser transportation infrastructure. As a result, the IDD dataset—which represents unstructured traveling situations—includes a large variety of vehicle classes, and shows adequate intra-class variability for vehicles—has been used to study vehicle recognition in this work. The suggested approach uses three neural networks—DenseNet21, SqueezeNet, and EfficientNet-B0—to extract features, which are then concatenated. An SVM classifier that uses a linear kernel has been trained using this fused collection of features. On a portion of the IDD dataset, the proposed methodology yields 89.73% recognition accuracy. In comparison with using the individual neural networks for feature extraction, there is a noted improvement in recognition accuracy of approximately 2.59%. The GTI dataset has also been evaluated, and all models achieved accuracy scores more than 99.4%.

D. Bhattacharya
Department of Electronics and Communication Engineering, Techno Main Salt Lake, EM-4/1, Sector V, Bidhannagar, Kolkata 700091, West Bengal, India

A. Bhattacharyya
Department of Computer Science and Engineering, Jadavpur University, 188, Raja Subodh Chandra Mallick Road, Kolkata 700032, West Bengal, India

M. Agrebi
Department of Computer Science, LAMIH UMR CNRS 8201, Université Polytechnique Hauts-de-France, Valenciennes, France

A. Roy
Department of Computer Science and Engineering, Dr. B. C. Roy Engineering College, Jemua Road, Fuljhore, Durgapur 713206, West Bengal, India

P. K. Singh (✉)
Department of Information Technology, Jadavpur University, Jadavpur University Salt Lake Campus, Plot No. 8, Salt Lake Bypass, LB Block, Sector III, Salt Lake City, Kolkata 700106, West Bengal, India
e-mail: pawansingh.ju@gmail.com

S. K. Biswas et al. (eds.), *Intelligent Computing Systems and Applications*, Lecture Notes in Networks and Systems 1307, https://doi.org/10.1007/978-981-96-3860-4_27

Keywords Automatic vehicle detection · Deep feature ensemble · DFE-AVD · EfficientNet · DenseNet · SqueezeNet · IDD dataset · GTI dataset

1 Introduction

Automatic vehicle detection (AVD) using machine learning has gained a lot of traction in the last few decades in the research community as well as in the real world. The primary motivation to work on such systems is four-fold—To prevent road accidents caused by human error, to assist disabled and elderly people, to manage traffic effectively to reduce commute times & for automatic car parking.

In India, about 4,49,002 vehicle accidents have been reported in the year 2019 leading to the death of 1,51,113 people [1]. 11% of deaths in the world attributed to accidents occur in India. In the United States, 33,244 vehicle crashes took place in 2019 leading to 36,096 deaths [2]. In 2019, it has been found that Indians residing in urban areas need to commute at least 2 hours daily for work [3]. Commute times have been steadily rising in the US for the last few years [4]. Traffic must be managed effectively to bring down commute times.

The standard way of classifying images using supervised machine learning involves feature extraction from the original image followed by feeding the extracted feature set into a machine learning model for classification. For training the classifier using machine learning, the extracted features from the images in training set and corresponding labels are used. The performance of such a system, to a very large extent, is determined by the quality of feature extraction techniques involved. Images fall into the category of high-dimensional data and all this information might not be required for the task at hand. Some of this data might be redundant, thus adversely affecting the performance of classifiers. Hence, it becomes necessary to extract relevant features before training the classification model.

There are multiple ways to obtain on-road information surrounding a car using sensors. In this context, sensors can be broadly classified into two types—active & passive sensors. Active sensors measure the travel time of reflected signals emitted from sensors to estimate object distance. In the past few years, object detection using data from active sensors has been widely studied. A downside to using these types of sensors is their excessive cost. In this work, our focus is on vision-based object detection which makes use of passive sensors such as cameras. Images offer richer information about the environment compared to active sensors. Higher resolution is also an added advantage. Easier installation, lesser expense, and accessibility make camera a far better alternative for obtaining on-road information. The images obtained via these cameras can be stored for later use—To enhance intelligent security systems as well as make use of the data to come up with strategies to minimize traffic congestion. The key factors to be taken into consideration to evaluate an AVD system, besides performance, are faster inference and usage of minimal computational resources. It must be noted that, despite all these advantages, vision-based object detection is still a very challenging task as described in the next subsection.

Deep learning techniques for feature extraction from image are more robust to changes than traditional computer vision techniques. They can also extract features which are beyond the limit of traditional techniques. The networks used for feature extraction in this paper are DenseNet21, SqueezeNet, and EfficientNet-B0. These networks extract features with very high speed. Thus, these networks have been preferred in this method over the traditional techniques as they extract more complex features and at a very high speed. Images are fed to the pipeline after which features are extracted by three CNNs and concatenated. This concatenated set of features are fed to SVM classifier with linear kernel which makes the prediction.

1.1 Challenges in AVD

Building an AVD using images is an inherently difficult task due to the nature of the data. Vehicles come in varying colors, shapes, and sizes. Traffic conditions, the presence of pedestrians and other objects, illumination differences, and weather conditions add to the complexity of a road scene.

This paper has been organized in the following manner—Sect. 2 highlights previous works on AVD, Sect. 3 covers the required theoretical background, Sect. 4 covers the proposed detection pipeline in this work, Sect. 5 presents experimental results, and Sect. 6 concludes this paper.

2 Related Work

In the research literature, the standard route for automatic vehicle detection is to use an image descriptor to extract features from images and then train models using these extracted features. Researchers have used numerous feature extraction techniques over the years like different variations of HOG, Gabor filters, and SIFT for vehicle detection. SVM is commonly used for the task of classification. But using these techniques has some disadvantages as the features are manually chosen. Often, such hand-crafted feature sets fail to encapsulate the entire content of an image. For complicated tasks, these algorithms fail to produce satisfactory results. Due to such limitations, researchers have moved more and more toward deep learning. Neural networks tend to identify high-level features from the input data. This eliminates the need for careful inspection of data and choosing suitable techniques for improved performance. Deep learning methods have produced better results than machine learning techniques, in practice, especially for image datasets.

The Indian Driving Dataset (IDD) [5] has been proposed to study autonomous navigation in unconstrained environments for few-shot learning, domain adaptation, and prediction of on-road behavior. Singh [6] performed a comparative analysis of models on both structure driving datasets such as Cityscapes and BDD and unstructured driving datasets such as IDD on semantic segmentation, instance segmentation, and object detection tasks. It has been observed that the DRN model [7] performed well for all the datasets that have been used. Nidamanuri [8] used Convolutional Neural Network (CNN) and Long-Short-Term-Memory (LSTM) models to build their Auto-Alert system for driving assistance. The IDD dataset has been used for their purpose because the authors claimed that the existing solutions don't work with chaotic traffic conditions. Kothandaraman [9] used unsupervised domain adaptation for proposing a semantic segmentation method that can recognize the class of each region in images and videos. The IDD dataset has been used to demonstrate the effectiveness of their method. Their model, BoMuDANet which uses pseudo-labeling and self-training can also identify unseen objects during inference. Baheti [10] proposed a novel architecture called Eff-UNet wherein the encoder is an EfficientNet which makes use of compound scaling effectively, and for the decoder network, UNet [11] is used to build the segmentation maps. This model held first place in the IDD lite segmentation challenge.

In the research literature, the GTI dataset has mostly been used for vehicle vs road classification. The GTI dataset contains only 2 classes and is very structured. It might not be very appropriate to train models to be used in unstructured traffic conditions. Therefore, we extend the study of vehicle recognition to the IDD dataset which has been captured in unstructured road scenes, thus making it a challenging task. In addition, the IDD datasets contain classes pertaining to different types of vehicles. The usage of this dataset, in the research literature, has mostly been restricted to detection, segmentation, localization, and domain adaptation—not classification.

3 Background

3.1 CNNs as Feature Extractors

Convolutional neural networks are the go-to for image classification, localization, and segmentation. Unlike traditional Image processing techniques which require hand-crafted features, ConvNets automatically learn the weights of the filters necessary to extract features for the given classification problem. In ConvNets, feature maps extracted from one layer are fed into the next layer. This helps the network learn the underlying structure of images hierarchically. This framework yields superior feature extractors over hand-crafted extraction techniques.

However, CNNs require large image datasets to be able to properly learn the weights for the kernel matrices to perform classification. A workaround for this is to use pre-trained models, i.e., models trained on large datasets like ImageNet [15] which has 1000 classes and over 14 million images. In practice, it works because these networks have learned to detect visual patterns (low level features like edges, curves, etc.) from images and such feature extractors can be used for other datasets as well. This helps researchers and practitioners to save computational resources and build on top of these pre-trained models. This is particularly useful for small datasets wherein the weights of these networks can be frozen and used as fixed feature extractors to train classifiers or by fine-tuning the final fully connected layers. Taking the features of multiple pre-trained models helps us to obtain robust & diverse feature sets for effective training of classifiers. In this work, DenseNet121 [12], EfficientNet-B0 [13], and SqueezeNet [14] have been used. The classification heads of all these networks have been removed prior to feature extraction.

DenseNet121: In this type of network architecture, a feed-forward approach is used to connect each layer to all other layers—This is unlike generic neural networks where each layer is connected to the next layer. For a certain layer, the feature maps generated by all the previous layers are used as input. This design choice has been found to curb the vanishing gradient problem & to facilitate feature propagation and feature reuse throughout the network. The architecture of DenseNets also aids in a reduction in the no. of parameters of the network. The authors claim this sort of architecture improves performance of networks because each layer has access to all the previous feature maps, which is the entirety of what the network has learned. Additionally, they also state that the feature maps can be thought of as the network's global state and each layer is responsible for contributing a certain no. of feature maps to the state. It has also been demonstrated that such networks are scalable to hundreds of layers such that no difficulty is experienced while optimizing the parameters of the networks.

SqueezeNet: The authors were primarily motivated to come up with a neural network architecture that provides close to state-of-the-art performance while having fewer memory requirements. This is because most of CNN research has been driven by improving the performance of these networks in terms of accuracy. SqueezeNet uses 50 times fewer parameters compared to AlexNet [16] while maintaining accuracy which is at par with it. The motivation for such work is as follows: Increased efficiency for distributed training of ConvNets, less bandwidth while exporting a new updated model from servers to autonomous cars and smaller models leading to easier deployment on devices with hardware constraints.

The authors also made use of model compression techniques to build a model that's 510 times smaller than AlexNet. The approach proposed by Han [17] called Dense-Sparse-Dense (DSD), which makes use of model compression as a way of regularizing while training the networks, has been used. SqueezeNet is mainly built up of "Fire" modules which consist of a "squeeze" sub-module made up of 1 x 1 filter followed by an "expand" sub-module which consists of a combination of 1 x 1 and 3 x 3 filters.

EfficientNet-B0: The authors conducted a systemic study on model scaling and recognized network width, depth, and resolution—On careful balancing of these factors, improvement in performance can be observed. A new method for scaling neural networks had been proposed which scales all the aforementioned dimensions uniformly. Experiments for scaling have been done on ResNet [18] and MobileNets [19]. Neural architecture search [20] has been used to create a network that serves as the baseline for upscaling networks. It has been found to achieve state-of-the-art accuracy scores & also perform well on transfer learning tasks—all of these while making sure that the no. of parameters is smaller by one order of magnitude.

3.2 Support Vector Machine

SVMs [21] are one of the most commonly used supervised learning methods. From the datapoints of both classes, we try to find the points which are closest to the hyperplane. These data points are known as support vectors. The distance between the support vectors and the separating hyperplane is called the margin. An SVM tries to maximize the margin to find the optimal hyperplane. If the data is linearly separable, this type of formulation would work just fine. For data that isn't linearly separable, the data needs to be projected to a higher dimensional space where it can be separated by a hyperplane. Kernels help to achieve this task. There are many types of kernels—Some of the most common ones are Linear Kernel, Polynomial Kernel, Gaussian Radial Basis Function (RBF) Kernel, etc.

4 Proposed Method

The image pixel values are divided by 255 to have their values in the [0, 1] range prior to rescaling and standardizing these images for feeding into the networks. Only the feature-extraction part of the network is kept, and the classification head is removed for all the networks. The neural networks have been used as fixed feature extractors hence all their weights have been frozen, and no fine-tuning was performed. The images are fed into the three networks consecutively, following which they were passed Global Average Pooling layers and 1024-dimensional, 1280-dimensional, 512-dimensional feature vectors are obtained respectively for DenseNet121, EfficientNet-B0, and SqueezeNet. For every image, all these feature vectors are concatenated into one long feature vector which is of 2816 dimensions. The concatenated feature vectors are standardized and this data is used to train an SVM with a Linear Kernel. The experiments are performed on both GTI and IDD datasets separately.

In order to evaluate the proposed feature ensemble technique, linear SVMs are trained for each of the pre-trained networks separately and accuracy score & weighted-mean F1 score were used. The entire pipeline has been summarized in Fig. 1.

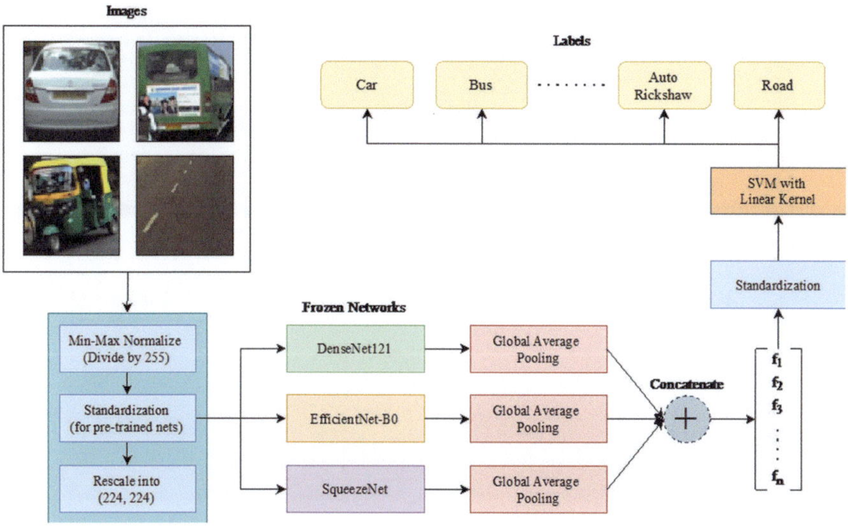

Fig. 1 Schematic diagram of our proposed deep feature ensemble method for AVD

5 Experimental Results

5.1 Datasets

GTI Dataset: Images are distributed among vehicle and non-vehicle classes in this dataset. The curators of the dataset stated that the position of rear portion of a vehicle relative to camera impacts framing the rear portion of a vehicle in the image hence each of the classes is sub-divided into 4 categories—front, left, right, and far. The dataset is representative of diverse driving conditions—sunny weather, cloudy days, poor lighting conditions, driving inside tunnels, and some images from cameras with poor resolution are also present. Figure 2 shows some samples of vehicles and non-classes in the dataset. The GTI dataset contains 7325 images out of which 4907 (2304 vehicles and 2603 non-vehicles) were used for training and 2418 (1121 vehicles and 1297 non-vehicles) for testing.

IDD Dataset: Unlike most autonomous navigation datasets which are obtained from well-structured driving environments, IDD images have been collected to reflect unstructured driving conditions. These images have been obtained from drive sequences shot on Indian roads. The dataset contains 10,004 images divided into 34 classes. Because of the huge size of the dataset and computational restrictions, in this work, experiments have been run on a small subset of it. In this subset, the vehicle categories which have been considered are autorickshaw, bicycle, bus, car, motorcycle, and truck. In addition, classes corresponding to person, rider, traffic sign, and vehicle fallback are also present. This makes the data truly representative of the

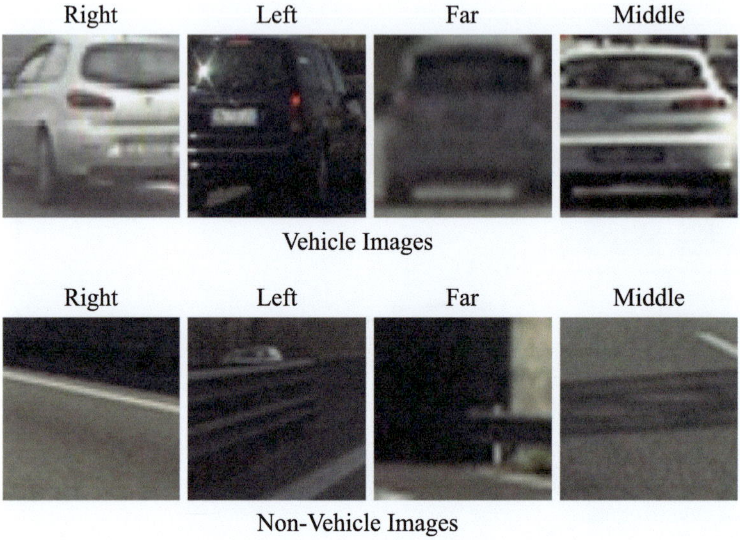

Fig. 2 Sample vehicle and non-vehicle images taken from GTI dataset

road scene in certain countries. We have additionally extracted 600 images of road patches because this class isn't annotated in the dataset. 2812 images are used for training and 2815 images are used for testing. In the extracted subset, the images are not uniformly distributed among various classes. 300 images of each autorickshaw, bus, car, motorcycle, person, rider, road, truck, and vehicle fallback classes are used for training whereas 300 each for testing. 30, 53, 29 images of animal, bicycle, and traffic sign classes are used respectively for training. 31, 54, 30 images of animal, bicycle, and traffic sign classes are used respectively for testing. Figure 3 shows samples from all the classes of the subset extracted from the IDD dataset.

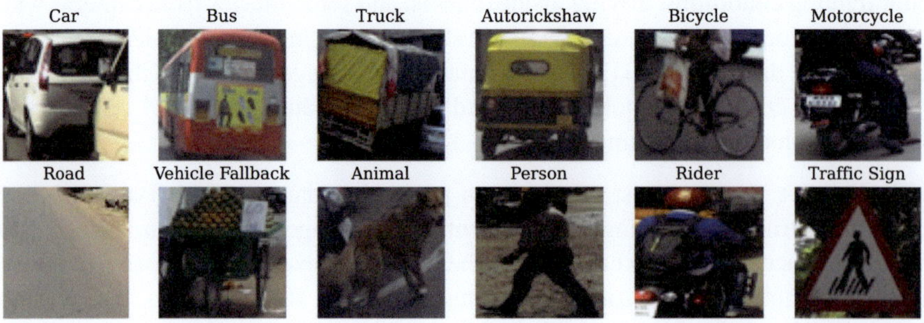

Fig. 3 Samples images taken from IDD dataset

5.2 Performance Evaluation

Table 1 shows the accuracy and weighted-mean F1 scores achieved by the proposed ensemble model as well as their candidate models on both IDD and GTI datasets. The ensemble model outperforms all the individual networks in terms of both the metrics for the IDD dataset. For the GTI dataset, very high accuracy is obtained for the individual networks as well as for the ensemble network—All the networks have comparable performances. Figure 4 shows the confusion matrix for the ensemble model trained and evaluated on the IDD dataset.

Table 1 Performance results obtained by the proposed DFE-AVD model on IDD and GTI datasets

Dataset	Feature extraction network	Accuracy	Weighted mean F1 score
IDD	DenseNet121	87.14	87.12
	EfficientNet-B0	85.90	85.89
	SqueezeNet	80.57	80.48
	DFE-AVD	**89.73**	**89.73**
GTI	DenseNet121	**99.88**	**99.88**
	EfficientNet-B0	**99.88**	**99.88**
	SqueezeNet	99.42	99.42
	DFE-AVD	99.83	99.83

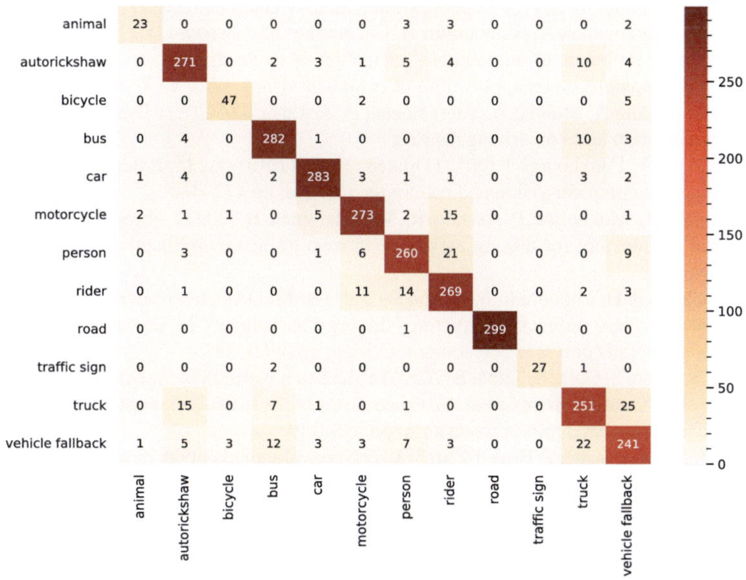

Fig. 4 Confusion matrix generated by the proposed DFE-AVD model on IDD dataset

6 Conclusion

A deep feature ensemble-based method has been proposed for automatic vehicle detection. The feature ensemble network makes use of the following pre-trained models—DenseNet121, EfficientNet-B0, and SqueezeNet. These networks have been specifically chosen such that the combined model has fewer memory requirements. In this case, the combined model takes less than 60 MB. The proposed method has been evaluated on the GTI dataset and a small subset of the IDD dataset. The proposed model has been found to outperform the individual networks for the IDD dataset where it achieves an accuracy of 89.73%. For the GTI datasets, the ensemble model gives comparable performance to the individual networks where it achieves 99.83% accuracy. Possible future directions to extend this work would be to test the proposed method on other vehicle image datasets in unstructured and harsh driving conditions.

References

1. Ministry of Road Transport and Highways, Government of India (2019) Road accidents in India 2019. New Delhi, India
2. Fatality Facts (2019). https://www.iihs.org/topics/fatality-statistics/detail/state-by-state
3. MoveInSync (2019) Travel time report Q1 2019 vs Q1 2018
4. Burd C, Burrows M, McKenzie B (2021) Travel time to work in the United States: 2019. American Community Survey Reports, United States Census Bureau, 2, p 2021
5. Varma G, Subramanian A, Namboodiri A, Chandraker M, Jawahar CV (2019) IDD: a dataset for exploring problems of autonomous navigation in unconstrained environments. In: 2019 IEEE winter conference on applications of computer vision (WACV). IEEE, pp 1743–1751
6. Singh D, Rahane A, Mondal A, Subramanian A, Jawahar CV (2021) Evaluation of detection and segmentation tasks on driving datasets
7. Yu F, Koltun V, Funkhouser T (2017) Dilated residual networks. In: Proceedings of the IEEE conference on computer vision and pattern recognition, pp 472–480
8. Nidamanuri J, Mukherjee P, Assfalg R, Venkataraman H (2021) Auto-alert: a spatial and temporal architecture for driving assistance in road traffic environments. Int J Intell Transp Syst Res 1–11
9. Kothandaraman D, Chandra R, Manocha D (2021) BoMuDANet: unsupervised adaptation for visual scene understanding in unstructured driving environments. In: Proceedings of the IEEE/CVF international conference on computer vision, pp 3966–3975
10. Baheti B, Innani S, Gajre S, Talbar S (2020) Eff-unet: a novel architecture for semantic segmentation in unstructured environment. In: Proceedings of the IEEE/CVF conference on computer vision and pattern recognition workshops, pp 358–359
11. Ronneberger O, Fischer P, Brox T (2015) U-net: convolutional networks for biomedical image segmentation. In: International conference on medical image computing and computer-assisted intervention. Springer, Cham, pp 234–241
12. Huang G, Liu Z, Van Der Maaten L, Weinberger KQ (2017) Densely connected convolutional networks. In: Proceedings of the IEEE conference on computer vision and pattern recognition, pp 4700–4708
13. Tan M, Le Q (2019) Efficientnet: rethinking model scaling for convolutional neural networks. In: International conference on machine learning. PMLR, pp 6105–6114

14. Iandola FN, Han S, Moskewicz MW, Ashraf K, Dally WJ, Keutzer K (2016) SqueezeNet: AlexNet-level accuracy with 50× fewer parameters and< 0.5 MB model size. arXiv:1602.07360
15. Russakovsky O, Deng J, Su H, Krause J, Satheesh S, Ma S, Huang Z, Karpathy A, Khosla A, Bernstein M, Berg AC, Fei-Fei L (2015) Imagenet large scale visual recognition challenge. Int J Comput Vis 115(3):211–252
16. Krizhevsky A, Sutskever I, Hinton GE (2012) Imagenet classification with deep convolutional neural networks. Adv Neural Inf Process Syst 25:1097–1105
17. Han S, Pool J, Narang S, Mao H, Gong E, Tang S, Elsen E, Vajda P, Paluri M, Tran J,Catanzaro B, Dally WJ (2016) Dsd: dense-sparse-dense training for deep neural networks. arXiv:1607.04381
18. He K, Zhang X, Ren S, Sun J (2016) Deep residual learning for image recognition. In: Proceedings of the IEEE conference on computer vision and pattern recognition, pp 770–778
19. Howard AG, Zhu M, Chen B, Kalenichenko D, Wang W, Weyand T, Andreetto M, Adam H (2017) Mobilenets: efficient convolutional neural networks for mobile vision applications. arXiv:1704.04861
20. Stanley KO, Miikkulainen R (2002) Evolving neural networks through augmenting topologies. Evol Comput 10(2):99–127. Pearson K (1901) LIII. On lines and planes of closest fit to systems of points in space. Lond Edinb Dublin Philos Mag J Sci 2(11):559–572
21. Boser BE, Guyon IM, Vapnik VN (1992) A training algorithm for optimal margin classifiers. In: Proceedings of the fifth annual workshop on computational learning theory, pp 144–152

Semantic Segmentation of Underwater Images with CNN Based Adaptive Thresholding

Sai Krishna Anand, Pranav Vigneshwar Kumar, Rohith Saji, Akhilraj V. Gadagkar, and B. R. Chandavarkar

Abstract Semantic segmentation remains a key research field in modern day computer vision and has been used in a myriad of applications across various fields. It can be extremely beneficial in the study of underwater scenes. Various underwater applications, like unmanned explorations and autonomous underwater vehicles, require accurate object classification and detection to allow the probes to avoid malicious objects. However, the models which work well for terrestrial images rarely work just as well for underwater images. This is because underwater images suffer from high blue light intensity as well as other ill-effects such as poor lighting and contrast. Trying to improve the model to account for bad image quality is not a great method as the model may misidentify noise as an image characteristic. In this paper, a unique CNN-based approach for post-processing image thresholding is proposed, on top of 3 models used for the semantic segmentation itself–Segnet, U-Net, and Deeplabv3+. The models' outputs are then subject to the CNN-based post-processing technique to binarize the outputs into masks, and provides improved segmentation results compared to the base models.

Keywords Semantic segmentation · Deep neural networks · Underwater applications · U-Net · Segnet · Deeplabv3+

1 Introduction

The process of semantic image segmentation entails the classification of individual pixels within an image [1]. It aims to assign each pixel to a certain class within the collection. When multiple objects of the same class are present in an image, each pixel of these objects is categorised and rendered with the identical colour. It is a crucial area and a significant research focus within computer vision. This notion is extensively

S. K. Anand · P. V. Kumar · R. Saji · A. V. Gadagkar (✉) · B. R. Chandavarkar
Wireless Information Networking Group (WiNG), Department of Computer Science and Engineering, National Institute of Technology Karnataka, Surathkal, India
e-mail: gadagkar.akhil@gmail.com

S. K. Biswas et al. (eds.), *Intelligent Computing Systems and Applications*, Lecture Notes in Networks and Systems 1307, https://doi.org/10.1007/978-981-96-3860-4_28

utilised across various sectors and is a subject of in-depth research, with practical applications in medicine [2], autonomous vehicles, and agricultural science. Research on the application of semantic segmentation in underwater scenarios is far less than in other disciplines. Enhancing the models utilised for terrestrial photographs is crucial for optimising the performance of real-time applications such as exploration probes and underwater vehicles. Traditional segmentation algorithms may underperform due to factors such as inadequate lighting and contrast, necessitating customisation for underwater photos through particular image enhancements [3].

The ocean is abundant in a variety of biological resources, energy, and metals. Underwater expeditions aim to investigate and study the abundant marine biological resources to meet different needs. Semantic segmentation [1] is highly advantageous for several underwater applications, particularly in underwater exploration missions involving probes and robotics. Segmenting underwater photos is crucial as it can help automate procedures, leading to more efficient study of the ocean's enormous areas [4].

Three models–Segnet [5], U-Net [2], and Deeplabv3+ [6]–are selected for this research in the field of semantic segmentation. Segnet utilises indices pooling to acquire a broader visual context for each pixel [7]. It is highly memory-efficient and is being examined for its speed and low memory usage. U-Net introduced skip connections, a feature that has been widely adopted in other models in research papers since then. The skip connections [2] enable the decoder to regain spatial information by establishing direct connections from the corresponding block in the encoder. Google built the cutting-edge model Deeplabv3+. Therefore, these three models were selected based on their past performance with both terrestrial and underwater images.

Previous studies have utilised a variety of post-processing techniques, drawing from image processing [8] and deep learning [9]. Various thresholding approaches such as Otsu and Gaussian Otsu [10] have been investigated. These methods offer distinct binarization strategies but are not effective as post-processing techniques because they optimise their own metric rather than the image metrics. A CNN model is suggested to predict the threshold directly from the segmentation model's output, rather than utilising CNNs to classify the image and then setting a fixed threshold. The threshold is utilised to convert the output into the final mask by binarizing it. This will be further examined in subsequent parts.

The rest of this paper is structured as follows. Section 2 includes a review of past methods for semantic segmentation and thresholding strategies to binarize model outputs for obtaining final masks. Section 3 provides a concise summary of the 3 models employed in the research and their distinctive properties. Section 4 presents a CNN-based post-processing strategy that utilises dynamic thresholding to enhance the segmentation results. Section 5 includes the experimental setup and results. Section 6 contains the conclusions.

2 Literature Review

In this section, past work in the field is explored and presented.

Xin and Huashan created an original post-processing enhancement framework and introduced a novel weighted composite filter (WCF) [8] to improve segmentation masks inside a unified framework. The weight ratio is adjusted to divide the WCF into a local and a global component. Additionally, in order to provide pertinent features, a guided image filter is constructed as a local filter that may reinstate boundary information.A minimum spanning tree-based filter is used as the global filter to determine global pixel similarity for picture matching. A post-processing enhancement framework has been developed that includes selection, normalisation, WCF, and argmax.

In their article, Massimo, Rajendra, Filippo, and Kristen provide a comprehensive examination of the innumerable methodologies employed within deep learning frameworks to enhance the network output outcomes or adequately prepare the input.Reference the source (r62). This work examines the effects of several pre- and post-processing strategies employed in deep learning models to address the highly intricate patterns found in histology images. Numerous of the methods discussed in this article are applicable to virtually any domain of image analysis; post-processing algorithms in particular.

Xiaolong, Zhidong, and Yuhan analyse traditional methods and datasets for segmentation, then explore several aspects of modern DNN-based techniques.Fangfang and Ming utilised DeepLabv3+ as the basis for their study, which is a state-of-the-art semantic segmentation network. The unsupervised colour correction method (UCM) module is integrated into the encoder structure of the framework initially to enhance image quality. Furthermore, the decoder structure incorporates two up-sampling layers to enhance the acquisition of target features and boundary information.

Jahidul, Chelsey, and Yuyang [11] developed one of the earliest extensive datasets for the purpose of performing semantic segmentation for underwater images. The dataset consists of nearly 1500 underwater photos along with their ground truth masks where each of the pixels are labelled with one of eight different classes. The different classes include: aquatic animals, marine invertebrates, wreckage and ruins, aquatic plants, human divers, robots, and the sea bed. They also present SUIM-Net, a fully convolutional model which employs an encoder-decoder network. It achieves a good balance between both performance and processing efficiency. SUIM-Net delivers rapid end-to-end inference while also making sure there is no drop in performance. This is especially important for usage in pipelines of real-time applications such as visually guided underwater robots.

Li and Huang studied the basic principles of the geometric active contour model [12] for addressing degradation issues in underwater image data, such as low resolution, low contrast, and distortion. The GACV adaptive picture segmentation algorithm has been enhanced. Based on the enhanced adaptive GACV image segmentation method, a mathematical model of underwater image segmentation was used to simulate underwater images. This was done to address the problem of degradation in underwater images caused by uneven light distribution and light absorption and dispersion in water.

3 Segmentation Models Overview

3.1 Segnet

SegNet is composed of three layers: encoder, decoder, and pixel-by-pixel classification [5] As illustrated in Fig. 1, the encoder network is composed of thirteen convolutional layers, which are identical to the initial 13 layers of convolution of the VGG-16 network employed for object categorization [13]. The fully connected layers at the deepest encoder output are eliminated to focus on preserving feature maps with higher resolution. Each of the thirteen layers comprising the decoder network corresponds to an encoder layer. A multiclass softmax layer [14] receives the output of the decoder for pixel-wise classification.

Translation invariance [15] is accomplished by SegNet during the subsampling [7] stage via max-pooling layers over tiny spatial shifts in the image. Integrating invariance with subsampling results in each pixel covering a larger spatial region. The window or picture context is referenced as [16]. These tactics enhance classification accuracy while decreasing the feature map's size. As a result, the image undergoes lossy representation, resulting in indistinct borders that are suboptimal for segmentation purposes.

SegNet utilises upsampling [17] in its decoder to maintain the same resolution between input and output images. To do this efficiently, it saves the max-pooling indices from the encoder network [16]. This results in a minor decrease in accuracy [18].

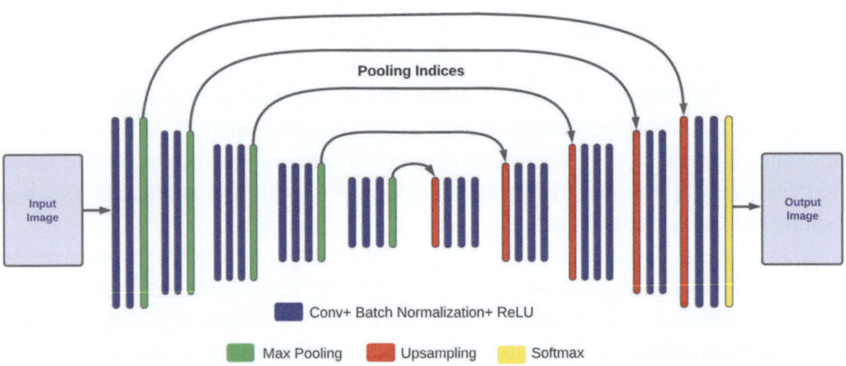

Fig. 1 SegNet architecture [5]

The last decoder layer generates a multi-channel feature map in RGB format, despite the fact that the initial encoder layer's input consists of 3 channels. The other decoders in the network generate feature maps with the same number of channels as their encoder inputs, making the final decoder layer unique. A trainable soft-max classifier [14] is applied to the high-dimensional feature representation to generate a 'K' channel image in this layer. Each of the 'K' classes is depicted in the channels of the output graphics, which carry probabilities. Each pixel's final class in the segmentation is determined by the channel with the highest probability.

SegNet's drawback is its limited capacity to learn various feature mappings compared to other models due to the absence of fully connected layers. This diminishes the SegNet's performance relative to other models in its category. Combining SegNet with ResNet can enhance segmentation accuracy for underwater photographs, but at the cost of increased processing demands.

3.2 U-Net

Fischer et al. [2] introduced the U-Net architecture. Initially designed for biological pictures, its use today extends well beyond the medical sector. The U-Net concept consists of two components: the encoder and decoder. U-Net architecture incorporates skip connections to get further information from the encoder network, distinguishing it from other encoder-decoder networks. Its name, U-Net, is derived from its symmetrical "U" shape. Skip connections were implemented to address various issues in various architectural designs [2, 19, 20].

During backpropagation, skip connections might cause the gradient to follow a different route. The efficacy of these yy routes has been empirically validated to enhance the convergence of the segmentation models. Skip connections in a neural network bypass some layers and immediately connect the output to other layers farther ahead in the network, instead of just the next layer. This process is primarily carried out in complex structures to preserve spatial information, as described below.

The encoder network comprises four blocks of convolution and max pooling layers [21, 22]. Each layer generates additional feature maps to extract information about the objects or instances present in the photos. The layers' form will be altered, causing the model to lose the spatial placement of the retrieved features. Through spatial upsampling of the feature maps, the decoder network retrieves these spatial features, or "where" the instances are located.

By joining the two maps, the appropriate encoder layer skip connections are included in the decoder network, which is a mirror image of the encoder network [21, 22]. This helps the decoder become more accurate at locating points in reference to the encoder's initial feature maps. To provide a more accurate output, the data is double convolutioned after the concatenation step is finished.

Skip connections are essential for the U-Net's superior performance, whether working with images from land or sea. The assessment was conducted on the U-Net model utilising the Fish4Knowledge dataset, which comprises segmentation masks and underwater images of fish [23]. The performance is defined by an F1 score of 0.92. Skip connections provide additional spatial information to the decoder network, allowing the U-Net to effectively adapt to underwater images. In order to improve contrast, luminosity, and additional attributes of images, one may directly apply pixel transformations and histogram equalisation enhancements.

3.3 Deeplabv3+

A group of Google researchers developed the sophisticated semantic segmentation model DeepLabv3+ [6] for terrestrial applications. DeepLabv3+ improves upon DeepLabv3 [24] by including an encoder-decoder architecture. An object boundary refinement module is included to improve the sharpness and clarity of the segmentation output, and DeepLabv3 serves as the encoder module. The model demonstrated promising outcomes, with an 82.1% test set performance on the Cityscapes[1] dataset and 89.0% on the PASCAL VOC 2012 [25] dataset.

The encoder module, seen in Fig. 2, is designed to utilise a backbone network like ResNet,[2] Xception,[3] or VGG [26] to extract features from the input image. The Atrous Spatial Pyramid Pooling (ASPP) [27] module receives the feature maps that were obtained from the backbone network.

The primary function of the ASPP module is to gather multi-scale contextual information. The representation of features from the backbone network is inputted into the ASPP module, where four parallel atrous convolutions with varied dilation rates are used to segment the object at various sizes. In the final phase of the encoder network, the output produced by the ASPP module is convolutioned by a factor of one.

The bilinear amplification of the encoder output's resolution by a factor of four constitutes the initial operation in the decoder component of the model. The upsampled output is merged with the 1x1 convolution of the low-level features. Following concatenation, the features are improved through a sequence of 3x3 convolutions, which are subsequently succeeded by a basic bilinear upsampling by a factor of 4. This will provide us the prediction for the semantic segmentation as referenced in [6].

DeepLabv3+ was designed for terrestrial images and does not function well when applied directly to underwater images. The issue lies in inadequate segmentation of the target boundary, imprecise edges, and classification mistakes.

---

[1] http://arxiv.org/abs/1604.01685.

[2] http://arxiv.org/abs/1512.03385.

[3] http://arxiv.org/abs/1610.02357.

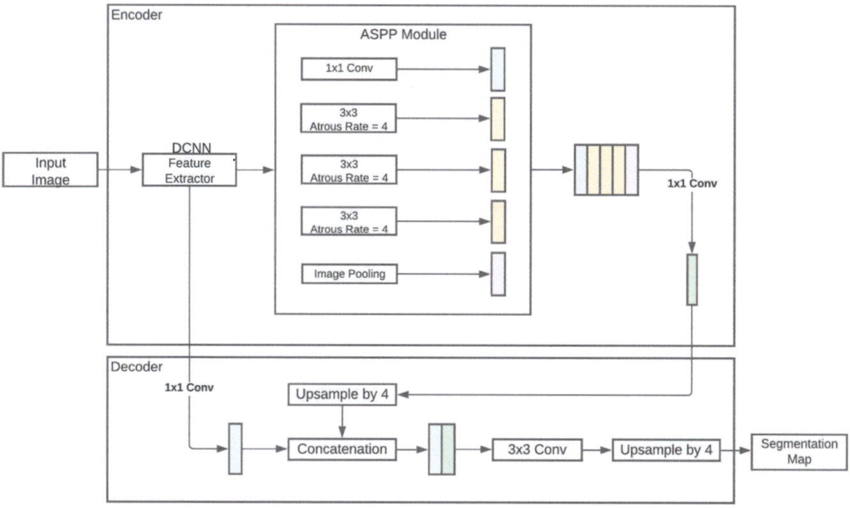

Fig. 2 DeepLabv3+ architecture [6]

4 CNN-Based Image Thresholding

The segmentation maps obtained as output by the different segmentation models require appropriate post-processing in order to produce accurate results. Post-processing [8, 9] can help in reducing the segmentation model's prediction errors. One of the most important post-processing steps is thresholding, which converts the model's output into a binary image. Improving the performance of the thresholding technique can have a positive impact on the results of the network output. Various post-processing techniques were discussed in the literature review and very few use convolutional neural networks as a means of thresholding the binary masks. Literature shows that the threshold is usually fixed to a static value for every mask and using CNNs for this purpose changes this to a dynamic thresholding technique.

For the task of thresholding, a CNN-based thresholding technique is proposed which takes an image as input and predicts a suitable threshold. A relatively simple and shallow neural network is considered so as to reduce the prediction time and ensure it does not affect the overall segmentation time. Minimal segmentation time is a crucial factor in real-time applications. The architecture, as shown in Fig. 3, consists of three blocks of convolution, batch normalisation [28] and max pooling layers. The number of filters in the convolution layer is increased progressively in the deeper layers of the network. The third and final block of convolution, batch normalisation and max pooling layers is followed by a two dense layers of 64 and 4 neurons respectively. Finally, the output layer consists of a single neuron which outputs the predicted threshold. Batch normalisation [28] and dropout [29] layers are included to prevent the model from over-fitting.

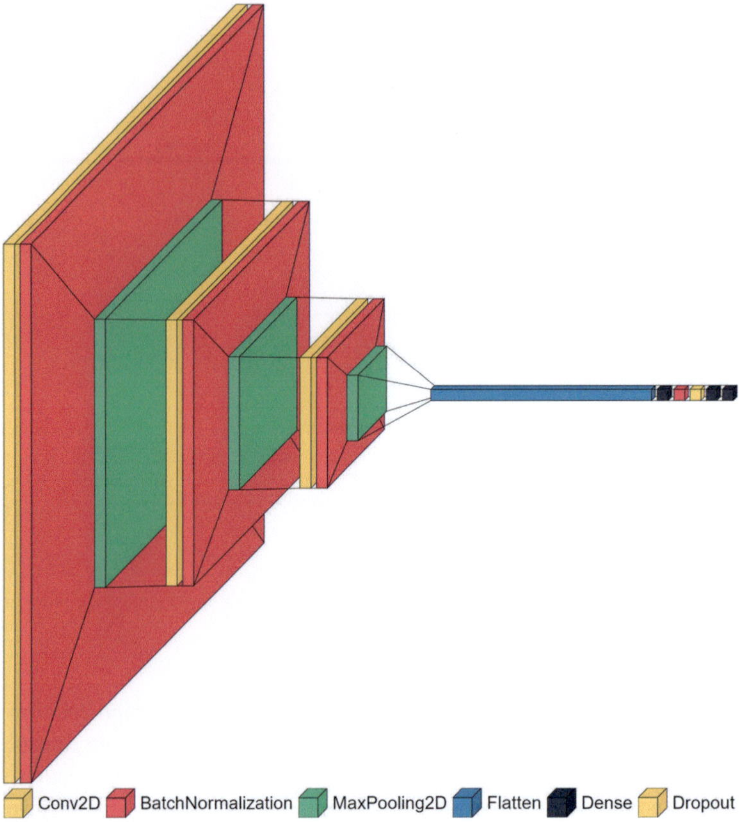

Fig. 3 CNN-based threshold model architecture

The rationale behind choosing a CNN-based thresholding technique over a basic global threshold is that it imparts intelligence into the process of choosing the right threshold. The model can adapt to the changes in the input image and learn the best possible threshold value. When compared to other traditional thresholding techniques like Otsu [10], the CNN-based method is simpler and more compute efficient.

5 Implementation and Results

The problem is formulated as an image-to-image translation from the input domain of underwater images to its target semantic mask in binary space for each of the classes. These individual binary images are then superimposed with different colours to obtain the final RGB mask. Since each pixel is to be classified into a binary value in every channel separately, cross-entropy loss [30] is used to supervise the training. Tensorflow's [31] libraries is used to implement the entire pipeline. The Adam optimizer is used for gradient descent along with a learning rate of 10-4, with a momentum value of 0.5 for all the models.

Various preprocessing methods–ICM, UCM, histogram equalisation, CLAHE, RGHS and Rayleigh Distribution–were performed on the images separately and multiple models were trained with each of these preprocessed sets of images. Out of the 6 methods attempted, ICM proved to provide the best results for underwater image segmentation. The 3 models, namely Segnet, U-Net and Deeplabv3+, are run with and without the preprocessed images and the best results have been tabulated later in this section. Data augmentation was performed with some standard libraries of Keras [32]. The rotation range was set to 0.2, the zoom range, width and height shift set to 0.05 and horizontal and vertical flip enabled. This ensured that the data flowed in batches directly from the augmentation module to the training pipeline.

The dataset used for this experiment is the SUIM dataset [33]. It commonly acts as a benchmark dataset for Underwater image segmentation. It contains of 8 different classes: (1) waterbody backdrop (BW), (2) aquatic plants/flora (PF), (3) human divers (HD), (4) robots and instruments (RI), (5) wrecks/ruins (WR) , (6) fish and other vertebrates (FV), (7) reefs and other invertebrates (RO), and (8) seafloor and rocks (SR). Each class can be denoted by a 3-bit number together with a corresponding colour. The models in this study forecast 5 of the 8 categories: HD, FV, RO, RI, and WR. The remaining items are considered to belong to a unified category.

The dataset is derived from a diverse collection of photos captured during collaborative human-robot experiments and explorations in several oceanic areas with distinct water compositions. The dataset comprises 1525 RGB photos for training and validation, plus an additional 110 images for testing and evaluating semantic segmentation models. The photos come in different resolutions: 1906×1080, 1280×720, 640×480, and 256×256. The data augmentation module enhances the model's performance by expanding its capability to process a larger variety of images beyond the initial batch of 1500 images in the dataset.

A range of thresholds are used for every image and the best threshold is calculated by comparing the ground truth mask with the output obtained from the segmentation model. These are saved for every image as the outputs with the predicted masks as input and the CNN model [31, 32] is trained on these masks. The model is trained for 200 epochs with a batch size of 4. The model essentially performs an image-to-scalar transformation an provides a threshold value between 0 and 1. This threshold is used by the segmentation model's predicted image and the final binarized image is stored as the output.

5.1 Evaluation Metrics

There are 2 metrics used in this paper to measure the quality of the segmentation:

1. **Intersection-Over-Union (IoU)**–The IoU metric [34] is used to find how much of the predicted image mask overlaps with the ground truth mask with respect to the total area covered by both the ground truth mask and the predicted mask. It is calculated using: *(Area of intersection of ground truth and predicted mask / Union of ground truth and predicted mask).*
2. **F1 Score**–Uses 2 metrics, precision and recall [35] which present information about the quality and the quantity of the produced results respectively. Calculated using precision (P) and recall (R) as

$$(2 * P * R)/(P + R) \tag{1}$$

5.2 Qualitative and Quantitative Analysis

The accuracy of the segmentation can be seen in the output images in Fig. 4. It can be seen that the U-Net and Deeplab models generate much more fine-grained contours and this loss is not very significant. The Segnet model generates much smoother contours in comparison to the other 2 models. The improvement in the Segnet and Deeplab is seen visually too, in the 2nd row of Fig. 4, with it being much closer to the ground truth label. The U-Net model, even though improved, displays this performance enhancement through further smoothening of contours in some images.

The performance of the segmentation models are measured using 2 main metrics–F1 score and mIoU score, both of which are shown in Table 1. The first half of the table shows the F1 scores of the base models as well as the F1 scores with the postprocessing, while the second half of the table displays the mIoU scores. From the results, it can be seen that U-Net and Deeplabv3+ perform significantly better than the Segnet model, even without postprocessing. The performance of Segnet greatly increases with postprocessing, with the F1 and mIoU scores increasing 9% and 10% respectively. The U-Net and Deeplabv3+ models improve as well, but not as much as the Segnet, owing to the fact that their base scores were already higher. Even though the Segnet's scores are lower, the model performs well considering that the Segnet does not save all its feature maps in order to perform space optimizations. This is very useful in real time applications, where slower models like Deeplabv3+ cannot be deployed due to its higher inference time.

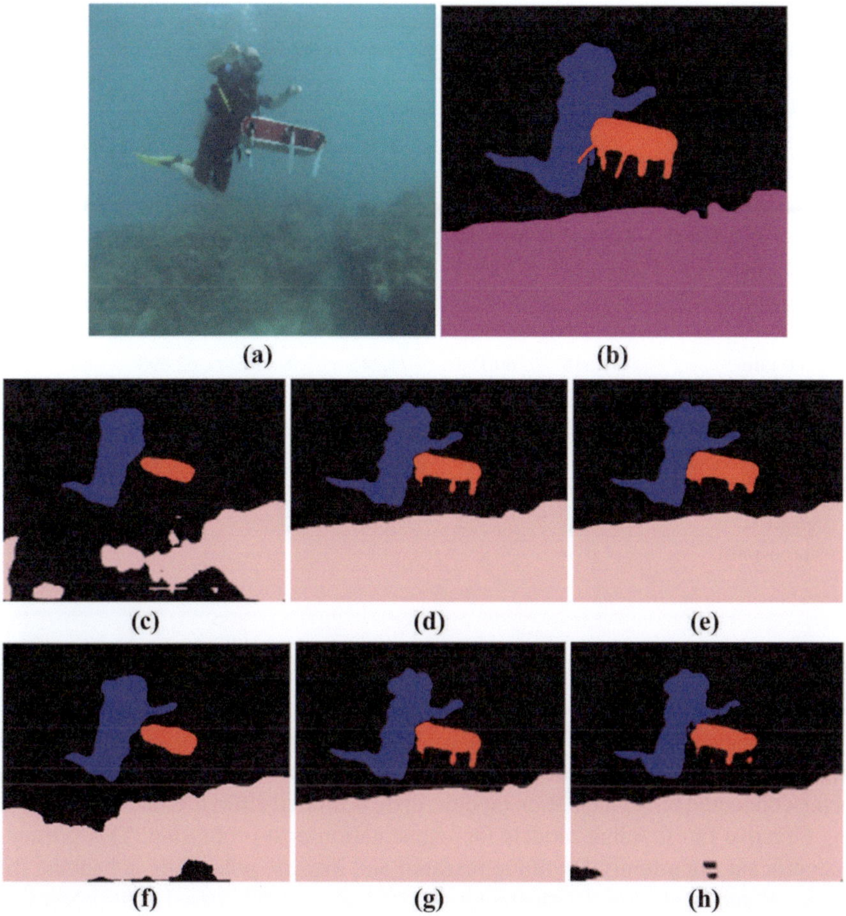

Fig. 4 Segmentation results of the models. **a** Original Image, **b** Ground Truth mask. **c** Base Segnet, **d** Base U-Net, **e** Base Deeplabv3+, **f** Segnet with postprocessing, **g** U-Net with postprocessing, **h** Deeplabv3+ with postprocessing

6 Conclusion and Future Work

Underwater applications rely heavily on semantic segmentation, particularly for unmanned or autonomous underwater vehicles. Inadequate segmentation outcomes can hinder the vehicles from effectively exploring and collecting data from the ocean's depths. This paper examined various state-of-the-art semantic segmentation approaches. The structures, distinctive traits, and innovations of the various models were analysed and showcased. The encoder-decoder architecture, fundamental to all the models studied, yields satisfactory outcomes even with underwater images. A CNN-based post-processing strategy was suggested to enhance the thresholding

Table 1 Experimental segmentation results. This table displays the F1 scores and the mIoU scores of the models [2, 5, 6]. The mean scores are presented in the table. RO, FV, HD, RI and WR represent 5 of the classes in the segmentation dataset

Model	RO	FV	HD	RI	WR	Combined
Segnet–F1	91.003	61.784	73.629	45.402	59.052	66.174
UNet–F1	94.848	78.475	94.748	81.801	87.383	87.451
Deeplabv3+–F1	95.886	83.625	91.879	78.704	86.506	87.320
Segnet Postproc–F1	92.688	68.356	82.516	57.204	76.156	75.384
UNet Postproc–F1	95.683	89.910	95.323	85.824	95.417	92.431
Deeplabv3+ Postproc–F1	97.862	83.180	94.420	79.072	92.815	89.469
Segnet–mIoU	91.515	60.108	71.246	58.431	62.698	68.799
UNet–mIoU	94.210	74.190	91.958	84.016	87.797	86.434
Deeplabv3+–mIoU	95.407	79.290	88.983	83.265	88.971	87.183
Segnet Postproc–mIoU	92.917	67.612	79.814	70.207	80.396	78.187
UNet Postproc–mIoU	94.890	84.463	91.646	88.289	95.740	91.006
Deeplabv3+ Postproc–mIoU	97.306	77.524	91.447	82.677	93.737	88.539

process with intelligence. This method offers dynamic thresholding and has potential for enhancement by improving the CNN model used for post-processing.

Future enhancements of this work could be categorised into two parts: the segmentation model and the post-processing model. Although considerable research is being conducted on segmentation models, there is limited effort focused on establishing effective thresholding models for segmentation post-processing. This presents a significant opportunity for future research and inquiry, potentially enhancing the accuracy and quality of segmentation outputs. Many CNN-based post-processing techniques in current research focus on classification and detection problems. There are limited post-processing tools available for segmentation tasks. The potential work in this field is limitless.

References

1. Jordan J (2020) An overview of semantic image segmentation. https://www.jeremyjordan.me/semantic-segmentation/
2. Ronneberger O, Fischer P, Brox T (2015) U-net: Convolutional networks for biomedical image segmentation
3. Xiaolong Liu ZD, Yang Y (2018) Recent progress in semantic image segmentation. [Online]. Available at: https://arxiv.org/ftp/arxiv/papers/1809/1809.10198.pdf
4. Liu F, Fang M (2020) Semantic segmentation of underwater images based on improved deeplab. J Marine Sci Eng 8(3). [Online]. Available at: https://www.mdpi.com/2077-1312/8/3/188
5. Badrinarayanan V, Kendall A, Cipolla R (2016) Segnet: A deep convolutional encoder-decoder architecture for image segmentation

6. Chen L-C, Zhu Y, Papandreou G, Schroff F, Adam H (2018) Encoder-decoder with atrous separable convolution for semantic image segmentation
7. Pathical S, Serpen G (2010) Comparison of subsampling techniques for random subspace ensembles. 1(08):380–385
8. Cheng X, Liu H (2020) A novel post-processing method based on a weighted composite filter for enhancing semantic segmentation results. Sensors 20(19) [Online]. Available at: https://www.mdpi.com/1424-8220/20/19/5500
9. Salvi M, Acharya UR, Molinari F, Meiburger KM (2021) The impact of pre- and post-image processing techniques on deep learning frameworks: A comprehensive review for digital pathology image analysis. Comput Biol Med 128:104129. [Online]. Available at: https://www.sciencedirect.com/science/article/pii/S0010482520304601
10. Yousefi J (2015) Image binarization using otsu thresholding algorithm 05
11. Islam MJ, Edge C, Xiao Y, Luo P, Mehtaz M, Morse C, Enan SS, Sattar J (2020) Semantic segmentation of underwater imagery: Dataset and benchmark. CoRR arXiv:2004.01241
12. Shelei L, Mengxing H (2018) Research of underwater image segmentation algorithm based on the improved geometric active contour models. International conference on intelligent autonomous systems (ICoIAS) 2018:44–50
13. Perumanoor TJ (2021) What is vgg16?–introduction to vgg16. https://medium.com/@mygreatlearning/what-is-vgg16-introduction-to-vgg16-f2d63849f615
14. Nwankpa C, Ijomah W, Gachagan A, Marshall S (2018) Activation functions: Comparison of trends in practice and research for deep learning
15. Biscione V, Bowers J (2020) Learning translation invariance in cnns
16. Fezan (2019) Understanding of semantic segmentation how segnet model work to perform semantic segmentation. https://medium.com/@fezancs/understanding-of-semantic-segmentation-how-segnet-model-work-to-perform-semantic-segmentation-5c426112e499
17. Dumitrescu Boiangiu CA (2019) A study of image upsampling and downsampling filters. Computers 8(4):30
18. Jordan J (2018) Evaluating image segmentation models. https://www.jeremyjordan.me/evaluating-image-segmentation-models/
19. Adaloglou N (2020) Intuitive explanation of skip connections in deep learning. https://theaisummer.com/skip-connections/
20. Drozdzal M, Vorontsov E, Chartrand G, Kadoury S, Pal C (2016) The importance of skip connections in biomedical image segmentation
21. Lamba H (2019) Understanding semantic segmentation with unet. https://towardsdatascience.com/understanding-semantic-segmentation-with-unet-6be4f42d4b47
22. Zhang J (2019) Unet explanation. https://towardsdatascience.com/unet-line-by-line-explanation-9b191c76baf5
23. NNa, MHP, SMH (2021) Semantic segmentation of underwater images using unet architecture based deep convolutional encoder decoder model, vol 03, pp 28–33
24. Chen L-C, Papandreou G, Schroff F, Adam H (2017) Rethinking atrous convolution for semantic image segmentation
25. Everingham M, Eslami SMA, Van Gool L, Williams CKI, Winn J, Zisserman A (2015) The pascal visual object classes challenge: A retrospective. Int J Comput Vis 111(1):98–136
26. Simonyan K, Zisserman A (2015) Very deep convolutional networks for large-scale image recognition
27. Chen L, Papandreou G, Kokkinos I, Murphy K, Yuille AL (2016) Deeplab: Semantic image segmentation with deep convolutional nets, atrous convolution, and fully connected crfs. CoRR arXiv:1606.00915
28. Christian Szegedy SI (2015) Batch normalization: Accelerating deep network training by reducing internal covariate shift. CoRR arXiv:1502.03167
29. Srivastava N, Hinton G, Krizhevsky A, Sutskever I, Salakhutdinov R (2014) Dropout: A simple way to prevent neural networks from overfitting. J Mach Learn Res 15(56):1929–1958. [Online]. Available: http://jmlr.org/papers/v15/srivastava14a.html

30. Zhang Z, Sabuncu MR (2018) Generalized cross entropy loss for training deep neural networks with noisy labels. [Online]. Available: https://arxiv.org/abs/1805.07836
31. Abadi M, Barham P, Chen J, Chen Z, Davis A, Dean J, Devin M, Ghemawat S, Irving G, Isard M, Kudlur M, Levenberg J, Monga R, Moore S, Murray DG, Steiner B, Tucker P, Vasudevan V, Warden P, Wicke M, Yu Y, Zheng X (2016) TensorFlow: A system for Large-Scale machine learning. In: 12th USENIX symposium on operating systems design and implementation (OSDI 16). Savannah, GA, USENIX Association, pp 265–283. [Online]. Available: https://www.usenix.org/conference/osdi16/technical-sessions/presentation/abadi
32. Keras documentation: Image data preprocessing. [Online]. Available: https://keras.io/api/preprocessing/image/
33. Robotics I, Vision Lab UM Suim dataset. [Online]. Available: http://irvlab.cs.umn.edu/resources/suim-dataset
34. Rosebrock A Intersection over union (iou) for object detection
35. Ting KM (2010) Precision and recall. Boston, MA, Springer US, pp 781–781. [Online]. Available: https://doi.org/10.1007/978-0-387-30164-8_652

Curved Shaped Defected Ground Structure: A Way to Achieve High Polarization Purity, Wide Bandwidth, and Stable Gain

Zonunmawii, Manoj Sarkar, L. Lolit Kumar Singh, Sudipta Chattopadhyay, and Abhijyoti Ghosh

Abstract In this investigation, a single-feed circular patch antenna (CPA) with a pair of curved dumbbell shape defected ground structure (CD-DGS) at the non-radiating periphery has been studied on RT-Duroid substrate material ($\varepsilon_r = 2.33$) for the enhancement of the co-polarization (CP) to cross-polarization (XP) isolation (polarization purity) and impedance bandwidth. A parametric study has been carried out by changing various parameters of the CD-DGS in order to find out an optimum structure that provides the best output in terms of polarization purity and impedance bandwidth without hindering the gain profile. The present investigation results around 12% impedance bandwidth with 26 dB CP-XP isolation. The proposed structure and classical conventional structure provide good impedance matching and stable radiation patterns. Therefore, the present investigation can be used fully in the modern wireless communication system where better XP performance is the major issue.

Keypoints Circular patch antenna · Curved dumbbell shape defected ground structure (CD-DGS) · Co-pole to cross-pole isolation · Impedance bandwidth · Gain

1 Introduction

In the modern era of the wireless communication field, the patch antenna is the most popular radiator which is used in the microwave field to find the potential application in modern communication because of its integration with an active device and having an excellent features like low weight, dual-polarization, easy to fabricate. Apart from this excellent feature it has some major demerits like narrow beamwidth, low gain, polarization purity issue, low impedance beamwidth, low power handling capacity. Polarization purity decreases while operating frequency increases [1, 2].

Zonunmawii · M. Sarkar · L. L. K. Singh · S. Chattopadhyay · A. Ghosh (✉)
Department of Electronics and Communication Engineering, Mizoram University, Aizawl, Mizoram, India
e-mail: abhijyoti_engineer@yahoo.co.in

S. K. Biswas et al. (eds.), *Intelligent Computing Systems and Applications*, Lecture Notes in Networks and Systems 1307, https://doi.org/10.1007/978-981-96-3860-4_29

The co-polarized electric fields are being generated from the radiating sides of the circular patch and radiates in the broadside direction. Few extents of orthogonal polarized fields called cross-polarization (XP) radiation took place which is theoretically investigated in [3].

Scientists and researchers have started working to address these issues, some handful investigations have been carried out to address the input and radiation performance of conventional CPA, like the different structures of defected patch structure (DPS), defected ground structure (DGS), and different shapes of shorting post technique. CPA with Arc-shaped [4], circular-shaped [5] DGS are reported with polarization purity of 5–7 dB but some impedance bandwidth-related issues have been come across in those documentations. Arc shaped DPS model has been reported in [6, 7] where polarization purity of around 20–23 dB is achieved. Different types of rectangular slots on circular patch surface have been documented in [8] where around 22% bandwidth and 3.24 dB Gain has been reported. Around 20 dB of XP suppression has been documented in [10–13] by modulating the feed structure. However, all these investigations led to complexity in the manufacturing process and more investigation is required to find a simpler technique for suppressing the XP radiation and provide improvement in the impedance bandwidth.

Therefore, in the present investigation, a curved dumbbell shape defect has been deployed on the ground plane at the non-radiating edge of the conventional patch on RT-Duroid substrate with dielectric constant $\varepsilon_r = 2.33$ and height ($h = 1.58$ mm) to address the limitation of the earlier investigation. This proposed structure is very simple and provides good polarization purity and better impedance bandwidth without hindering the basic radiation profile (Fig. 1).

2 Theoretical Background and Parametric Study

The conventional CPA radiated linearly polarized electric field along the broad side direction of the patch. It generally occurred due to the fringing field that resides at the radiating edges of the patch which is the dominant mode TM_{11} [2, 9]. Nevertheless, a few handfuls amount fringing fields of the dominant mode are existing at the non-radiating edge of the conventional CMA as shown in Fig. 2.

The orthogonal resonance field of weak TM_{11} mode in CPA along with higher-order orthogonal excitation mode TM_{21} is the key factor of producing high XP radiation from the non-radiating edges of the CMPA [9]. The field distribution of higher-order filed in TM_{21} mode is orthogonal to the fundamental dominant mode TM_{11} as shown in Fig. 2. A closed inspection reveals that the radiation from non-radiating edges which is mainly responsible for XP radiation. Therefore, a symmetrical pair of curved dumbbell shape DGS have been incorporated on the ground plane to eliminate such radiation without hindering its dominant TM_{11} mode radiation.

The parametric studies have been done using a commercially available software package (high-frequency structure simulator; HFSS v.14) [14] to find out the best possible proposed structure antenna geometry. At first, a circular patch antenna with

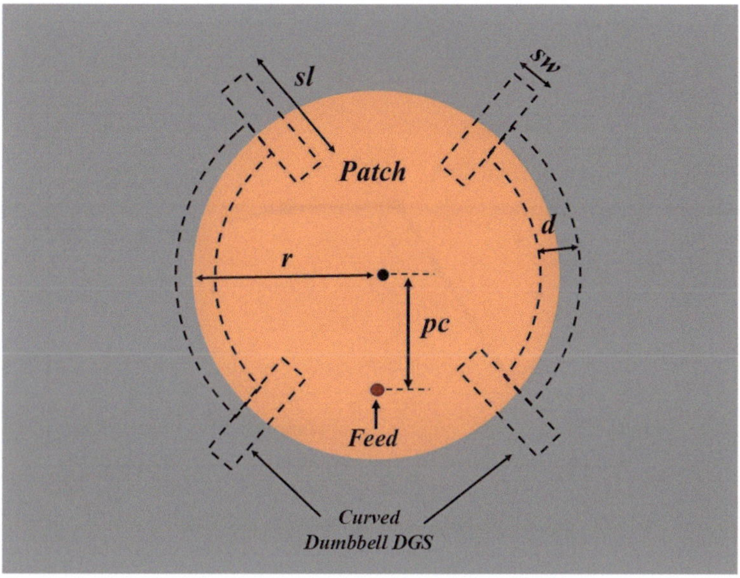

Fig. 1 Top view of the proposed structure with curved dumbbell shape DGSs at the non-radiating edges and parametric study

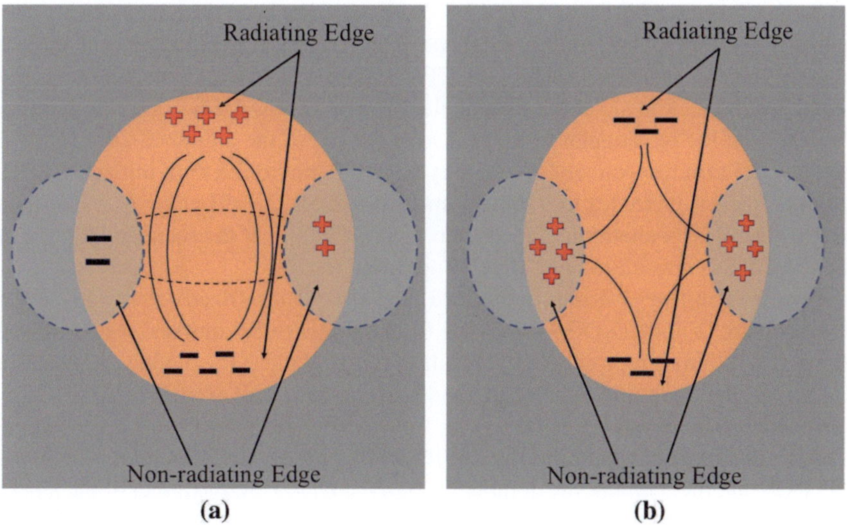

Fig. 2 Electric field and current distribution over the patch surface. **a** Fundamental dominant TM_{11} mode. **b** Next higher-order TM_{21} mode

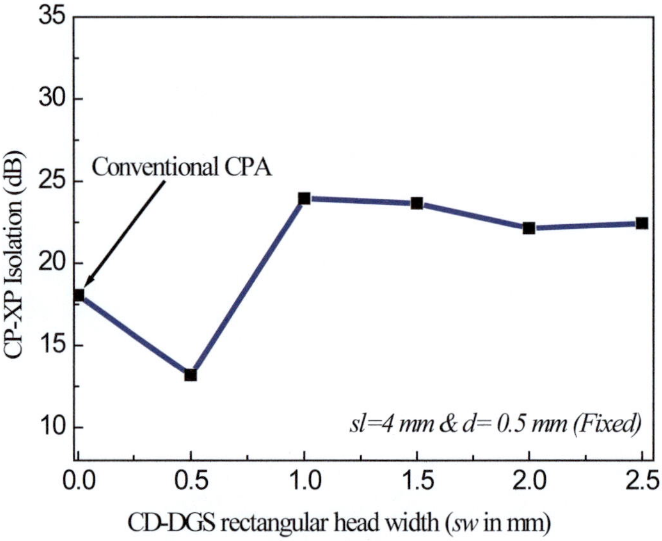

Fig. 3 Simulated plot of CP-XP isolation with respect to variation of CD-DGS rectangular head slot width (*sw*). (Keeping $d = 0.5$ mm, $sl = 4$ mm fixed)

radius (r) 7 mm on RT-Duroid ($\varepsilon_r = 2.33$) substrate has been designed. After that a pair of symmetric curved dumbbell shape DGS have been deployed on the ground plane at the non-radiating sides of the CPA. Figure 3 shows the polarization purity plot when the width (sw) of the rectangular head of the CD-DGS is varying keeping the length of the rectangular head (sl) and width (d) of the dumbbell arc fixed at 4 mm and 0.5 mm respectively. As sw vary from 0.5 mm to 2.5 mm, initially CP-XP decrease then it increases to a maximum isolation of 23.9 dB at $sw = 1$ mm then again, the CP-XP isolation decreases. So, the width (sw) of the rectangular head of the CD-DGS is considered as 1 mm for optimum results.

Now, keeping $sw = 1$ mm (optimized) and $sl = 4$ mm (fixed), the width of the dumbbell arc (d) of CD-DGS is varying from 0.5 mm to 2.5 mm and it is observed that at $d = 0.5$ mm it provides best result in terms of CP-XP isolation as shown in Fig. 4. So, the optimum value of the width of the dumbbell arc (d) of CD-DGS is considered as 0.5 mm.

After finalizing the optimum value of the width (sw) of the rectangular head of the CD-DGS and the width of the dumbbell arc (d) of CD-DGS, the effect of the length (sl) of the rectangular head of the CD-DGS structure has been studied by varying the sl from 2 mm to 5 mm and documented in Fig. 5. It has been seen that the polarization purity decreases while increasing sl up to 3.5 mm but after that polarization purity improves without hampering other radiation profile parameters. Conventional CPA ($sl = sw = d = 0$ mm) provide the co-pole to cross-pole isolation around 18 dB whereas the proposed structure with $sl = 4$ mm, $sw = 1$ mm (optimized), and $d = 0.5$ mm (optimized) provides co-pole to cross-pole isolation of 26 dB.

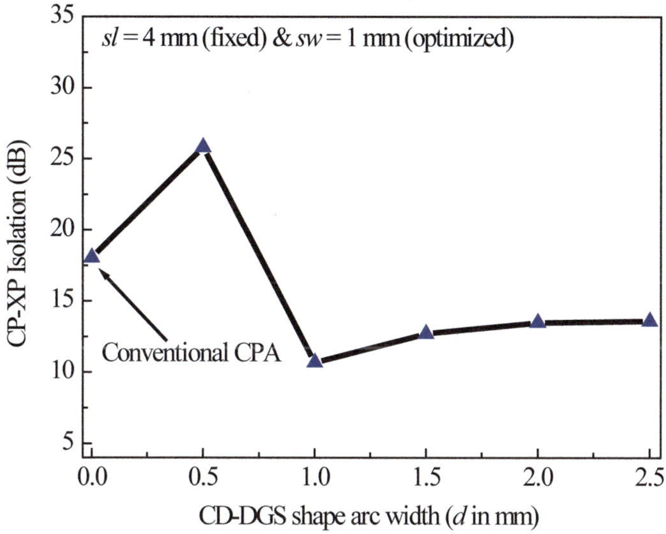

Fig. 4 Simulated plot of CP-XP isolation with respect to variation of dumbbell arc (d) of CD-DGS. (sw = 1 mm (optimized), sl = 4 mm (fixed))

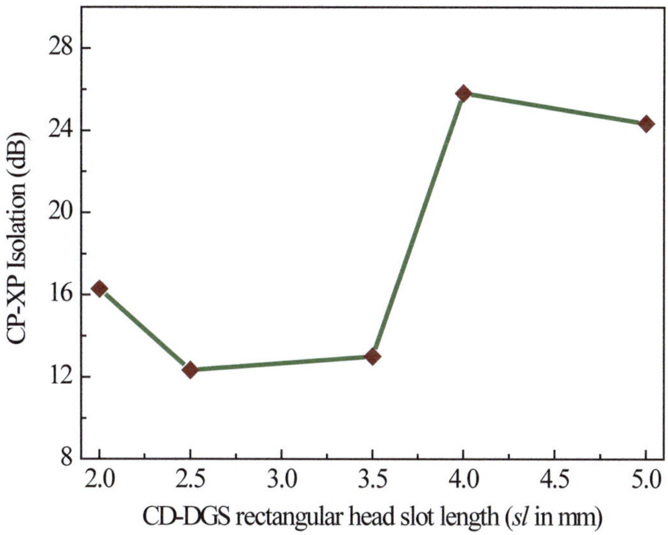

Fig. 5 Simulated plot of CP-XP isolation with respect to variation of CD-DGS rectangular head slot length (sl). (sw = 1 mm (optimized), d = 0.5 mm (optimized))

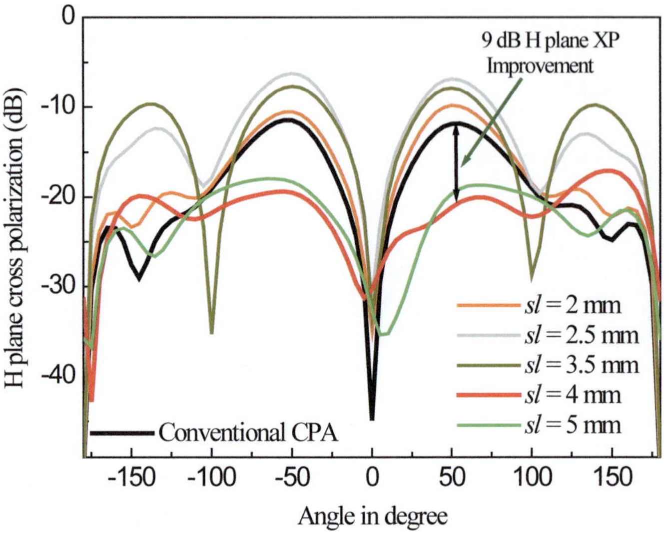

Fig. 6 H plane cross-polarization radiation with respect to CD-DGS rectangular head slot length (*sl*)

The CP-XP isolation shown in Fig. 6 further confirms the improvement of polarization purity by the proposed curved dumbbell shape defected ground structure (CD-DGS) incorporated CPA that is discussed earlier.

Figure 7 clearly shows the gain profile of the CD-DGS incorporated CPA. It is clearly evident that the incorporation of curved dumbbell shape defected ground structure (CD-DGS) at the non-radiating edge does not hamper the co-polarization gain performance. The gain profile is quite the same as conventional CPA and stable also.

3 Optimum Structure

At the beginning, a conventional circular microstrip patch antenna has been designed having radius (r) 7 mm on the top of RT-Duroid substrate (ε_r= 2.33) with height (h) 1.75 mm over the ground of 70*70 mm^2. After that a symmetrical pair of curved dumbbell shape DGS structure has been deployed on the ground plane at the non-radiating edges where the width of the arc of CD-DGS (d) is 0.5 mm, the width (sw) of the rectangular head of the CD-DGS is 1 mm, and the length (sw) of the rectangular head of the CD-DGS is 4 mm. This is the way the final optimum structure has been carried out as shown in Fig. 1.

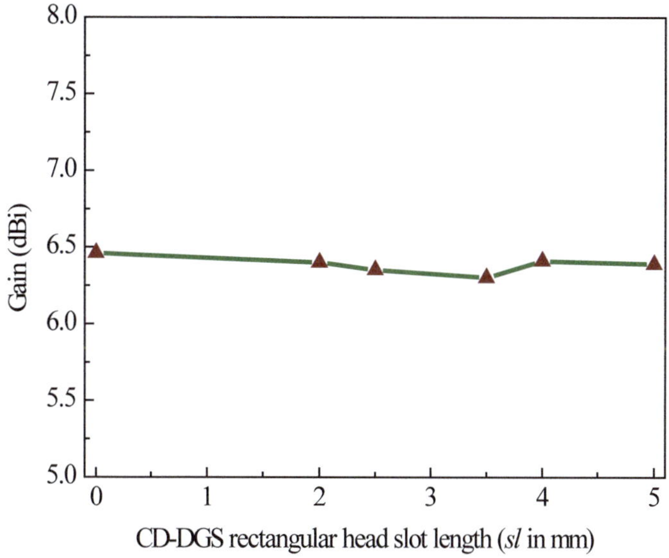

Fig. 7 Variation of gain of the proposed CD-DGS structure with respect to CD-DGS rectangular head slot length (*sl*)

4 Simulated Result and Discussion from Optimized Structure

The reflection coefficient profile of any antenna structure shows how efficiently matching is done. Proper matching is very much required to get the optimum performance from any antenna structure. The comparison of the reflection coefficient (S_{11}) profile of conventional and proposed structures has been revealed in Fig. 8 with the help of [14]. Reflection coefficient profile is clearly evident that the convention CPA has narrow impedance bandwidth around 4% while the CPA with CD-DGS provides an impedance bandwidth of 12% without effecting the frequency spectrum which is quite good as compared to a conventional CPA and both structures provide very good impedance matching.

The complete E plane co-polar (CP) and cross polar (XP) radiation profile of conventional and proposed structure have been revealed in Fig. 9. The E plane CP radiation of the proposed structure is quite same as conventional CPA, where both the structures attain 6.41 dBi gain and XP radiation is below −30 dB which is quite good and convincing. Figure 9 also provides a clear evident that the incorporation of the CD-DGS at the non-radiating edge does not affect the CP radiation of the E plane.

The H plane radiation pattern of the proposed structure and conventional has been recorded in Fig. 10. In this present model, the H plane cross-polarization radiation has been surprisingly improved as compared to the conventional CPA. Without hindering

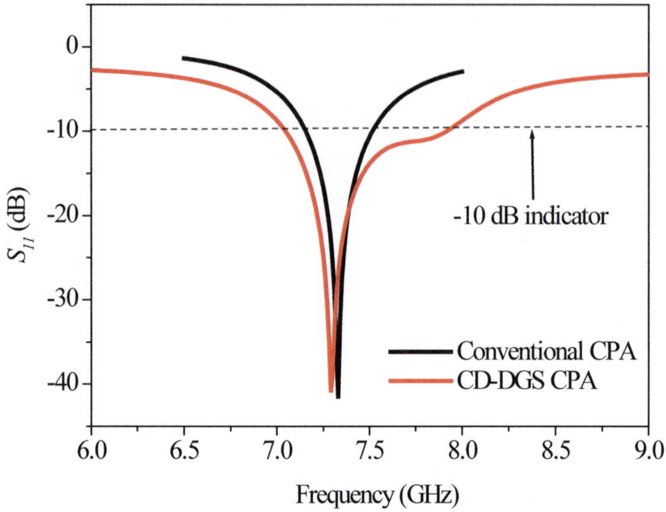

Fig. 8 Simulated S_{11} profile of conventional and proposed CD-DGS integrated CPA

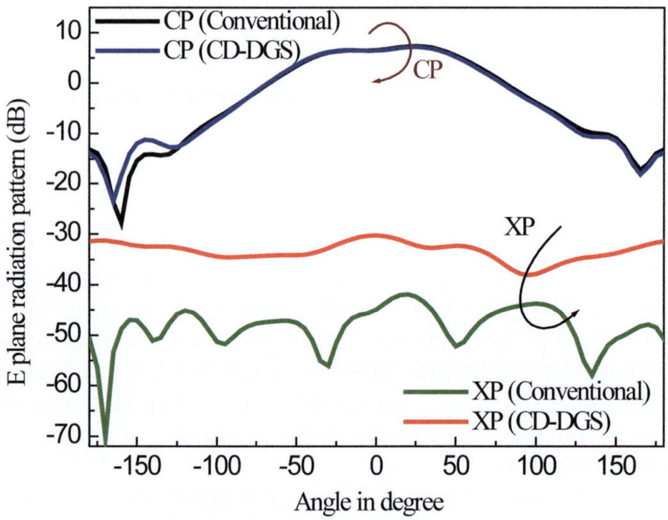

Fig. 9 E plane radiation pattern of the conventional and proposed structure

the broad side co-polarization radiation deployment of CD-DGS in the H plane peak XR radiation decreased by almost 9 dB where the CP profile of conventional and proposed structures remain constant. The peak value of XP of the proposed structure and conventional CPA is −20.4 dB and −11.45 dB respectively.

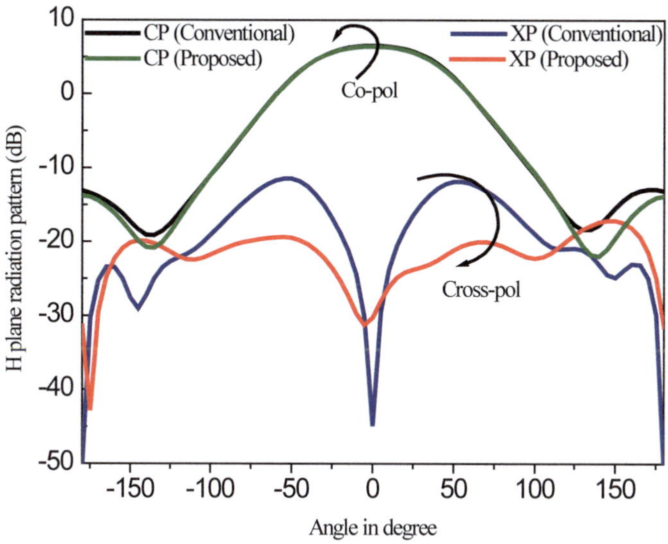

Fig. 10 Normalized H plane radiation pattern of the conventional and proposed structure

5 Conclusion

A pair of curved dumbbell shape DGS structures has been deployed at the non-radiating edge of conventional circular patch antenna on RT-Duroid substrate to improve the polarization purity and impedance bandwidth without hindering the other parameter of the basic radiation pattern. The optimum structure provides 25.8 dB polarization purity with 12% impedance bandwidth while for conventional CPA polarization purity and bandwidth are only 18 dB and 4% respectively. Around 8 dB improvement in polarization purity and 8% improvement in impedance matching is achieved with the investigated structure. The incorporation of CD-DGS does not have much effect on the CP gain. This present investigation can be useful where high polarization purity, wide bandwidth and a stable gain is required.

References

1. James JR, Hall PS (eds) (1992) Handbook of microstrip antenna, vol 1. Peter Peregrinus, London
2. Garg R, Bhartia P, Bahl I, Ittipiboon A (2001) Microstrip antenna design handbook. Artech House, Norwood, USA
3. Lee KF, Luk KM, Tam PY (1992) Cross polarization characteristics of circular patch antennas. Electron Lett 28(6):587–589
4. Guha D, Kumar C, Pal S (2009) Improved cross-polarization characteristics of circular microstrip antenna employing arc-shaped defected ground structure (DGS). IEEE Antennas Wirel Propag Lett 8:1367–1369

5. Guha D, Biswas M, Antar YMM (2005) Microstrip patch antenna with defected ground structure for cross polarization suppression. IEEE Antennas Wirel Propag Lett 4:455–458
6. Ghosh A, Ghosh SK, Ghosh D, Chattopadhyay S (2016) Improved polarization purity for circular microstrip antenna with defected patch surface. Int J Microw Wirel Technol (Cambridge University Press) 8(1):89–94
7. Ghosh A, Ojha S, Rana Y, Anand A, Kumar S, Chattopadhyay S (2014) Improved cross polarization performance of circular microstrip antenna with arc shaped defected patch surface. In: 1st international science and technology congress 2014(IEMCONGRESS'2014). Elsevier Publication, Kolkata, India, pp 87–92
8. Mondal K, Murmu L, Sarkar PP (2017) Investigation on compactness, bandwidth and gain of circular microstrip patch antenna. In: 2017 devices for integrated circuit (DevIC), pp 742–746. https://doi.org/10.1109/DEVIC.2017.8074050
9. Ghosh SK, Ghosh D, Chattopadhyay S (2016) Improved polarization purity for circular microstrip antenna with defected patch surface. Int J Microw Wirel Technol 8(1):89–94
10. Petosa A, Ittipiboon A, Gagnon N (1999) Suppression of unwanted probe radiation in wideband probe-fed microstrip patches. Electron Lett 35:355–357
11. Schejbal V, Kovarik VA (2006) Method of cross-polarization reduction. IEEE Antennas Propag Mag 48:108–111
12. Li P, Lai HW, Luk KM, Lau KL (2004) A wideband patch antenna with cross-polarization suppression. IEEE Antennas Wirel Propag Lett 3:211–214
13. Chen ZN, Chia MYW (2003) Broad-band suspended probe-fed plate antenna with low cross-polarization level. IEEE Trans Antennas Propag 51:345–346
14. HFSS: High frequency structure simulator. Ver. 13. Ansoft Corp., USA

Analysis of Hate Speech Recognition Using SVM and Naïve Bayes Classifier

Archana Chaudhari and Rahul Borole

Abstract Sentiment Analysis is a popular topic for scientific and statistical research and market updates. Sentiment analysis is part of data Science which deals with the sentiments of the customer or user reviews. Data science is the branch in which the data is collected and after some computations on the data, analysis is done. Sentiments analysis is also called as Text/Data analysis. Social media is often seen to witness hate speech. Hate is spread in different ways based on the caste, sex, race, religion, country, etc. Due to these hateful comments, there are chances of occurrence of stressful and harmful situations. Various machine learning approaches can be used to detect hate speech. To detect such hate content from social media, classification algorithms are used for sentiment analysis. Classification algorithms classify the given data between different classes. In this work, an analysis of hate speech recognition using Support Vector Machines (SVM) and Naïve Bayes algorithm classifier is presented.

Keywords Machine learning · Hate content · Offensive content · Sentiment analysis · Sentiments · SVM · Naïve Bayes · Classification

1 Introduction

The internet is witnessing a surge in the proliferation of hurtful comments and negative reviews across diverse platforms, such as social media and online discussion forums. The anonymity afforded to users enables them to express their opinions freely. Identifying and tackling hate poses a considerable challenge due to the absence of a universally accepted definition of hate. The perception of hate often hinges on individual perspectives regarding a comment. Recognizing the potential harm, many

A. Chaudhari (✉) · R. Borole
Department of Instrumentation and Control Engineering, Vishwakarma Institute of Technology, Pune 411037, India
e-mail: archana.chaudhari@vit.edu

R. Borole
e-mail: rahul.borole19@vit.edu

S. K. Biswas et al. (eds.), *Intelligent Computing Systems and Applications*, Lecture Notes in Networks and Systems 1307, https://doi.org/10.1007/978-981-96-3860-4_30

online platforms have taken steps to combat hate speech, promptly deleting such comments when identified.

Several definitions attempt to capture the essence of hate, reflecting the subjective nature of this phenomenon. Some of the definitions are mentioned here:

Encyclopedia of the American Constitution: "Hate speech is speech that attacks a person or group on the basis of attributes such as race, religion, ethnic origin, national origin, sex, disability, sexual orientation or gender identity."

Facebook: "We define hate speech as a direct attack on people based on what we call protected characteristics—race, ethnicity, national origin, religious affiliation, sexual orientation, caste, sex, gender, gender identity and serious disease or disability. We also provide some protections for immigration status. We define attack as violent or dehumanizing speech, statements of inferiority, or calls for exclusion or segregation."

Twitter: "Hateful conduct: You may not promote violence against or directly attack or threaten other people on the basis of race, ethnicity, national origin, sexual orientation, gender, gender identity, religious affiliation, age, disability or serious disease."

Sentiment analysis, a crucial aspect of data science, revolves around working with various forms of data extracted from platforms like social media, online shopping websites, networking sites and people's reviews of products or services. This data primarily manifests as text, encompassing a broad spectrum of content, including political or religious views and personal experiences.

Within sentiment analysis, different approaches yield outcomes. One prominent method involves a lexicon-based approach where a dictionary of words, or lexicon, assigns predefined sentiment scores to words, enabling the computation of the overall sentiment score of a given text. On the other hand, machine learning approach leverages the predictive power of algorithms based on past data outcomes. Various algorithms, such as Naïve Bayes, Support Vector Machines (SVM), N-gram, K-Nearest Neighbours (KNN), are instrumental in achieving sentiment analysis within the machine learning framework. Each algorithm contributes uniquely to the understanding and prediction of sentiment of textual data.

In the proposed work, hate speech detection using SVM and Naïve Bayes algorithm approach is proposed, and its performance is analysed.

The research paper is organized as follows: Section 1 presents the introduction to the topic. Section 2 discusses the literature review for the research work. Section 3 presents the analysis of hate speech recognition using SVM and Naive Bayes classifier is demonstrated. Section 4 discusses the experimental results and performance of the classifiers. Section 5 presents the conclusions of the research work.

2 Literature Review on Related Work

Sentiment analysis is a very popular topic for research and development. There are many research and experiments done in this field of sciences. Sentiment analysis uses statistical tools to explain and describe the data. Sean MacAvaney et al. [1] discuss about the challenges which are faced in hate detection. It stated various definitions of hate from various sources. It also proposed using SVM for hate detection. Christos Troussas et al. [2] discuss about the Naïve Bayes classifier for Facebook statuses. The data gathered was from Facebook users having 5000 positive sentiments and 5000 negative sentiments. The authors stated that using the Naïve Bayes algorithm gives higher accuracy.

Anna Schmidt and Michael Wiegand [3] proposed a survey for hate speech detection. The survey includes the features of hate speech, classification methods and the challenges for hate speech detection. The classification methods stated are Support Vector Machines (SVM) and Recurrent Neural Networks (RNN).

An application of text mining is document classification. Various algorithms are used for this purpose, but Naïve Bayes is used widely because of its simplicity in training and classifying stage. Pre-processing and feature selection are two factors for good mining quality and performance of quality mining depends on the dataset. This is discussed by Ting et al. [4].

Oauth protocol is used for authentication for accessing tweets using Twitter API. The classifier used is Naïve Bayes classifier. Authors Vadivukarassi et al. [5] state that increasing the number of features increases the accuracy of classifier.

Putri et al. [6] discuss about the SMOTE technique used for the balancing the imbalanced dataset. It is over sampling technique used for under sampled datasets. Different classification algorithms were tested using SMOTE technique and without. Comparison result showed that the SMOTE Naïve Bayes method gave better accuracy compared to others. SVM shown better results in non-SMOTE method.

Vandan Mujadia et al. [7] collected the dataset in three languages and used algorithms like SVM, Random Forest and LSTM. Authors conclude that the traditional Machine Learning algorithms provide stronger results and feature identification than neural network basked approaches. Similar experiment was conducted by Inna Vogel and Meghana Meghana [8] and concluded that the model of SVM using TF-IDF is good classifier for English Language, whereas for Spanish, Bi-directional Long Short Term Memory (Bi-LSTM) is better classifier. SVM classifier is better for other types of datasets too. Munir Ahamd et al. [9] used dataset of self-driving cars and Apple products from Twitter. The results of SVM shown were dependent on input data. To improve the performance of the model, using larger and different datasets is suggested. Luo et al. [12] provided the solution of data sparseness. They used vector space model for text representation. In other words, SVM classifier. Study found that multiple feature dimensions are not proportional. Authors Nurulhuda Zainuddin and Ali Selamat [13] used SVM for two datasets. Concluded that TF-IDF plays important role in extracting the classical features in dataset.

A detailed study about Naïve Bayes classifier is presented by Rish [10]. Author discusses different cases like dataset with noise, dataset without noise. Ways to reduce information loss are provided. Naïve Bayes is strong algorithm if used effectively. Islam et al. [11] investigated the performance of Naïve Bayes classifier and K-nearest neighbour classifier. The result of Bayesian inference depends strongly on prior probabilities, which must be available in order to apply the method directly. Since the Bayes theorem provides a principled way to calculate the posterior probability of each hypothesis given the training data, we can use it as the basis for straightforward learning algorithm that calculates the probability for each hypothesis then outputs the most probable.

For multilingual hate speech detection, Michele Corrazza et al. [14] identified that RNN based LSTM is well performing across different languages. They evaluated the model for different languages like English. German and Italian. Another study by Ika Alfina et al. [15] of hate speech detection in Indonesian Language. The comparison between Naïve Bayes and SVM. Author concluded that the word n-gram feature was superior than the character n-gram. Overall SVM results were lower than other compared algorithms.

SVM is extremely strong performer, but this algorithm capture ‘sparse’ and ‘dense’ features in document for classification. This makes the detection of relations in sentence hard. On other hand, Long Short-Term Memory (LSTM) algorithm of type Recurrent Neural Networks (RNN) is used to capture the dependencies in the sentences. For classification, dataset having three classes, namely, ‘strong hate’, ‘weak hate’ and ‘no hate’ was taken. After the results, Fabio Del Vigna et al. [16] found that the accuracy of SVM was slightly greater than LSTM.

Zeerak Waseem and DrikHovy [17] collected a dataset of tweets and annotated in different classes as sexist, racist, neither, etc. To check a tweet is offensive or not, some rules are provided by author. The classification algorithm used was N-grams. Thomas Davidson et al. [18] weighted the entries in dataset using Term-Frequency Inverse Document Frequency (TF-IDF) method. Using 5-fold cross validation, author tested various algorithms like logistic regression, linear SVM, decision trees, Naïve Bayes, etc. Author states that automatic classification methods can achieve relatively higher accuracy.

Ayme Arango et al. [19] discuss about the problem of overestimation of performance of approaches for hate speech detection on twitter. Deep learning models mainly have two issues, they overfit data easily and they are difficult to explain. These issues must be considered while using deep learning models.

3 Analysis of Hate Speech Detection Using SVM and Naïve Bayes Classifiers

The proposed method classifies the hate speech using Naïve Bayes or SVM algorithm. Figure 1 represents the overview of the classifier method. It basically contains 3 phases, pre-processing, Training and application phase. In pre-processing phase, data pre-processing computations are performed on the data to make it usable. In this phase, all the constraints which do not signify any information are removed. The next phase is the training phase in which the processed data is trained under different algorithms to solve the problem. In the proposed method, for training phase SVM and Naïve Bayes are used. The application phase classifies any random input into different classes.

Sentiment Analysis is carried out in the following stages:

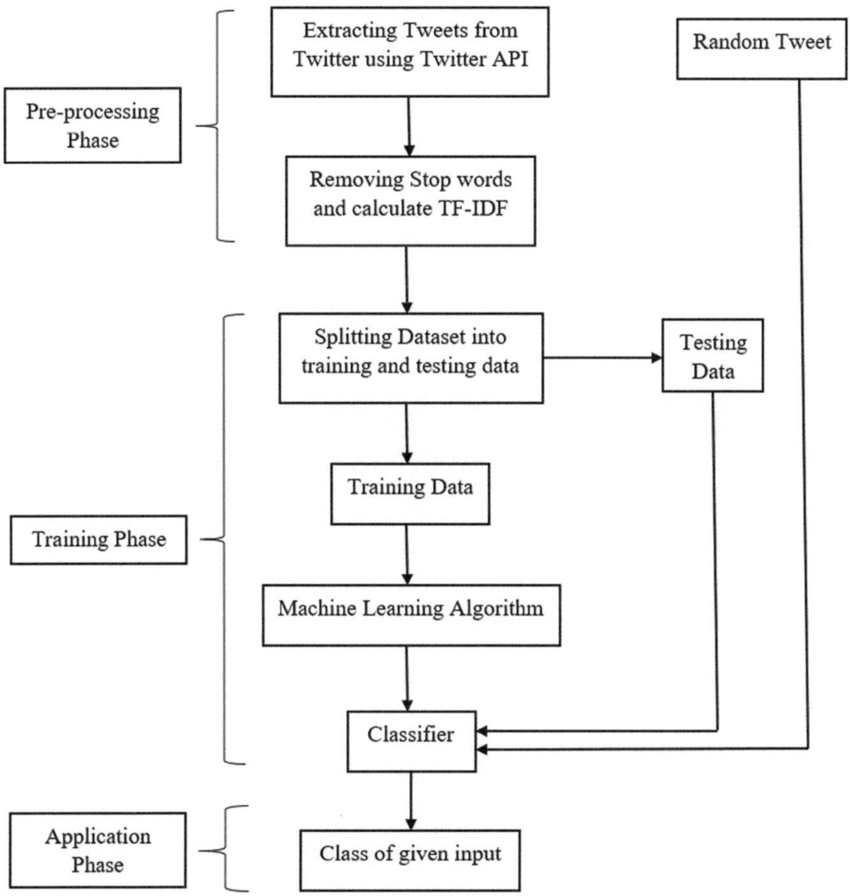

Fig. 1 Flowchart for classifier method

- The interested field for analysis is selected.
- The dataset for the given field is created/taken.
- Pre-processing is done on the dataset to clean the dataset and make it usable.
- The next step is to develop a machine learning model for sentiment analysis. This can be achieved by various algorithms which are available.
- The dataset is first vectorized.
- The stop words which do not contribute to sentiments are removed.
- The dataset is then further divided into training and testing dataset.
- The model is trained using the selected machine learning algorithm.
- Test the model using testing data and evaluate the performance metrics.

The training process of the model involves the utilization of two algorithms: Support Vector Machines (SVM) and the Naïve Bayes algorithm. Prior to training, certain computations are conducted on the dataset, which consists of words or sentences. To convert alphabetical data into numerical form, TF-IDF vectorization method is employed. This method transforms words or sentences into floating-point data and incorporates two key components: Term frequency (TF) and Inverse Document Frequency (IDF). TF represents the frequency of a specific word in the document, indicating its importance. IDF, on the other hand, is the weight of the term, aiming to reduce the significance of terms.

The IDF is calculated using the formula:$idf_i = \log\left(\frac{n}{df_i}\right)$where, idf_i is the inverse document frequency and df_i is the document frequency

And the TF-IDF score is computed as follows:$W_{i,j} = tf_{i,j} \times idf_i$

Support Vector Machines (SVM) is a classification-based algorithm that maps the data to a high-dimensional space for effective categorization. It determines a linear or polynomial separator, known as a hyperplane, between two classes. SVM is chosen for its ability to enhance predictive accuracy without overfitting, especially when dealing with a large number of data points. The selection of the kernel, which is a mathematical function used for data point transformation, depends on experimentation with the dataset to achieve optimal accuracy.

The Naïve Bayes algorithm is fast-working, classification-based algorithm for categorizing categorical variables. It operates on the principles of Bayes Theorem, calculating the probability of an event occurring given the occurrence of another event. The fundamental assumption of Naïve Bayes is that the features in the dataset are mutually independent. It calculates the conditional probability of the features, from which the posterior probability is derived using the formula:

$$P(A|B) = P(A) * P(B|A)/P(B)$$

$$P\left(a_1, a_2 \ldots\ldots a_n|v_j\right) = (\arg\max)v_j \varepsilon V \Pi P\left(a_i|v_j\right)$$

$$v_{NB} = (\arg\max)v_j \varepsilon V \Pi P\left(a_i\middle|v_j\right)$$

The Naïve Bayes prediction is determined by maximizing the posterior probability, classifying the feature into the appropriate class.

4 Results and Discussion

4.1 Database Description

To perform sentiment analysis, a dataset (Table 1 shows metadata) of tweets from different users is taken. This dataset has 56,745 tweets from different users along with the toxicity of tweet. This dataset has both positive and negative sentiments.

Second dataset (Table 2 shows metadata) used is also dataset having tweets from various users. This dataset has 3 classes, offensive text, hateful text and neither. It has 25,295 entries.

Table 1 Metadata of dataset 1

Name	HateSpeechDataset.csv
File type	CSV (Comma Separated Values)
File extension	.csv
File size	4.94 MB
Row count	56,745
Column count	2
Column 1	Toxicity
Type of column 1	Categorical
Column 2	Tweet
Type of column 2	Text
Source	Kaggle datasets

Table 2 Metadata of dataset 2

Name	OffensiveData.csv
File type	CSV (Comma Separated Values)
File extension	.csv
File size	2.42 MB
Row count	25,295
Column count	7
Column 6	Class
Type of column 6	Categorical
Column 7	Tweet
Type of column 7	Text
Source	Kaggle datasets

4.2 Performance Metrics

After training and testing the model, evaluation of model is necessary. For the evaluation of performance of model there are several performance metrics which can be calculated. The performance metrics which can be considered are:

- ROC AUC Score
- Accuracy Score
- Confusion Matrix.

ROC AUC Score is the area under the curve of True-positive rate against False-Positive rate. ROC-AUC is the measure of the model to distinguish between given two classes. The higher the AUC, the better model can distinguish between two classes.

Accuracy Score is the metric used for evaluation of classification models. Accuracy is the ratio of correct predictions to total number of predictions done. Higher the accuracy, higher the correct predictions are done.

Confusion Matrix (Table 3) is an evaluation metric used for evaluation of classification models. Confusion matrix can have two or more classes. The confusion matrix for binary classification is given as:

Where TP—True Positives, FP—False Positive, FN—False Negative, TN—True Negative.

Using the given confusion matrix, we can calculate the statistical terms like accuracy, precision, specificity, recall and F measure.

$$Accuracy = \frac{TN + TP}{Total\ Observations} \tag{6}$$

$$Precision = \frac{TP}{TP + FP} \tag{7}$$

$$Specificity = \frac{TN}{TN + FP} \tag{8}$$

$$Recall\ or\ Sensitivity = \frac{TP}{TP + FN} \tag{9}$$

Table 3 Confusion matrix for binary classification

	Actual values		
Predicted values		Negative (0)	Positive (1)
	Negative (0)	TN	FP
	Positive (1)	FN	TP

Table 4 Confusion matrix using SVM model

	Actual values		
Predicted values		Negative (0)	Positive (1)
	Negative (0)	6401	164
	Positive (1)	430	4354

4.3 Experimental Results

Hate Speech dataset is divided into training and testing dataset. Training dataset is *80%* and testing dataset is *20%*. After testing and training the classification model using Support Vector Machines (SVM) and Naïve Bayes algorithm, we calculate the performance metrics of the model. Confusion matrix for SVM is shown in Table 4.

ROC AUC Score for the given model is calculated in Python. The value got as outcome for SVM is *0.9425* and for Naïve Bayes is *0.9180*. This value describes that of 100 entries of classification, 94 and 91 values are predicted in correct classes by respective algorithms. Confusion matrix for Naïve Bayes is shown in Table 5.

Accuracy score is calculated, and values are *0.9476* and *0.9205* respectively. As the accuracy is higher, the more correct the predicted value is.

Confusion matrix for the given trained model for SVM model is represented by Table 4. Table 5 demonstrates the confusion matrix using Naive Bayes model.

Offensive dataset is divided into training and testing dataset. Training dataset is 80% and testing dataset is 20%. After testing and training the classification model using Support Vector Machines (SVM), we calculate the performance metrics of the model. Table 6 represents confusion matrix using SVM for offensive data. The Accuracy Score calculated for this model is 0.9120.

Support Vector Machines (SVM) usually increases the accuracy of model because it overcomes the problem of over fitting. Naïve Bayes algorithm is effective for larger

Table 5 Confusion matrix using Naïve Bayes model

	Actual values		
Predicted values		Negative (0)	Positive (1)
	Negative (0)	6130	435
	Positive (1)	467	4317

Table 6 Confusion matrix using SVM model for offensive data

	Actual values			
Predicted Values		Class 1	Class 2	Class 3
	Class 1	54	190	27
	Class 2	25	3703	106
	Class 3	4	84	766

datasets compared to other datasets. This algorithm can also be used for the different types of datasets which can be easily accessed through the internet to check the behaviour of both algorithms.

5 Conclusions

In the research work, two machine learning algorithms Support Vector Machines (SVM) and Naïve Bayes are implemented for hate speech detection and classification. After the experiments, SVM model was found to have greater accuracy compared to Naïve Bayes. But Naïve Bayes also gave considerable accuracy. For the multi class classification, Naïve Bayes cannot be used because it is usually used for binary classification. For the multi class classification using Naïve Bayes, it doesn't give any results for third class.

References

1. MacAvaney S, Yao H-R, Yang E, Russell K, Goharian N, Friender O (2019) Hate speech detection: challenges and solutions. Plos One J
2. Troussas C, Virvou M, Espinosa KJ, Llaguno K, Caro J (2013) Sentiment analysis of Facebook statuses using Naive Bayes classifier for language learning
3. Schmidt A, Wiegand M (2017) A survey on hate speech detection using natural language processing. In: Fifth international workshop on natural language processing for social media
4. Ting SL, WH Ip, Tsang AHC (2011) Is Naïve Bayes a good classifier for document classification? Int J Softw Eng Its Appl
5. Vadivukarassi M, Puviarasan N, Aruna P (2017) Sentimental analysis of tweets using Naïve Bayes. World Appl Sci J 35. IDOSI Publications
6. Putri TTA, Sriadhi S, Rahmadani R, Hutahaean HD (2020) A comparison of classification algorithms for hate speech detection. IOP Conf Ser: Mater Sci Eng
7. Mujadia V, Mishra P, Sharma DM (2019) IIIT-Hyderabad at HASOC 2019: hate speech detection. In: International FIRE
8. Vogel I, Meghana M (2021) Profiling hate speech spreaders on Twitter: SVM vs. Bi-LSTM. In: Conference and labs of the evaluation forum
9. Ahmad M, Aftab S, Ali I (2017) Sentiment analysis of tweets using SVM. Int J Comput Appl
10. Rish I (2001) An empirical study of the Naïve Bayes classifier. Watson TJ Research Centre
11. Islam MJ, Wu QMJ, Ahmadi, M Sid-Ahmed MA (2007) Investigating the performance of Naïve Bayes classifiers and K-nearest neighbor classifiers. In: International conference on convergence information technology
12. Luo F, Li C, Cao Z (2016) Affective-feature-based sentiment analysis using SVM classifier. In: International conference on computer supported cooperative work in design
13. Zainuddin N, Selamat A (2014) Sentiment analysis using support vector machine. In: International conference on computer, communication and control technology
14. Corazza M, Menini S, Cabrio E, Tonelli S, Villata S (2020) A multilingual evaluation for inline hate speech detection. ACM Trans Internet Technol
15. Alfina I, Mulia R, Fanany MI, Ekanata Y (2017) Hate speech detection in the indonesian language: a dataset and preliminary study

16. Del Vigna F, Cimino A, Dell'Orletta F, Petrocchi M, Tesconi M (2017) Hate me, hate me not: hate speech detection on Facebook. In: Italian conference on cybersecurity
17. Waseem Z, Hovy D (2016) Hateful symbols or hateful people? Predictive features for hate speech detection on Twitter. NAACL-HLT
18. Davidson T, Warmsley D, Macy M, Weber I (2017) Automated hate speech detection and the problem of offensive language. In: International AAAI conference on web and social media
19. Arango A, Pérez J, Poblete B (2019) Hate speech detection is not as easy as you may think: a closer look at model validation. In: SIGIR
20. Rottger P, Vidgen B, Nguyen D, Waseem Z, Margetts H, Pirrehumbert JB (2021) HATE-CHECK: functional tests for hate speech detection models arXiv:2012.15606v2 [cs.CL]. Accessed 27 May 2021
21. Florio K, Basile V, Polignano M, Basile P, Patti V (2020) Time of your hate: the challenge of time in hate speech detection on social media. Article in applied sciences
22. Aluru SS, Mathew B, Saha P, Mukherjee A (2020) Deep learning models for multilingual hate speech detection. arXiv:2004.06465v3 [cs.CI]. Accessed 9 Dec 2020

Networking and Cryptography

Mitigation of Distributed Denial of Service Attack in a Software Defined Network Using Flow Entry Based Open Flow Methodology

K. Bavani, M. P. Ramkumar, G. S. R. Emil Selvan, and M. Mahalakshmi

Abstract Denial of Service (DDoS) attack has been used as a weapon by many hackers to cause a network failure thereby preventing the users from using the network service. This attack is perpetrated to create a financial loss, to gain money, to seek revenge, etc. Software-Defined Network (SDN) has grown widely to gain the hope of many researchers and network administrators to meet the need for maintenance of the data centers. Today's world mostly prefers SDN due to its flexibility, cost-efficient, manageability, etc. SDN has a network architecture that changes its traffic pattern every time whenever a change is done in its central architecture. The separation of the planes in SDN ensures the security of entire network devices and their functions thereby protecting the network. So, the attackers target the controller to bring it down leading to a network failure. In the existing methods, there are many mitigation schemes that help to mitigate the attack. But still, the attacker tries a new way to perpetrate the DDoS attack. The proposed approach uses Flow entry-based OpenFlow methodology to mitigate the DDoS attack at the early stage.

Keywords DDoS attack mitigation · OpenFlow protocol · SDN security · Software defined network

K. Bavani (✉)
Department of Computer Science and Engineering, Sathyabama Institute of Science and Technology, Jeppiaar Nagar, India
e-mail: bavaniyathra@gmail.com

M. P. Ramkumar · G. S. R. Emil Selvan · M. Mahalakshmi
Department of Computer Science and Engineering, Thiagarajar College of Engineering, Madurai, India
e-mail: ramkumar@tce.edu

G. S. R. Emil Selvan
e-mail: emil@tce.edu

S. K. Biswas et al. (eds.), *Intelligent Computing Systems and Applications*, Lecture Notes in Networks and Systems 1307, https://doi.org/10.1007/978-981-96-3860-4_31

Table 1 Disparity of SDN with traditional and conventional network

Characteristics	Conventional network	Traditional network	SDN
Network management	Difficult	Difficult	Easy
Performance	Static	Static	Dynamic
Configuration	Manual	Manual	Automatic
Innovation	Limited design and operations	Limited design and operations	Improved design and operations with software upgrade
Programmable network	No	No	Yes

1 Introduction

1.1 Software Defined Network

In this era, there is a dynamic change in the networking fields as the elaborating ideas in the technical fields are in advance. Therefore, programmable networks are in need to tackle these ideas in the networking field. The control plane in this network act as the controller for the entire network which is otherwise termed as SDN controller.

This controller employs OpenFlow protocol (also known as OpenFlow controller) which is used to enable communication with various networking devices in the network. To make the controller communicate with other networking devices, the southbound interface (OpenFlow) controls these networking devices and makes them taciturn and utilize them for the purpose of forwarding the data packets [1]. SDN is rapidly growing in both the industrial and research community fields [2]. Although it is preferred in today's world, it has several challenges that are to be solved. The comparison of SDN with Traditional and Conventional network is shown in Table 1.

1.2 Overview of SDN Framework and Its Integrant

The networking framework of SDN provides the programmability of the network. The controller in SDN performs a centralized control [3] over the entire network where the data plane that is the forwarding plane connect the networking devices to bring out all the active data path elements. The control plane provides an ease to users or the net- work administrators by allocating the resources at virtual level. Therefore, this network is precisely virtualized. The architecture of SDN is shown in Fig. 1. It consists of three primary layers known as the data plane, control plane and the application plane. The deployment sections in SDN includes SDN controller, southbound API, northbound API, and network applications.

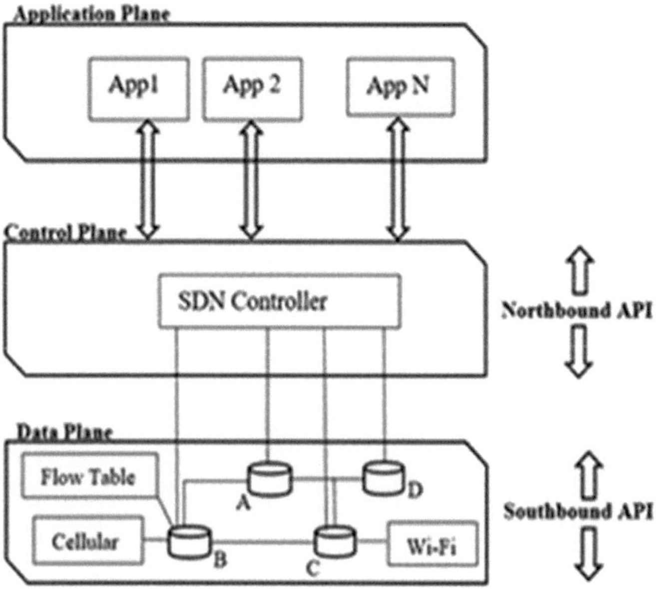

Fig. 1 SDN architecture

1.3 Data plane

The packet forwarding devices are present in the data plane. When these devices need to communicate with the controller, they do so with the help of the OpenFlow interfaces [4]. As the forwarding devices enable this interface, it provides a forwarding table which has 3 fields that includes rules, actions, counters or stats.

1.4 Southbound API

The southbound API act as a link or channel between the forwarding network devices and the control plane. It allows the controller to manage the network behavior and monitor the flow table entries of all the switches. They have many mutual interfaces that connects the controller. They are OpenFlow, POF [5], OpFlex [6], and OpenState [7]. The southbound API mostly prefers the OpenFlow protocol interface.

1.5 Controller

The device discovery and the network topology helps the controller to extract the lower level information and provide it to the application plane. Distinctive set of con trollers are available with unique fundamental architecture [8]. The centralized controllers such as POX, NOX, Beacon, Maestro, Ryu NOS and Floodlight are considered to be best in the data centre networks.

1.6 Northbound API

The northbound API is used for the purpose of developing various applications. To set the instructions in the forwarding network devices, the northbound API extracts the low-level information of the southbound API.

1.7 Application Plane

The application plane has high level instructions which will be given to the controller, where these instructions are the flow rules that are executed on the forwarding devices [9]. OpenFlow Protocol and its Function.

OpenFlow protocol uses the SDN controller to control and manage switches in the network. The switches contain the flow tables for the data packets shown in Fig. 2. The switches compare the data packet's flow entries in the flow table [10] during the arrival of data packets into the network. The communication between the controller and the switches are enabled with the help of the OpenFlow Protocol [11].

When a new data packet enters into the network, it checks for the match in the flow table [12]. Since the flow tables are empty at the initial phase, no match is found. So, the data packets are sent [13] to the controller to make the further conclusion either to forward them to the switch and create a flow entry in the flow table or drop them [14]. The action to create a flow entry is sent to the switches and the data packets are thus sent to the destination hosts [15].

1.8 SDN Security

Securing the network by ensuring the CIA triad is the major goal of the network. The SDN architecture has a special characteristic of having a centralized controller. Many network administrators [16] thought whether the presence of centralized controller in the SDN leads to a network failure or secure the network in case of attack [1]. Since, the controller is centralized in the network, it controls and monitors all the net-

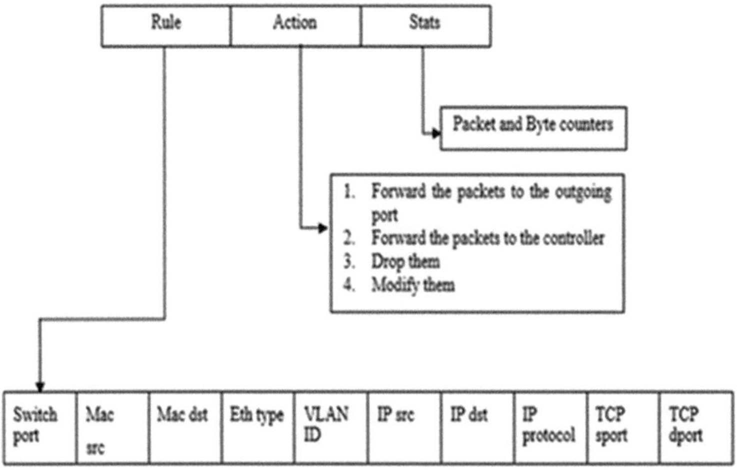

Fig. 2 OpenFlow table

work resources and its activities [17]. Also, in SDN, the controller is responsible for the flow initiation that takes place whenever the data packets enter into the network [18].

The network behavior is monitored with the help of the controller [19]. If a controller goes down, then the entire network resources fail. Therefore, the perpetrator gains the control of the entire network and performs malicious activities. Multiple controllers can be introduced in the SDN to prevent a network failure, but multiple controllers lead to a problem of gaining access and controlling the network resources.

1.9 DDoS Attack

The attacker or cyber terrorists creates a virus program and penetrates it into many web pages. When the network users visit those web pages, the virus program gets automatically downloaded [20] without the awareness of the user and gains complete control of the user's device. Similarly, the attacker creates botnet [20] and gives command to the botnets to perform attack on a target web server or network. This is called as Distributed Denial of Service (DDoS) attack. If the attacker uses any one machine or his own machine to attack the target web server or network, then it is called as Denial of Service (DoS) attack.

2 Existing Approaches of Mitigation of DDoS Attacks in SDN

The Software Defined Network has a physical separation of forwarding data plane from the control plane thereby which the controller acts as a centralization and controls the entire network devices and their functions. In, they proposed a new metric called as information distance metric and also proved that this metric helps in mitigating the DDoS attack at its early stage.

Entropy is used as an important parameter to detect and mitigate DDoS attacks in SDN. They used entropy for DDoS attack detection and rate limits for DDoS attack mitigation purpose. In certain set of rules are defined in the controller so that if the traffic packets were suspected to be attack packets, then they are removed from entering the network and dropped. They have proposed various mitigation mechanisms which includes blocking the ports that is once the attack packets are detected, their source of attacking port is found and blocked.

3 Problem Definition

The fundamental architecture of SDN keeps on changing to various design entities whenever the network administrators update the SDN. So, the attacker tries to bring down the controller of the SDN therefore leading the network to a failure. It is important to protect the controller from overloading and also protect the switch from breakage. The detection mechanism is achieved successfully at a higher accuracy but at the time of mitigation, the capacity of processing the traffic packets is not achieved by the switch. Either the switch breakage occurs or the switch overloads the controller and offloads it leading to a network failure. The attacker tries to increase the flow of attack traffic packets along with the legitimate traffic packets only after some time and not at the initial stage. So, the switch cannot withhold all of these attack packets suddenly therefore leads to a breakage and causes the network failure.

3.1 Objective

To mitigate the DDoS attack in SDN at its early stage using Flow entry based OpenFlow methodology.

4 Proposed Approach

Here, a statistical tactic is suggested to obtain the fine-grained patterns of data. In the statistical approach flows, window size, number of windows, flow threshold and flow timeout are proposed.

To compute the statistical data, collection of data packets is done with the help of windows. It is assumed that the window size to be 50 and take number of windows as 5. The reason to choose the window size as 50 is, for mitigation of DDoS attack in the early stage the window size should not be too large and if the window size is too small then it leads to computational overhead. So, the window size 50 balances the problem on both sides. So, for each window there are 50 packets. These 50 packets are grouped into various flows according to the destination ip address they must reach.

The controller then forwards this flow of data packets to the switch, where the switch creates the flow entry for that flow of data packets and updates the counter value to 1. For choosing the flow timeout value, the time taken for the first flow of first window to enter the network switch is computed and this first flow timeout value is set as the mitigating value. If it exceeds the mitigating value, then the entire flow of data packets is suspected to be malicious and are dropped from the OpenFlow table. This method is set to prevent the high number of data packets (which mostly contains attack packets) entering the network switch and reducing its processing capacity. Once, the DDoS attack is detected in the network, the switch and the controller are under the threat where they may get crash or overload anytime.

When one flow of data packets enters the network and if any one packet in that flow is suspected to be malicious then the entire flow is dropped or deleted from the flow table in the network switch. The threshold value for the flow rate is set. The flow timeout is set for 4 s for a window. For a window of time 4 s, the number of new data flow entry is computed and compared with the flow rate threshold value. If the flow entry count exceeds the flow rate threshold value then all the data packets of that flow entry are dropped till the end of the flow time. Then a new window of data flows enters into the network and they are counted and compared. If the initial flow timeout value in a window is set to 4 s, then the mitigation mechanism works better at huge DDoS attack. For the implementation, to create a network topology mininet tool is used. This tool is most widely used for the purpose of DDoS attack detection and mitigation. The controller used here is pox controller. Scapy is used for generating the traffic packets. The simulation is run in Ubuntu 18.04 Operating System. The network topology created is shown in Fig. 3.

The system layout of the proposed approach is shown in Fig. 4. The representation in Fig. 5 explains the detection of attack with the help of proposed threshold value and the mitigation process starts after the detection at the early stage. Once the mitigation process starts, the attack rate starts decreasing to a certain level.

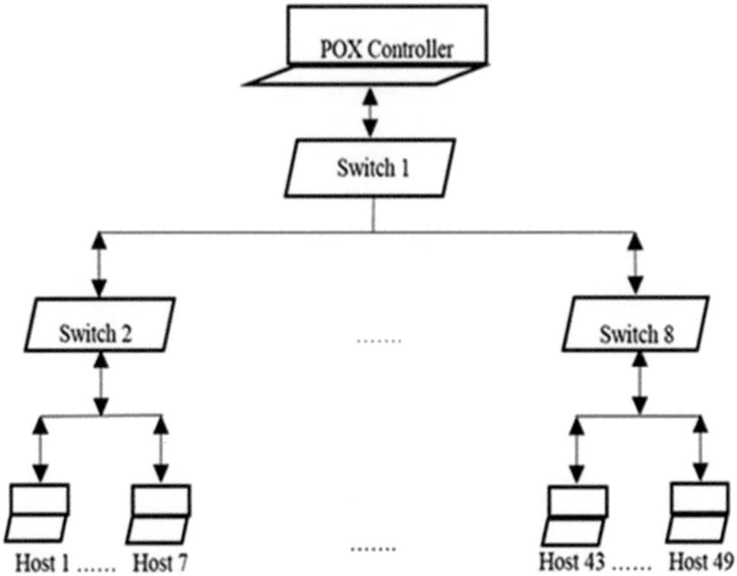

Fig. 3 Network topology

In Fig. 6, there is no increase in the attack rate during the mitigation process. This is due to the constant flow rate without any increase in the attack packets. Here, the controller drops the flow of data packets that it suspects to be malicious with the help of the counter value. So, in Fig. 6 it is clear that the attack rate decreases during the mitigation process and the DDoS attack has been prevented. The performance analysis for mitigation of DDoS attack in SDN using OpenFlow statistics methodology is shown in Table 2. They are tested in sending various range of traffic packets to the network and the controller's operation is monitored. The window size is set from 10 to 100 packets in the sequential number of test and the controller's behavior is monitored. Using Open- Flow Statistic methodology, the attack traffic has been successfully lowered.

The controller's behavior in detecting and mitigating the DDoS attack at the early stage is monitored. In Figs.7, 8, 9 and 10 the mitigation of DDoS attack for 10, 20, 50 and 100 packets/sec using OpenFlow statistics methodology is shown respec tively. Using the OpenFlow statistics methodology, it is shown that the controller succeeds in lowering the attack traffic in the network.

In Figs. 11, 12, 13 and 14 the mitigation of DDoS attack for 10, 20, 50 and 100 packets/sec without using OpenFlow statistics methodology is shown respectively. Here, the controller gets overloaded and fails to lower the attack traffic.

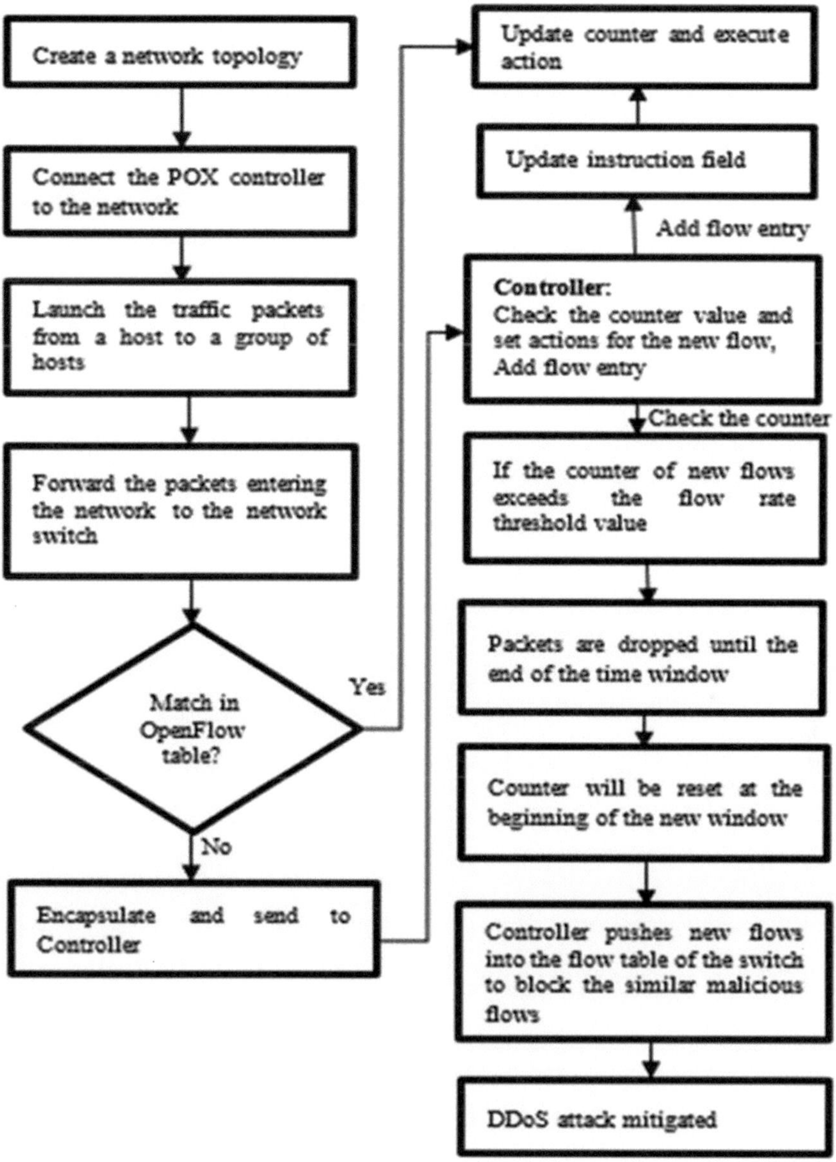

Fig. 4 System design

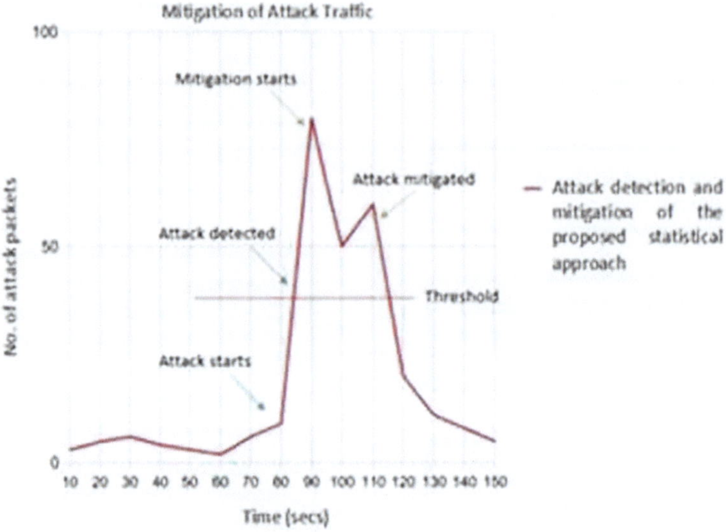

Fig. 5 Mitigation of DDoS attack at a standard attack flow

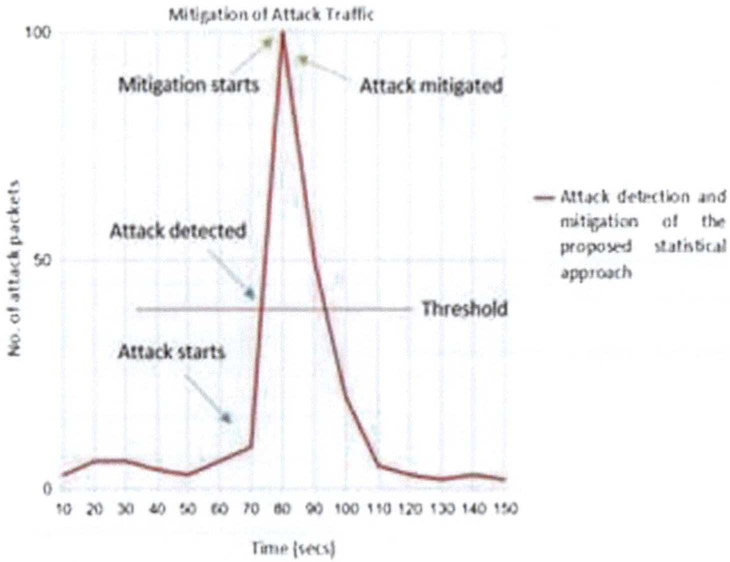

Fig. 6 Mitigation of DDoS attack at an increased attack flow

Table 2 Performance analysis of mitigation with and without using OpenFlow statistics

Methodology		
Packets/ Sec	Using OpenFlow statistics methodology	Without using OpenFlow statistics methodology
10	Operational (succeeds in lowering the attack traffic)	Operational (succeeds in lowering the attack traffic)
20	Operational (succeeds in lowering the attack traffic)	Operational (succeeds in lowering the attack traffic)
50	Operational (succeeds in lowering the attack traffic)	Partially Operational (Fails in lowering attack traffic completely)
100	Operational (succeeds in lowering the attack traffic)	Crash (Either network switch crashes or controller overloads)

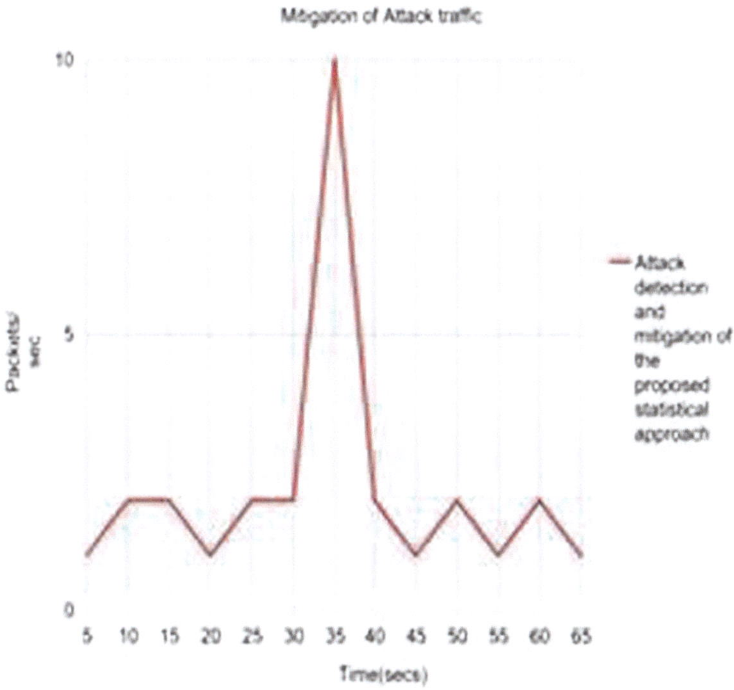

Fig. 7 Mitigation of DDoS attack for 10 packets/sec using OpenFlow statistics methodology

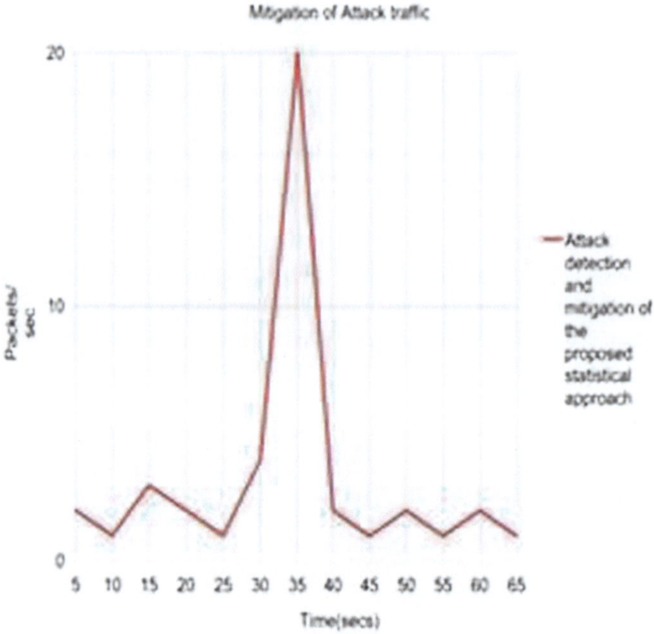

Fig. 8 Mitigation of DDoS attack for 20 packets/sec using OpenFlow statistics

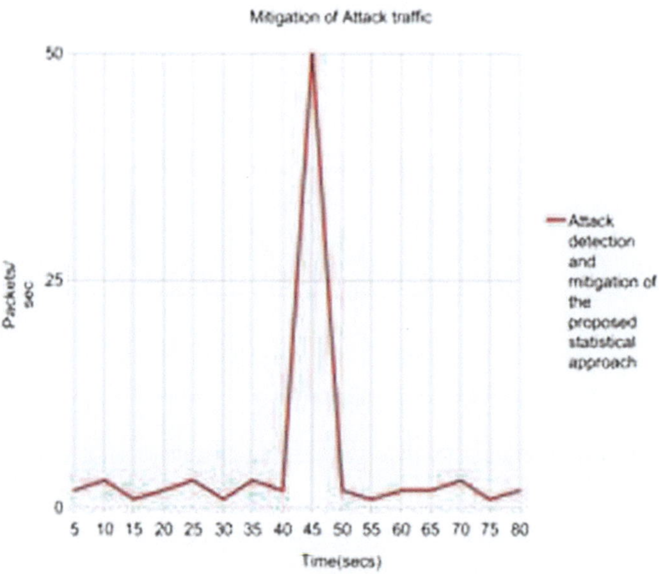

Fig. 9 Mitigation of DDoS attack for 50 packets/sec using OpenFlow statistics methodology

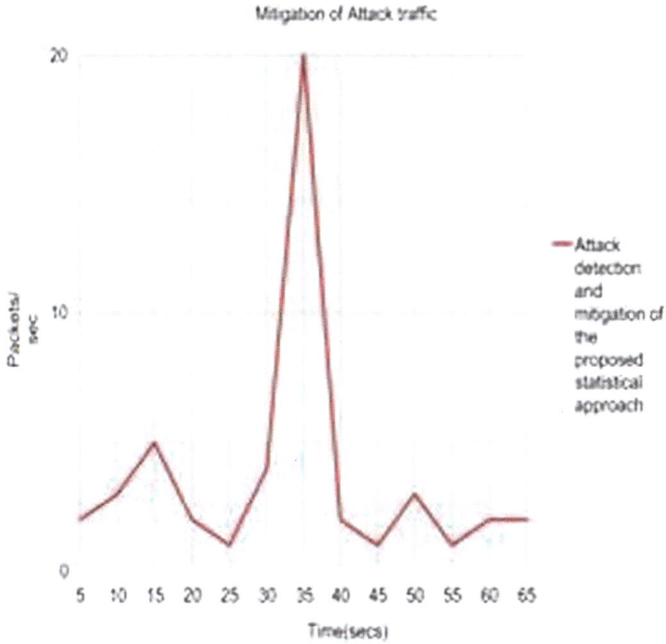

Fig. 10 Mitigation of DDoS attack for 100 packets/sec using OpenFlow statistics methodology

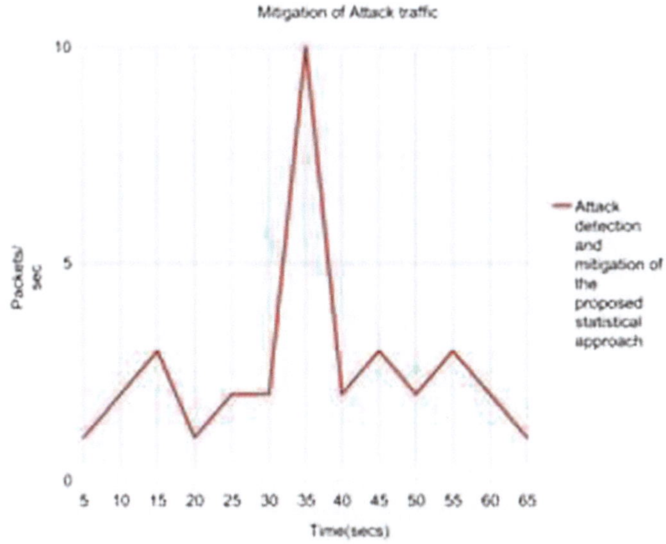

Fig. 11 Mitigation of DDoS attack for 10 packets/sec without using OpenFlow statis tics methodology

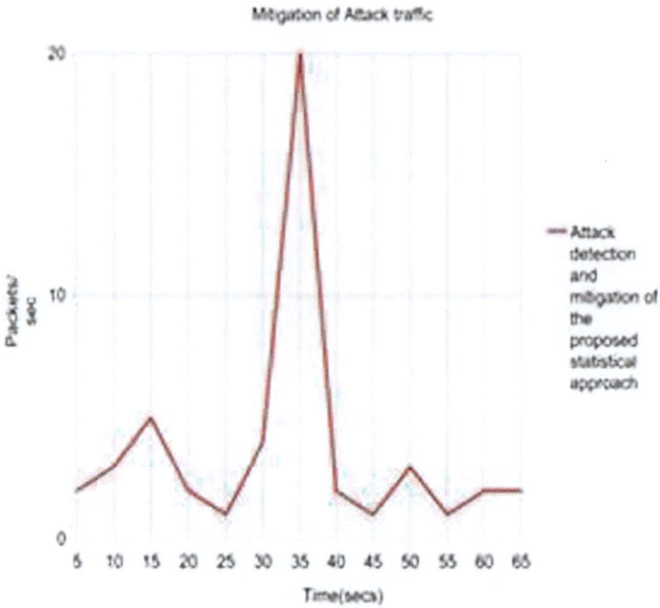

Fig. 12 Mitigation of DDoS attack for 20 packets/sec without using OpenFlow statistics methodology

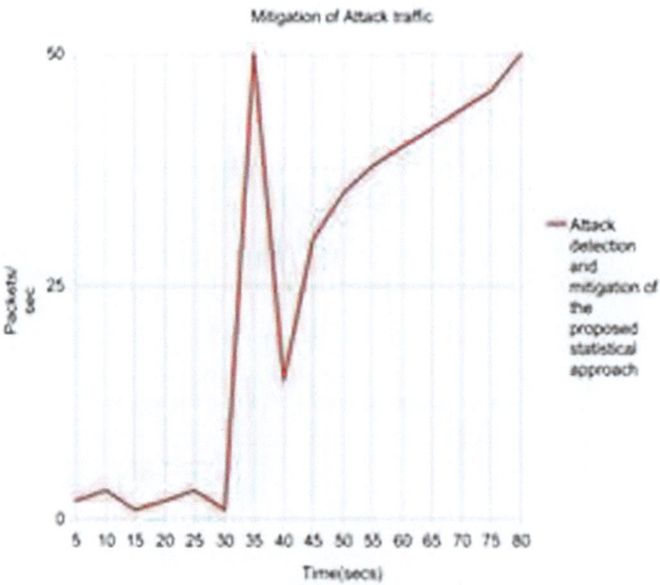

Fig. 13 Mitigation of DDoS attack for 50 packets/sec without using OpenFlow statistics methodology

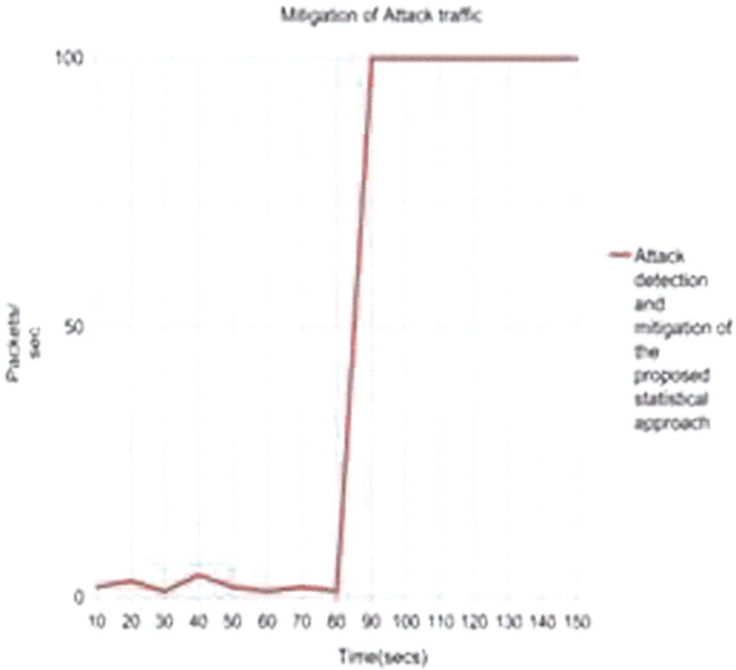

Fig. 14 Mitigation of DDoS attack for 100 packets/sec without using OpenFlow statistics methodology

5 Conclusion

The proposed approach successfully mitigates the DDoS attacks at its early stage and also protects the switch and the controller from getting crashed or overloaded. If the attacker tries to increase the amount of attack packets at any time, this mechanism works successfully in mitigating the DDoS attack in SDN. The proposed method was designed only for a single controller network environment, so the attacking paths can not be identified. Also, the proposed approach cannot detect the attack against the entire network. So, in future, the multi-controller network environment can be designed to detect the DDoS attack against and protect the network devices from the DDoS attack without overloading the controller.

References

1. Shaghaghi A, Kanhere SS, Kaafar MA, Bertino E, Jha S (2018) Gargoyle: a network- based insider attack resilient framework for organizations. In: 2018 IEEE 43rd conference on local computer networks (LCN), Chicago, IL, pp 553–561
2. Goransson P, Black C, Culver T (2016) Software defined networks: a comprehensive approach, Morgan Kaufmann, 2nd ed
3. Akhunzada A, Ahmed E, Gani A, Khan MK, Imran M, Guizani S (2015) Securing soft- ware defined networks: taxonomy, requirements, and open issues. In IEEE Commun Mag 53(4):36–44
4. Dargahi T, Caponi A, Ambrosin M, Bianchi G, Conti M (2017) A survey on the security of stateful SDN data planes. IEEE Commun Surv and Tutor 19(3):1701–1725
5. Abdou A, Van Oorschot PC, Wan T (2018) Comparative analysis of control plane security of SDN and conventional networks. IEEE Commun Surv and Tutor 20(4):3542–3559
6. Kreutz D, Ramos FM, Verissimo P, Rothenberg CE, Azodolmolky S, Uhlig S (2015) Software-defined networking: a comprehensive survey. Proc IEEE 103(1):14–76
7. Braga R, Mota E, Passito A (2010) Lightweight DDoS flooding attack detection using NOX/ OpenFlow. In: 2010 IEEE 35th conference on local computer networks (LCN), Piscataway, New Jersey, pp 408–415
8. Badotra S, Nagpal D, Panda SN, Tanwar S, Bajaj S (2020) IoT-Enabled healthcare network with SDN. In: 2020 8th International conference on reliability, infocom technologies and optimization (trends and future directions) (ICRITO), Noida, Uttar Pradesh, pp 38–42
9. Kloti R, Kotronis V, Smith P (2013) OpenFlow: a security analysis. In: 2013 21st IEEE international conference on network protocols (ICNP), IEEE, Goettingen, Germany, pp 1–6
10. Li Q, Zou X, Huang Q, Zheng J, Lee PP (2018) Dynamic packet forwarding verification in SDN. IEEE Trans Dependable Secur Comput 16(6):915–929
11. Kandoi R, Antikainen M (2015) Denial-of-service attacks in OpenFlow SDN networks, IFIP/ IEEE international symposium on integrated network management (IM). IEEE, Ot- tawa, ON, Canada, pp 1322–1326
12. Kia M (2015) Early detection and mitigation of DDoS attacks in software defined networks. Ph.D. Thesis, Ryerson University, Toronto, Canada
13. Hausheer D, Hohlfeld O, Schmid S, Gu G (2018) Security and performance of software-defined networks and functions virtualization. Comput Netw 138(1):15–17
14. Ubale T, Jain AK (2018) Taxonomy of DDoS attacks in software-defined networking environment. In: International conference on futuristic trends in network and communication technologies, Springer, vol 958, no 1, pp 278–291 (2018)
15. Ahmad I, Namal S, Ylianttila M, Gurtov A (2015) Security in software defined networks: a survey. IEEE Commun Surv and Tutor 17(4):2317–2346, Forth- quarter
16. Bhushan K, Gupta B (2018) Distributed denial of service (DDoS) attack mitigation in software defined network (SDN)-based cloud computing environment. J Ambient Intell Humaniz Comput 10(5):1–13
17. Bavani K., Ramkumar MP, Emil Selvan GSR (2020) Statistical approach based detection of distributed denial of service attack in a software defined network. In: 2020 6th International conference on advanced computing and communication systems (ICACCS), IEEE, Coimbatore, Tamilnadu, pp 380–385
18. Sahoo KS, Panda SK, Sahoo S, Sahoo B, Dash R (2019) Toward secure software-defined networks against distributed denial of service attack. J Supercomput 75(8):4829–4874
19. Rebecchi F, Boite J, Nardin PA, Bouet M, Conan V (2019) DDoS protection with stateful software-defined networking. Int J Network Manage 29(1):20–42
20. Ahalawat A, Dash SS, Panda A, Babu KS (2019) Entropy based DDoS detection and mitigation in OpenFlow enabled SDN. In: 2019 International conference on vision towards emerging trends in communication and networking (ViTECoN), IEEE, Vellore, Tamilnadu, pp 1–5

A Review on Different Types of Attacks on RSA and ECC

Heisnam Sursita Devi, Ningthoukhongjam Tutu Raja, Yambem Jina Chanu, and Kh. Manglem Singh

Abstract With the advancement of technologies, there is a growing number of cryptosystems. Among them, RSA and ECC are two efficient Asymmetric key cryptosystems. As a result, it is suitable for use while using devices with low memory, like smart cards, mobile phones, and so on. However, during the cryptography process, there are moments when information about the private key is leaked. So, understanding different types of vulnerable attacks on RSA and ECC cryptosystems will help scientists, researchers, and developers in selecting the most appropriate design parameters of these systems. The study briefly overviews the different types of attacks on RSA and ECC.

Keywords Attacks on RSA and ECC · Power attack · Timing attack · Factorization attacks

1 Introduction

In the current era of digital technology, communication has evolved as essential to daily existence as food, housing, and clothing. It is everyone's lifeblood. To enable safe communication across an insecure medium, it is difficult to supply the necessary tools. The CIA triad, which refers to the three key security objectives of confidentiality, integrity, and availability, is a term used to describe the value of information. These security objectives can be achieved in various ways, but cryptography is by far the most common and all-encompassing method.

The RSA System was created by Leonard Max, Adi Shamir, and Ronald Linn Rivest in 1978 [1]. This system's discrete exponential problems and solid foundation in number theory make it popularly used in public key (Asymmetric) cryptosystems. Huge integers that are the product of two large prime integers are difficult to factor, which gives RSA its security. The cryptosystem is most frequently used to protect user information and guarantee the integrity of digital data [1].

H. S. Devi (✉) · N. T. Raja · Y. J. Chanu · Kh. Manglem Singh
NIT Manipur, Imphal, Manipur, India
e-mail: sitaheisnam@gmail.com

S. K. Biswas et al. (eds.), *Intelligent Computing Systems and Applications*, Lecture Notes in Networks and Systems 1307, https://doi.org/10.1007/978-981-96-3860-4_32

Neil Koblitz and Victor S. Miller developed elliptic curve cryptography (ECC) in 1985 [2]. It is a new class of public key algorithms that operate with smaller keys than RSA while maintaining equivalent strength and operating more quickly [2].

Numerous researchers have examined the RSA and ECC systems for vulnerabilities since their initial disclosure. Even though many exciting attacks have resulted from years of research, none of them has been very catastrophic. Instead, they largely serve to highlight the risks associated with improper usage of RSA and ECC. This paper's main goal is to shed light on very few of these attacks and outline the important mathematical techniques used to carry out these attacks. Throughout the research, we use "Alice" and "Bob" to represent two random parties who want to communicate with each other, and "Eve" to represent a malevolent attacker who wants to interfere with the communication between Alice and Bob.

The paper is organized as follows. The process involved in RSA is described in Sect. 2. In the next Section, various attacks on RSA are described in brief. Then, we briefly describe ECC in Sect. 4. In Sect. 5. Side-channel attacks on ECC are discussed, and in the next section, we present a comparative. In Sect. 7 Fault analysis attacks on ECC are presented and in Sect. 8 some of the common applications of RSA and ECC are presented in brief. Lastly, we concluded the paper in Sect. 9.

2 RSA Process

An RSA cryptosystem involves three processes, namely, key generation, encryption, and decryption algorithms are explained in the following.

2.1 RSA Key Generation

We began by considering

$$n = p.q \tag{1}$$

Here, p and q are two prime numbers. n is called the RSA modulus. Then the sender Bob computes the Euler totient function.

$$\varphi(n) = (p-1)(q-1). \tag{2}$$

An odd integer e is the selected such that it is coprime to $\varphi(n)$, that is, $gcd\,(e, \varphi(n)) = 1$.

The pair of integers (n, e) is the public key. e is called the encryption exponent [3]. Then Bob computes another integer d such that

$$1 < d < \varphi(n)\, d \cdot e \equiv 1 \bmod \varphi(n). \quad (3)$$

Assuming M *is a* message that is sent by Alice to Bob. It is an integer, and $M \in N$. To encrypt M, Alice computes $c = M^{e} mod\, n$. And to decipher the ciphertext, Bob computes $M = c^{d} mod\, n$ [4].

Algorithm 2.1 (RSA key generation)
{
Input: Two prime numbers $p\ and\ q$.
Output: $e\ and\ d$.
Select $p\ and\ q$ two large prime integers such that $p\ and\ q$ are not equal
Compute $n = p \cdot q$;
Compute $\varphi(n) = (p-1) \cdot (q-1)$;
Select e such that $gcd\ (e, \varphi(n))$;
Find d such that $e \cdot d \equiv 1\ mod\ \varphi(n)$;
Public key= (e, n);
Private key =d;
Return Public key and Private key;
}

2.2 RSA Encryption

Those who are aware of the public key (e, n) can encrypt a plaintext M with $0 \leq M < n$ producing the encrypted text c shown below:

$$c = M^{e} mod\, n \quad (4)$$

Encryption is done in blocks with a block size of n blocks, each of which has a binary value less than n, that is, block size $\leq log_2(n) + 1$. If the block size is i, then $2^i < n \leq 2^{i+1}$ [4].

Algorithm 2.2 (RSA encryption)

Input: Public key (e, n), Plaintext M such that $M < n$ and $gcd(M, n) = 1$;
for each triplet of the form (e, n, M)
    Compute $c = M^{e}\ mod\ n$;
end
return Ciphertext c;

2.3 RSA Decryption

The decryption is given as follows:

$$M = c^{d} mod\, n \quad (5)$$

for any integer M with $0 \leq M < n$.

The encryption and the decryption are inverse operations.

Since

$$e \cdot d \equiv 1\, mod\, \varphi(n),$$

So we have

$$e \cdot d = k \cdot \varphi(n) + 1$$

for some integer $k. \geq 1$

$$\begin{aligned}(M^e)^d &\equiv M^{k\varphi(n)+1} mod\, n \\ &\equiv M^{k\varphi(n)} M\; mod\, n \\ &\equiv 1^k M\; mod\, n \\ &\equiv M\; mod\, n\end{aligned}$$

Hence proved.

The decryption algorithm is described below:

Algorithm 2.3 (RSA decryption)

Input: Private key d, Public modulus n and Ciphertext c;
for each triplet of the form (d, n, c)
    $M = c^d\ mod\ n$;
end for;
return decipher text M;

3 Attacks on RSA

The majority of attacks in this category benefit from the unique characteristics of the RSA function. Usually, they get an advantage of system abuse, incorrect selection of the private or public exponent, the relationship between the ciphertexts, poor padding, and many other things.

3.1 *Chosen Ciphertext Attack*

In this attack, the cryptanalyst selects a ciphertext to retrieve information, at least in parts, and obtains the decryption using an unidentified key. Let us say Alice sends Bob the message m encrypted as ciphertext $c = m^e$ mod n. The following actions are taken by Eve to decode the original file when she intercepts c and send it to Bob.

1. Eve choose a random integer $x \in Z$.
2. She computes $y = c \cdot x^e modn$.
3. She sends y to Bob for decryption and gets $z = y^d mod\ n$.

Eve finds the plaintext m because

$$z = y^d mod\ n = \left(c \cdot x^e\right)^d mod\ n = \left(c^d \cdot x^{ed}\right)$$
$$mod\ n = \left(c^d \cdot x\right)$$
$$mod\ n = (m \cdot x) mod\ n$$

Thus finds, $M = z \cdot x^{-1} mod\ n$.

It resembles a chosen plaintext attack in many ways, although it targets a decryption function rather than an encryption mechanism. An instance of a chosen plaintext attack is that the attacker has the option to select plaintexts and read the ciphertexts that go with them. In comparison to chosen plaintext assaults, the attack is thought to be less useful in practical applications. However, the difficulty of chosen plaintext and chosen ciphertext attacks does not directly correlate. A cipher could be susceptible to one attack but not to another, or vice versa. A key scenario in public key cryptography is the selected ciphertext attack, in which the attacker always has access to known plaintext and even chosen plaintext situations because the encryption key is known to all parties.

3.2 *Factorization Attacks*

In a factorization attack, p and q factors are sought after using the knowledge of n. The attacker can circumvent the system's security by posing as the key owners and use the stolen cryptographic data to decrypt sensitive information. This attack is directed at an RSA cryptography library, which is used to generate RSA Keys. Attackers can do this by gaining the target's public key, which they can then use to get the private keys to various security tokens, smartcards, and motherboard chipsets [5].

However, the factorization attack on RSA will be successful if we get additional information about the other value, such as the totient function $\varphi(n)$.

From Eqs. (1) and (2) we get

$$n - \varphi(n) + 1 = pq - (p-1)(q-1) + 1 = pq - pq + p + q - 1 + 1 = p + q$$

So, if we know $n = pq$ and $n - \varphi(n) + 1 = p + q$, then we can form a quadratic equation that has roots p and q.

$$x^2 - (p+q)x + pq = x^2 - (n - \varphi(n) + 1)x + n = (x-p)(x-q) = 0.$$

3.3 Hastad's Broadcast Attack

When the same plaintext to be encrypted is delivered to different persons after encryption using multiple moduli, but with the same and low encryption exponent, Hastad's broadcast attack may arise. Small public key e values, like 3, are regarded cryptographically unsafe since an attacker may readily decrypt the encrypted content [6]. Alice may send different encrypted texts of the same message text m using public keys (e, n_i) of the same small exponent e and different moduli n_i to different people as below:

$$\begin{aligned} c_1 &= m^e mod n_1 \\ c_2 &= m^e mod n_2 \\ &\vdots \\ c_i &= m^e mod n_i \end{aligned}$$

where $gcd\left(n_i, n_j\right) = 1$ for $i \neq j$.

An attacker can compute a solution to the following system

$$\begin{aligned} x &= c_1 mod\ n_1 \\ x &= c_2 mod\ n_2 \\ &\vdots \\ x &= c_i mod\ n_i \end{aligned}$$

The solution of x is $c\ mod n_1 n_2 \ldots n_i$. The original message could be computed by finding the eth root of c. The plaintext can be deduced in polynomial time, $log(n)$.

3.4 Franklin and Reiter Related Message Attack

Franklin and Reiter discovered a small public exponent attack that is based on the encryption of two related messages m_1 and m_2 with the shared modulus n and very small encryption exponent e, such that $m_1 = f(m_2) mod n$, where $m_1 \neq m_2$ and $f = ax + b$ with $b \neq 0$, producing the ciphertexts c_1 and c_2 [7].

Suppose the ciphertext c_1 is sent by Alice to Bob. Eve, posing as Bob, conveys to Alice that he did not receive the message, so he requested to resend it. This time, Alice slightly alters the message with a different timestamp and resends the ciphertext c_2 to Bob and also to Eve. Now, two ciphertexts of the two related plaintexts are in the hands of Eve. Franklin and Reiter [8] showed how to recover the plaintext from the known information (n, e, c_1, c_2, f) in time polynomial in $log(n)$ as following:

$$Since,\ c_1 = m_1^e\ mod\ n = (am_2 + b)^3 mod\ n = a^3m_2^3 + 3a^2m_2^2b + 3am_2b^2 + b^3$$
$$a^3m_2^3 - 2b^3 + 3b\left(a^2m_2^2 + am_2b + b^2\right) \tag{6}$$

$$Therefore,\ 3b\left(a^2m_2^2 + am_2b + b^2\right) = c_1 - a^3c_2 + 2b^3, Since\ c_2 = m_2^3\ mod\ n$$

Multiplying both sides by $(am_2 - b),\ we\ get$

$$(am_2 - b)\left(c_1 - a^3c_2 + 2b^3\right) = 3b(am_2 - b)\left(a^2m_2^2 + am_2b + b^2\right)$$
$$3b\left((am_2)^3 - b^3\right),\quad Since$$
$$(am_2)^3 - b^3 = (am_2 - b)\left(a^2m_2^2 + am_2b + b^2\right) = 3b\left(a^3c_2 - b^3\right)$$

Since Eq. (6).
Therefore, we get

$$m_2 = \frac{1}{a}\left(\frac{3a^3bc_2 - 3b^4}{c_1 - a^3c_2 + 2b^3} + b\right) = \frac{b}{a}\left(\frac{c_1 + 2a^3c_2 - b^3}{c_1 - a^3c_2 + 2b^3}\right) modn.$$

We get m_1 from $m_1 = am_2 + b.m_1 = am_2 + b.$

To obtain the result, all the calculations are computed in polynomial time $log(n)$. We can see that, when $e = 3$, it is sure that we cannot send messages that are linearly related. In the case of $e \neq 3$, the computation may result in a non-linear form, and the attack fails in such a case.

3.5 Short Pad Attack

For a more secure, it is better to pad the message before encryption, which transforms the plaintext m_2 in the plaintext $m_1 = m_2 + b$, where b is usually a random number. Even in such a case, it is possible to recover the original message if the encryption exponent that is $e = 3$ and the $b's$ absolute value is negligible, as shown in [9]. The value of b should satisfy $|b| < n^{\frac{1}{9}}$. The value of m_2 from the previous section is

$$m_2 = b\left(\frac{c_1 + 2c_2 - b^3}{c_1 - c_2 + 2b^3}\right) modn \tag{7}$$

Putting the value of m_2 from Eq. (6) in the above, we get

$$0 = b^9 + (3c_2 - 3c_1)b^6 + \left(3c_2^2 + 21c_2c_1 + 3c_1^2\right)b^3 + (c_2 - c_3)^3 modn$$

We can find the value of m_1 and m_2 for related messages after solving the monic polynomial of degree 9 for the value of b.

3.6 Cycling Attack

The cycling attack is regarded as one of the earliest attacks on the RSA cryptosystem [10]. This attack encrypts the ciphertext repeatedly, as the name suggests. The ciphertext of any plaintext is a permutation of all the letters in the range from 0 to $n-1$ [5]. Eve will get the plaintext if she continuously encrypts the ciphertext c. However, she has never seen the plaintext, she does not know when to stop. Therefore, she should find the smallest integer x that satisfies

$$c = c^x mod\ n \tag{8}$$

where $x = e \cdot d = k \cdot \varphi(n) + 1$ for some positive integer k. We can find both d and the Euler totient function $\varphi(n)$ of the modulus n.

So, the decryption exponent $d = \frac{x}{e} = \frac{k \cdot \varphi(n)+1}{e}$.

3.7 Short Message Attack

In the brief message attack, it may be assumed that the attacker is familiar with some plaintext blocks. If this presumption is accurate, the attackers can try encrypting each block of plaintext data to see if it produces the known cipher text.

It is advised that in order to prevent this short message attack, it should pad the plain text before encrypting it.

The RSA cryptosystem appears ineffective if the message m and the encryption exponent e are comparatively very much smaller than the modulus n, that is, $c = m^e < n$. The original message m can be recovered by taking the *eth* root of the ciphertext c. For protection against such an attack, the encryption exponent e should be large, the message m should be large, or both should be large.

Short message attack [4] can be thwarted using Optimal asymmetric encryption padding (OAEP).

3.8 Common Modulus Attack

When the identical message is encrypted using two separate coprime encryption exponents and the one modulus to produce two different ciphertexts [11], The message is decrypted using the standard modulus attack and the plaintext. In another word, it is possible to retrieve the original plaintext from the attacker Eve if Alice encrypts the same plaintext with two different encryption keys e_1 and e_2 with $gcd(e_1, e_2) = 1$, and the same modulus n, producing the ciphertexts $c_1 = m^{e_1} modn$ and $c_2 = m^{e_2} modn$. The recovered plaintext is given by

$$m = c_1^a \cdot c_2^b mod n \tag{9}$$

where $a \cdot e_1 + b \cdot e_2 = 1$, and the values of a and b can be obtained by using the extended Euclidean algorithm. We can compute m in polynomial time $log(n)$ [12].

3.9 Wiener's Low Decryption Exponent Attack

Wiener's attack takes advantage of using a small decryption exponent d to break the RSA cryptosystem under the conditions $q < p < 2q$ and $d < \frac{1}{3}n^{\frac{1}{4}}$, where $n = pq$ [13]. d can be recovered from the congruence $e \cdot d \equiv 1 mod \varphi(n)$, which is equal to $e \cdot d - k \cdot \varphi(n) = 1$, where the Euler totient function $\varphi(n) = (p-1)(q-1)$, and k is an integer. We have

$$\left|\frac{e}{\varphi(n)} - \frac{k}{d}\right| = \frac{1}{d \cdot \varphi(n)}$$

Since $n = p \cdot q > q^2$, so we have $q < \sqrt{n}$ and $0 < n - \varphi(n) = p + q < 2q + q - 1 < 3\sqrt{n}$, So, we have

$$\left|\frac{e}{n} - \frac{k}{d}\right| = \left|\frac{e \cdot d - k \cdot n}{d \cdot n}\right| = \left|\frac{1 + k(\varphi(n) - n)}{d \cdot n}\right| < \frac{3 \cdot k \cdot \sqrt{n}}{d \cdot n} < \frac{3 \cdot k}{d \cdot \sqrt{n}}$$

Since $k < d$, so we have $3 \cdot k < 3 \cdot d < n^{\frac{1}{4}}$, and hence

$$\left|\frac{e}{n} - \frac{k}{d}\right| < \frac{1}{d \cdot n^{\frac{1}{4}}}$$

Finally, since $3 \cdot d < n^{\frac{1}{4}}$, we have

$$\left|\frac{e}{n} - \frac{k}{d}\right| < \frac{1}{3 \cdot d^2}$$

This implies that $\frac{k}{d}$ is a convergence of the ongoing fraction expansion of the fraction $\frac{e}{n}$. The total number of convergents of $\frac{e}{n}$ is $O(n)$.

Let $\frac{a}{b}$ be the convergent of $\frac{e}{n}$. If it is correct for $\frac{k}{d}$, then the totient function can be computed as

$$\varphi(n) = (p-1)(q-1) = \frac{be - 1}{a}$$

Now $p + q = pq - (p-1)(q-1) + 1$ and $q - p = \sqrt{(p+q)^2 - 4pq}$.
If both $p + q$ and $q - p$ are positive, then $\frac{a}{b}$ is the correct guess for $\frac{k}{d}$.

Durfee improved the Wiener's attack to show that a private exponent $d < n^{.292}$ rendering the system insecure [14].

3.10 Partial Key Exposure Attack

The complete private exponent d can be recovered, and the modulus n can be factored if the encryption exponent e is small and a fraction of d, say $\frac{N}{4}$ (where N is the number of bits of n) least significant bits of d is available.

We know

$$ed - k\varphi(n) = ed - k(p-1)(q-1) = ed - k(n-s+1) = 1,$$

where $s = p + q$ From Eq. (2).

$$\therefore ed_0 \equiv 1 + k(n-s+1) mod 2^{\frac{N}{4}},$$

where d_0 is $\frac{N}{4}$ least significant bits of $d$and $d \equiv d_0 mod 2^{\frac{N}{4}}$

We know that

$$p^2 - sp + n \equiv 0\, mod\; 2^{\frac{N}{4}},\; 2^{\frac{N}{4}}\; Since,\; p^2 - sp + n = p^2 - (p+q)p + n = p^2 - p^2 - pq + n = 0$$

We solve the above equation to obtain a candidate value for $pmod 2^{\frac{N}{4}}$. We can find the corresponding $q_0 \equiv \frac{n}{p_0}$ value.

Let $r \geq 2^{\frac{N}{4}}$ and $p_0 \equiv pmodr$. Then we can find $q_0 = q \equiv \frac{n}{p_0} modr$. We can create a polynomial

$$f(x, y) = (rx + p_0)(ry + q_0) - n$$

The solution (x_0, y_0) will reveal the factorization of n. The running time is $O(en^{\varepsilon})$ for some ε.

3.11 Timing Attack

An attacker can find the decryption exponents working backward if he can get the ciphertext and the amount of time taken to decipher it by the recipient's computer [15]. This attack is derived from the modular exponentiation algorithm. It is computed bit by bit with the constraint of one squaring whenever the corresponding bit in the decryption exponent results in 0, or else with one squaring and one multiplication whenever the corresponding bit gives 1. This indicates that whenever the corresponding bit is 1, the time taken in each iteration gets longer. The attacker thus finds the value of the decryption exponent d from the timing difference.

To stop the timing attack, until now there exist two methods. The first method adds a random delay in the exponentiation during the encryption for the bit 0 so that the same amount of time is taken for both bit 0 and bit 1. The second is the blinding, as recommended by Rivest, before the decryption process starts, each ciphertext is multiplied by a random number.

3.12 Power Attack

Power attack is a derivation of the timing attack with the exception that the attacker can determine whether more power is used to decrypt data if the bit value is 1. The same methods used to thwart the timing attack can be used in the power attack.

4 Elliptic Curve Cryptography (ECC)

ECC is a replacement class of public key algorithm which is able to be implemented using smaller keys than RSA while keeping similar strength with faster speed. Discrete logarithms provide ECC's strength over an elliptic curve group. It's related that it was employed by mathematicians to go looking out the formula for arc length of an ellipse.

A Weierstrass equation is used to describe an elliptic curve E over a field K:

$$E : y^2 + a_1xy + a_3y = x^3 + a_2x^2 + a_4x + a_6 \tag{10}$$

in which a_1, a_2, a_3, a_4, a_6 are constant and $\Delta = 0$. Here, Δ represents the E's discriminant. Equation (10) may be simplified with the aid of using making use of an alternate of coordinates. If characteristic of K is not identical to 2 or 3, then E can be written as

$$y^2 = x^3 + ax + b \tag{11}$$

where a, b are constant. If characteristic of $K = 2$, then E can be written as

$$y^2 + xy = x^3 + ax^2 + b \tag{12}$$

if E is non-super singular.

ECC Key Generation

A crucial step in a cryptographic system is the key generation. Where, in ECC, it is required to generate two keys a public and a private key. In an open communication, usually, with the help of the recipient's private key, the recipient will be able to decode the communication that was encrypted by the sender using the recipient's public key.

Now, a number 'd' is to be selected within the range of 'n'. With the help of the following equation, we can determine the public key

$$Q = d * P \tag{13}$$

where P is a point on the curve, 'Q' is the public key, and 'd' is the private key.

ECC Encryption

Let 'm' be the message being sent to the recipient. In order to map this message on the elliptic curve. Let's say 'm' has the point 'M' on the elliptic curve 'E'. And randomly 'k' is selected in between (1– (n–1)).

Two cipher texts, namely, C_1 and C_2 will be generated.

$$C_1 = k * P$$
$$C_2 = M + k * Q$$

C_1 and C_2 are the cipher text.

Decryption

The message 'M' that was sent to us must be returned.

$$M = C_2 - d * C_1.$$

4.1 *Scalar Multiplication*

An abelian group can be defined as a pair of a point (x, y) together with the point which is present at infinity on the curve E. Point multiplication is considered to be the result of the operation P, where k is a scalar and point $P \in E(K)$ and both of them are provided. The dilemma of Elliptic Curve Discrete Logarithm Problem (ECDLP) involves determining k where two specified positions P and Q are given in such a way that $Q = kP$ is the foundation for the security of ECC.

When performing elliptic curve scalar multiplication (ECSM), k information can be leaked in numerous ways. The attacker wants to use physical attacks to recover the whole bit stream of k. Physical attacks are classified into two primary types: a) side-channel analysis (SCA) and b) fault analysis (FA). We will quickly go through the SCA (passive attacks, also known as known SCA) and FA against an ECC implementation.

5 Side Channel Attacks on ECC

A cryptographic system can be attacked using the extra information or side effects of elliptic curve cryptography technology. Additional data about power usage, electromagnetic radiation, and time can lead to side-channel attack. SPA (Simple Power Analysis), DPA (Differential Power Analysis), SEMA (Simple Electromagnetic Analysis), and DEMA are examples of side-channel assaults (Differential Electromagnetic Analysis).

We will quickly go through some of the SCA (also known as passive attacks) against an ECC in this section.

5.1 Power Analysis Attack (PAA)

Paul Kocher first presented the power analysis method in 1996, and then the PAAs in 1999 [15]. These assaults have grown to pose a serious threat to tamper-resistant technology. By looking at the power usage traces, PAA gives attackers the ability to discover the cryptographic device's secret key or some of its information. Particularly for mobile devices like WSNs, smart cards, mobile phones, RFIDs, etc., this poses a severe concern [14].

5.2 Simple Power Analysis Attack (SPAA)

By examining the power utilization from a single measure trace and taking advantage of side-channel leakage information, the SPAA seeks to discover the secret key d [16].

5.3 Differential Power Analysis Attack (DPAA)

In DPAA, the attacker uses statistical approaches to pull the secret information out of the obvious differences in power usage generated by numerous data and operational computations [16].

5.4 Refine Power Analysis (RPA)

Refine power analysis attack (Goubin-type DPA) attacks solely by concentrating on the presence of a point P_0 on E(K), where K and $P_0 \neq O$ agrees that 0 is present in one of the coordinates. By estimating the power taken of a chosen message and some selected locations on the elliptic curve, RPA might be able to determine the next bit of the scalar [17].

5.5 Zero Point Analysis (ZPA)

The idea behind the zero-point analysis approach is that even if a point doesn't have a coordinate with a zero value, the auxiliary register might. It is a variation on the RPA. As a result of this attack, all spots that consume no power are visible [18].

5.6 Doubling Attack (DA)

Doubling attack (DA) [14] relies on two queries P and $2P$. By using comparable point doubling (PDBL) operations to compute dP and d, the DA can identify situations in which the same operation is performed twice ($2P$). There are two variants of DA: regular and relative DA (relative doubling attack suggested by Yen et al. in [15]). Relative DA employs a completely new method to generate the key bit.

5.7 Template Attack

A fully controllable device is necessary for the template attack and proceeds in two phases, which is very similar to differential power analysis assault (two-round technique: template building and matching). The attacker characterizes the noise and builds a precise model of the intended signal source during the template construction phases (also known as the profiling phases). The actual attack takes place in the second stage. Template attacks take advantage of the so-called differential SCA kind of leakage since they are data-dependent. The attack presupposes that the variance contains the majority of the side-channel data [19].

5.8 Carry-Based Attack (CBA)

CBA targets the ECSM's defenses rather than the ECSM itself. The CBA is dependent on propagation that happens when performing repeated sub-word additions of long integers [19].

5.9 Address-Bit DPA

The relation of the register address and the key is examined by the address-bit attack (ADPA) [37]. Itoh et al. [25] are the ones to apply the first ADPA to ECC. For instance, an algorithm implementation. Regardless of the key bit's value, 1 in [25] accomplishes point addition and doubling, However, the doubled point's address is completely reliant on k_i. As a result of this, k_i can be retrieved if the attacker can detect the difference of the data that is read from R [0] and R [1].

5.10 Simple Electromagnetic Analysis Attack (SEMA)

In SEMA attack, the attacker straight away determines parts of the game key by using the information from a single electromagnetic radiation measurement. While an algorithm incorporates a conditional branch that causes a brand-new radiation pattern every time the department is taken, an SEMA assault is regularly done. On an implementation without protection, an SEMA attack on a single shot can find all the key bits. It is far from impossible to detect electromagnetic radiation from a clever card, as verified by means of Quisquater and Samyde at [25].

There are two varieties of radiations, intentional and accidental, according to Agrawal et al. [26, 27]. Direct contemporary flows purpose the first type. The second type is introduced by distinct couplings, FM and AM modulations, and so on. Exploring unintentional radiations is in which EM actually benefits above other facet channel assaults [26, 27]. EM leakage genuinely incorporates a number of channels. The SEMA attack is applied to the EC processor described in [28, 29], which employs a conditional branch algorithm for EC factor multiplication. The conditional department inside the set of rules became what made the SEMA assault successful.

5.11 Differential Electromagnetic Analysis Attack (DEMA)

A DEMA attack makes use of the property that special statistics processing calls for an extraordinary amount of strength and radiates an exclusive subject. Multiple measurements are employed to eliminate background noise in the DEMA attack, and the secret key d is obtained via statistical analysis. De Mulder et al. in [23] present a DEMA attack on an EC processor's FPGA (Field Programmable Gate Array) implementation. The attackers employ a simulated version of the attacked device.

This version is used to estimate the expenses of a device's electromagnetic radiation when compared to the device's actual measured electromagnetic radiation. Statistical procedures are used to compare data. Among others, the most common are the deviation from the mean check and the correlation analysis. The version predicts the amount of internal channel leakage within a particular runtime for the correlation analysis. These forecasts correspond to actual electromagnetic radiation.

6 Comparative Side Channel Attacks on ECC

The difference between a simple SCA and a differential SCA is called a comparative SCA [37]. The reuse of values is determined by comparing two segments of the same or separate leakage traces. The double attack [38] is the first attack of this type to be officially recorded.

The concept of the doubling attack is that even if the attacker is blind to the operation being performed, he can nevertheless detect when the same operation is performed twice. When multiplying two points, such as $2P$ and $2Q$, the intruder might not be aware of the values of P and Q, but he can determine if $P = Q$. It is possible to recover all the bits of k by differing two power traces, one for kP and one for $k(2P)$.

7 Fault Analysis Attacks on ECC

In addition to side-channel analysis, an attacker can purposefully interfere with the cryptographic hardware to discover the secret. A faulty clock, a loss of power, and other tools can be used to cause faults on the victim device. The division of fault attacks against the ECC into three types is based on the scalar recovery mechanism. The three types are:

(a) Safe-error based analysis
(b) Weak-curve-based analysis.
(c) Differential fault analysis.

7.1 Safe-Error Analysis

It was Yen and Joye [30, 31] who first introduced the idea of safe-error. They discussed the C safe-error and M safe-error as two different sorts of safe-error.

C safe-error: The bogus procedures that are routinely included to increase SPA resistance are used in this attack. As an illustration, consider the add-and-double-always algorithms [32, Alg. 1], where step 3 dummy addition allows for safe-error. When the dummy point addition is being performed, the enemy can create momentary errors. The end outcomes will be incorrect if the scalar bit $k_i = 1$. Aside from that, there is no impact on the outcome. By ensuring that the results are accurate, the adversary can regain k_i.

M safe-error: This attack takes advantage of some memory blocks being present are erased incorrectly. Yen and Joye [30] were the ones who first suggested this assault on RSA. But that also holds true for ECSM.

7.2 Weak Curve-Based Analysis

Biehl et al.'s study on a weak curve fault attack against an ECC implementation was published in 2000 [33]. The crucial finding is that the addition formulae do not employ the variable a_6 from the definition of E in Eq. (1). Accurate answers were produced by the addition equations for the curve E that deviates from E exclusively in a_6.

An ECC processor can be deceived by the adversary by using a point $P \in E(F)$, where E a vulnerable curve exists in terms of cryptography. The attacker can determine k once ECDLP E has been solved. Fault induction is usually necessary when moving a scalar multiplication from a strong curve E to a weak curve E. By employing flawed points [33], invalid curves [35], and twist curves [34], the adversary leverages faults to strike a weak curve.

Invalid point attacks: In this attack, a starting point for scalar multiplication can be found at the point P on the weak curve E. If kP is executed without first verifying P's accuracy, no defects must be introduced. The attacker will attempt to shift the point P once it has been verified by the ECC processor. To accomplish this, an attacker would need to create a defect at a particular period.

Invalid curve attacks: By relaxing the restrictions on the insertion of a defect, Ciet and Joye in [33] improved the attack. They demonstrate how any flaws, including those in the parameters for the field representation and curve $a1, a2, a3, and\ a4$ may result in the scalar multiplication being carried out on a weak curve.

Twist curve-based FA: Many cryptographically robust curves have weak twist curves, as Fouque et al. [34] discovered in 2008. When a scalar multiplication is performed without making use of the y-coordinate, the results for points on the given curve E and its quadratic twist are accurate, but the outcomes for weak twists can leak k. A random $x \in Fp$ on a prime field-defined elliptic curve Fp has a 50% likelihood of being a point on E or its twist. The likelihood that a random fault on P's x-coordinate will strike a location on the (weak) twist curve is therefore 50%.

7.3 Differential FA

Using the difference between the accurate findings and the flawed results, the Differential Fault Attack (DFA) can infer some details about the scalar. Muller Biehl et al. in [33] published the first DFA on an ECSM. In order to describe this attack, they employ a right-to-left multiplication technique. The values of Q and R at the conclusion of the i th iteration are shown by the symbols Q_i and R_i, respectively. Let k(i)

= *k* / 2i. With flaws induced, let Q_i represent the value of Q. Listing from the Most Significant Bits (MSB) to the Least Significant Bits (LSB), the attack reveals k.

1. ECSM is run once, and the correct result (Q_{l-1}) is collected.
2. Repeat the ECSM to cause a one-bit flip to occur Q_i, where l − m ≤ i < l and m are small.
3. Recall that $Q_{l-1} = Q_i + (k(i)2i)P$ and $Q_{l-1} = Q_i + (k(i)2i)P$.

In order to create Q_i and Q_i, the attacker then makes an attempt all conceivable k(i) ∈ {0, 1,.., 2 m − 1}. The difference between the two $\{Q_i, Q_i\}$ values with the proper value of k(i) is only one bit. The left-to-right multiplication algorithm can also be attacked by a DFA. Furthermore, it is stated that a problem introduced at random moments during an ECSM is sufficient for the attack [33].

7.4 *Sign Change FA*

Blomer et al. in [36] presented sign change fault (SCF) attacks in 2006. A scalar is intended for those implementations when the non-adjacent form is used for the encoding. A point's sign only affects its y-coordinate when utilizing curves constructed over the prime field. P is always a valid point because the SCF attack doesn't call for operations of the elliptic curve to stray from the initial group $E(Fp)$.

8 Applications

For a large message, the RSA method is highly sluggish. This is why it is used to encrypt and authenticate short messages, as well as to produce digital signatures. The certificate for the remote server to which we have connected is validated using RSA. Once our browser has validated the certificate chain (using RSA signature functionality), it will most likely utilize RSA to accomplish a secure key exchange with the server, allowing us to connect securely with a distant server through HTTPS.

Asymmetric encryption solves the issue of a secure exchange of ciphers by allowing users to freely share a public encryption key to everyone and keeping their private key protected within their company.

Elliptic curve cryptography is used in public key encryption, the generation of digital signatures, factorization, and primality testing.

9 Conclusions

In this research, we provide a thorough analysis of attacks against asymmetric key cryptography, RSA, and ECC. This paper makes the following contribution: First, we have gone through the various forms of RSA attacks, and then we'll go over ECC attacks briefly. RSA and ECC have been the subject of extensive research for many years, and while some interesting attacks have been discovered, no severe attacks have yet been discovered. This article should prompt further research on ECC and RSA in the near future in order to implement countermeasures.

References

1. Rivest RL, Shamir A, Adleman L (1978) A method for obtaining digital signatures and public-key cryptosystems. Commun ACM 21(2):120–126
2. Singh AK (2014) A review of elliptic curve based signcryption schemes. Int J Comput Appl 102(6):8887
3. Forouzan BA, Mukhopadhyay D (2010) Cryptography and network security, Tata McGraw Hill, 2nd edition
4. Cavallar S et al (1999) Factorization of RSA-140 using the number field sieve. In: international conference on the theory and application of cryptology and information security. Springer, Berlin, Heidelberg
5. Hastad J (1985) On using RSA with low exponent in a public key network. In: lecture notes in computer science
6. Jason Hinek M (2010) Cryptanalysis of RSA and its variants. Chapman Hall
7. Yacobi O, Yacobi Y (2006) A new related message attack on RSA. Essays in Memory of Shimon Even, pp 187–195
8. Coppersmith, D (1997) Small solutions to polynomial equations, and low exponent RSA vulnerabilities. J Cryptol
9. Simmons GJ, Norris MJ (1977) Preliminary comments on the MIT public-key cryptosystem. Cryptologia
10. Simmons GJ (1983) A privacy protocol using the RSA crypto algorithm. Cryptologia
11. Salah IK, Darwish A, Oqueli S (2006) Mathematical attacks on RSA cryptosystem. J Comput Sci 2(8):665–671
12. Wiener MJ (1990) Cryptanalysis of short RSA secret exponents. IEEE Trans Inform Theory 36:553–558
13. Boneh D, Durfee G (2000) Cryptanalysis of RSA with private keys less than. IEEE Trans Inf Theory 46(4):1339–1349
14. Kocher P (1996) Timing attacks on implementations of Diffie-Hellman. RSA, DSS, and other systems, Science 1109:104–113
15. Houssain H, Badra M (2012) Power analysis attacks on ECC: a major security threat. (IJACSA) Int J Adv Comput Sci Appl
16. Kocher P, Jaffe J, Jun B (1999) Differential power analysis. In: proceeding advances cryptology—CRYPTO'99, Santa Barbara, CA, vol 1666, pp 388–397
17. Goubin L (2003) A refined power-analysis attack on elliptic curve cryptosystems. In: proceedings of PKC 2003, LNCS 2567. Springer Berlin / Heidelberg, pp 199–211
18. Akishita T, Takagi T (2005) Zero-value register attack on elliptic curve cryptosystem. IEICE Transactions 88-A (1):132–139
19. Fouque P, R´eal D, Valette F, Drissi M (2008) The carry leakage on the randomized exponent countermeasure. In: cryptographic hardware and embedded systems—CHES, ser. LNCS. Springer, vol 5154, pp 198–213

20. Junfeng F, Verbauwhede I (2012) An updated survey on secure ECC implementations: attacks, countermeasures and cost. In: Naccache D (ed): Quisquater Festschrift, LNCS 6805, Springer-Verlag Berlin Heidelberg, pp 265–282
21. Junfeng F, Guo X, De Mulder E Schaumont P, Preneel B, Verbauwhede I State-of-the-art of secure ECC implementations: a survey on known side-channel attacks and countermeasures. 978-1-4244-7812-5/10/$26.00 c 2010 IEEE
22. Loiseau A, Lecomte M, Fournier JJA Template attacks against ECC: practical implementation against Curve25519. 978-1-7281-7405-1/20/$31.00 c 2020 IEEE
23. De Mulder E, Buysschaert P, Ors SB, Delmotte P (2005) Electromagnetic analysis attack on an FPGA implementation of an elliptic curve cryptosystem. EUROCON
24. Kadir SA, Sasongko A, Zulkifli M (2011) Simple power analysis attack against elliptic curve cryptography processor on fpga implementation. In: 2011 international conference on electrical engineering and informatics. Bandung, Indonesia
25. Quisquater JJ, Samyde D (2001) Electromagnetic analysis (EMA): measures and countermeasures for smard cards. In: proceedings of the international conference on research in smart cards: smart card programming and security (E-smart), Attali I, Jensen T, eds, LNCS. Springer-Verlag, vol 2140, pp 200–210
26. Agrawal D, Archambeault B, Rao JR, Rohatgi P (2002) The EM side-channel(s): attacks and assessment methodologies. In: proceedings of the 4th international workshop on cryptographic hardware and embedded systems (CHES), Kaliski Jr BS, Koc CK, Paar C, eds, LNCS. Springer-Verlag, vol 2523, pp 29–45
27. Agrawal D, Archambeault B, Chari S, Rao JR, Rohatgi P (2003) Advances in side-channel cryptanalysis. RSA Laboratories Cryptobytes. Spring, vol 6, no 1, pp 20–32
28. Chari S, Rao JR, Rohatgi P (2002) Template attacks. In: proceedings of the 4th international workshop on cryptographic hardware and embedded systems (CHES), Kaliski Jr BS, Koc CK, Paar C eds, LNCS. Springer-Verlag, vol 2523, pp 172–186
29. Ors SB, Batina L, Preneel B, Vandewalle J (2005) Hardware implementation of an elliptic curve processor over GF(p) with montgomery modular multiplier. Int J Embed Syst
30. Yen S-M, Joye M (2000) Checking before output may not be enough against fault-based cryptanalysis. IEEE Trans Comput 49(9):967–970
31. Joye M, Yen SM (2002) The montgomery powering ladder. In: international workshop on cryptographic hardware and embedded systems. Springer, Berlin, Heidelberg
32. Coron JS (1999) Resistance against differential power analysis for elliptic curve cryptosystems. In: international workshop on cryptographic hardware and embedded systems. Springer, Berlin, Heidelberg
33. Biehl I, Meyer B, Müller V (2000) Differential fault attacks on elliptic curve cryptosystems. In: annual international cryptology conference. Springer, Berlin, Heidelberg
34. Fouque PA et al (2008) Fault attack on elliptic curve montgomery ladder implementation. In: 2008 5th workshop on fault diagnosis and tolerance in cryptography. IEEE
35. Ciet M, Joye M (2005) Elliptic curve cryptosystems in the presence of permanent and transient faults. Des Codes Crypt 36(1):33–43
36. Blömer J, Otto M, Seifert JP (2006) Sign change fault attacks on elliptic curve cryptosystems. In: international workshop on fault diagnosis and tolerance in Cryptography. Springer, Berlin, Heidelberg
37. Homma N, Miyamoto A, Aoki T, Satoh A, Shamir A (2008) Collision-based power analysis of modular exponentiation using chosen-message pairs. In: international workshop on cryptographic hardware and embedded systems. Springer, Berlin, Heidelberg, pp 15–29
38. Giraud C (2006) An RSA implementation resistant to fault attacks and to simple power analysis. IEEE Trans Comput 55(9):1116–1120

Genetic Algorithm-Based Energy-Efficient Charging Scheme for Rechargeable Wireless Sensor Network

Aaditya Lochan Sharma, Subash Harizan, and Biswaraj Sen

Abstract Due to its vast range of applications in various fields, Wireless Sensor Networks (WSNs) have received a lot of attention in the modern technological world. Limited energy source of the sensor impacts the lifetime of the network. Various techniques and approaches have been proposed in this regard. However, these techniques only extend the lifetime of the network to some extent. The evolution of Wireless Energy Transfer (WET) technology in recent years has resulted in the deployment of Wireless Rechargeable Sensor Networks (WRSN), in which sensors are equipped with rechargeable batteries that can be recharged using Wireless Mobile Charger (WMC). However, scheduling of WMC is one of the challenging issues and to recharge the sensor nodes is also a well-referenced research topic. To determine the ideal charging path for the WMC, a Genetic Algorithm (GA)-based energy-efficient on-demand charging scheme with a novel mutation operation is proposed in this work. The proposed novel mutation operation aids in improving performance and accelerating convergence with the goal of finding the optimal charging path for the WMC.

Keywords Wireless rechargeable sensor networks · Scheduling · Mobile charge · WET

1 Introduction

WSNs have acquired a lot of attraction in recent years due to their wide range of applications such as industry, agriculture, healthcare, military, environmental/ earth sensing, disaster management, and so on [1–5]. Each sensor has a battery that

A. L. Sharma (✉) · S. Harizan · B. Sen
Department of Computer Science and Technology, Sikkim Manipal Institute of Technology, Sikkim Manipal University, Sikkim, India
e-mail: aadityalochansharma@gmail.com

B. Sen
e-mail: biswaraj.s@smit.smu.edu.in

S. K. Biswas et al. (eds.), *Intelligent Computing Systems and Applications*, Lecture Notes in Networks and Systems 1307, https://doi.org/10.1007/978-981-96-3860-4_33

serves as its power source. As the batteries have limited energy once the battery runs out of charge, the sensor node dies, limiting the network's ability to operate indefinitely. Several techniques and approaches have been proposed to reduce energy consumption by sensor nodes in order to extend network lifetime [6, 7]. However, these solutions merely prolonged the network's limited lifetime. Some techniques suggested replacing the sensor node batteries or the sensor node itself [8]. However, this is not feasible for harsh environment.

For an on-demand charging challenge, the authors in [9] presented a deep reinforcement learning technique. The WMC can charge the SNs numerous times using this approach. A new metric known as charging reward is used, which allows WMC to respond to charging demand in real time while also transferring maximum energy to SNs. The authors in [10] have proposed a hybrid meta-heuristic technique for optimizing the charging schedule of multiple WMCs. The ideal charging schedule is determined using a combination of cuckoo search and genetic algorithm, with the goal of reducing charging delay and overall travel distance. The authors in [11] described a method in which a drone is employed as a WMC. The drones have a limited flight range, and multiple charging pads have been placed across the network. The base station finalizes the charging schedule after receiving the charging requests and provides SNs and pad coordinates to the drone.

Many techniques have been proposed to solve the problem of sensor node energy restoration, such as energy harvesting from the environment (solar, wind, etc.) [12]. Wireless energy transfer (WET) technology has been found to be a revolutionary approach. WET uses magnetic resonance coupling or radio frequency to wirelessly transmit energy from one device to another [13]. The WET technology was then combined with WSN to create the Wireless Rechargeable Sensor Network (WRSN), which has been intensively researched and studied in recent years.

The geographic location of the nodes is ignored in FCFC (First Come First Serve) [14], and requests are serviced according to their arrival time. The authors in [15] propose a broker node recharging scheme for surveillance-based wireless sensor networks, with the purpose of minimizing sensor node waiting time and WMC overhead. In CRS [16], the network is divided into grids of equal size, each containing a number of SNs, one of which is chosen as the cluster head. Data is collected and sent to the WMC by the cluster head. This scheme takes into account several WMC, with each WMC being assigned to a section of the grid. ACO (Ant Colony Optimization) algorithm is used to determine the WMC's optimal schedule. Tomar and Jana [17] combine two prominent MADM (Multi-Attribute Decision-Making) algorithms, TOPSIS (Technique for Order of Preference by Similarity to Ideal Solution), and AHP (Analytic Hierarchy Process) to provide a novel WMC charging scheduling scheme. Several factors are considered, including SN residual energy, distance to WMC, energy consumption rate, and neighborhood energy weightage as well as WMC's limited energy. The relative weights of the network properties are first determined using AHP. The attributes of each requesting node are then assessed, and TOPSIS is used to determine the most appropriate node to be charged.

The authors of [18] propose a novel pricing strategy based on spatial and temporal discretization as well as a cost-effective approximation method. A three-dimensional network region is considered in the network model, with an unmanned aerial vehicle (UAV) serving as the WMC. The authors in [19] proposed a bi-objective charge scheduling optimization problem with multiple WCVs that takes into account different spatio-temporal restrictions of WMC mobility and sensor charging. The WMCs are assigned to different regions using an offline algorithm. For heterogeneous WRSNs with multiple WMC, the authors developed an on-demand, multi-node, and partial charging scheme in [20]. The problem is known as the Charging Schedule Generation Problem (CSGP). To address this issue, a unique partial charging system is proposed that combines NSGA-II (Non-dominated Sorting Genetic Algorithm) and AHP-TOPSIS to optimize the WMC charging path. The authors of [21] propose an Importance-different Charging Scheduling (IDCS) technique based on matroid theory to maximize charging utility while minimizing data loss.

2 Related Model and Problem Formulation

In this section, we present the network model, problem definition, and LP formulation of the problem for the WRSN. We list the notations along with their description in Table 1.

Table 1 Notations

Notation	Meaning
$\boldsymbol{BS}$	Base station
$\boldsymbol{SN}$	Sensor node
$\boldsymbol{N}$	Number of SNs
$\boldsymbol{N_r}$	Number of SNs sending charging request
$\boldsymbol{s_{ei}}$	Initial energy of SN s_i
$\boldsymbol{s_{eri}}$	Residual energy of SN s_i
$\boldsymbol{s_{eci}}$	Energy consumption rate of SN s_i
$\boldsymbol{WMC}$	Wireless mobile charger
$\boldsymbol{E_C}$	Energy spent by WMC to charge node
$\boldsymbol{E_T}$	Energy spent by WMC in traveling
$\boldsymbol{S_{MC}}$	Speed of WMC
$\boldsymbol{dist_{bf}}$	Euclidean between BS and first SN to be charged
$\boldsymbol{dist_{lb}}$	Euclidean distance between last recharged SN and BS
$\boldsymbol{dist_{ij}}$	Euclidean distance between nodes s_i *and* s_j
$\boldsymbol{Th}$	Energy threshold value beyond which the SN sends the charging request
$\boldsymbol{S_{tour}}$	Set of tours
$\boldsymbol{\beta_{ji}}$	Boolean variable

2.1 Network Model

In this work, we assume an obstacle-free two-dimensional WRSN with N number of rechargeable SN, a BS, and a wireless mobile charger WMC with infinite energy. We expressed the WRSN as a weighted graph $G = (N \cup BS, L, S_{ei})$, where L denotes the link between any two SNs, D represents the Euclidean distance between any two SNs, and S_{ei} denotes the initial energy of the SN. The SNs are randomly deployed in the network area. SN depletes its energy for sensing, sending, and receiving of data. Once the energy of the SN falls below the define threshold value (Th) they send charging request to BS by single-hop or multi-hop communication. The charging request of the SN contains its location coordinate, residual energy (S_{eri}), and energy consumption rate (S_{eci}). Here, we adopted the same energy consumption model as given in [14]. On receiving the charging request from SNs, BS schedules the WMC according to planned path to recharge the requested SNs (Nr). Note, within this charging schedule if any SN sends the charging request, then they have to wait for the next charging schedule to be recharged. After recharging the Nr WMC returns back to BS.

2.2 Problem Definition

In the proposed work, problem definition can be defined as follows. Once the energy of the SN falls below the defined threshold value, they can send charging request to BS. On receiving the charging request, BS schedules the WMC to recharge the requested SNs. However, the energy consumption rate of the SNs is different making it a very challenging issue to replenish the energy of the requested SNs based on their location coordinates and energy level. Therefore, our aim is to find an optimal path for the WMC so that it can efficiently recharge the requested SNs. WMC returns back to BS after recharging the requested SNs and waits for the next request.

2.3 Linear Programming Problem (LPP) Formulation

In this section, an LPP is formulated to find an optimal recharging path for the WMC. Let $dist_{ij}$ be the Euclidean distance between nodes S_i and S_j and $\beta_j\beta_j$ be the Boolean variable defined as follows:

$$\beta_{ji} = \begin{cases} 1, & \text{if node } S_j \text{ directly visited after node } S \\ 0, & \text{otherwise} \end{cases} \quad (1)$$

The LPP to find the optimal path for the WMC is given as

$$Minimize \sum_{i=1}^{N_r+1} \sum_{j=1,j\neq i}^{N_r+1} dist_{ij}\beta_{ji} \tag{2}$$

subject to

$$\sum_{i=1}^{N_r+1} \beta_{ji} = 1, \forall i, j = 1 \, to \, N_r, \, and \; i \neq j \tag{3}$$

$$\sum_{j=1}^{N_r+1} \beta_{ji} = 1, \forall j, i = 1 \, to \, N_r, \, and \; i \neq j \tag{4}$$

$$\sum_{i=1}^{N_r+1} \sum_{j=1,j\neq 1}^{N_r+1} \beta_{ij} \leq |S| - 1, \; \forall S \; CN_r + 1 \tag{5}$$

$$\sum_{i=1}^{N_r+1} (E_{Ci} + E_{Ti}) < E_{MC} \tag{6}$$

Here, Eq. 2 represents the objective function. Constraint 3 represents that WMC must visit only one node from another single node. Constraint 4 represents that WMC must visit only one node after visiting node S_i. All the requested SNs are fully connected is ensured by constraint 5. Finally, constraint 6 ensures that the energy required by WMC to travel and to recharge the requested SNs should not exceed its maximum energy.

3 Proposed Work

3.1 An Overview of Genetic Algorithm

The Genetic Algorithm (GA) [22–24] is widely studied evolutionary algorithm to solve various optimization problems. Initially, GA starts with the generation of possible solutions called chromosomes of some size (say P). From initial population, a set of chromosomes (parents) are selected based on some selection operation like tournament selection, roulette wheel selection, etc. and a new population is created. Thereafter, parent chromosomes undergo crossover operation to generate two child chromosomes. The crossover operation is followed by the mutation operation. In mutation operation, a randomly selected gene value of the generated child chromosomes is altered with another valid value in search of better chromosome. Here, mutation operation does not guarantee a better or improved chromosome.

Finally, the derived fitness function is used to evaluate the mutated child chromosome. The mutated child chromosomes with better fitness value than the parent will

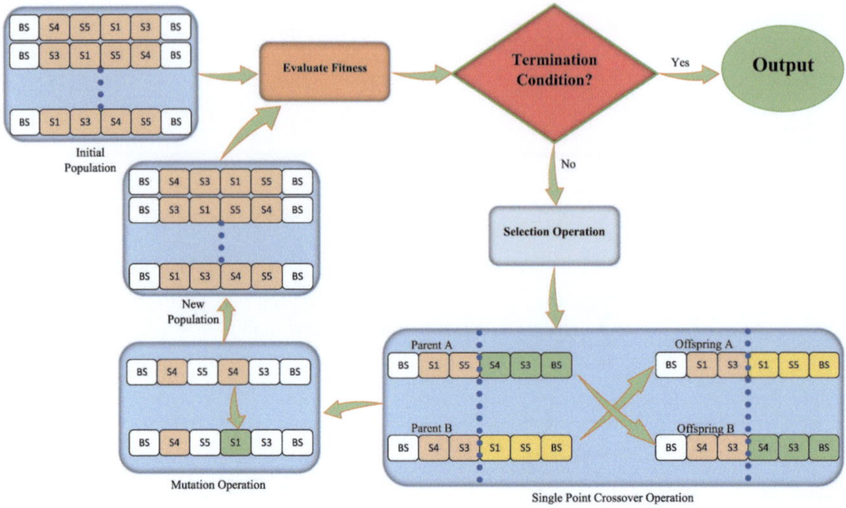

Fig. 1 Genetic algorithm flow diagram

replace the parent from the population else they are discarded. The process of selection, crossover, and mutation will iterate till the termination criteria are achieved. The complete procedure of the GA is shown in Fig. 1, where the genes of the chromosomes are represented by SNs and BS.

In the proposed work, an improved mutation operation is applied instead of traditional operation to ensure the validity of the chromosome. The procedure of the proposed GA to obtain the optimal path for WMC is described below.

3.2 *Chromosome Representation*

In this work, each individual chromosome represents the complete charging path for the requested SNs. Therefore, we have taken the length of the chromosome same as the number of requested SNs, i.e., Nr. Here, each individual gene value represents the unique identity of the requested SN.

Let us consider a WRSN with six SNs S = {s1, s2, s6} and one WMC as shown in Fig. 2. We assume that out of six SNs, four SNs s1, s3, s4, and s5 send charging request to the BS. However, to decide the length of the chromosome for the corresponding network, we have added BS as a gene value at the front and end parts of the chromosomes which consists of requested SNs. Note that scheduling of the WMC always starts from BS and also ends at BS. Thus, the length of the chromosome is equal to the number of charging request and BS (at the front and end of the chromosome), i.e., 6. Figure 3 represents the corresponding chromosome for the given network scenario in Fig. 2. From given Fig. 3, it can be observed that gene value at position 4 is s1 (unique ID of SN 1), which implies that SN s1 will be charged by WMC at third position, i.e., after recharging SNs s4 and s3. Similarly, SNs placed at individual gene value will be charged according to the gene position value, i.e., SN at first gene position will be charged first, SN at second gene position will be charged second, and so on.

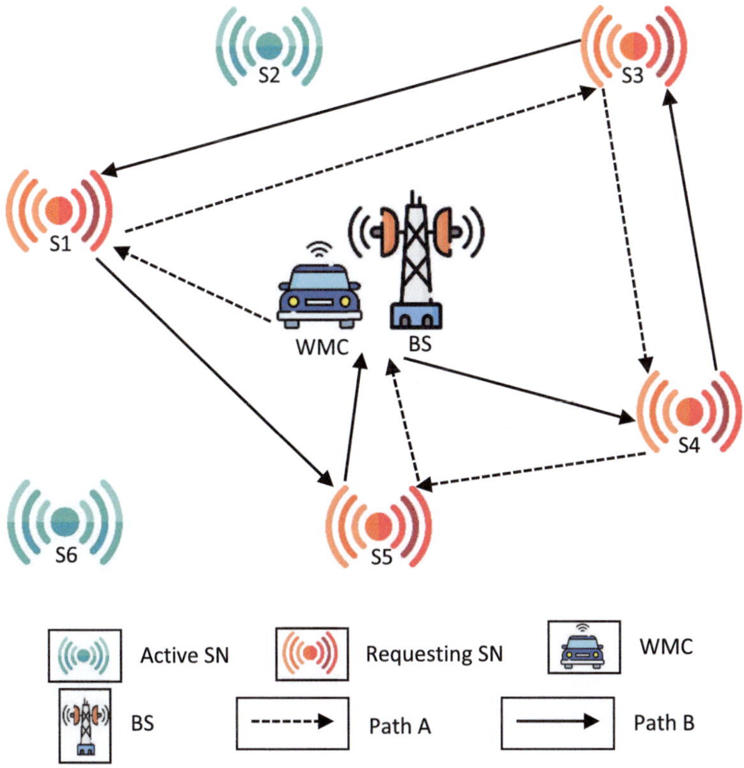

Fig. 2 Network scenario

Fig. 3 Chromosome representation for the corresponding network scenario shown in Fig. 2

3.3 Initial Population Initialization

A set of randomly generated chromosomes of size P represents the initial population. As given, Algorithm 1 is used to generated the initial population.

Algorithm 1: Initial-Population-Generation

Input: (1) Number of requested SNs, N_r.

(2) Size of initial population P, i.e., number of chromosomes.

Output: Initial population with P number of chromosomes.

1: *InitialPopulation* = ∅

2: *Chromosome* = $\{g_i|\forall i, 1 \leq i \leq N_r\}$

3: **for** $j = 1$ to P

4: **for** $i = 1$ to N_r

5: g_i=*randomly generates the coordinates of individual* N_r

6: **end for**

7: *InitialPopulation* = *InitialPopulation* ∪ *Chrom*

8: **end for**

9: Return *InitialPopulation*

3.4 Derivation of Fitness Function

The quality of the chromosome is assessed by the derived fitness function. In the proposed work, our aim is to find the optimal path for the WMC to recharge the requested SNs. Here, the priority to recharge the requested SNs is based on the path length of the WMC and residual energy of the requested SNs. The SNs with minimum traveling distance and minimum residual energy are given more priority to recharge. Therefore, we derived the fitness function considering all these three factors as follows.

Objective 1 (minimization of path length of the WMC): Total path length of the WMC to recharge all the Nr numbers of SNs consists of three parts. First part is the distance between BS and the first node to be recharged represented as $\sum_{i=1}^{N_r-1} dist_{(i,i+1)}$. Second part is the distance between first node to the rest of the node to be recharged which is represented as $dist_{bf}$ $dist_{bf}$. Lastly, third part is the distance between last recharged node and the BS represented as $dist_{lb}$. Note that we consider that WMC starts from BS and after recharging all requested SNs return back to BS. Therefore, the first objective can be denoted as follows:

$$Objective\ 1 : Minimize\ O_1 = dist_{bf} + \sum_{i=1}^{N_r-1} dist_{(i,i+1)} + dist_{lb} \tag{7}$$

Objective 2 (maximization of residual energy of the requested SNs): Let us consider S_{eri} and S_{eci} be the residual energy and energy consumption rate of the SN S_i. As our second objective is to maximize the residual energy of the SNs which can be represented as follows:

$$Objective\ 2 : Maximize\ O_2 = \sum_{i}^{N_r} \frac{S_{eri}}{S_{eci}} \tag{8}$$

In the proposed work, we used weight sum approach (WSA) [6] to derive an efficient fitness function, so that trade-off can be built with these conflicting objectives. In WSA, a single scalar objective function is derived by multiplying each objective function with weight value and adding them together as follows:

$$Fitness = \left\{ \left(\sum_{i=1}^{N_r-1} dist_{(i,i+1)} + d_{lb} \right) + w_2 \times \left(1 - \sum_{i}^{N_r} \frac{S_{eri}}{S_{eci}} \right) \right\} \tag{9}$$

where w_1 and w_2 represent the weight value such that $\sum w_i = 1$ and $0 \leq w_i \leq 1$, $\forall i = 1$ to 3. Here, fitness function consists of three objectives O_1, *and* O_2, where objective function with higher range value will dominate the overall fitness function. Therefore, values of all objectives are normalized within the same scale of (0,1) to provide equal priority to all three objectives. However, the charging priority of the requested SNs is based on the distance from the current node, residual energy, and energy consumption rate. Therefore, our main objective is to minimize the fitness function given as

$$Objective : Minimize\ Fitness \tag{10}$$

Chromosome with the minimum fitness values is considered as the better charging schedule for the WMC. However, objectives are conflicting in nature.

3.5 Selection Operation

In the selection operation, two parent chromosomes are chosen from the population for crossover operation. In the proposed work, chromosomes are randomly selected so that all chromosomes have equal opportunity to take part in crossover operation.

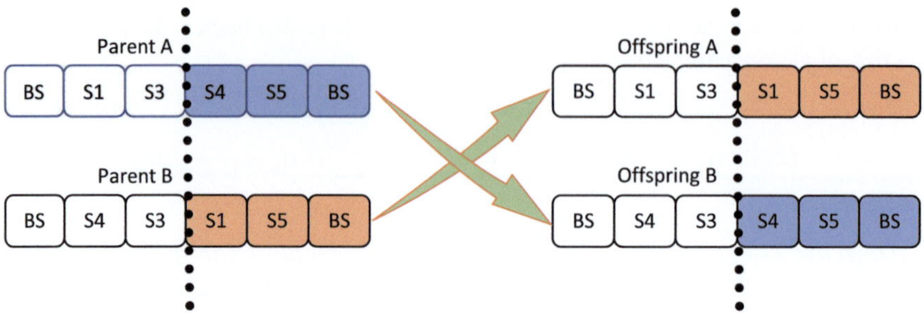

Fig. 4 Crossover operation

3.6 *Crossover Operation*

Selection operation is followed by crossover operation. There are various types of crossover operations like one point, two point, etc. Here, we used single point crossover as shown in Fig. 4. In this operation, two parent chromosomes interchange their information beyond a randomly selected single point of reference to produce new offspring.

3.7 *Improved Mutation Operation*

In mutation, a gene from the chromosome is selected randomly or based on some probability and replaced by the other valid values to produce a better chromosome. Note that mutation never ensures better chromosomes. In the proposed work, a novel mutation scheme is proposed in contrast to the conventional mutation operation. Initially, we identify the repeated gene value in the chromosome. Then, these repeated gene values are replaced with a valid gene value which is part of chromosome but is discarded during mutation. The proposed mutation operation will generate the valid chromosome, i.e., valid recharging path for the WMC as depicted in Fig. 5. From given Fig. 5, it can be observed that in offspring chromosome gene value s4 (SN 4) is repeated. Therefore, this gene value is replaced by gene value s1 which is discarded or removed during crossover operation.

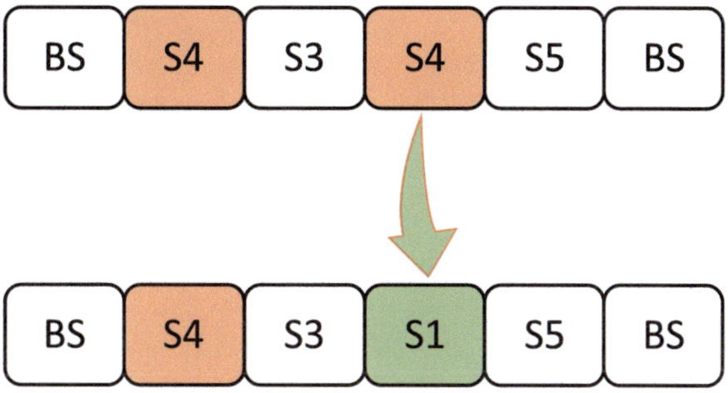

Fig. 5 Improved mutation operation

4 Experimental Results

We have executed an extensive simulation using MATLAB and C programming language. We have considered an obstacle-free WRSN with 300 m x 300 m square with randomly deployed 100 SNs. Out of the 100 SNs varying number of SNs request for charge. Here we have considered 10, 15, 20, 25, 30, 35, 40 number of SNs requesting for charge. Here we have taken chromosome population from size 10 to 100. The proposed algorithm is then used to find the optimal path for the WMC. The results of the algorithm are then compared with greedy algorithm. Fitness value represents the quality of the chromosome, higher the fitness value better the solution.

We have evaluated the fitness value for varying number of iterations. With the increase in the number of iterations the fitness value also increases but after a certain number of iterations there is no increase in the fitness value and the graph becomes linear as illustrated in Fig. 6. This happens because no offspring with a fitness value better than the fitness value of the chromosomes in the new population can be produced any more. As a result, we have discovered the best option. When compared to the greedy algorithm, the proposed genetic algorithm produces better results, as illustrated in Fig. 6.

We have also evaluated the fitness value with varying population size. With the increase in population size, the fitness value improves because a larger number of genes means a larger variety of potential combinations, which could result in a better chromosome. After a certain population size (80), the graph becomes linear because although we may get new offspring, the fitness value is not changing. When compared to the greedy algorithm, the proposed genetic algorithm provides a greater fitness value for the respective population size, as illustrated in Fig. 7.

One of the main objectives of this work was to reduce the total distance traveled by the WMC. Thus, we have evaluated the distance traveled by the WMC with respect to varying number of requested SNs. As shown in Fig. 8, when compared with greedy algorithm, the proposed genetic algorithm provides shorter path for the WMC to travel and recharge the SNs. As can be observed from Fig. 8, a higher number of requesting SNs do not always imply that the WMC travels a greater distance because the requesting SNs may be positioned distant from the BS, increasing the total distance, or aligned near the BS, reducing the total distance.

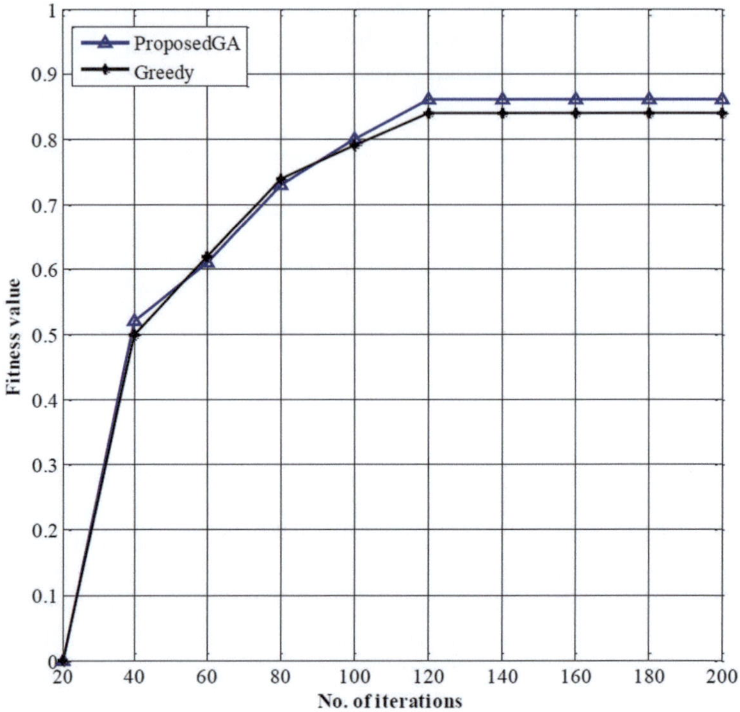

Fig. 6 Fitness values by varying number of iterations

We have also evaluated the energy consumed by the WMC for varying number of requesting SNs (same as Fig. 8). The WMC consumes 0.25 Joule of energy for traveling 1 unit of distance. Thus, the total energy consumed by the WMC is the

$$\begin{aligned}\text{Total Energy Consumed} &= \text{Energy consumed for travelling} \\ &\quad + \text{Energy transferred to sensor.}\end{aligned}$$

We have also executed algorithm to calculate total energy consumed by WMC to recharge the requested SNs by varying the number of requested SNs as shown in Fig. 9. From Fig. 9, it can be observed that WMC consumes less energy by scheduling proposed algorithm as compared to greedy algorithm.

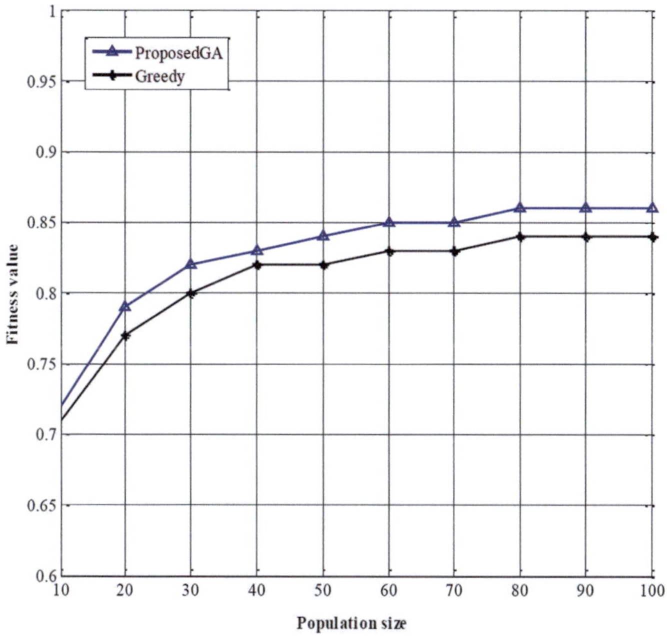

Fig. 7 Fitness value by varying number of population size

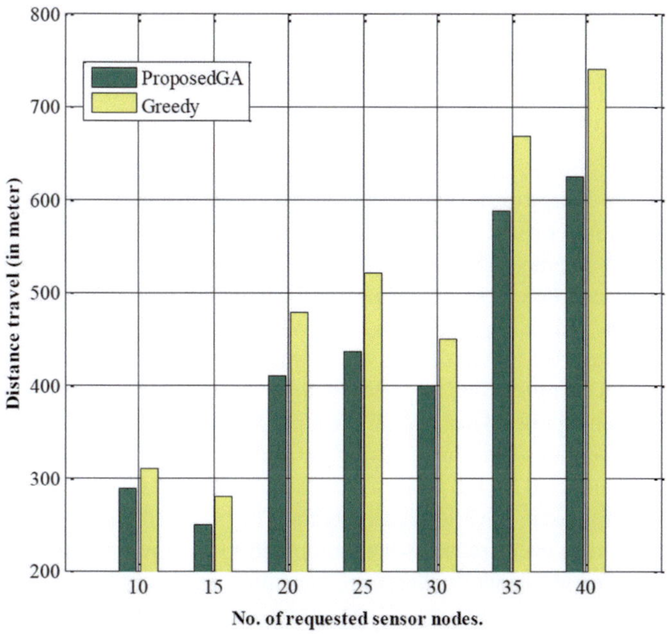

Fig. 8 Distance traveled by WMC by varying number of requested sensor nodes

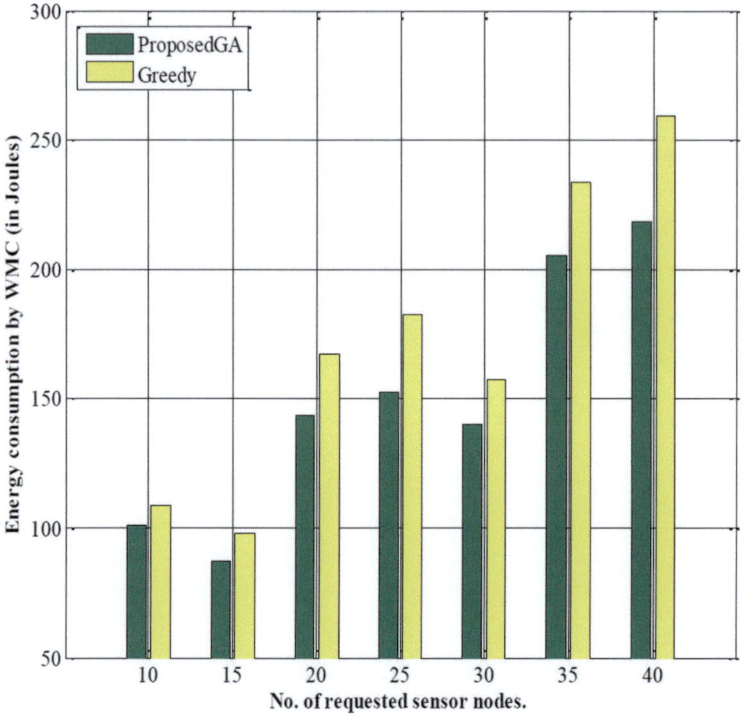

Fig. 9 Energy consumed by WMC for varying number of sensor nodes

5 Conclusion

Due to the finite energy of the battery, the lifetime of conventional WSNs is limited. Some applications, on the other hand, require the network's availability for an extended period of time. WET and WSN have been combined to create a new technology known as wireless rechargeable sensor network (WRSN). SNs in WRSNs are rechargeable and can be charged by any WMC. In this paper, we have proposed an energy-efficient WMC scheduling technique based on genetic algorithm. Here traditional mutation operation is replaced with an improved mutation operation. The problem was then rephrased in terms of linear programming (LPP). The quality of the chromosomes/solutions is evaluated using the derived fitness function. An extensive simulation is conducted for the proposed work. Simulated results are compared with greedy algorithm based on various parameters. The obtained results outperform the greedy algorithm in every regard when compared to it. However, multiple WMCs are not considered in this work, therefore, our future work will be in this regard.

References

1. Bal M (2014) Industrial applications of collaborative wireless sensor networks: a survey. In: 2014 IEEE 23rd international symposium on industrial electronics (ISIE). IEEE, pp 1463–1468
2. Feroz Khan AB, Anandharaj G (2019) A cognitive key management technique for energy efficiency and scalability in securing the sensor nodes in the IoT environment: CKMT. SN Appl Sci 1(12):1–7
3. Madhu A, Sreekumar A (2014) Wireless sensor network security in military application using unmanned vehicle. Int J Electron Commun-Tion Eng 51:58
4. Sedighian Kashi S (2019) Area coverage of heterogeneous wireless sensor networks in support of internet of things demands. Computing 101(4):363–385
5. Wang Q, Cui Z, Wang L (2021) Charging path optimization for wireless rechargeable sensor network. Peer-To-Peer Netw Appl 14(2):497–506
6. Harizan S, Kuila P (2019) Coverage and connectivity aware energy efficient scheduling in target based wireless sensor networks: an improved genetic algorithm based approach. Wireless Netw 25(4):1995–2011
7. Kulshrestha J, Mishra MK (2018) Energy balanced data gathering approaches in wireless sensor networks using mixed-hop communication. Computing 100(10):1033–1058
8. Tong B, Wang G, Zhang W, Wang C (2011) Node reclamation and replacement for long-lived sensor networks. IEEE Trans Parallel Distrib Syst 22(9):1550–1563
9. Cao X, Xu W, Liu X, Peng J, Liu T (2021) A deep reinforcement learning-based on-demand charging algorithm for wireless rechargeable sensor networks. Ad Hoc Netw 110:102278
10. Chawra VK, Gupta GP (2021) Hybrid meta-heuristic techniques based efficient charging scheduling scheme for multiple Mobile wireless chargers based wireless rechargeable sensor networks. Peer-To-Peer Netw Appl 14(3):1303–1315
11. Chen J, Yu CW, Ouyang W (2020) Efficient wireless charging pad deployment in wireless rechargeable sensor networks. IEEE Access 8:39056–39077
12. Ren X, Liang W, Xu W (2014) Data collection maximization in renewable sensor networks via time-slot scheduling. IEEE Trans Comput 64(7):1870–1883
13. Kurs A, Karalis A, Moffatt R, Joannopoulos JD, Fisher P, Soljacic M (2007) Wireless power transfer via strongly coupled magnetic resonances. Science 317(5834):83–86
14. Wang Q, Kong F, Wang M, Wang H (2017) Optimized charging scheduling with single mobile charger for wireless rechargeable sensor networks. Symmetry 9(11):285
15. Gharaei N, Al-Otaibi YD, Rahim S, Alyamani HJ, Khani NAKK, Malebary SJ (2021) Broker-based nodes recharging scheme for surveillance wireless rechargeable sensor networks. IEEE Sens J 21(7):9242–9249
16. Gupta GP, Chawra VK (2020) Coverage-aware recharge scheduling scheme for wireless charging vehicles in the wireless rechargeable sensor networks. In: data management, analytics and innovation. Springer, Singapore, pp 663–671
17. Tomar A, Jana PK (2021) A multi-attribute decision making approach for on-demand charging scheduling in wireless rechargeable sensor networks. Computing 103(8):1677–1701
18. Lin C, Yang W, Dai H, Li T, Wang Y, Wang L, Zhang Q (2021) Near optimal charging schedule for 3-D wireless rechargeable sensor networks. IEEE Trans Mob Comput
19. Kumar R, Mukherjee JC (2021) On-demand vehicle-assisted charging in wireless rechargeable sensor networks. Ad Hoc Netw 112:102389
20. Priyadarshani S, Tomar A, Jana PK (2021) An efficient partial charging scheme using multiple mobile chargers in wireless rechargeable sensor networks. Ad Hoc Netw 113:102407
21. Ouyang W, Obaidat MS, Liu X, Long X, Xu W, Liu T (2021) Importance-different charging scheduling based on matroid theory for wireless rechargeable sensor networks. IEEE Trans Wireless Commun 20(5):3284–3294
22. De Jong K (1988) Learning with genetic algorithms: an overview. Mach Learn 3(2):121–138

23. Mirjalili S (2019) Genetic algorithm. In: evolutionary algorithms and neural networks. Springer, Cham, pp 43–55
24. Goldberg DE (2002) The design of innovation: Lessons from and for competent genetic algorithms, vol 1. Kluwer Academic Publishers, Boston

Enrichment of Impedance Bandwidth and Co-Polarization to Cross-Polarization Isolation of Circular Patch Antenna with Rectangular DGS Structure

Zonunmawii, Manoj Sarkar, L. Lolit Kumar Singh, Sudipta Chattopadhyay, and Abhijyoti Ghosh

Abstract A coaxial probe feed single element circular patch antenna (CPA) with rectangular shape defected ground structure (DGS) at the radiating edge on RT-Duroid substrate (εr = 2.33) has been studied in this paper. Consequential enhancement of impedance bandwidth and co-polarization to cross-polarization isolation (polarization purity) has been observed from the proposed structure. The proposed structure provides around 31.5% impedance bandwidth and 24 CP-XP isolation which is quite good as compared to conventional circular patch antenna. The classical CPA structure and proposed structure show good impedance matching and stable radiation pattern within the operational frequency. The defected ground structure geometry is very simple and easy to manipulate; therefore, this present structure might be ideal for the application in the wireless communication system where high bandwidth, good polarization purity, and the stable radiation pattern are required.

Keywords Circular patch antenna · Rectangular defected ground structure · Impedance bandwidth · Gain · Co-polarization · Cross-polarization

1 Introduction

In the modern era communication, patch antenna is most useful radiator due to its ease of integration with active device like (MMIC, HMMIC, and MEMS), low cost, easy fabrication process, and lightweight. Besides this advantage it has major disadvantage, like narrow bandwidth, low co-polarization gain, and poor polarization purity. Co-polarization (CP) radiation to cross-polarization (XP) radiation isolation commonly known as polarization purity (PP) decreases as operating frequency of

Zonunmawii · M. Sarkar · L. L. K. Singh · S. Chattopadhyay · A. Ghosh (✉)
Department of Electronics and Communication Engineering, Mizoram University, Tanhril, Aizawl, Mizoram, India
e-mail: abhijyoti_engineer@yahoo.co.in

S. K. Biswas et al. (eds.), *Intelligent Computing Systems and Applications*, Lecture Notes in Networks and Systems 1307, https://doi.org/10.1007/978-981-96-3860-4_34

the antenna increases [1, 2]. Conventional CPA radiates linearly polarized fields in broad side direction in its dominant TM_{11} mode. However, few extents of orthogonal polarization, called cross-polarized (XP) radiation take place at first higher order mode, i.e., TM_{21} mode and those were theoretically studied in [3].

Scientists and researchers in antenna field have proposed different techniques like defected ground structure (DGS), defected patch structure (DPS), shorting technique, to improve the input and radiation performance of Conventional CPA. CPA with circular DGS [4], arc-shaped DGS [5] is documented to attain polarization purity of 5–7 dB but impedance bandwidth is very narrow in those reports. Few defected patch surface (DPS) models are reported in [6, 7] where polarization purity around 20–25 dB has been reported. Around 22% bandwidth and 3.24 dBi gain has been reported with different type of rectangular slots on circular patch surface [8]. Polarization purity around 25 dB is reported in [9] by fully shorting the non-radiating edges of the conventional CPA but impedance bandwidth is very narrow in the range of 7–9%. In another structure [10], rectangular strip with discrete shorts is reported to achieve 27 dB polarization purity with only 8% impedance bandwidth.

Different type of cavity enclosed CPA has been investigated in [11, 12] for bandwidth enhancement, where 7–8% impedance bandwidth is documented in [11] and around 20% impedance bandwidth is observed in [12] without much improvement in polarization purity. In all the above reports, enhancement of either impedance bandwidth or polarization purity has been reported. The simultaneous improvement of these parameters has not been reported till now.

Therefore, in the present investigation, a pair of rectangular defects have been deployed at the radiating edge of the conventional CPA on RT-Duroid substrate with dielectric constant (ε_r) 2.33 and height (h) 1.58 mm in order to counsel the limitation of the earlier structure and for the concurrent improvement in impedance bandwidth and polarization purity. The investigated antenna provides 24 dB CP-XP isolation and 31.5% impedance bandwidth. So simultaneous enrichment of impedance bandwidth and polarization purity is achieved in comparison with a conventional CPA.

2 Theoretical Background and Parametric Study

The conventional CPA is basically an open resonator. Any type of modification by placing defects below the patch or on the ground plane definitely hampers the tradition cavity boundary conditions and changes the field between ground plane and patch. This changes in the filed distribution also influences the input as well as the radiation characteristics of the modified structure. The defect size and shape critically decide the behavior of the cavity fields so the selection of the defect dimensions is very much important to get the best performance from the modified antenna structure.

The bandwidth of the antenna is inversely proportional to the quality factor and quality factor is inversely proportional to the different losses occurred in the antenna. So, the bandwidth increases with the increase in loss which increases as the defect size increases.

With the above theoretical concept, a pair of rectangular DGSs has been incorporated on the ground plane at the radiating sides of a conventional CPA with radius (r) 7 mm. The length (sl) of the rectangular DGS is kept as the same as the diameter of the CPA. Then the width (sw) of the rectangular DGS is varied to get the optimum dimension of the DGS.

First, DGS with width length 0.5 mm ($sw = 0.5$ mm) is placed. Improvement of the impedance bandwidth of rectangular DGS (RDGS) integrated CPA is observed as compared to the conventional CPA. The conventional CPA ($sl = sw = 0$ mm) provides a impedance bandwidth of 4% where the impedance bandwidth increases to 11% with RDGS structure with length 14 mm and width 0.5 mm ($sl = 14$ mm and $sw = 0.5$ mm) as shown in Figs. 2 and 3. Figure 3 also shows that due to the RDGS the resonant frequency corresponding to the lowest order mode moves towards the upper side in the frequency spectrum. The effective permittivity of the substrate changes due to the incorporation of the RDGS (Umesh electromagnetics). The width (sw) of the RDGS increases further up to 1.6 mm. As the width (sw) increases, the impedance bandwidth increases up to $sw = 1.5$ mm, then, for $sw = 1.6$ mm, it decreases. Both Figs. 2 and 3 confirm this. The impedance bandwidth with $sw = 1.5$ mm is 31.5% which is much higher than the conventional CPA.

Along with the impedance bandwidth, a study on polarization purity is also performed. Figure 4 shows the variation of XP in H plane as the width (sw) varies from 0.5 to 1.6 mm keeping the RDGS length (sl) fixed at 14 mm. Initially, the polarization purity decreases after placing the RDGS but later it improves and provides best polarization purity with $sw = 1.5$ mm. The polarization purity is only 15 dB in case of classical CPA, whereas, with modified structure, it is around 24 dB.

With the above parametric studies on impedance bandwidth and polarization purity, it is confirmed that the RDGS integrated CPA structure provides optimum performance with $sl = 14$ mm and $sw = 1.5$ mm.

3 Optimum Structure

At first, conventional circular microstrip patch antenna has been designed with radius (r) 7 mm on RT-Duroid material with dielectric constant (ε_r) 2.33 and height (h) 1.58 mm. After that a pair of rectangular DGS, slot has been deployed at the radiating edge where the length (sl) of the rectangular DGSs is kept same as the diameter of the circular patch, i.e., 14 mm. From the parametric studies, it is seen that the rectangular DGSs with width (sw) 1.5 mm provides the best output in terms of impedance bandwidth and polarization purity. This way the final proposed structure has been built as shown in Fig. 1.

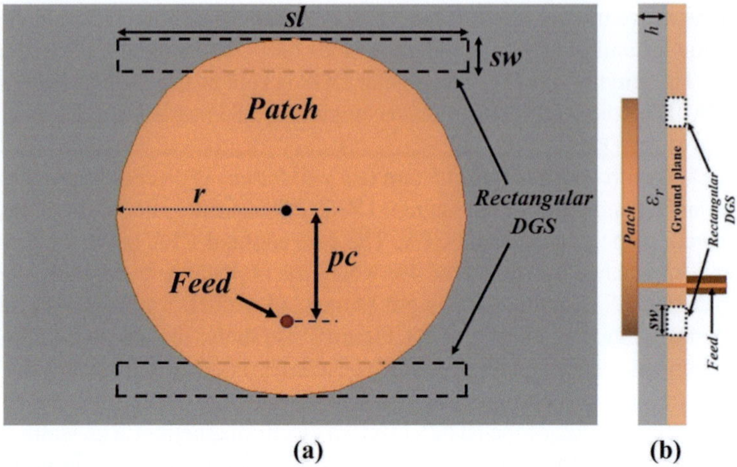

Fig. 1 Schematic representation of proposed rectangular defective ground structure (DGS) integrated CMPA **a** top view **b** side view

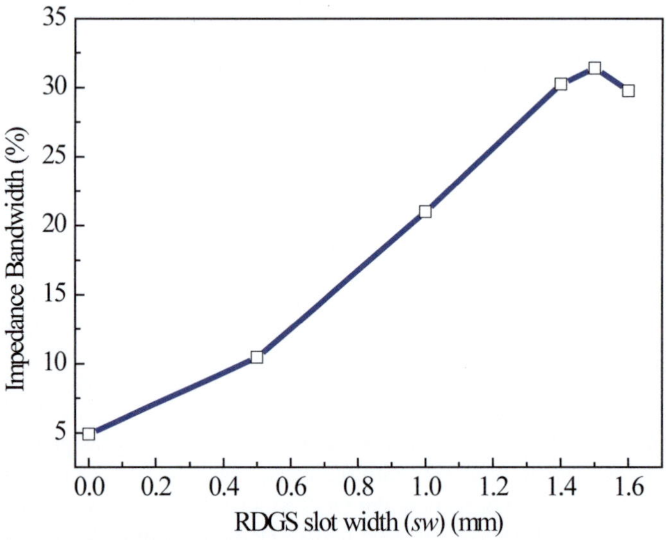

Fig. 2 Variation of impedance bandwidth (%) as a function of RDGS width with RDGS length (*sl*) fixed at 14 mm

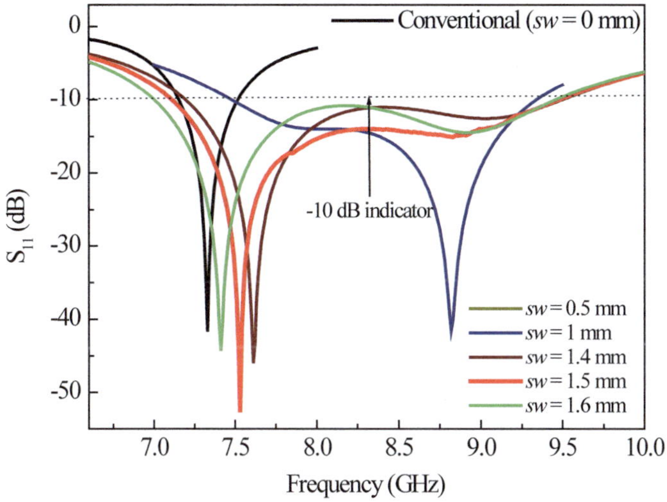

Fig. 3 Reflection coefficient profile for conventional CPA and RDGS incorporated CPA with RDGS length (*sl*) fixed at 14 mm

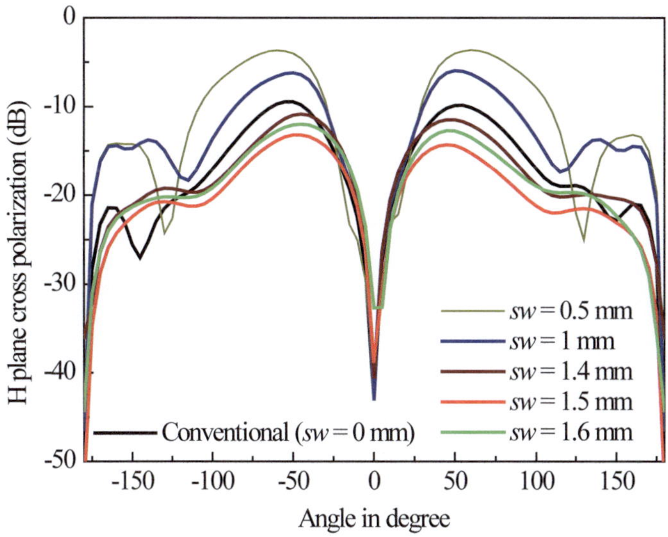

Fig. 4 Variation of impedance bandwidth (%) as a function of RDGS width (*sw*) with RDGS length (*sl*) fixed at 14 mm

4 Results and Discussion

This section discusses the output obtained with the help of [13] from the optimized proposed structure. Figure 5 depicted the reflection coefficient (S_{11}) profile of the conventional CPA and proposed CPA with rectangular DGS structure. Conventional CPA resonates at 7.27 GHz, whereas proposed antenna with rectangular DGS resonates at 7.53 GHz; so, it clearly evident that due to the deployment of rectangular DGS structure at the radiating edge the resonance frequency has been shift towards higher frequency spectrum but it quite same as conventional patch and both structures provide very good impedance matching. It is very much clear that the conventional CPA has narrow impedance bandwidth in the order of 4%, whereas the proposed modified structure provides an impedance bandwidth of 31.5% which is much higher as compared to the conventional CPA.

The E plane normalized radiation profile has been revealed in Fig. 6. It is evident from the figure that E plane XP is below −37 dB for both conventional CPA as well as the proposed antenna, which is quite convincing. The co-polarization radiation in E plane has not been affected much due to the incorporation of rectangular DGSs at the radiating edges.

Comparison of proposed model and conventional CPA H plane radiation pattern has been revealed in Fig. 7. The peak H plane co-polar gain is 6.52 dBi and 6.69 dBi in case of conventional CPA and proposed antenna, respectively. The peak XP level in H plane is −9.45 dB and −18.2 dB in case of conventional CPA and proposed antenna, respectively. The deployment of rectangular DGSs at the radiating edge do not affect the co-polarization radiation in H plane, whereas the peak XP radiation decreases by almost 9 dB. This clearly improves the PP by a factor of 9 dB in case of modified rectangular DGS integrated CPA.

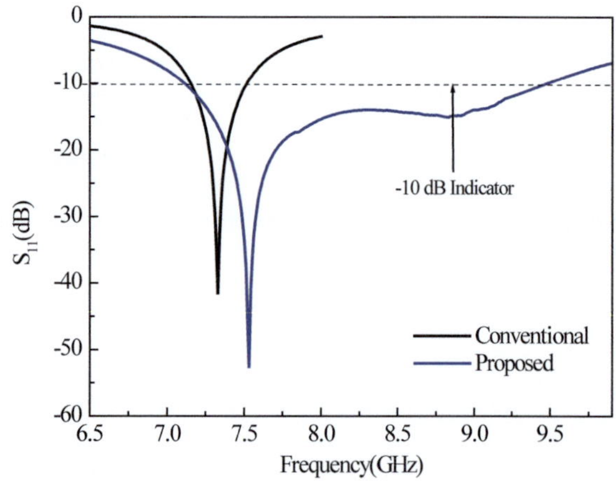

Fig. 5 Simulated reflection coefficient profile of conventional and proposed rectangular DGS integrated CPA

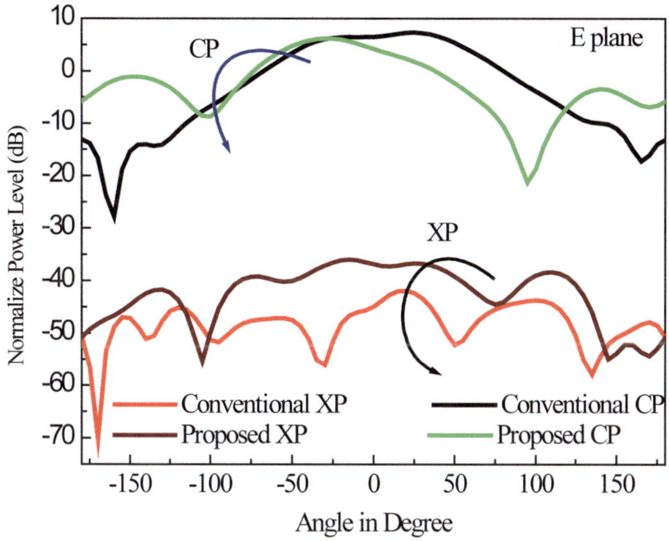

Fig. 6 Normalized E plane radiation pattern of conventional and proposed rectangular DGS integrated CPA

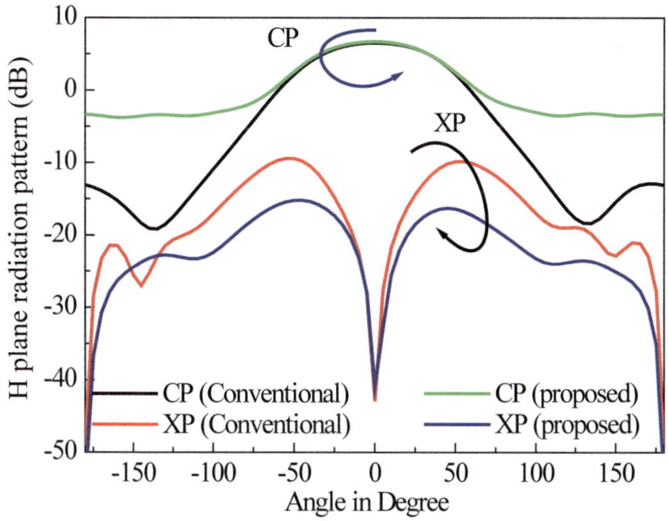

Fig. 7 Simulated conventional and proposed structure radiation pattern in H plane

5 Conclusion

A pair of rectangular defects has been deployed on the ground plane at the radiating edge of conventional circular patch antenna in order to provide enrichment in the impedance bandwidth and polarization purity. 31.5% impedance bandwidth and around 24 dB CP-XP isolation have been achieved from the proposed structure, while, for conventional CPA, impedance bandwidth and polarization purity are only 4% and 15.9 dB, respectively. Enrichment of around 27.5% in the impedance bandwidth and 8 dB polarization purity as compared to a conventional CPA is achieved using the proposed structure. This investigation can be useful in the modern wireless communication field, whereas wide impedance bandwidth and high polarization purity is required.

References

1. James JR, Hall PS (eds) (1992) Handbook of microstrip antenna, vol 1. Peter Peregrinus, London
2. Garg R, Bhartia P, Bahl I, Ittipiboon A (2001) Microstrip antenna design handbook. Artech House, Norwood, USA
3. Lee KF, Luk KM, Tam PY (1992) Cross polarization characteristics of circular patch antennas. Electron Lett 28(6):587–589
4. Guha D, Biswas M, Antar YMM (2005) Microstrip patch antenna with defected ground structure for cross polarization suppression. IEEE Antennas Wireless Propag Lett 4:455–458
5. Guha D, Kumar C, Pal S (2009) Improved cross-polarization characteristics of circular microstrip antenna employing arc-shaped defected ground structure (DGS). IEEE Antennas Wirel Propag Lett 8:1367–1369
6. Ghosh A, Ghosh SK, Ghosh D, Chattopadhyay S (2016) Improved polarization purity for circular microstrip antenna with defected patch surface. Int J Microw and Wirel Technol 8(1):89–94
7. Ghosh A, Ojha S, Rana Y, Anand A, Kumar S, Chattopadhyay S (2014) Improved cross polarization performance of circular microstrip antenna with arc shaped defected patch surface. In: 1st International science and technology Congress 2014 (IEMCONGRESS'2014), Elsevier, pp 87–92
8. Mondal K, Murmu L, Sarkar PP (2017) Investigation on compactness, bandwidth and gain of circular microstrip patch antenna. In: 2017 devices for integrated circuit (DevIC), pp 742–746
9. Zonunmawii, Ghosh A, Singh LLK, Chattopadhyay S, Sim CYD (2021) Reduced-surface-wave-inspired circular microstrip antenna for concurrent improvement in radiation characteristics. IEEE Antenna and Wave Propag Lett 20(5):858–862
10. Zonunmawii, Ghosh A, Singh LLK, Chattopadhyay S (2022) Cavity field modulation with modulating circular patch antenna surface: a key to realize reduced horizontal radiation and omni-present improvement in radiation performance. IEEE Access 10:18434–18444
11. Karmakar NC (2002) Investigations into a cavity backed circular patch antenna. IEEE Trans Antenna Propagat. 50:1706–1715
12. Singh AK, Gangwar RK, Kanaujia BK (2014) Cavity backed annular ring microstrip antenna loaded with concentric circular patch. In: the 8th European conference on antennas and propagation (EuCAP 2014), pp 2155–2158
13. HFSS: High frequency structure simulator, Ver. 13. Ansoft Corp., USA

Miniaturization of Rectangular Microstrip Antenna Using Shorting Posts for Low Frequency Wireless Underground Sensor Networks

Sarath Kumar Annavarapu, L. Lolit Kumar Singh, Sudipta Chattopadhyay, and Abhijyoti Ghosh

Abstract For Electromagnetic wave to propagate underground low frequencies are preferred. Design of a miniaturized planar antenna at such low frequency specifically in P band (250–500 MHz) is much suitable for wireless underground sensor networks. Miniaturization at such low frequency is cumbersome task. In this work Microstrip patch antenna with copper shorting technique is presented. Miniaturization with number and position of vias are examined and presented. The observation paves a patch to the miniaturization of antennas with judicious placement of vias and its number. The results shows that frequency variations of Microstrip patch antenna by changing number and position of shorting vias without altering the physical dimensions.

Keywords Shorting vias · Wireless underground sensor networks · Low frequency antenna

1 Introduction

Wireless Underground Sensor Networks (WUSN) play important role in soil monitoring, habitat monitoring, oil, and other underground components [1]. At High frequency Electromagnetic waves get absorbed by the surface of the earth and are not useful for underground propagation. Frequencies in the range of Mega Hertz (MHz) are preferred for the same. Designing a Microstrip Patch Antenna (MPA) at such low frequency is a tedious task as the size of the antenna becomes larger. So, miniaturization techniques play important role. A compact and broadband microstrip antenna with shorted patch, stacked shorted patch and slot loading patch is presented in [2]. Several works [3–8] reported the use of shorting technique to reduce the

S. K. Annavarapu · L. L. K. Singh · S. Chattopadhyay · A. Ghosh (✉)
Department of Electronics and Communication Engineering, Mizoram University, Aizawl, Mizoram, India
e-mail: abhijyoti_engineer@yahoo.co.in

S. K. Biswas et al. (eds.), *Intelligent Computing Systems and Applications*, Lecture Notes in Networks and Systems 1307, https://doi.org/10.1007/978-981-96-3860-4_35

size of microstrip patch antenna. In [3] miniaturization techniques such as material loading, reshaping the antenna, shorting and folding techniques, slots and defected ground structure are reported. In [4], a simple folding operation of a wall shorted patch resulting in a stacked shorted patch is designed for ISM band. Specifically, in [5] slit etching technique is performed to reduce the patch size which is working in the frequency of 436 MHz for CubeSat payload. In [6], H-shaped microstrip antenna is studied through transmission line model and analysis is done by Method of Moments. Miniaturization and tuning of patch antenna using a stacked superstrate layer is reported in [7] in which short circuit post are used for miniaturization. In the present work, the shorting vias are explored in a different way in terms of number and position on the patch for low frequency applications in P band.

2 Antenna Design

Conventionally exciting a MPA at a particular frequency depends on the length of the patch. It is established fact that resonant frequency is inversely proportional to the length of the patch. MPA of length x width (375 mm $\times$ 450 mm) is required for the frequency of 362 MHz. Such an antenna is not feasible for any applications. As such to design an antenna at frequency ranges of MHz requires larger length. In this paper, Miniaturization of the MPA using shorting vias are explored in the novel way without altering the length of the patch. A Quarter Wave Patch (QWP) can be obtained by use of shorting vias which reduces the patch size to half. This QWP will produce the electric field same as that of the conventional MPA with the judicious placement of shorting vias. The shorting vias can be modeled as inductance parallel to the reference model of a Patch [8].

2.1 Variation of Number of Shorting Vias

To analyze the effect of number of shorting vias [9] on the patch a QWP is taken as shown in Fig. 1a. The number of shorting vias is varied from 24 to 3 and the variation in the frequency is depicted in Fig. 1b. The parameters used in the antenna are mentioned in Table 1. The results from Fig. 1b justify that Miniaturization [10] of the structure can be obtained by varying the number of shorting vias instead of altering the length. Simulation results depict the clear frequency shift from 447 to 362 MHz.

The Miniaturized antenna obtained with 3 shorting vias shown in Fig. 2 is taken as Reference structure and is further analyzed for the varying the position of the shorting vias.

Fig. 1 Variation in the frequency with number of shorting vias **a** QWP antenna with shorting vias **b** Reflection coefficients comparison for different number of shorting vias

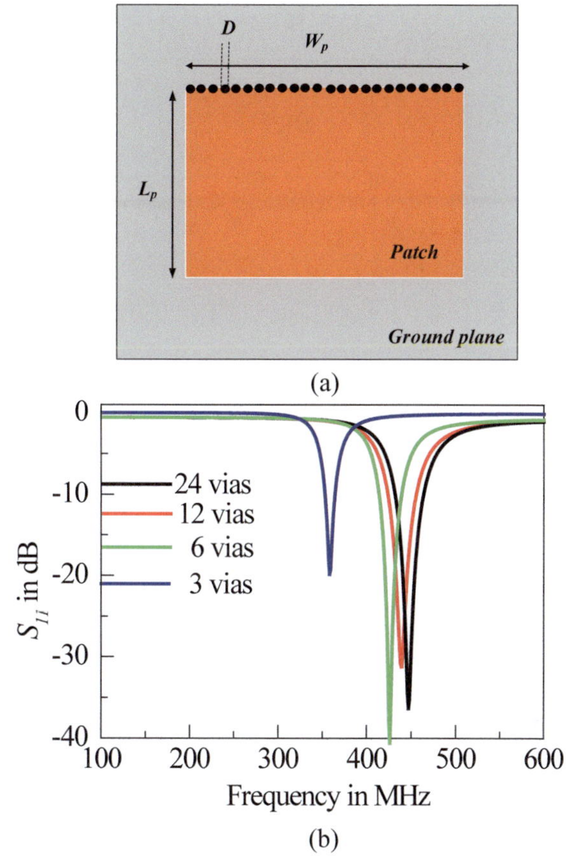

Table 1 Parameters of antenna

L (mm)	W (mm)	L_p (mm)	W_p (mm)	D (mm)
225	286	155	186	2

2.2 *Variation of Shorting Vias Position*

The length and width of the conventional antenna is made to 155 mm × 186 mm on air substrate, which is less than half the size of the conventional antenna at the same frequency [11]. Then shorting vias of diameter ($D = 2$ mm) are placed as shown in Fig. 2. The analysis of RA is done by dragging the shorting vias by p mm form the edge of the patch as shown in Fig. 3.

From Fig. 3b, it is evident that as the position of shorting vias are moved down the edge, frequency increases. The reason for increasing in frequency can be interpreted as follows. When the shorting vias are at the edge of the patch, the shorted edge

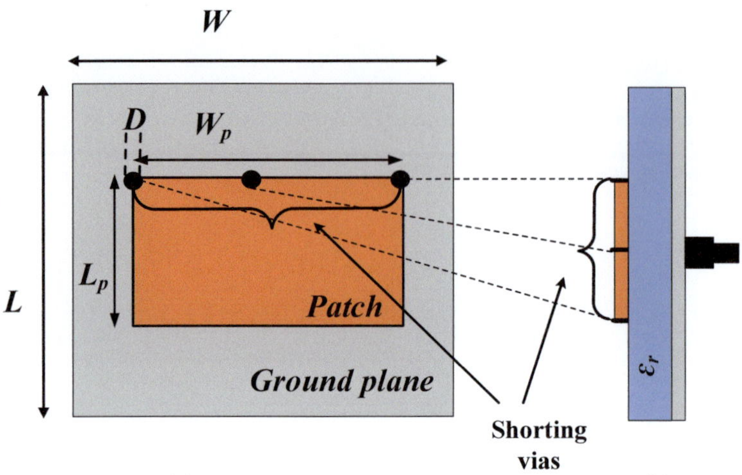

Fig. 2 Proposed miniaturized antenna with three shorting vias **a** Top View **b** cross-sectional view

will not radiate. In this scenario patch is radiating through non-shorted edge. As the shorting vias are moved down form the edge, the second edge of the patch also starts radiating, which increases the frequency of operation. The E filed radiation from the shorted edges of the patch with 3 shorting vias for positions $p = 0$ mm and $p = 20$ mm is shown in the Fig. 4.

The frequency variations with number and position of shorting vias are consolidated and is shown in Fig. 5.

3 Results and Discussions

The simulation is performed using High-Frequency Structure Simulator (HFSS). The use of shorting vias in terms of number and their position for the purpose of frequency shifting is thoroughly discussed. Observing the results of the previous section, it can be confirmed that miniaturization can be obtained by reduction number of shorting vias as depicted in Fig. 1b. For a structure with the n number of shorting vias *($n =$ 24, 12, 6, 3)*, miniaturized version can be obtained for the value of $p = 0$ mm as shown in Fig. 3b. Varying in the value of p in turn varies the operating frequency. Figure 5 depicts the cumulative results for similar length of patch with variation in the number and position of shorting vias.

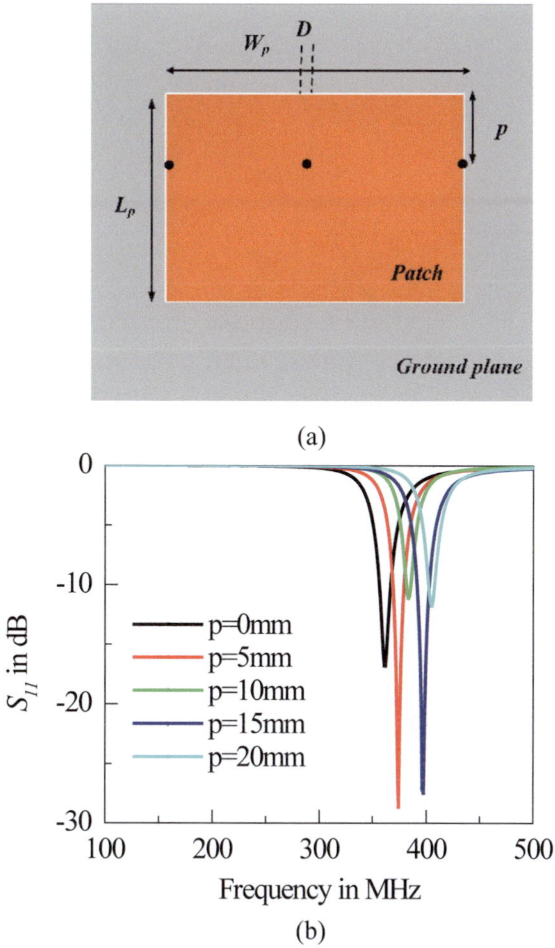

Fig. 3 Variation in the frequency with position of shorting vias **a** RA with position of slots at p **b** S_{11} comparison for various values of p

Figure 6 shows the gain plots of the miniaturized structures with different values of n. The values of Co pol (CP) and Cross pol (XP) of E plane and H plane are plotted. From the above results it is evident that patch of same length with varying the position and number of shorting vias with reasonable gain, which can be useful of wireless Underground applications, can be designed.

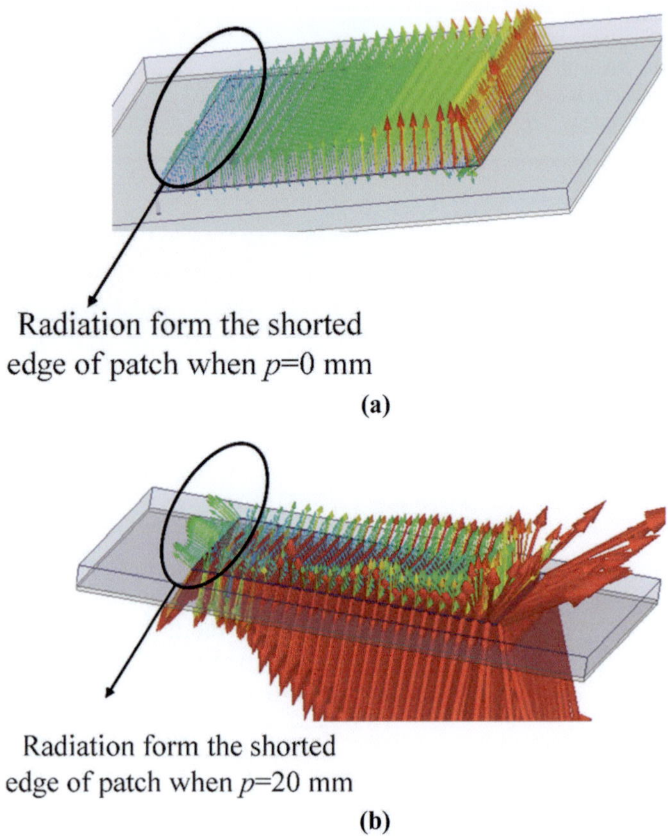

Fig. 4 E filed at the shorted edges of the patch **a** slots at $p = 0$ mm **b** slots at $p = 20$ mm

Fig. 5 Variation in frequency with number and position of shorting vias

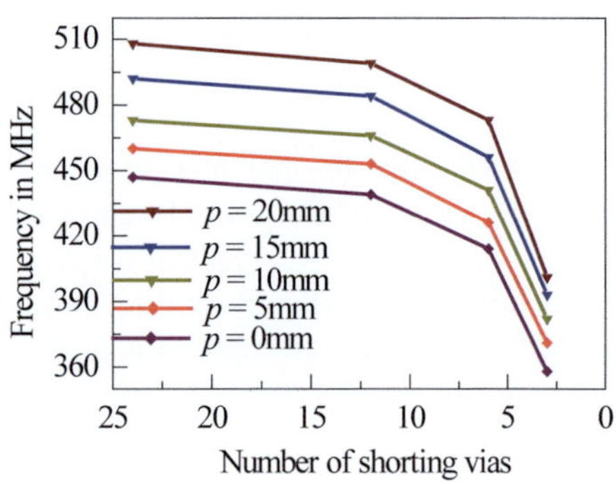

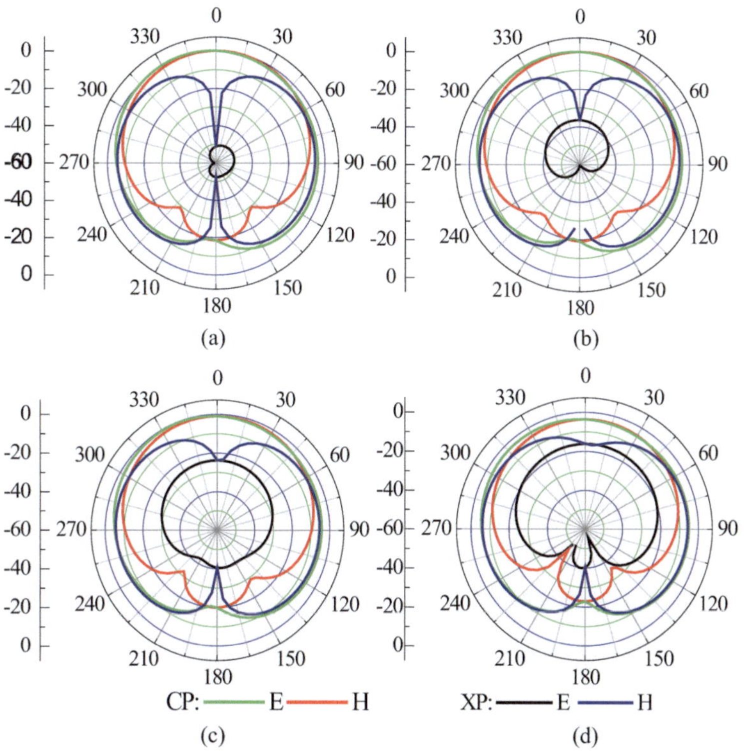

Fig. 6 Radiation pattern of the miniaturized QWP **a** with 24 shorting vias **b** with 12 shorting vias **c** with 6 shorting vias **d** with 3 shorting vias

4 Conclusion

Microstrip patch antenna miniaturization using shorting vias are discussed. The concept of shorting vias is applied to the low frequency antennas. The decrease in number of shorting vias decreases the operating frequency. Variation in the Operating frequency with position *p* from the edge of the patch is presented. The proposed structure at 362 MHz is useful in the propagation of underground Electromagnetic wave. Further simulation shows that the frequency can be varied by varying the position and number of the shorting vias. This can be applied and useful of all the ranges of frequencies.

References

1. Mustafa AAG, Radosveta S Wireless underground sensor networks: channel modelling and operation analysis in the terahertz band. In Int J Antennas Propag. https://doi.org/10.1155/2015/780235
2. Priyanka K, Rajesh N (2014) Review and survey of compact and broadband microstrip patch antenna. In: IEEE International conference on advances in engineering and technology research. 978-1-4799-6393-5/14
3. Muhammad UK, Mohammad SS, Raj M (2015) Microstrip patch antenna miniaturisation techniques: a review. In: IET microwaves, antennas and propagation, vol 9, no 9, pp 913–922
4. Li R, Dejean G, Tentzeris MM, Laskar J (2004) Development and analysis of a folded shorted-patch antenna with reduced size. In: IEEE transactions antennas propagation, no 52.2, pp 555–562
5. Kakoyiannis CG, Constantinou P (2008) A compact microstrip antenna with tapered peripheral slits for CubeSat RF payloads at 436 MHz: miniaturization techniques, design, and numerical results. In: IEEE international workshop on satellite and space communications, pp 255–259
6. Anguera J, Boada L, Puente C, Borja C, Soler J (2004) Stacked H-shaped microstrip patch antenna. In: IEEE transactions antennas propagation, vol 4, no 52, pp 983–993
7. Bokhari SA, Zurcher JF, Mosig JR, Gardiol FE (1996) A small microstrip patch antenna with a convenient tuning option. In: IEEE transactions antennas and propagation vol 44, no 11, pp 1521–1528
8. Porath R (2000) Theory of miniaturized shorting-post microstrip antennas. In: IEEE transaction on antennas and propagation, vol 48, no 1, pp 41–47
9. Chatterjee S, Ghosh K, Paul J, Chowdhury SK, Chanda D, Sarkar PP (2013) Compact microstrip antenna for mobile communication. In: microwave optical technology letters, vol 55, no 5, pp 954–957
10. Filippo C, Agostino M (2012) Closed-form analysis of reflection losses in microstrip reflect array antenna. In: IEEE transaction in antenna and propagation, vol 60, no 10, pp 4650–4660
11. HFSS: High frequency structure simulator, Ver. 14. Ansoft Corp., USA

A Comprehensive Review on Result Extraction and Analysis in Underwater Network Simulators: A UnetStack Perspective

Akhilraj V. Gadagkar, B. R. Chandavarkar, and Harendra Singh Kushwaha

Abstract Simulation of Underwater Acoustic Sensor networks (UASNs) enables researchers to test and verify the protocols or techniques developed quickly and inexpensively. Many underwater network simulation platforms are available such as NS-2 Miracle, Aqua-Net, DESERT, SUNSET, and UnetStack. After conducting a simulation, any simulation platform typically generates a simulation trace or log file. This file has to be processed further to analyze the network performance. However, this requires writing additional scripts or programs, making processing trace files laborious, time-consuming, and prone to error. Thus, a tool that can automate this task of processing the simulation traces and extracting the required result for network performance analysis would help the researchers to focus on developing and validating their work. A few automated tools are available for specific simulators, but no such automated tool is available for the UnetStack simulator. UnetStack is a popular industry standard underwater network simulation platform used in research and development. The community edition is freely and publicly available for research and academia. This work reviews the process and the tools for result extraction and analysis. Furthermore, the work discusses details on available methods in UnetStack for extracting and analyzing the results with their limitations and the scope for building an automated trace analysis tool and finally gives concluding remarks.

Keywords Underwater network simulation · Simulation trace · Performance analysis · UnetStack

A. V. Gadagkar (✉) · B. R. Chandavarkar · H. S. Kushwaha
Wireless Information Networking Group (WiNG), Department of CSE, NITK, Surathkal, Mangalore, India
e-mail: gadagkar.akhil@gmail.com

S. K. Biswas et al. (eds.), *Intelligent Computing Systems and Applications*, Lecture Notes in Networks and Systems 1307, https://doi.org/10.1007/978-981-96-3860-4_36

1 Introduction

Underwater Acoustic Sensor Networks (UASNs) are being developed with increasing demand to respond to various applications in the underwater world. Underwater sensor networks typically consist of static sensor nodes that are deployed in a targeted region or mobile nodes such as the Autonomous Underwater Vehicles (AUVs) to perform collaborative monitoring and information acquisition tasks [1]. Communication with acoustic signals is a reliable approach for sending and receiving messages in an underwater environment. Ongoing enhancements in the functionality of underwater acoustic modems have made it feasible to have a wide range of applications using submerged acoustic networks [2–5]. In underwater network research, realizing a comprehensive testbed consisting of a reasonable network topology to assess the performance of an algorithm or a protocol implemented and is not always feasible as it incurs high cost, time, and requires abundant human resource. This practical difficulty highlights the requirement for a simulation platform that can accurately mimic an underwater scenario while maintaining generality.

Network simulation provides a swift and economical way to determine the network activity and its behavior under different operational circumstances. Prior to the deployment of the network, simulation results enable to measure performance, diagnose issues, ascertain the causes, and rectify problems. Nodes in underwater networks face several technical hurdles [6]. Furthermore, as there are many underwater network simulators available, choosing the right one based on the study requirements is critical for validating and analyzing results. Some popular, open-source, free-to-use underwater network simulators are Aqua-Mate, SUNSET, DESERT, and UnetStack. Out of these, UnetStack is a popular industry standard underwater network simulation platform. The community edition is freely and publicly available for research and academia, with UnetStack3 being the current latest version available. UnetStack is flexible and efficient, supporting both real-time and discrete simulation modes. Further, UnetStack allows simulation code to be ported directly to any UnetStack compatible modem without the overhead of cross compilation. Thereby, UnetStack becomes a practical selection for most of the underwater network research and development [7].

Typically, the underwater simulators capture the events that occur in the network during the simulation and, as a result, generate simulation trace or log file as the output of the simulation done. Further, extracting the relevant fields or data from the trace or log files is necessary to analyze the network activity and performance. However, this requires one to write additional scripts or programs, which makes examining results for performance analysis to be laborious and time-consuming. The results obtained might contain errors if the implemented script or program has bugs. This brings the demand for a tool that can automate extracting, analyzing fields from output files, and presenting the performance analysis results, which can significantly reduce the effort and time. The trace analyzer should be able to take the *trace* and/or *log* files generated by the simulator as input and extracts the essential data, which it then displays more efficiently. Thereby enabling researchers to focus on developing and testing the protocols or methods to be implemented for their underwater network, rather than examining output files of simulations.

Various network simulators have their trace analyzer tool, the NS-2 has an NS-2 visual trace analyzer, NS-3 has Netanim, and the Gnuplot is a trace analyzer. As of now, no such trace analyzer is available for the UnetStack. As a result, this work discusses the UnetStack simulation results concerning the output files generated, exploring the file format and the meaning of the fields, and also the limitations posed by the file. Extracting relevant data from the output file is significant in obtaining the simulation results. Therefore, the features that are currently accessible in UnetStack3 also are discussed. Further, the pseudocodes present the calculations for general performance measures obtained as output from the trace analysis tool.

1.1 Various Underwater Simulators

NS2-MIRACLE NS2-MIRACLE is a framework that expands the functionality of the NS-2 network simulator. It comes with a dynamic engine that has potential to carry out cross-layer communications enabling the layers of the network stack to contain multiple modules simultaneously. Additionally, it enables different IP, MACs, link layers, or physical layers to use within the same node. The collection of libraries in NS2-MIRACLE paves a way to effortlessly generate and simulate communication systems in NS-2. The modular nature makes the code to be ported, re-used, or expanded easily, thus enabling researchers to redefine and extend the network protocols in a dynamic and modular manner. Further, it also supports inter-protocol and inter-layer communication by allowing cross-layer message exchange and allows more than one wireless interface per mobile node to use.

Aqua-Net/Mate Aqua-Net Mate is a virtual channel simulator that works in real time. Aqua-Net Mate has been created from scratch, making it much easier for researchers to customize underwater network protocols. Cross-layer design [8] is supported by Aqua-Net Mate. Original Aqua-Net was developed for a real-world system and lacked simulation capabilities, which led to the development of Aqua-Net Mate. The main features of Aqua-Net Mate are that it is very expandable, highly adjustable, and user-friendly that it supports real-time scheduling, and that it has a uniform customizable interface.

SUNSET The SUNSET is an open-source simulation platform with a set of tools for testing and developing protocols for underwater networks. It is developed based on the NS-2 simulator and its expansion NS2-MIRACLE [9]. It implements the MAC and routing protocols for underwater wireless sensor networks. It is capable of replicating and adhering to communication protocols. It may authorize and evaluate acoustic devices such as modems self-governing submerged and surface vehicles constantly and can also be leveraged to link devices together over the internet. SUNSET allows an entirely new manner of conducting underwater communication research.

SUNSET's real-time scheduler allows it to imitate real-world scenarios having extensive durations (in real time) in a short amount of time. However, it may be challenging to pinpoint time synchronization and accuracy issues because all events are executed in a single thread sequentially.

DESERT DESERT is another open-source tool for simulating underwater networks. Its goal is to become a useful tool for replicating, following, and interpreting underwater network protocols for research [10]. Through the network layer, data link layer, and physical layer, it supports the application and transport layers. Executes a fundamental correspondence, organizing engineering that allows disparate hubs to communicate in a submerged environment reliably. Compared to SUNSET, DESERT implementation is simple but less efficient in real-world tests. DESERT has constraints, including excessive energy and resource usage, despite having a basic design that allows for smooth transition flexibility between simulation, emulation, and field-level testing.

UnetStack UnetStack is a configurable agent-based underwater network platform that provides quick design, simulation, testing, and deployment for software-defined underwater networks. In 2004, UnetStack was created as part of the Unet project at the National University of Singapore's Acoustic Research Lab. UnetStack serves as the foundation for an underwater network simulator.

UnetStack is built on the Fjage agent framework [11] and does not use a traditional protocol stack with layers. In UnetStack, the typical stack includes a set of software agents that represent several layers of the network stack and execute predefined services. This methodology is known as service-oriented architecture [12], allows information to be shared, services to be delivered, and negotiation of behaviors between different agents. Since service-oriented agent architecture is highly extendable, researchers can create, test, and add new functions to a process that would not have been possible with typical layer-based network architecture. Furthermore, as the stack is scalable, adding new agents and services or replacing existing agents to match the developmental requirement can be done. UnetStack allows to define custom agents to implement network protocols. Agents can be developed in Groovy or other supported languages, as UnetStack gives API bindings for numerous languages such as Java, Groovy, Python 3, Julia, C, and JavaScript. The protocol developed can be simulated and the performance of the implemented protocol can be tested by deploying and simulating in different scenarios and obtain the results.

Simulation in the UnetStack performed can be in real-time or discrete-event simulation. UnetStack simulator (also known as "UnetSim") is an UnetStack component. UnetSim allows you to simulate a large underwater network on even a single machine. A simulation scenario is specified in UnetSim using a Groovy domain-specific language (DSL) script. Groovy is a dynamic and flexible Java virtual machine (JVM) language. UnetStack supports the dynamic implementation of Groovy codes, allowing researchers to develop scripts and modify and tune their protocols without recompiling their code. Further, the shell provides the interface to make simulations interactive. In addition, modifying the script can be done if necessary to gather and display network performance statistics [11].

2 Simulation Results in UnetStack

Simulation results are extracted from trace and log files, for any simulation UnetStack typically generates the trace and the log file which should be further processed to extract the desired results. This section presents details on the log file and two types of trace files generated as output files for a simulation.

2.1 Output Files of a Simulation in UnetStack

Log File The *logs/log-0.txt* file comprises detailed text logs and uses the Java logging framework. The simulation script and the agents can log additional data in various log levels to this file. Simulation scripts or the agents can do custom logging using the methods provided to log at different levels, thereby giving an adaptable and adjustable approach to logging in simulation for later investigation. A common concentrate of the logs in *logs/log-0.txt* is displayed in Fig. 1.

Figure 1 shows an instance of the *logs/log-0.txt*, wherein the first column represents the timestamp which is in the Unix epoch time, the second column represents the loglevel, the third column defines the fully qualified class name, and the last column designates the log message.

New Trace File Starting from UnetStack 3.3, the UnetStack simulator generates trace files in rich JSON format with default name *logs/trace.json*. Before UnetStack 3.3, in the previous versions, the trace files were being generated in NS-2 NAM trace format with default name *logs/trace.nam* which contained traces for events captured by the simulated modem implemented as the physical agent. However, as agents in the stack became more refined and associations between agents became more complex, this led to the need for a sophisticated trace plan to examine protocols with multi-agent interactions. To cater to these needs, the authors of UnetStack came up with a JSON format trace file. The authors signify that the JSON trace format gives

```
1632466366151|INFO|org.arl.unet.UnetBoot@1:doInvoke|fjage Build: fjage-1.9.0/2901c246/8-04-2021_21:48:05
1632466366347|INFO|org.arl.unet.UnetBoot@1:doInvoke|UnetBoot: [sim]
1632466366622|INFO|org.arl.fjage.remote.EnumTypeAdapter@1:<clinit>|Groovy detected, using GroovyClassLoader
1632466366668|INFO|org.arl.fjage.remote.MessageAdapterFactory@1:<clinit>|Groovy detected, using GroovyClassLoader
1632466366838|INFO|org.arl.fjage.remote.MasterContainer@1:openTcpServer|Listening on port 1100
1632466366958|INFO|org.arl.fjage.connectors.TcpServer@11:run|Listening on port 1100
1632466367574|INFO|org.arl.fjage.connectors.WebServer@1:add|Adding web context: /fjage
1632466367578|INFO|org.arl.fjage.connectors.WebServer@1:add|Adding web context: /
1632466367579|INFO|org.arl.fjage.connectors.WebServer@1:add|Adding web context: /web
1632466367599|INFO|org.arl.fjage.connectors.WebServer@1:add|Adding web context: /logs
1632466367618|INFO|org.arl.fjage.connectors.WebServer@1:add|Adding web context: /ws
1632466367850|INFO|org.arl.fjage.connectors.WebServer@1:start|Started web server on port 8080
1632466367852|INFO|org.arl.fjage.remote.MasterContainer@1:addConnector|Listening on ws://127.0.1.1:8080/ws
1632466367858|INFO|org.arl.fjage.connectors.WebServer@1:add|Adding web context: /fjage/shell/ws
1632466369498|INFO|org.arl.fjage.remote.MasterContainer@1:init|Initializing agents...
1632466369500|INFO|org.arl.fjage.shell.ShellAgent/sim@25:init|Agent unetsh#-6ruvz8 init
1632466369599|INFO|org.arl.fjage.remote.MasterContainer@1:init|Agents ready...
1632466369601|INFO|org.arl.fjage.remote.MasterContainer@1:start|Starting container...
```

Fig. 1 logs/log-0.txt

```
{"version": "1.0","group":"EventTrace","events":[
 {"group":"SIMULATION 1","events":[
  {"time":1632207689088,"component":"arp::org.arl.unet.addr.AddressResolution/A","threadID":"c4ae9103-476e-473e-b67b-9403fc5bda7f","stimulus
  {"time":1632207689088,"component":"arp::org.arl.unet.addr.AddressResolution/B","threadID":"54510ba1-867a-4f7c-b97f-f4e2884514e2","stimulus
  {"time":1632207718845,"component":"remote::org.arl.unet.remote.RemoteControl/A","threadID":"4bb3e3bd-b46f-4a2a-af74-4484a9a8bf1c","stimul
  {"time":1632207718856,"component":"transport::org.arl.unet.transport.SWTransport/A","threadID":"7ccfbac1-cfce-4a22-b59d-3e84d6489e45","sti
  {"time":1632207718856,"component":"remote::org.arl.unet.remote.RemoteControl/A","threadID":"4bb3e3bd-b46f-4a2a-af74-4484a9a8bf1c","stimul
  {"time":1632207718858,"component":"transport::org.arl.unet.transport.SWTransport/A","threadID":"7ccfbac1-cfce-4a22-b59d-3e84d6489e45","sti
  {"time":1632207718890,"component":"router::org.arl.unet.net.Router/A","threadID":"78d223f2-3b54-465a-a9a7-a97061e5fd0f","stimulus":{"claz
  {"time":1632207718893,"component":"uwlink::org.arl.unet.link.ReliableLink/A","threadID":"78d223f2-3b54-465a-a9a7-a97061e5fd0f","stimulus":
  {"time":1632207718893,"component":"router::org.arl.unet.net.Router/A","threadID":"78d223f2-3b54-465a-a9a7-a97061e5fd0f","stimulus":{"claz
  {"time":1632207718899,"component":"uwlink::org.arl.unet.link.ReliableLink/A","threadID":"78d223f2-3b54-465a-a9a7-a97061e5fd0f","stimulus":
  {"time":1632207718900,"component":"mac::org.arl.unet.mac.CSMA/A","threadID":"f64b7355-3544-41aa-b440-a8ddca3a4be4","stimulus":{"clazz":"o
  {"time":1632207719241,"component":"mac::org.arl.unet.mac.CSMA/A","threadID":"f64b7355-3544-41aa-b440-a8ddca3a4be4","stimulus":{"clazz":"o
  {"time":1632207719253,"component":"uwlink::org.arl.unet.link.ReliableLink/A","threadID":"78d223f2-3b54-465a-a9a7-a97061e5fd0f","stimulus":
  {"time":1632207719269,"component":"phy::org.arl.unet.sim.HalfDuplexModem/A","threadID":"44e6f977-486f-4a56-b48e-e7d08a7502d2","stimulus":{
  {"time":1632207719272,"component":"phy::org.arl.unet.sim.HalfDuplexModem/A","threadID":"44e6f977-486f-4a56-b48e-e7d08a7502d2","stimulus":{
  {"time":1632207719323,"component":"phy::org.arl.unet.sim.HalfDuplexModem/A","threadID":"44e6f977-486f-4a56-b48e-e7d08a7502d2","stimulus":{
  {"time":1632207720683,"component":"phy::org.arl.unet.sim.HalfDuplexModem/B","threadID":"72ce46e5-f896-4803-91a5-b6fb28c18afe","stimulus":{
  {"time":1632207720694,"component":"uwlink::org.arl.unet.link.ReliableLink/B","threadID":"d822e397-b422-49a5-92fa-216db4638192","stimulus":
  {"time":1632207720695,"component":"phy::org.arl.unet.sim.HalfDuplexModem/B","threadID":"4b57260f-9457-4abf-9060-fdce27c93a84","stimulus":{
  {"time":1632207720695,"component":"uwlink::org.arl.unet.link.ReliableLink/B","threadID":"d822e397-b422-49a5-92fa-216db4638192","stimulus":
  {"time":1632207720696,"component":"phy::org.arl.unet.sim.HalfDuplexModem/B","threadID":"4b57260f-9457-4abf-9060-fdce27c93a84","stimulus":{
  {"time":1632207720697,"component":"remote::org.arl.unet.remote.RemoteControl/B","threadID":"4f1b477d-bbdb-4fd5-b391-0ee489ce6721","stimul
  {"time":1632207720750,"component":"phy::org.arl.unet.sim.HalfDuplexModem/B","threadID":"4b57260f-9457-4abf-9060-fdce27c93a84","stimulus":{
  {"time":1632207722351,"component":"phy::org.arl.unet.sim.HalfDuplexModem/A","threadID":"f3cc8f39-a8c0-4967-907d-350c8ea20fe0","stimulus":{
  {"time":1632207722352,"component":"uwlink::org.arl.unet.link.ReliableLink/A","threadID":"78d223f2-3b54-465a-a9a7-a97061e5fd0f","stimulus":
  {"time":1632207722354,"component":"router::org.arl.unet.net.Router/A","threadID":"78d223f2-3b54-465a-a9a7-a97061e5fd0f","stimulus":{"claz
  {"time":1632207722355,"component":"transport::org.arl.unet.transport.SWTransport/A","threadID":"7ccfbac1-cfce-4a22-b59d-3e84d6489e45","sti
  {"time":1632207726892,"component":"mac::org.arl.unet.mac.CSMA/A","threadID":"f64b7355-3544-41aa-b440-a8ddca3a4be4","stimulus":{"clazz":"o
 ]}
]}
```


Fig. 2 logs/trace.json

a much profuse trace that provides ease of analysis, employing any programming language having JSON support. A small extract from a typical *trace.json* file is shown in Fig. 2.

In Fig. 2, each entry in the file represents the activities of the agent in the course of simulation. The leftmost column represents the time stamp for the event, the column to the next denotes the agent which is involved in activity on a node with its fully qualified class name. Further, the *threadID* which uniquely identifies the iteration, then *stimulus* provides the name of the request class with its fully qualified class name, then *messageID* uniquely represents the message, and *performative* indicates the message type (which can be REQUEST, AGREE, REFUSE, etc.). The sender field represents the sender or the forwarder, the recipient field represents the intermediate node or the receiver, for the responses given by the agents the response field contains the details related to it.

Legacy Trace File Since the initial version of UnetStack, the UnetStack simulator generated trace file with default name *logs/trace.nam*. This trace file followed the NS2 NAM trace format with the notion of making it easy to understand and handle. The *logs/trace.nam* contains traces for different events about each of the packets carried in the network during simulation. It captures events from the physical agent, implemented as a simulated `HalfDuplexModem` and thereby generates traces for events that occur when a packet arrives at the physical agent.

Although the *logs/trace.nam* is easy to understand and analyze, there is a significant limitation of the *logs/trace.nam*. Unlike *logs/trace.json* which contain traces for all the interactions of all the agents in the stack. The simulation traces generated in the *logs/trace.nam* are limited only to the physical agent. Since the *logs/trace.nam* captures only the events that occur in the physical agent, the traces generated here are limited to the activities that happen only in the physical layer. An extract from the trace file is shown in Fig. 3. Furthermore, the newer version of simulator can be

```
# BEGIN SIMULATION 1
n -t 8.005000 -s 3 -x 0.000000 -y 0.000000 -Z 0.000000 -a 3
+ -t 8.005000 -s 3 -d 2 -i 40839989 -p 0 -x {3.0 2.0 -1 ------- null}
- -t 8.005000 -s 3 -d 2 -i 40839989 -p 0 -x {3.0 2.0 -1 ------- null}
n -t 8.005000 -s 1 -x 0.000000 -y 0.000000 -Z 0.000000 -a 1
n -t 8.005000 -s 2 -x 0.000000 -y 0.000000 -Z 0.000000 -a 2
n -t 8.005000 -s 4 -x 0.000000 -y 0.000000 -Z 0.000000 -a 4
r -t 9.005000 -s 3 -d 2 -i 40839989 -p 0 -x {3.0 2.0 -1 ------- null}
r -t 9.005000 -s 3 -d 1 -i 40839989 -p 0 -x {3.0 2.0 -1 ------- null}
r -t 9.005000 -s 3 -d 4 -i 40839989 -p 0 -x {3.0 2.0 -1 ------- null}
+ -t 42.042000 -s 1 -d 2 -i 254433913 -p 0 -x {1.0 2.0 -1 ------- null}
- -t 42.042000 -s 1 -d 2 -i 254433913 -p 0 -x {1.0 2.0 -1 ------- null}
r -t 43.042000 -s 1 -d 2 -i 254433913 -p 0 -x {1.0 2.0 -1 ------- null}
r -t 43.042000 -s 1 -d 4 -i 254433913 -p 0 -x {1.0 2.0 -1 ------- null}
r -t 43.042000 -s 1 -d 3 -i 254433913 -p 0 -x {1.0 2.0 -1 ------- null}
  :
  :
```

Fig. 3 logs/trace.nam [14]

Table 1 Description of events for the traces in *logs/trace.nam*

Event	Event description
+	Indicates enqueue of a packet
d	Indicates drop of a packet
–	Indicates dequeue of a packet
r	Indicates reception of a packet

configured to fall back to legacy trace file by adding following lines in the simulation script:

- *trace = new NamTracer()*
- *trace.open('logs/trace.nam').*

Various events that could be either **+**, **–**, **d**, or **r** according to the extraction from the *logs/trace.nam* file are shown in Fig. 3 and their descriptions are shown in Table 1.

2.2 Limitation

The primary objective of any simulator is to provide the performance metrics of the simulation performed in such a way so that the results can be analyzed in a short time and with less effort. In general, simulators produce one or more trace files, relevant information from the trace file is obtained in form of a graph or other formats using a trace analysis tool. (For instance, with NS3 graphs are plotted using the tool called GnuPlot.) However, UnetStack currently does not have any tool available for result extraction. Hence, it requires the researchers to put substantial effort and time to process trace files and obtain necessary results. The key challenge in building an

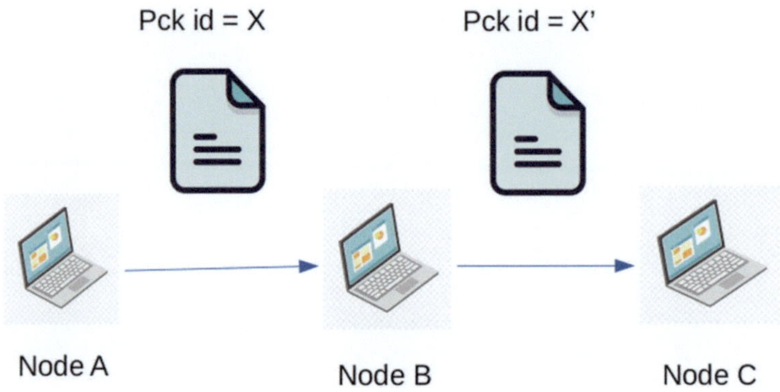

Fig. 4 Change in the packet ID at the intermediate node

result analysis tool for UnetStack lies in uniquely identifying a packet in the course of transmission.

The aforesaid problem encounters in a multi-hop transmission, where the packet ID gets altered at intermediate node. The reason for this is the fact that an intermediate node when forwarding the packet received from its predecessor node generates a new packet ID treating it as a new transmission. Thus, resulting in generation of the new packet ID for the same packet forwarded. In Fig. 4, it can be noticed that the data originated from the source (Node-A) reaches the intermediate node (Node-B) with packet ID as X. The intermediate node forwards the same packet to sink (Node-C) with packet ID changed to X^l. This change in the packet ID at intermediate node results in the challenge to measure the performance metrics as tracing the packet transmission considering the packet ID is not feasible approach. Assuming a situation where a single packet from a single point of source is transmitted, then though the packet ID changes the intermediate node calculation of metrics can be done ignoring the packet ID. However, when there are multiple sources even this does not work and in practical network there will be multiple packets flowing through multiple sources. Therefore, it is really hard to uniquely identify packets from the trace file. Furthermore, if one uses the *logs/log-0.txt* to derive performance metrics, even this approach has the problem of consistency and uniformity as this requires user to log additional information to the *logs/log-0.txt*. Thus, it makes the process of building an automated analysis tool a challenging task.

3 Analyzing Performance in UnetStack

This section discusses the features provided by UnetStack for calculating performance measures. Further the method for using log file (*logs/log-0.txt*) with custom logging for performance calculation is also discussed here.

3.1 *Analyzing Protocol Performance with Discrete-Event Simulation*

Discrete-event simulation enables one to conduct large number of simulations in a small duration, which allows Monte Carlo testing of the protocols developed. The discrete-event mode simulation of four-node ALOHA network to analyze the performance of ALOHA is demonstrated in with a Groovy script in [14] which demonstrates how throughput varies as a function of load.

The result of the simulation depicts the simulated and theoretical throughput measure as shown in Fig. 5.

As per the simulation statistics generated for the simulation, number of frames transmitted is 1456, frames received are 984, with 0.2026 being the offered load. Thereby, throughput estimated is 0.1367. ALOHA performance curve theoretically expressed as $y = x \exp(-2x)$. Therefore, repeated simulation of ALOHA with increasing load, when offered load reaches 0.5, results in a matching maximum throughput value with the theoretical performance.

3.2 *Automated Trace Analysis Tool*

The recent version of UnetStack outputs the simulation trace file by default in the JSON format. For automated trace extraction and analysis, an experimental JSON trace file analysis tool named `VizTrace` is demonstrated in the article published

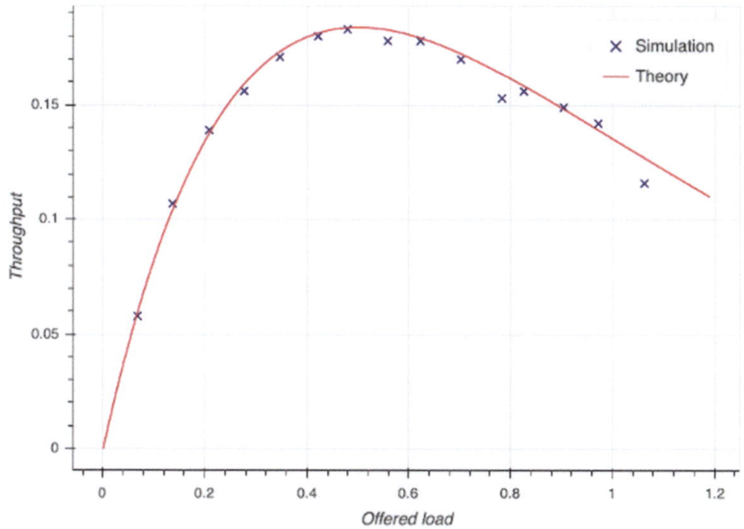

Fig. 5 Simulated and theoretical ALOHA performance [14]

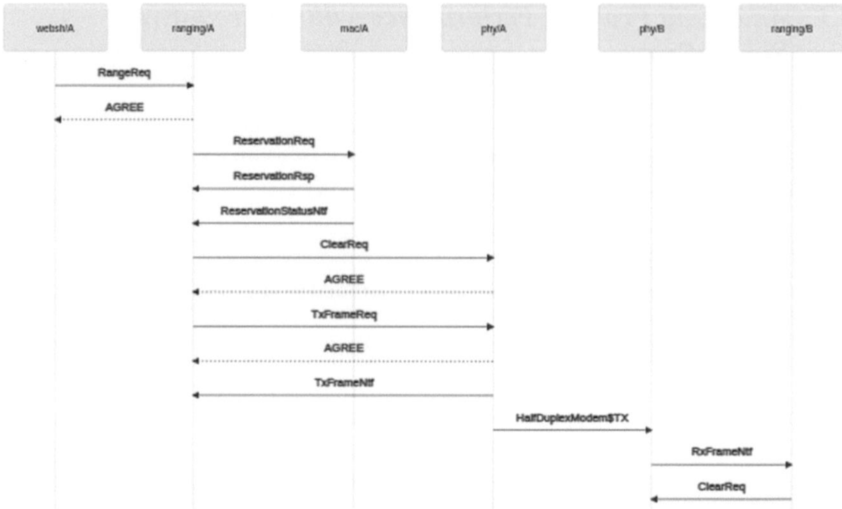

Fig. 6 Sequence diagram describing agent interactions [15]

UnetStack blog [15]. `VizTrace` is written in Julia. It can produce sequence diagram from a JSON trace file. To know the working of the tool, create a basic two-node network and measure the distance between nodes A and B, pass the *logs/trace.json* to the tool, and it will draw a sequence diagram as shown in Fig. 6.

3.3 Performance Calculation Using Custom Logging in logs/log-0.txt

The log file generated for a simulation provides a way to calculate the performance metrics. Here, the data required to calculate necessary performance metric is made to be logged to the *logs/log-0.txt* from a simulation script or from a custom defined agent. Logging custom data at the info loglevel is done using either the *log.info()* or *log.fine()* methods which logs at the info or the fine loglevel. Further, a program or script is implemented to extract necessary fields or the data from the log file and perform calculation of required performance metrics and write the calculated values to a file. Thereafter, a program implemented for plotting reads the file for the calculated values and generates the required plots showing the results for performance metrics.

4 Scope of the Work

Generating the results of simulation and presenting them in an effective manner is vital which can reduce the time and effort. The purpose of an automated trace analysis tool is to provide effective way to present the simulation result which aids in visualizing and measuring the performance metrics. The tool should contain an graphical user interface which is flexible to use and satisfies most of the user requirements, thereby parsing the trace files manually is avoided. Therefore, the productivity will be more enabling the user to focus more on simulating and analyzing the network.

4.1 Design of the Automated Trace Analysis tool

The design of the proposed automated trace analysis tool is done with three layers. An initial layer which can be called as the input layer, an intermediate layer which is the processing layer, and the third layer to be the output layer. A file containing results of simulation is fed to the input layer which then parses it. The processing layer parses the traces recorded in the trace files generated during simulation thereby calculating the parameters like delay, PDR, etc. The output layer functionality will be to represent the calculated performance parameters with suitable graphs and other representations making the inference easy. As a result, the trace analysis tool performs extraction of required information, calculating the required parameters with an aim to present the results.

4.2 Trace Analysis Tools for Different Simulators

For analyzing the trace files, NS-2 visual trace analyzer is an open-source analyzer tool for NS-2 simulator which is implemented with C# and .NET framework. This tool requires the .NET framework and thus can be run on Windows platform only. For other platforms, such as Linux and Mac, these tools have not been tested. Further, a similar implementation can be seen in Java which can be executed on different operating system platforms and this tools accepts a Tcl, trace file as input, it then parses the file and generates the required metrics. There is another tool called the GnuPlot that can be used with NS2 to plot graphs. With NS3, in addition to the default visualization tool a new visualizer named Netanim is available as a native package of NS3 simulator. UnetStack currently lacks a good trace analyzer tool that can provide a detailed representation of underwater environments with all real-time parameters. Even yet, only a small amount of work is being done to evaluate the *logs/log-0.txt* and *logs/trace.json* files with more complex features. As a result, it is required to develop a trace analyzer for simulations in UnetStack that includes advanced features such as extraction, analysis, and plotting of most of the network performance metrics such as packet loss, packet delay, end-to-end latency, and packet received.

5 Concluding Remarks

Underwater network simulation provides an efficient and cost-effective approach for testing and validating the research work. However, the task of processing the simulation traces for extracting and analyzing the desired network performance metrics is an additional overhead which incurs significant time and effort, which signifies the need for trace analysis tool which can automate result extraction and analysis. This work reviews need of underwater simulation and tool for trace analysis with different simulation platforms and components. Further, highlighting the features of UnetStack as an ideal choice for most of the underwater simulations the work discusses the existing approaches and facilities that are available in UnetStack for result extraction and analysis and scope for developing an automated trace analysis tool that can be more generic and accurate.

References

1. Ismail N-SN, Hussein LA, Ariffin SHS (2020) Analyzing the performance of acoustic channel in underwater wireless sensor network (UWSN). In: 2010 Fourth Asia international conference on mathematical/analytical modelling and computer simulation. IEEE, pp 550–555
2. Dol HS, Casari P, Van Der Zwan T, Otnes R (2016) Software-defined underwater acoustic modems: historical review and the NILUS approach. IEEE J Ocean Eng IEEE 42:722–737
3. Demirors E, Sklivanitis G, Santagati GE, Melodia T, Batalama SN (2018) A high-rate software-defined underwater acoustic modem with real-time adaptation capabilities. In: IEEE Access 6:18602–18615
4. Renner C, Golkowski AJ (2016) Acoustic modem for micro AUVs: design and practical evaluation. In: Proceedings of the 11th ACM international conference on underwater networks & systems. association for computing machinery, New York, NY, United States. Shanghai China, pp 1–8
5. Sherlock B, Tsimenidis CC, Neasham JA (2015) Signal and receiver design for low-power acoustic communications using m-ary orthogonal code keying. In: OCEANS 2015-Genova. IEEE, pp 1–10
6. Ilyas M, Mahgoub I (2004) Handbook of sensor networks: compact wireless and wired sensing systems. CRC Press
7. Chandavarkar BR, Gadagkar AV (2020) A framework for residual energy model in unetstack simulator for underwater sensor networks. In: 2020 5th International conference on computing, communication and security (ICCCS). IEEE, pp 1–8
8. Shiwen M (2010) Chapter 8: fundamentals of communication networks. In: Cognitive radio communications and networks. Academic Press, Oxford, pp 201–234
9. Petrioli C, Petroccia R, Potter JR, Spaccini D (2018) The SUNSET framework for simulation, emulation and at-sea testing of underwater wireless sensor networks. In: Ad Hoc networks, vol 34. Elsevier, pp 224–238
10. Masiero R, Azad S, Favaro F, Petrani M, Toso G, Guerra F, Casari P, Zorzi M (2012) DESERT underwater: an NS-miracle-based framework to design, simulate, emulate and realize test-beds for underwater network protocols. In: 2012 Oceans-Yeosu. IEEE, pp 1–10
11. Chitre M, Bhatnagar R, Soh W-S (2014) UnetStack: an agent-based software stack and simulator for underwater networks. In: 2014 Oceans-St. John's. IEEE, pp 1–10
12. Akyildiz IF, Pompili D, Melodia T (2016) State-of-the-art in protocol research for underwater acoustic sensor networks. In: Proceedings of the 1st ACM international workshop on Underwater networks, New York, NY, United States. ACM, pp 7–16

13. Torres D, Friedman J, Schmid T, Srivastava MB, Noh Y, Gerla M (2015) Software-defined underwater acoustic networking platform and its applications. In: Ad Hoc networks, vol 34. Elsevier, pp 252–264
14. UnetStack Handbook. https://unetstack.net/handbook/unet-handbook-preface.html
15. Unet Blog. https://blog.unetstack.net/whats-new-in-UnetStack-3.3

Implementation of Next Hop Selection in Underwater Routing Using TOPSIS

Afrah Nayeem, B. R. Chandavarkar, and Pradeep Nazareth

Abstract Underwater Acoustic Sensor Networks (UASNs) are a prominent research area in academia and industry. UASNs are used in a wide range of applications, including ocean monitoring, environmental monitoring, and underwater surveillance. However, UASNs face a number of challenges like high ocean interference and noise, high propagation delay, low bandwidth, and limited battery life. In addition, most underwater applications demand efficient routing. Selecting the optimal next hop is important to ensure reliable and efficient packet delivery. This paper proposes an attribute fetching and node ranking mechanism to be used in a routing protocol. The source node obtains the attributes from its neighbors and ranks the nodes in decreasing order of priority. The priorities can be used to choose the best next hop in a routing protocol. The proposed mechanism is implemented in UnetStack, an agent-based underwater simulator.

1 Introduction

Water covers about 70% of the Earth's surface. A wealth of unexplored resources exists underwater. However, the underwater environment is too complicated to explore. Underwater acoustic Sensor Networks are becoming very popular. Underwater acoustic sensor networks (UASNs) can be applied in both off-shore and on-shore fields, like environmental monitoring, patrolling coastal waters, assisted navigation, deep-sea mining, and disaster prevention. However, underwater networks pose tough challenges like low acoustic speed, poor network connectivity, high deployment cost, low data rate, and greater energy consumption [1]. We can overcome the above challenges by exploiting routing protocols [2] that efficiently forward the data packets from the source to the sink. Choosing the best next hop is critical to efficient and reliable packet delivery in routing.

A. Nayeem · B. R. Chandavarkar · P. Nazareth (✉)
Wireless Information Networking Group (WiNG), Department of Computer Science and Engineering, National Institute of Technology Karnataka, Surathkal, Mangalore, India
e-mail: pradeep.nitk2017@gmail.com

S. K. Biswas et al. (eds.), *Intelligent Computing Systems and Applications*, Lecture Notes in Networks and Systems 1307, https://doi.org/10.1007/978-981-96-3860-4_37

This paper proposes a node ranking mechanism for the best next hop selection in a routing protocol. The source node obtains link quality, energy attribute, and depth from its neighbors. The node then uses the Technique for Order of Preference by Similarity to Ideal Solution (TOPSIS) algorithm to rank its neighbors based on the three criteria [3]. The source node selects the best neighbor as the next hop and forwards the packet to it. The mechanism has been implemented and simulated in UnetStack, an agent-based network stack for underwater communication [4].

The rest of this paper is organized as follows. Section 2 presents the related work. Section 3 discusses the proposed method in detail. Section 4 presents the simulation of the mechanism in UnetStack. The results of the simulation are analyzed in Sect. 5. Finally, the conclusion is presented along with suggestions for future work in Sect. 6.

2 Related Work

This section focuses on the literature work that used the TOPSIS or relevant techniques in the various protocols in the area of networking. Azad and Sharma [5] designed a fuzzy TOPSIS-based cluster head selection technique. The candidate node with the highest TOPSIS rank in that cluster is selected as the cluster head. The protocol, in which the candidate node with the higher residual energy and neighbor node density, & lower distance to the base station, is selected as cluster head. The nodes choose to be associated with the cluster head that is nearest to them. The node connects directly with the base station if it is closer to it than the cluster head. The cluster heads use TDMA to gather data from their cluster members, aggregate it, and send it to the base station.

Jain and Rani [6] designed a novel ring-based cross-layer routing protocol for the IoT for smart city applications. The protocol proposes an IoT-WSN [7] protocol for real-time applications. The nodes follow a ring system, and the base station is assumed to be located at the center. The next hop candidates must be from the next ring or within the same ring in the transmission range. Routing is based on the shortest path, using the TOPSIS algorithm to choose the best node at each hop. The optimal nodes are selected based on the following criteria: remaining energy, the distance of selected nodes from the base station, the Euclidean distance between mth selected node, and the next optimum nth node in the transmission range, and the number of neighboring nodes.

Saloni and Deepthi [8] designed an optimized dual-hop routing protocol based on grid clustering. In this routing protocol, the network is split into many equal-sized grids. The node with the maximum residual energy in each grid is chosen as the grid head. The sensor nodes belonging to the same grid are only allowed to communicate with each other and with the grid head. The grid heads can communicate with the other grid heads. The selected grid head relays the data collected from the nodes to the base station. The optimal next hop is chosen by applying the TOPSIS on all the grid heads. The three criteria are each grid head's remaining energy, distance from the sink, and the number of neighboring grid heads. The grid heads closer to the base station forward the aggregated data directly to the base station, while the farther grid heads employ dual-hop communication.

A fuzzy-logic-based cluster head selection and TOPSIS-based mobile sink path planning were proposed by Sunil and Prabhat [9]. The cluster heads are selected by the scheme using fuzzy logic, and the nodes associate with the cluster head that is nearest to them. The new cluster heads are unknown to the sink. To discover the identities of every cluster head, the mobile sink follows a circular route. Whenever the cluster heads are in communication range, they transmit data to the mobile sink. The three factors—remaining energy, node degree, and edge cost—are used by the TOPSIS technique to rank each cluster head. In accordance with the TOPSIS rank, the mobile sink visits the cluster heads and gathers the data from the cluster heads.

A fuzzy TOPSIS-based cluster head selection strategy was presented by Khan and Bilal [10], Sorin et al. [11]. According to this protocol, the cluster heads gather data from each cluster member, aggregate it, and deliver it to the sink. The sink broadcasts its location to the sensor nodes. Every sensor node broadcasts a 'Hello' message carrying the node attributes and updates its neighborhood tables. The cluster heads are then selected using fuzzy TOPSIS. The five main criteria considered in choosing the cluster head are the residual energy of the node, the energy consumption rate, the distance between the node and the sink, the node density, and the average distance to each neighbor. The cluster heads broadcast advertisements to their neighbors. The node chooses to associate with the closest cluster head. Re-clustering is performed only if the threshold value is smaller than the neighbors to avoid overhead.

3 Proposed Mechanism

In the proposed method, every node periodically broadcasts requests for attributes of neighbors. The neighbors respond to the request with their link quality, residual energy, and depth. The node applies the TOPSIS algorithm, calculates the TOPSIS score, and ranks the neighbors accordingly. The node can directly forward the packet to the best node or form clusters based on the specific routing protocol. The complete mechanism has been depicted in Fig. 1.

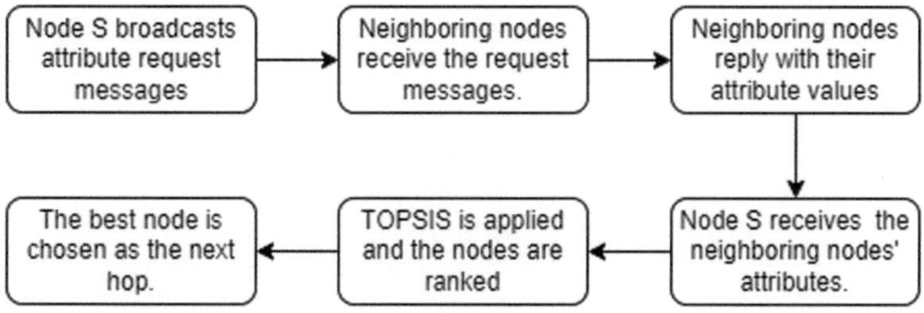

Fig. 1 Attribute fetching and node ranking mechanism

3.1 Obtaining Attributes from the Neighbors

In the proposed method, every node transmits attribute requests regularly. Neighbors respond to the node by indicating their depth, link quality, and residual energy. Link quality significantly influences communication [12]. The quality is measured either by transmissions or by underwater characteristics. Another major factor is the limited battery power of nodes. Batteries cannot be easily recharged underwater [13]. Hence, the protocol must ensure fair and judicious use of energy.

The wireless link quality is estimated in terms of packet delivery probability using the signal-to-noise ratio (SNR) approach [14]. The path loss over a distance d for a signal of frequency f due to large-scale fading is given as $A(d, f) = d^k a(f)^d$, where k is the spreading factor, and a(f) is the absorption coefficient. The spreading factor k denotes the propagation geometry. For a typical scenario, k is given as 2. The absorption coefficient a(f) is computed using Thorp's formula [15]. Therefore, the average SNR over distance d is estimated using Eq. 1:

$$\Gamma(d) = \frac{E_b/A(d, f)}{N_0} \tag{1}$$

In a nonfading additive white Gaussian noise (AWGN) channel, E_b and N_0 are constants that indicate the average transmission energy per bit and noise power density.

3.2 Ranking of Neighboring Nodes Using TOPSIS

After the source node obtains the attributes from its neighbors, the TOPSIS algorithm ranks them in decreasing order of priority. The priorities can be utilized to choose the best next hop in a routing protocol.

TOPSIS [3] is a multi-criteria decision-making technique that chooses the best alternative from many alternatives based on the similarity to the ideal solution. The selected alternative is closest to the best solution and farthest from the worst solution. The alternatives and their attributes are represented in a decision matrix. A_i denotes an alternative i and C_j denotes the criterion j. x_{ij} denotes the value of attribute j for the alternative i. The alternatives are nothing but the nodes, and the criteria are the node attributes: link quality, depth, and energy.

$$\begin{array}{c} \begin{matrix} C_1 & C_2 & C_3 \end{matrix} \\ \begin{pmatrix} x_{11} & x_{12} & x_{13} \\ x_{21} & x_{22} & x_{23} \\ \cdots & \cdots & \cdots \\ x_{m1} & x_{m2} & x_{m3} \end{pmatrix} \begin{matrix} A_1 \\ A_2 \\ A_3 \\ A_m \end{matrix} \end{array}$$

Step 1: The x_{ij} values are normalized using sum normalization as follows:

$$X_{ij} = \frac{x_{ij}}{\sum_{i=1}^{m} x_{ij}} \tag{2}$$

Step 2: The weights of the attributes are calculated using the entropy method as shown in equations (3) and (4). The X_{ij} values are further multiplied with their respective weights to give v_{ij}.

$$E_j = -\frac{\sum_{i=1}^{m}(x_{ij} * ln(x_{ij}))}{ln(m)} \tag{3}$$

$$w_j = \frac{1 - E_j}{\sum_{j=1}^{3}(1 - E_j)} \tag{4}$$

Step 3: The positive ideal solution A^+ and the negative ideal solution A^- are computed. The distance of an alternative A_i from the positive ideal solution and the negative ideal solution is measured as d_i^+ and d_i^-.

$$A^+ = v_1^+, v_2^+, v_3^+ = (max_i v_{ij}| \quad i = 1, 2, ..., m), j = 1, 2, 3 \tag{5}$$

$$A^- = v_1^-, v_2^-, v_3^- == (min_i v_{ij}| \quad i = 1, 2, ..., m), j = 1, 2, 3 \tag{6}$$

Step 4: The TOPSIS score of the alternative is calculated (7).

$$score = \frac{d_i^-}{d_i^+ + d_i^-} \tag{7}$$

Step 5: The alternatives are then ranked on the basis of their TOPSIS scores.

4 Simulations

The proposed algorithm has been implemented in the UnetStack3 simulator. The first part of this section briefly describes the UnetStack tool. The second part presents the simulated topology, simulation parameters, and the two custom Unet agents that implement the proposed algorithm.

4.1 UnetStack

UnetStack [4] was created in 2004 at the National University of Singapore's Acoustic Research Lab as part of the Unet project. UnetStack is an agent-based stack and an underwater network simulator that can be used to create, simulate, deploy, and test underwater networks quickly. The default stack comes with a set of software agents representing various layers of the traditional network stack. UnetStack is built on top of the fjage framework that defines the agents and services in the stack, as shown in Fig. 2. Agents are the fundamental building blocks of UnetStack, exchanging messages and providing well-defined services.

When an agent offers a service, it advertises the service for other agents to query. Physical service, datagram service, routing service, node information service, and remote access service are some of the essential services provided by UnetStack. Moreover, new agents and services may be included, or existing agents can be extended to fulfill the requirements. UnetStack simulator is a UnetStack component. UnetSim is a popular and effective simulator designed for underwater network simulations. An underwater network with plenty of nodes can be simulated on a single computer using UnetSim. It can run a real time or discrete event simulation of an underwater network.

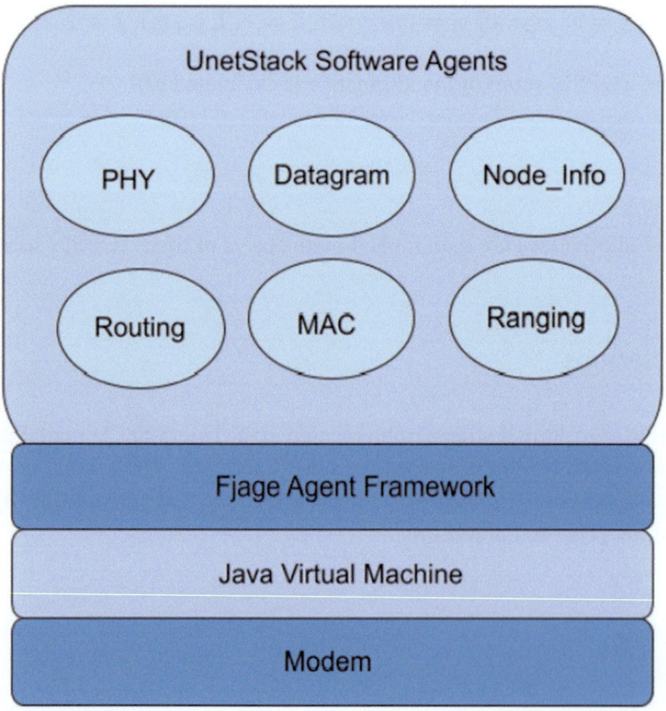

Fig. 2 UnetStack architecture

4.2 Implementation Using UnetStack

The topology shown in Fig. 3 has been simulated in UnetStack. The simulation runs in real-time mode. The simulation's channel property configures the physical channel settings, such as communication range and sound speed. Each node is described by the node name, location, ports, and network stack (agents to be loaded). Table 1 and Fig. 3 illustrate the simulation parameters and the simulation map, respectively.

Two custom agents have been defined to implement the Protocol: the attribute agent and the TOPSIS agent. The attribute agent responds to each incoming attribute

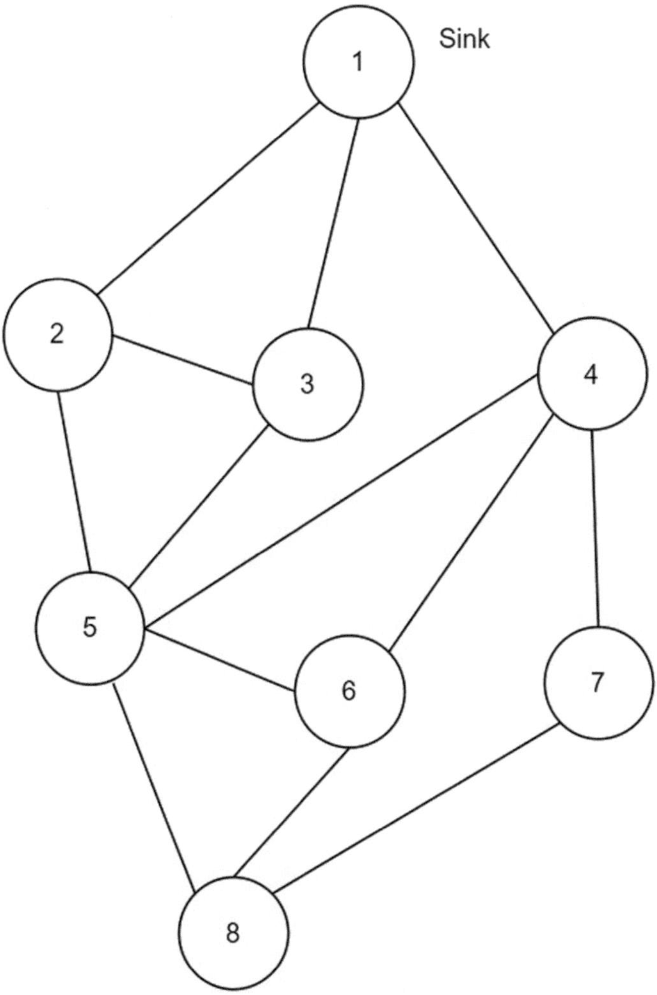

Fig. 3 The simulated topology

Table 1 Simulation parameters

Parameters	Values
Number of nodes	8
Communication range	500 m
Simulation time	5 min
Transmission distance	Euclidean distance between the nodes
Simulation mode	RealTimePlatform
Sound speed	1500 mps

request datagram with an attribute reply datagram. The sender's location coordinates are included in the attribute request. These position coordinates are used by the neighbor to determine the distance to the sender, which is then used to estimate the link quality using the parameters in Table 2. The neighbor's depth, link quality, and energy attribute are all included in the attribute reply. The attribute request is sent as a broadcast message, whereas the attribute reply is sent as a unicast message. Two different protocol numbers, Protocol.USER and Protocol.DATA, are used to distinguish between the attribute request and the attribute reply.

The Attribute agent is responsible for the following tasks:

1. Broadcast attribute request messages periodically
2. Accept attribute request messages from other nodes
3. And respond to the sender with the values of its attributes

The attribute agent broadcasts attribute requests periodically using a ticker behavior. However, if all of the nodes broadcast the attribute requests at the same time, there would be a lot of collisions. To avoid collisions, a random delay of 0-2 s is included. The startup() method subscribes to notifications from PHYSICAL service agents. When a notification arrives, the processMessage() method is invoked. The agent verifies if the received datagram is of protocol USER in the processMessage() method. The NODE INFO agent provides the agent with the node's position coordinates, and the agent estimates the link quality, energy attribute, and depth. The link quality is calculated using Eq. 1 and the parameters in Table 2. The agent then returns to the sender a datagram with protocol.DATA containing an array of the three attributes.

Table 2 Link quality parameters

Link quality parameters	Values
Eb	40 J
N0	1 J
k	2
f	18 KHZ
a(f)	3 dB/km

The TOPSIS agent is responsible for the following tasks:

1. Receive the attribute reply messages from the neighbors
2. Update that neighbor's attribute values in the hash map
3. Apply TOPSIS algorithm and calculate TOPSIS scores
4. Rank the neighbors according to their TOPSIS scores.

The topsis agent of a node stores the attribute values of its neighbors in a hash map. The hash map contains a key-value pair. The key is the address of the neighboring node, while the value is an array comprising the node's link quality, energy attribute, and depth. The startup() function in the Topsis agent subscribes to the Physical service to be notified of the attribute reply messages. When a notification arrives, the processMessage() method is invoked, and the sender's attributes are updated in the hash map. Consider node 4 in the topology; its neighbors are 6, 7, 5, 1. The hash map at node 4 is as follows:
6 : [44.0 , 81.0 , 475.0],
7 : [8.0 , 52.0 , 915.0],
5 : [113.0 , 62.0 , 350.0],
1 : [24.0 , 67.0 , 200.0].

The TOPSIS function is periodically called by the agent. The hash map is passed to the TOPSIS function. It computes each surrounding node's TOPSIS score and returns the nodes sorted by their scores. The TOPSIS agent at node 4 sorts the nodes in the following order: [5: 0.81, 6: 0.34, 7: 0.21, 1: 0.14]. The events that occur during the simulation are recorded in the log file for further analysis.

5 Results and Analysis

This section presents the simulation and demonstrates the implemented model in UnetStack. The given topology is simulated in real-time mode for 5 min, and the log file is analyzed. Consider the log output of node 4 shown in Table 3. The attribute agent at node 4 received an attribute request from 1 and responded to it with its attribute values. Then it broadcasted a request message. The TOPSIS agent at 4 received attributes from neighbors 5, 6, 1, 7 and ranked them based on their TOPSIS score.

The nodes obtain the neighbor attributes, update the hash map, and produce the TOPSIS ranks as shown in Table 4. Consider node 4, at timestamp 1, the priority order of the neighbors is 5, 6, 7, 1. At timestamp 2, the priority order of the neighbors becomes 5, 6, 1, 7. At timestamp 3, the priority order of the neighbors becomes 7, 5, 6, 1. The shifting of node priorities is caused by shifting criterion values. The link quality, depth, and energy attributes are the criteria used to rank nodes. While a node's depth and link quality remain constant, the energy attribute changes. This results in a change in the node ranking.

Table 3 Log at node 4

Agent	Event
attribute@4	4 received attribute request from 1
attribute@4	4 replied with its attributes 18,98,500 to 1
attribute@4	4 broadcasted attribute requests
topsis@4	4 received attributes [113, 69, 350] from 5
topsis@4	4 received attributes [44, 87, 475] from 6
topsis@4	4 received attributes [24, 24, 200] from 1
topsis@4	4 received attributes [8 , 96 , 915] from 7
topsis@4	Applying TOPSIS on : [5:[113.0, 69.0, 350.0], 7:[8.0, 96.0, 915.0], 1:[24.0, 24.0, 200.0], 6:[44.0, 87.0, 475.0]]
topsis@4	Node Rank: [5 : 0.81 , 6: 0.35 , 7:0.23 , 1: 0.14]

Table 4 Output of TOPSIS agent

Rank of neighboring nodes (in order)			
Node	Nodes at timestamp 1	Nodes at timestamp 2	Nodes at timestamp 3
1	2, 4, 3	4, 2, 3	4, 2, 3
2	5, 1, 3	5, 1, 3	5, 3, 1
3	1, 5, 2	5, 2, 1	5, 2, 1
4	5, 6, 7, 1	5, 6, 1, 7	7, 5, 6, 1
5	8, 6, 4, 2, 3	6, 2, 8, 4, 3	6, 4, 2, 3, 8
6	5, 8, 4	5, 8, 4	8, 5, 4
7	4, 8	8,4	8, 4
8	5, 6, 7	5, 6, 7	5, 6, 7

6 Conclusion and Future Work

This paper proposed an attribute fetching and node ranking mechanism for the best next hop selection in a routing protocol. The source node obtains its neighbors' link quality, energy attribute, and depth. The TOPSIS score is computed, and the neighbors are ranked in the decreasing order of priority. The priorities can be used to choose the best next hop in a routing protocol. The implementation has been done in UnetStack, and the simulation results have been logged and verified. Further, the work can be extended by considering additional attributes based on the underwater application. Additionally, the proposed mechanism can be tested in a real-time test-bed to analyze the actual performance.

References

1. Akyildiz I, Pompili D, Melodia T (2005) Underwater acoustic sensor networks: research challenges. Ad Hoc Netw 3:257–279. https://doi.org/10.1016/j.adhoc.2005.01.004
2. Li N, Martínez J-F, Meneses Chaus JM, Eckert M (2016) A survey on underwater acoustic sensor network routing protocols. Sensors 16(3). https://doi.org/10.3390/s16030414. https://www.mdpi.com/1424-8220/16/3/414
3. Yoon HC (1981) Multiple attribute decision making: Methods and applications. Springer
4. Chitre M, Bhatnagar R, Soh W-S (2014) UnetStack: an agent-based software stack and simulator for underwater networks, pp 1–10
5. Azad P, Sharma V (2013) Cluster head selection in wireless sensor networks under fuzzy environment. ISRN Sensor Network
6. Rani J (2017) A novel approach for smart cities in convergence to wireless sensor networks. Sustain Cities Soc
7. Bajaj K, Sharma B, Singh R (2020) Integration of WSN with IoT applications: a vision, architecture, and future challenges, pp 79–102. https://doi.org/10.1007/978-3-030-38516-3_5
8. Saloni Kakkar D (2018) Topsis optimized dual-hop routing protocol for homogeneous wireless sensor networks with grid-based clustering
9. Prabhat S (2021) Multi-criteria decision-based path planning for data collection in fuzzy-cluster based large sensor networks
10. Khan R, Bilal Y (2016) Fuzzy-topsis based cluster head selection in mobile wireless sensor networks. J Electric Syst Inf Technol
11. Sorin N, Dzitac S, Dzitac I (2016) Fuzzy topsis: a general view. Proc Comput Sci 91:823–8. https://doi.org/10.1016/j.procs.2016.07.088
12. Nazareth P, Chandavarkar BR (2020) Link quality-based routing protocol for underwater acoustic sensor networks, pp 1–6. https://doi.org/10.1109/ICCCNT49239.2020.9225289
13. Ovaliadis K, Savage N, Kanakaris V (2010) Energy efficiency in underwater sensor networks: a research review. J Eng Sci Technol Rev. https://doi.org/10.25103/jestr.031.25
14. Stojanovic M (2006) On the relationship between capacity and distance in an underwater acoustic communication channel. Association for Computing Machinery, New York, NY, USA, WUWNet'06, pp 41–47. https://doi.org/10.1145/1161039.1161049
15. Sehgal A, Tumar I, Schönwälder J (2009) Variability of available capacity due to the effects of depth and temperature in the underwater acoustic communication channel, p 1. https://doi.org/10.1109/OCEANSE.2009.5278268

Cluster Formation for Underwater Routing in UnetStack3

Pradeep Nazareth, B. R. Chandavarkar, and Anusha P. Das

Abstract Underwater Acoustics Sensor Networks (UASNs) are utilized in a range of underwater applications, including sea habitat monitoring, offshore research, and mineral exploration. Due to the underwater current, low bandwidth, high water pressure, fluctuations in link quality between nodes, propagation latency, and error probability, underwater communication is challenging. Because of these difficulties, data transmission in UASNs is unreliable during routing. One strategy to improve routing speed is to use an opportunistic routing technique. The sender will transmit the data to the set of neighbours in opportunistic routing such that at least one neighbour can receive and forward the data. The main processes in opportunistic routing include evaluating the adjacent nodes, picking the group of neighbours, and coordinating among the selected nodes to transfer the received data. The optimum next-hops during routing are picked. The numerous properties of neighbouring nodes are analysed and the neighbouring nodes are used for forming clusters that are utilised to choose the best next-hops. In this paper, a novel approach for sensor node clustering technique for UASNs is proposed. Here, it is assumed that the neighbours of a sender node are already ranked. A suitable algorithm like TOPSIS (Technique for Order of Preference by Similarity to Ideal Solution) is employed to search for the best next-hops and determines a set of candidates to be considered for cluster formation. The protocol has been implemented and simulated in UnetStack3, an agent-based network stack for underwater communication.

Keywords Underwater acoustic sensor networks · Cluster members · Cluster Head · Euclidean distance · Opportunistic routing protocol · UnetStack

---

P. Nazareth (✉) · B. R. Chandavarkar · A. P. Das
Wireless Information Networking Group (WiNG), Department of Computer Science and Engineering, National Institute of Technology Karnataka, Surathkal, Mangalore, India
e-mail: pradeep.nitk2017@gmail.com

S. K. Biswas et al. (eds.), *Intelligent Computing Systems and Applications*, Lecture Notes in Networks and Systems 1307, https://doi.org/10.1007/978-981-96-3860-4_38

1 Introduction

Oceans encompass a vast portion of the earth [1]. Oceans are home to a diverse range of animals and other species. They are the primary source of food, transportation, and other resources for numerous lifeforms. As society and the economy have progressed, more people have begun to pay attention to maritime conditions. Human activities associated with industry and urban expansion tourism have a significant influence on marine condition frameworks. Human interest in marine exploration was limited in the early days due to the extreme environment of oceans, such as wide expanses and high water pressure.

Sensor networks were quite popular in the research field in the twentieth century. These networks were originally solely utilised in terrestrial areas. With the passage of time, advancements in oceanic modem technology allowed wireless sensor networks to be used in underwater settings [2]. Sensor network deployment and management are more difficult in this setting than in land-based scenarios. Many applications need the monitoring and communication of an undersea environment. These applications include a broad range of topics, including climate monitoring, undersea environment, marine ecology, natural disasters, navigation, and picture transmission to remote locations on land, as well as offshore investigation and surveillance.

The routing performance is impacted by the dynamic links between nodes, resulting in low communication efficiency. Opportunistic Routing (OR) can be utilised to increase the reliability of communication [3]. In traditional routing, only a single next-hop forwarder node is entrusted with the task of packet forwarding. OR is a novel routing paradigm for UASNs that chooses a set of candidate nodes and the incoming packet will be forwarded to the cluster formed from this candidate set. OR takes advantage of wireless sensor networks' broadcasting capabilities. Sensor networks have improved in efficiency, throughput, and reliability owing to OR [4]. Many energy-saving approaches have been established in UASNs employing opportunistic routing to extend the network lifetime [5].

An appropriate routing protocol must be developed to meet the issues outlined above. The suggested method employs an algorithm to find the optimal next-hops by choosing a group of candidates that are eligible to participate in the process of cluster formation. A suitable method is used to determine an ideal measure of nodes required for routing and give rankings to the nodes in the selected candidate set based on the computed measures. Clustering refers to the grouping of sensor nodes into logical groups. These nodes collect information and transmit it to the base station. The cluster head is picked, and the remaining nodes in the candidate set are referred to as cluster member nodes. The cluster head is in charge of all inter and intra-cluster communication. Cluster heads will be chosen based on various parameters like energy, link quality, etc. of the nodes.

In this paper, we propose a novel approach for sensor node clustering technique for UASNs to increase the throughput, and reliability of the underwater networks and address the above mentioned challenges. It is assumed that the neighbours of a sender node are already ranked using an appropriate algorithm like TOPSIS to search for the best next-hops and determine a set of candidates to be considered for cluster formation. The proposed approach has been implemented and simulated in UnetStack, an agent-based network stack for underwater communication testing and deployment.

The remaining part of the paper is as follows: Section 2 discusses related work, Sect. 3 contains an elaboration on the proposed cluster formation approach, Sect. 4 contains details of results and analysis of the clustering approach, followed by conclusion in the Sect. 5.

2 Related Work

Clustering is the process of grouping sensor nodes to forward packets received. Each sensor node at the ocean's bottom monitors numerous events in the oceanic environment, such as temperature and pressure. One of the primary goals of clustering is to prevent redundant data transmission by receiving nodes and to solve the hidden-node problem [6, 7].

The techniques of cluster creation utilised in various routing protocols. Depth-based routing (DBR) protocol has been developed for UASN communication to efficiently route data to the proper destination to enable energy-efficient routing. The depth of a sensor node is measured using evidence data from lower nodes transmitted to receiver nodes. These lower nodes have a depth sensor installed inside it which is used to measure depth. For energy-efficient deep routing, depth and residual energy information are used, although the end-to-end delay is significant in DBR [8].

Another protocol developed for an underwater environment is energy efficient DBR (EEDBR) [9]. In this protocol, a sensor node picks its neighbour node based on how much energy it has left and how close it is to the sink node, then saves the identifier of the neighbour node in the routing table. Thereafter, each node uses its routing tables to send data packets to the next node with the most remaining energy and the smallest sink depth.

In WSNs and UWSNs, the Low Energy Adaptive Cluster Head Selection (LEACH) method is most typically utilised [10]. The main goal of this method is to minimise the amount of energy used by sensor networks. The paper [11] described a solution for underwater sensor networks that uses a low-energy clustering structure. When a member node chooses a cluster to join, it takes into account the cluster head's location, as well as energy intake and consumption between member nodes and the cluster head. The UMOD-LEACH protocol is a modification of the Leach protocol, and the improved method achieves 30% better outcomes in an underwater environment than (LEACH) [12]. It employs the principle of localization as well as time division multiple access (TDMA).

In the paper [13], another adaptive clustering routing method (ACUN) for underwater networks was presented. It employs multipath routing based on two primary parameters: the (CH) node's remaining energy and the distance between (CH) and sink. The Leach-MAC method discussed in the paper [14], is a variation of this approach that is more efficient in WSNs. Also, [15] proposes a region-based (R-Leach) and modified weighted (W-Leach).

Hydrocast is depth-based, sender-based routing protocol [16]. It selects the subset of forwarding nodes that maximizes progress while limiting channel interference. It is stateless and avoids costly distributed localization. The advantage of the Hydrocast routing protocol is, that it avoids the hidden-node problem. In that direction, the sender or forwarding node selects the candidate nodes by forming the clusters. One of the clusters will be selected as the candidate forwarding set. As the first step of selecting a candidate forwarding set, the node computes the Normalized Advancement (NADV) of its neighbors present at a lower-depth. The NADV computed by the Eq. 1:

$$NADV_n = d_n \times P_n \tag{1}$$

where d_n and p_n are the progress and packet delivery probability (PDP) of the neighbor node n respectively. Packet delivery probability is dependent on the frequency, the distance between sender and receiver, and size of the packet [17]. To form the cluster, sender S will select the neighboring node C which is having highest NADV. The node S will identify common nodes in, its own neighboring set and neighbors of C at the distance $< 0.5 \times R$, where R is the maximum communication range of the node. The clustering process will continue by considering the highest NADV value among other nodes, which are not clustered. Likewise, all nodes must be clustered. Finally one among multiple clusters needs to be selected as a candidate forwarding set by considering the highest Expected Packet Advancement (EPA). The EPA of the cluster can be computed by the Eq. 2.

$$EPA(Cluster_k) = \sum_{i=1}^{k} d_{n_i} P_{n_i} \prod_{j=0}^{i-1} (1 - P_{n_j}) \tag{2}$$

Nodes in the $Cluster_k$ are ordered according to their priorities, where $n_1 > n_2 > n_3, .. > n_k$.

Stateless Routing Protocol (SORP) is a hybrid strategy that combines receiver and sender-based approaches [18]. When a node gets a data packet in the SORP protocol, it first updates its neighbouring table and then determines whether or not it is eligible to participate in packet forwarding. If it is ineligible, the node simply discards the received packet. The nodes merely read the header of the incoming packet to decide whether to accept or reject it as soon as possible to save energy. In this protocol, the selected candidate forwarding set must satisfy the following three conditions:

- The receiving node must be placed at a lower depth in the upper hemisphere of the current forwarding node transmission coverage region.
- The receiving node must be a standard node. It must not be a node that is trapped or void. Any data packet received by the detected trapped and void nodes should be dropped until their state is altered.

- The receiving node should be within the forwarding area, i.e. its relative distance to the best candidate node should be less than a certain value. Since they are in the suppression region, receiving nodes outside of this area must simply drop the received packets.

3 Proposed Cluster Formation Approach

Every node in the topology forms its cluster. In the proposed cluster formation approach, it is assumed that the neighbours of every node for which a cluster is to be generated are already ranked using their scores. It is also assumed that the scores of the neighbours are determined using a suitable algorithm that considers attributes like link quality, energy, depth, etc. These neighbour scores are used to form a cluster for every node in the topology.

3.1 *Clustering Process*

In the clustering process, the sender node X has its list of neighbours and their scores or ranks. Whenever node X acts as a relay node, it has to forward the packets to nodes in its cluster. To identify the cluster for node X, the novel proposed approach is employed. Figure 1 represents the flowchart that depicts the sequence of steps of the processes involved in a cluster formation. The corresponding steps have been elaborated in the following subsections.

3.1.1 Cluster Head Election

The cluster head (CH) is chosen based on the scores of the nodes in the cluster. The CH is elected as the neighbour node with the highest score i.e. highest rank in the cluster, while the remaining nodes with lesser scores become candidates for cluster members (CM). The CHs work together to collect data, process it, and transmit it to the sink node.

Let S denote the set containing scores of all the neighbours of sender node X. Also, let n_x denote a neighbour of sender node X. Let C be the cluster set that includes the nodes in the cluster. The selected CH must not already be a part of the cluster C. The CH is determined using the following equation:

$$CH = \{n_x : score(n_x) > i;\ \forall\, i \in S, n_x \notin C\} \tag{3}$$

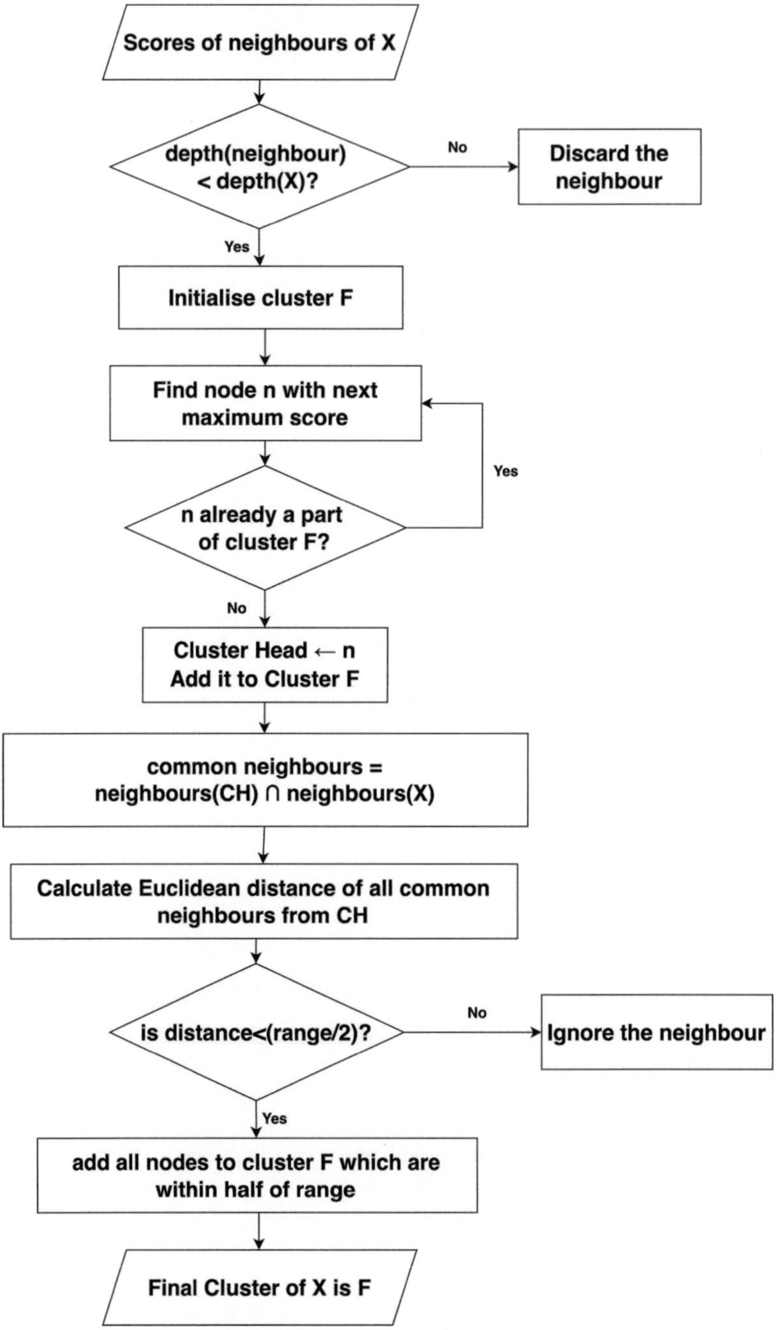

Fig. 1 Flowchart for cluster formation

3.1.2 Election of Cluster Members

CH will select its Cluster Members (CMs). Sender node X must have its neighbour information. Node X must know neighbours of the CH which are present in 50% of the communication range. The CH has to specify which nodes are within half of its communication range.

The CH picks its CMs based on their distance from the cluster head, whether it is a neighbour of the sender node X, and how close it is to the sink node. Therefore the neighbours within half the communication range and are common neighbours with X and has smaller sink depth are chosen as CMs. This approach to consider neighbours with smaller depth is also proposed in the DBR protocol discussed in Sect. 2 [9].

Cluster Head (CH) will select its Cluster Members (CMs). The sender node X, for which cluster is to be formed, must have its neighbour information like scores and coordinates. The CH has to specify which nodes are within 50% of the communication range. Only half of the communication range is considered because there is the possibility of two or more than two data receiving nodes may be out of coverage range from each other. It results in duplicate packet transmission and hidden-node problem [19].

Let C denote the cluster of node X, N_X represents neighbours of sender node X and $N_{CH_{(1/2)}}$ represents neighbours of CH within half the communication range.

$$C = [CH \cup \{N_X \cap N_{CH_{(1/2)}}\}]. \tag{4}$$

4 Results and Analysis

Deploying underwater networks is a difficult task. It must take into account a variety of conditions and constraints, such as low bandwidth, high propagation delay, significant packet loss, and so on. For the successful deployment of underwater networks in the face of such constraints, it is essential to use optimized network protocols with low overhead and significant cross-layer information sharing which are specially designed for use in such underwater networks. Traditional layered network stacks provide adequate separation of concerns, but the resulting protocols are suboptimal. Cross-layer optimization addresses this flaw by allowing direct interaction between layers [20]. UnetStack employs a different approach. Here, the stack is made up of software agents that deliver well-defined services. This method is also known as Service-Oriented Architecture (SOA) [21]. SOA allows for good separation of concerns while allowing information to be shared, services to be rendered, and behaviours to be negotiated between various agents. The resultant network stack is flexible, allowing for the rapid design, simulation, testing, and deployment of software-defined underwater networks. UnetStack can be used on daily use computers in discrete-event simulation mode to model underwater networks and test protocol performance. Moreover, as the stack is scalable, new agents and services may be included or existing agents can be extended to fulfill the development requirements.

The agents in UnetStack interact to provide a complete underwater networking solution. UnetStack employs the open source fjage framework to implement this agent-based approach. UnetStack is built on top of the fjage framework, that defines the agents and services in the stack. Agents exchange messages to communicate with each other. A message in UnetStack can be a request, response or a notification. Notification is a broadcasted message related to a particular topic. These are received by any agent that has subscribed to that topic.

A UnetStack simulator used to design underwater network simulations. An underwater network consisting of multiple nodes is simulated on a computer using UnetStack. It can run a real-time simulation or a discrete event simulation of an underwater network. UnetStack is simple to set up, use, and learn. Once an agent has been designed and tested in the simulator, it may be deployed and tested underwater using any UnetStack-compatible modem.

Using a simulation script, an 8-node topology is defined in UnetStack. The simulation's channel property is used to configure physical channel settings such as communication range and sound speed. Each node is described by the node attributes including node name, location, ports, and the network stack (agents to load) are specified. Table 1 represents the simulation parameters. Figure 2 illustrates the simulation map, respectively. The simulation is run in a real time mode.

To implement the protocol, a custom agent named *ClusteringAgent* has been defined. The startup() function subscribes to PHYSICAL service agent notifications. Using a ticker behaviour, the *ClusteringAgent* generates clusters and makes neighbour information requests on a regular basis. Using this gathered neighbour information, the *ClusteringAgent* computes the clusters for each sender node in the topology. The events that occur during the simulation are recorded in the log file for further analysis.

Figure 1 represents the flowchart for Cluster Formation. The simulation was done in UnetStack. An eight node topology was used for the same. The algorithm was run on intervals of 60 s. The following results and analysis have been obtained:

Table 1 Simulation parameters

Parameters	Values
Number of nodes	8
Communication range	1000 m
Simulation time	1 min
Transmission distance	Euclidean distance between the nodes
Simulation mode	RealTimePlatform
Sound speed	1500 mps

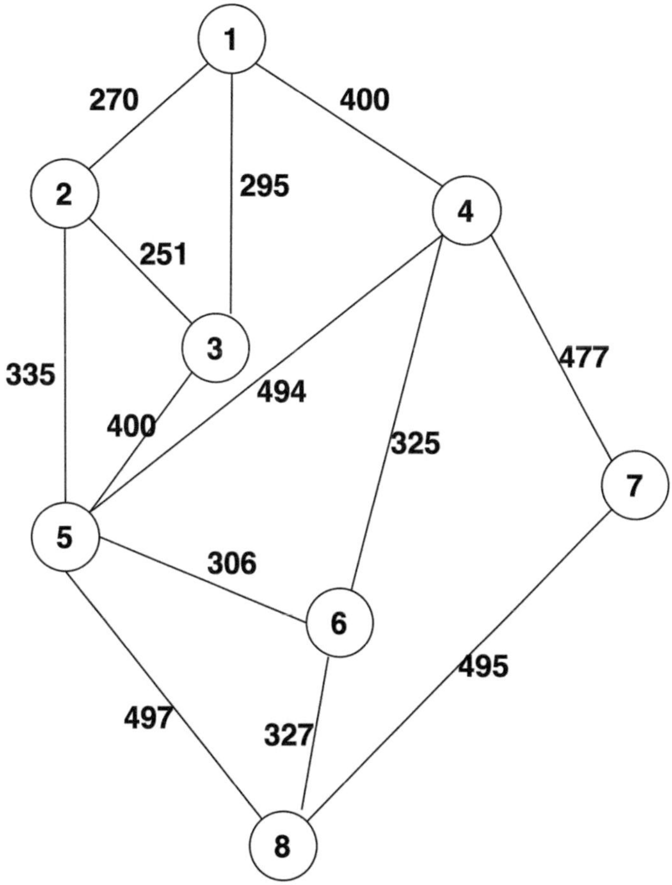

Fig. 2 Topology used for simulations

- The cluster of each node was changing. Upon analysis, it was found that the attribute, energy of each neighbour is variable in nature because of which the resulting cluster is not the same at every iteration.
- All the resultant clusters contain nodes only at lower depths. This would help transmission of signals to the surface sink efficiently.

Figure 2 shows the 8-node Topology used for simulation. The corresponding relevant results obtained in the log file are extracted as shown below:

Table 2 depicts the log outputs for the simulation on the eight nodes. Table 3 shows the log outputs for the cluster formation for sender node 4. The steps involved for the cluster formation of node 4 is described below:

The above Table 3 is explained as follows. As shown in Fig. 2, *sender node* 4 has neighbours 1, 7, 6, 5. It can access its neighbour information like their *coordinates*

Table 2 Cluster formation logs

Cluster formation logs at different timestamps

Sender node	Time stamp 1	Time stamp 2
1	Scores:[4:0.80, 2:0.34, 3:0.20]	Scores:[4:0.81, 2:0.32, 3:0.19]
1	Final cluster:[3]	Final cluster:[3]
2	Scores:[5:0.76, 1:0.48, 3:0.33]	Scores:[5:0.76, 1:0.48, 3:0.33]
2	Final cluster:[3]	Final cluster:[3]
3	Scores:[5:1.00, 2:0.26, 1:0.04]	Scores:[5:1.00, 2:0.26, 1:0.04]
3	Final cluster:[]	Final cluster:[]
4	Scores:[5:0.66, 6:0.47, 1:0.36, 7:0.26]	Scores:[5:0.63, 6:0.49, 7:0.27, 1:0.27]
4	Final cluster:[5, 6]	Final cluster:[5, 6]
5	Scores:[8:0.87, 6:0.68, 4:0.67, 2:0.26, 3:0.18]	Scores:[8:0.87, 6:0.68, 4:0.67, 2:0.26, 3:0.18
5	Final cluster:[2, 3]	Final cluster:[2, 3]
6	Scores:[5:0.94, 8:0.16, 4:0.03]	Scores:[5:0.94, 8:0.16, 4:0.02]
6	Final cluster:[5]	Final cluster:[5]
7	Scores :[4:0.83, 8:0.17]	Scores:[4:0.83, 8:0.17]
7	Final cluster:[4, 8]	Final cluster:[4, 8]
8	Scores:[5:0.86, 6:0.44, 7:0.15]	Scores:[5:0.85, 6:0.46, 7:0.15]
8	Final cluster:[6, 5]	Final cluster:[6, 5]

Table 3 Cluster formation for Sender node 4

Log results for Sender node 4
$ Node 4 has neighbours: [5, 6, 1, 7]
$ Clustering agent can access scores: [5:0.66, 6:0.47, 1:0.36, 7:0.26]
$ Coordinates before removing higher depth nodes: [1:[225.0, 750.0, 200.0], 7:[500.0, 335.0, 915.0], 6:[275.0, 250.0, 475.0], 5:[0.0, 300.0, 350.0]]
$ depth of sender node: 500
$ Coordinates after removing higher depth nodes: [1:[225.0, 750.0, 200.0], 6:[275.0, 250.0, 475.0], 5:[0.0, 300.0, 350.0]]
$ Cluster Head1: 5
$ 4 has final cluster list: [5, 6]

and *scores*. The *Clustering Agent* then discards all the neighbours that are at a higher *depth*. Here, *depth* of sender node 4 is 500. Node 7 has a *depth* of 915 hence this node gets discarded from the set of candidates for cluster formation. The neighbour nodes remaining are 1, 6, 5. Among these 3 neighbour nodes, 5 has the highest *score* which is 0.66. So, *node* 5 becomes the *Cluster Head*. The cluster head then generates its corresponding cluster. The common neighbours of sender node 4 and cluster head which are within half the communication *range* of the cluster head is chosen as *Cluster Members*. Here, for a node to be eligible to be a cluster member, it must be within distance 500 from the sender node 4. Thereby, *node* 6 is at a distance of 325 which is within half the communication range. Hence the *cluster* formed for sender node 4 is [5, 1] with node 5 is the *Cluster Head*.

5 Conclusion

The underwater acoustic channel has unique features such as low bandwidth, high propagation delay and also the sensor nodes are constrained of resources such as power, memory space, and so on. Hence the efficient utilization of the resources is the major criteria to be addressed. Hence in this paper, a novel approach for sensor node clustering technique for UASNs is proposed and it is assumed that the neighbours of a sender node are already ranked. The proposed model has been successfully implemented and simulated in UnetStack, an agent-based network stack for underwater communication and has achieved good results. Opportunistic Routing protocols can help overcome various UASN difficulties, such as network congestion and energy efficiency. The comparison of the results obtained with the results of other algorithms also helps to understand how the cluster formation is able to perform better in less time and with less energy since only the necessary nodes transmit the data. Based on the scores, the nodes can form clusters. Any neighbourhood transmission will be overheard by the nodes, and hence it will be able to discard the packet if it has previously been forwarded. thereby mitigating the hidden-node terminal problem.

References

1. Nayyar A, Puri V, Le D-N (2019) Comprehensive analysis of routing protocols surrounding underwater sensor networks (uwsns): proceedings of icdmai 2018, vol 1 pp 435–450. https://doi.org/10.1007/978-981-13-1402-5_33
2. Ayaz M, Abdullah A (2009) Hop-by-hop dynamic addressing based (h2-dab) routing protocol for underwater wireless sensor networks, pp 436–441. https://doi.org/10.1109/ICIMT.2009.70
3. Rahman MA, Lee Y, Koo I (2017) Flcor: fuzzy logic-based cooperative opportunistic routing for underwater acoustic sensor networks, pp 116–120. https://doi.org/10.1145/3057039.3057074
4. Coutinho R, Boukerche A, Vieira L, Loureiro A (2016) Geographic and opportunistic routing for underwater sensor networks. IEEE Trans Comput 65(2):548–561. https://doi.org/10.1109/TC.2015.2423677
5. Menon VG, Midhunchakkaravarthy D, Sujith A, John S, Li X, Khosravi MR (2022) Towards energy-efficient and delay-optimized opportunistic routing in underwater acoustic sensor networks for iout platforms: an overview and new suggestions. Comput Intell Neurosci
6. Noh Y, Lee U, Lee S, Wang P, Vieira LF, Cui J-H, Gerla M, Kim K (2015) Hydrocast: pressure routing for underwater sensor networks. IEEE Trans Veh Technol 65(1):333–347
7. Coutinho R, Boukerche A, Vieira L, Loureiro A (2016) Geographic and opportunistic routing for underwater sensor networks. IEEE Trans Comput 65(2):548–561. https://doi.org/10.1109/TC.2015.2423677
8. Yan H, Shi Z, Cui J (2008) Dbr: depth-based routing for underwater sensor networks. Lect Notes Comput Sci 4982:72–86. https://doi.org/10.1007/978-3-540-79549-0_7

9. Abdul Wahid H-JJ, Lee S, Kim D (2011) Eedbr: Energy-efficient depth-based routing protocol for underwater wireless sensor networks, pp 223–234
10. Braman GR, Umapathi A (2014) A comparative study on advances in leach routing protocol for wireless sensor networks: a survey. Int J Adv Res Comput Commun Eng
11. Li X, Wang Y, Zhou J (2012) An energy-efficient clustering algorithm for underwater acoustic sensor networks. In: 2012 International conference on control engineering and communication technology, pp 711–714
12. Alhazmi AS, Moustafa AI, Aldosari FM (2018) Energy aware approach for underwater wireless sensor networks scheduling: Umod_leach. In: 2018 21st Saudi computer society national computer conference (NCC), pp 1–5
13. Wan Z, Liu S, Ni W, Xu Z (2019) An energy-efficient multi-level adaptive clustering routing algorithm for underwater wireless sensor networks. Cluster Comput 22. https://doi.org/10.1007/s10586-018-2376-8
14. Batra PK, Kant K (2016) Leach-mac: a new cluster head selection algorithm for wireless sensor networks. Wirel Netw 22:49–60
15. Rashid MM, Al-Jamali NAS (2019) Modified w-leach protocol in wireless sensor network. J Eng 25(3):68–80. https://doi.org/10.31026/j.eng.2019.03.06. https://joe.uobaghdad.edu.iq/index.php/main/article/view/j.eng.2019.03.06
16. Noh Y, Lee U, Lee S, Wang P, Vieira LF, Cui J-H, Gerla M, Kim K (2015) Hydrocast: pressure routing for underwater sensor networks. IEEE Trans Veh Technol 65(1):333–347
17. Domingo MC (2008) Overview of channel models for underwater wireless communication networks. Phys Commun 1(3):163–182
18. Darehshoorzadeh A, Boukerche A (2015) Underwater sensor networks: a new challenge for opportunistic routing protocols. IEEE Commun Mag 53(11):98–107. https://doi.org/10.1109/MCOM.2015.7321977
19. Ghoreyshi SM, Shahrabi A, Boutaleb T (2017) Void-handling techniques for routing protocols in underwater sensor networks: survey and challenges. IEEE Commun Surv Tutor 19(2):800–827
20. Baldo N, Miozzo M, Guerra F, Michele R, Zorzi M (2010) Miracle: the multi-interface cross-layer extension of ns2. EURASIP J Wirel Commun Netw. https://doi.org/10.1155/2010/761792
21. Papazoglou M (2003) Service-oriented computing: concepts, characteristics and directions. In: Proceedings of the fourth international conference on web information systems engineering. https://doi.org/10.1109/WISE.2003.1254461

Comparative Analysis of Machine Learning Algorithms for Phishing Website Detection

Omkar Gore, Archana Chaudhari, Juilee Gaikwad, Smitesh Gaychor, and Devesh Gugale

Abstract Phishing is the most straightforward method of obtaining sensitive information from unsuspecting consumers. Phishers want to get their hands on sensitive data such as bank account numbers, usernames, and passwords. People working in cybersecurity are now searching for reliable and effective methods for detecting phishing websites. The goal of this research is to apply machine learning to distinguish between legitimate and phishing URLs by extracting and analysing numerous components. Attackers deceive Internet users by impersonating a legitimate website in order to obtain personal information. Sending emails to respectable corporations or enterprises can also be used. The accuracy of the prediction can be improved using machine learning methods. Phishing site URLs have certain characteristics, in the proposed work, these characteristics are investigated for the detection of phishing websites. The study examines machine learning methods and how well they function to identify phishing websites. Experiments conducted by using a hybrid strategy demonstrate improved accuracy for phishing website detection.

Keywords Phishing website · Logistic regression · Data classification · Naive Bayes · Machine learning · Tokenisation · Feature extraction

1 Introduction

Cybersecurity researchers have recently become highly concerned about phishing as it is not very difficult to create fake websites that look legitimate. Even though intelligent experts can recognise scammer websites, but all users certainly cannot, which results in some users becoming victims of phishing assaults. Some of the attacker's objectives is to maliciously steal bank account information and login information for

O. Gore (✉) · A. Chaudhari · J. Gaikwad · S. Gaychor · D. Gugale
Department of Instrumentation and Control Engineering, Vishwakarma Institute of Technology, Pune, India
e-mail: omkar.gore19@vit.edu

A. Chaudhari
e-mail: archana.chaudhari@vit.edu

S. K. Biswas et al. (eds.), *Intelligent Computing Systems and Applications*, Lecture Notes in Networks and Systems 1307, https://doi.org/10.1007/978-981-96-3860-4_39

their unethical benefits. Businesses in the United States lose $2 billion every year as a result of phishing attacks on their customers. The third Microsoft Computing Safer Index Report, which was published in February 2014, estimates that phishing-related attacks could have an annual global economic effect of up to $5 billion. Since there is very less user awareness in this domain, the attackers are becoming very powerful. It is difficult to prevent phishing attacks since they feed on users' vulnerabilities, yet it is crucial to increase security with phishing detection systems.

Phishing is the practise of imitating the characteristics and features of emails in order to make them appear identical to the original. It appears to be comparable to the authorised sources. This email appears to have come from a legitimate corporation or organisation, according to the user. As a result, the user is forced to use the links in the phishing email to access the phishing website. The websites which do phishing are designed to imitate the look of a legitimate company's own website. Phishers compel users to fill up personal information by sending them scary notifications or account validation messages, which push them to fill out the required information, which they can then misuse. They set up the circumstance in such a way that the user has no other choice but to visit their spoofed website.

The goal of this paper is to identify phishing websites. In the final phishing detection rate, some significant features such as URL, domain identity, security, and encryption criteria can be used to identify the phishing website. On this massive dataset, applying machine learning techniques can improve the approaches perform better. Many studies have independently validated these procedures and then compared their results to those of other methods. The research presents a hybrid strategy for improving data accuracy. To improve accuracy, four vivid machine learning algorithms are used: random forest, Naive Bayes, logistic regression, and decision tree technique.

2 Related Work

Many studies have been conducted in the topic of phishing detection in the past. We obtained data from a variety of sources and thoroughly examined it, which aided us in driving our own approaches in the development of a more safe and accurate system.

The authors of this paper [1] suggested a technique depending on machine learning algorithm that uses a collection or set of hybrid characteristics to classify websites. The main feature set is extracted using the cumulative distribution gradient technique, while the secondary features are extracted with the help of the data perturbation ensemble technique. Random forest in conjunction with ensemble learner, the method used to train the classifier, correctly identifies phishing websites with a precision of 94.6%.

Using a machine learning technique, the proposed method [2] recognised phishing websites. The feature set contained six features depending on the website structure, which were picked after the authors conducted a comparative analysis. The classifier was trained using a support vector machine, and it was successful in distinguishing websites into two categories: legitimate and phishing. For the website classification, the presented model has an accuracy of 84%.

The authors of this paper [3] created a browser plugin to detect phishing websites. The system performed comparative research to train the classifier using a different range of machine learning algorithms and techniques, including random forest, along with support vector machine, and also the algorithm of K-Nearest Neighbour (K-NN). A content-based strategy for extracting the JavaScript and HTML functionalities of the websites was included in the feature set. The dataset that was derived from the downloads of UCI-Machine Learning Repository included a 22-feature classification approach for detecting phishing websites.

This research [4] provides a simplified feature classification phishing website detection approach. To analyse the gathered features, Support Vector Machine (SVM) and Logistic Regression methods were employed. Nineteen features were chosen and used for classification out of a total of 30 features found. The dataset was obtained from the machine learning repository at UCI Irvine, and big data was used to create the model. With an accuracy of 95.62%, Support Vector Machine (SVM) surpassed the other two techniques.

This paper [5] examines how different elements of legitimate and phishing URLs can be extracted and analysed in order to demonstrate how machine learning can be used to detect phishing URLs. Decision tree, random forest, and support vector machine algorithms were used to identify phishing websites, with accuracy ratings of 97.11, 97.14, and 96.51, respectively. The study aims to identify phishing URLs and to identify the best machine learning approach by evaluating the accuracy rate, false positive rate, and false negative rate of each algorithm.

The study's authors offered a URL-based anti-phishing machine learning algorithm in their paper [6]. They used 14 elements of the URL to analyse the effectiveness of their technique and assess whether the website was malicious or legitimate. The suggested method was trained using over 33,000 URLs of phishing and genuine websites in support vector machine and Nave Bayes classifiers. The learning process was the emphasis of the phishing detection method. They discovered 14 distinct characteristics that distinguish phishing websites from real websites. When websites trained on SVM classification algorithm were found, the results of their experiment had a precision of above 90%.

The paper [7] uses URL-based features to train models that seek to mitigate zero-day attacks with the help of software proposals which analyse the URL of a website to distinguish legitimate from fraudulent websites. Based on the observations, the random forest classifier had an F1-score of 97%, a precision of 97%, and a recall of 99%. This model simply analyses the URL and makes no use of any other resources, the model suggested in this study is rapid and effective.

The fundamental premise of this paper [8] is to carry out the frameworks with efficiency, precision, and cost-effectiveness. To complete the task, four ML-managed classification models were used. The four classification models are K-NN, kernel-SVM, random forest classifier, and decision tree. A labelled dataset is included in the supervised classification, which is used to train the models. To analyse patterns and learn from them, cybersecurity products and security systems can use machine learning algorithms to effectively prevent assaults that repeat continuously and quickly take action responding to changing behaviour. It can help users in being better in terms of security and be more proactive in order to prevent risks and responding to active attacks in real time.

The authors [9] devised a method for distinguishing between dangerous and benign websites. The values of the true positives were found to be greater than the rate of false positives in this study. Another study [10] proposed a convolutional neural network (CNN) to identify the URL of a phishing website. Researchers who did the experiments used a sequential strategy to acquire information present in URLs in this investigation. On benchmark datasets, it obtained accuracies of 98.58%, 95.46%, and 95.22%, respectively.

This paper [11] addressed this issue by applying machine learning techniques as well as a novel dataset for discovery of phishing and its detection, which contains 5000 real websites and more than around 5000 phishing pages. Various machine learning methods were tested for achieving the appropriate results and after that random forest and multilayer perceptron models were selected after that. To improve the efficiency of the models, vivid feature extraction and selection techniques were applied on dataset. The experiment's best result was obtained by using 20 of the obtained features and applying them to the random forest algorithm. 98.11% of the time, the prediction was correct.

Convolutional Neural Networks (CNNs) have been suggested by the authors [12] as a method of identifying phishing URLs. To compare several conventional machine learning models and deep learning models, multiple feature sets including handcrafted, character embedding, character level TF-IDF, and character level count vector characteristics are used. The suggested model in the study beat existing phishing URL models, according to the studies, with accuracy rates of 95.02% on the provided dataset and 98.58%, 95.46%, and 95.22% on benchmark datasets.

In classifying phishing attempts, the authors of this paper [13] provide a classification model. This methodology includes feature extraction from websites as well as website classification. During feature extraction, UCI yielded 30 characteristics. These features were categorised using Support Vector Machine (SVM), Naive Bayes (NB), and Extreme Learning Machine (ELM). The Extreme Learning Machine (ELM), which surpassed SVM and NB in accuracy with 95.34%, used six activation functions. The results were attained using the MATLAB software.

In this paper [14], they propose a phishing detection method based on lexical analysis of URLs and machine learning classifiers. The tests were performed on a dataset that had 1,056,937 tagged URLs at the time of testing (phishing and legitimate). This dataset was processed to provide 22 unique features, which were then further reduced to a smaller set using a variety of feature reduction techniques. The findings show that Support Vector Machines (SVMs) are superior to random forest, gradient boosting, neural network, and SVM classifiers, which achieved the highest accuracy in detecting the studied URLs with a rate of 99.89%.

In this research, the authors [15] created a Google Chrome add-on that uses machine learning methods to identify the content of phishing websites. Here 22 features were collected from the UCI-Machine Learning Repository for this dataset. The algorithms K-NN, SVM, and random forest were used to compare accuracy in terms of F1-score, precision, recall, and all other metrics. The chrome extension for random forest, which was developed using Python, HTML, JavaScript, and CSS, obtained the highest rating. This plugin's alleged harmful site list, which is continuously expanding, is a drawback.

The authors [16] suggested a strategy that uses a URL identification method based on the random forest algorithm to detect phishing sites. The three steps of the show are parsing, heuristic data classification, and performance analysis. Only eight features are selected for parsing out of a total of 31. The random forest approach had a 95% accuracy rate.

The logistic regression (LR) method offers an effective detection for independent variables gathered for two classes. However, it has limitations when it comes to some elements, such as feature repetition and incompatibility, as well as outlier values having bad prediction impacts. On the other hand, the support vector machine (SVM) method can be utilised because it is better suited for a high number of independent variables [17]. This method works well for nonlinear situations, but it has drawbacks when dealing with huge, noisy datasets.

The authors of this research [18] described a revolutionary machine learning technique for detecting phishing websites. Here they have compared five machine learning techniques—Decision Tree (DT), Random Forest (RF), Gradient Boosting (GBM), Generalised Linear Model (GLM), and Generalised Additive Model (GAM). Comparisons were made between the top three algorithms, decision tree, random forest, and GBM. With 98.4% accuracy, 98.59 recall, and 97.70 precision, the random forest algorithm delivers the best accuracy, recall, and performance results.

The authors of [19] described a phishing URL approach that used a blacklist and unique lexical components. They gathered 18 common lexical features and a collection of URLs from URL repositories using a crawler. They employed advanced machine learning methods including classification and under/oversamples. The automated methods outperform other ML algorithms already in use. Because lexical features were challenging to apply in practical contexts, the study concentrated on content-related features. The experiment's results outperformed those of pre-existing classification schemes.

The study [20] examined various deep learning and machine learning techniques for identifying URLs by analysing unique URL components. The authors examined a number of supervised learning techniques for phishing URL detection that made use of a vocabulary, WHOIS attributes, traffic rank data, PageRank, and page importance factors. They examined the impact of training data diversity on classifier accuracy. The study's classification methods include SVM, K-NN, RFC, and ANN.

3 Methodology

The techniques for using machine learning to identify phishing websites are covered in this part, along with the suggested framework. Figure 1 presents the proposed framework for phishing website detection. For extraction of lexical features, tokenisation is implemented in the proposed work. Figure 2 describes the process of tokenisation. Tokenisation is the process of replacing sensitive data with one-of-a-kind identification symbols that retain all of the material's critical information while without jeopardising its security. A token is a bit of data that serves as a substitute for another, more valuable piece of data. The further subsections elaborate the proposed method.

3.1 Dataset Description

One of the most important aspects of our research is the dataset. The dataset for our suggested model, which we acquired from Kaggle, is nothing more than a table collecting information about phishing and authentic websites. Kaggle is one of the most well-known public repositories, and has a massive dataset collection appropriate for training machine learning models. There are 5,49,346 entries and 2 columns in

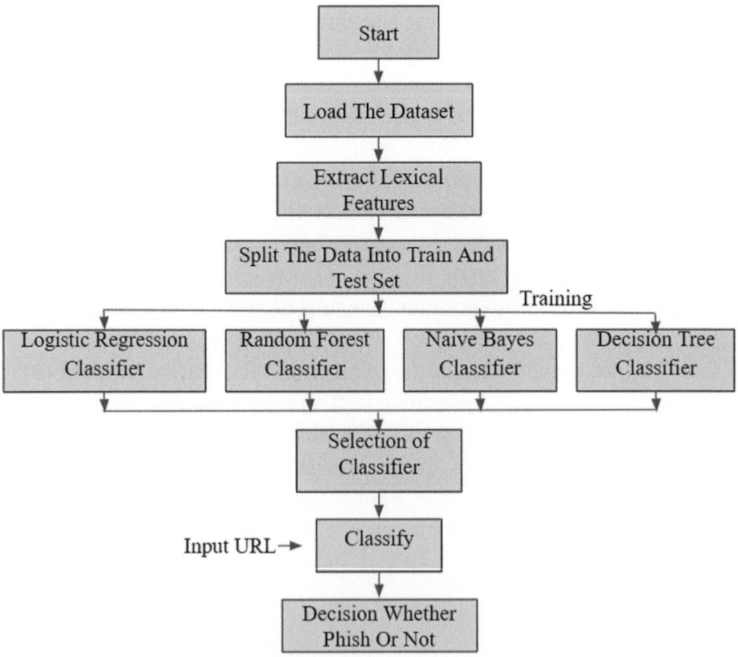

Fig. 1 Flow chart of the proposed system

Fig. 2 Flow chart of the tokenisation

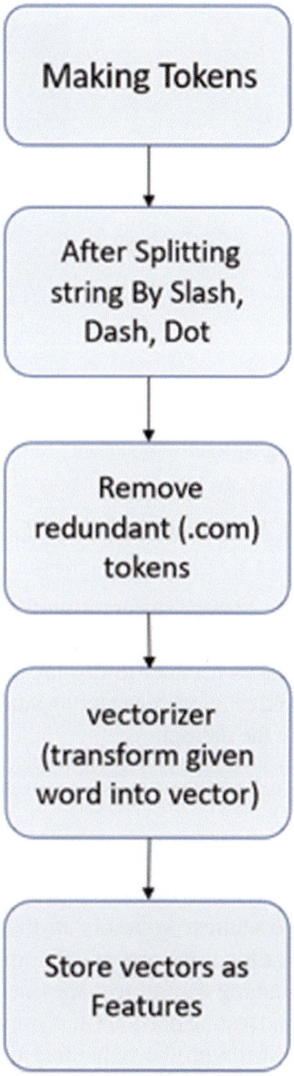

that dataset. The URL is in the first column, while the label is in the second. The prediction column, which has two categories, is in the label column. The first is a good URL, which means it contains no dangerous material, and the second is a poor URL.

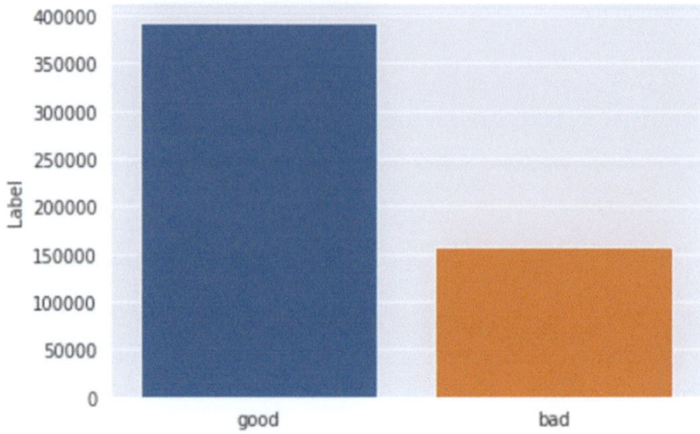

Fig. 3 Ratio of good and bad URLs

3.2 Distribution of Good and Bad URL

We created different guidelines for various features based on the examination of good and bad websites. With the use of pictorial representation, data visualisation makes it easier to display data, no matter how complex it is, and to analyse trends and relationships among variables. Figure 3 presents the ratio of good and bad URLs in the dataset.

3.3 Data Pre-processing

To address volatility in the values among distinct independent variables, feature scaling is the process of normalising or standardised the independent variables of the training dataset to a specified range. It's crucial to separate the dataset into training and testing portions. It is important to train a model with a fraction of the total dataset, test it with the remaining data, and then evaluate its performance. We divided the dataset into an 80:20 ratio as follows: 80% of the dataset was used for training and 20% for testing using a stratified sample technique. The Scikit-Learn module of the Python computer language was used to generate the train-test split.

3.4 Apply Machine Learning

This paper proposes to use machine learning technique after the data is pre-processed. Four machine learning categorization techniques—decision tree, random forest,

Naive Bayes, and logistic regression—have been selected because of their ability to detect phishing websites.

Steps for implementing the proposed method are as follows:

Step1: Import the required libraries and the dataset for phishing websites.

Step2: Remove missing data from the data by pre-processing it.

Step3: Divide the dataset in half, allocating 80% of the data to the training set and 20% of the data to the test set.

Step4: Pick a machine learning algorithm to use.

Step5: Create a classifier model for the specified machine learning approach using the training set.

Step 6: Test the classifier model using the specified machine learning algorithm using the test set.

3.4.1 Decision Tree Algorithm

The decision tree method is simple to understand and apply. The decision tree begins with choosing the root, or the best splitter, among the categorization-relevant features. Up until the leaf node, the algorithm will keep growing the tree. In a tree representation, each leaf node of the tree corresponds to the class label, and each internal node of the tree corresponds to an attribute. A training model created by decision trees is then utilised to forecast a target value or class. By learning straightforward decision rules inferred from prior data, a Choice Tree is used to generate a training model that may be used to predict the class or value of the target variable (training data). We start at the base of the tree when using Decision Trees to forecast a class label for a record. Comparison of the values of the record's attributes with the root attribute. On the basis of the comparison, we move to the next node by following the branch that corresponds to that value. The confusion matrix for the decision tree algorithm is shown in Fig. 4. The confusion matrix represents the true positives, false positives, true negatives, and false negatives for the classification. Also the classification report for the same model has been represented in Fig. 5 which contains Precision, Recall, F1-Score, and Support.

3.4.2 Random Forest Algorithm

A forest is created by the random forest algorithm out of a group of decision trees. Random forest is a popular supervised machine learning technique for addressing classification and regression challenges. Using the average for regression and the majority vote for classification, it builds decision trees from a variety of samples. One of the important characteristics of the random forest algorithm is that it can handle datasets with both continuous and categorical variables, as in regression and classification. It offers better outcomes for classification challenges. A high level of detection accuracy is achieved by using several trees. Trees are produced by using the bootstrap procedure. The dataset's characteristics and samples are chosen at

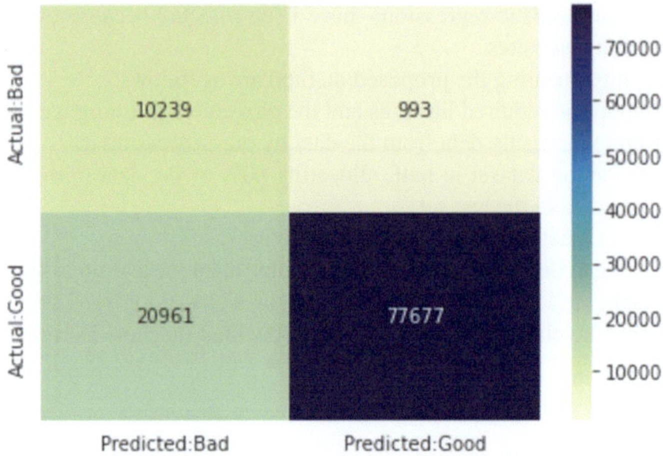

Fig. 4 Confusion matrix for decision tree

CLASSIFICATION REPORT

	precision	recall	f1-score	support
Bad	0.33	0.91	0.48	11232
Good	0.99	0.79	0.88	98638
accuracy			0.80	109870
macro avg	0.66	0.85	0.68	109870
weighted avg	0.92	0.80	0.84	109870

Fig. 5 Classification report for decision tree

random from a replacement tree to create the bootstrap technique. Similar to the decision tree approach, the random forest algorithm finds the optimal splitter for classification from a set of randomly chosen features. To find the optimum splitter, the random forest method also makes use of the Gini index and information gain techniques. This procedure will keep going until a forest of trees has been generated at random. The method will calculate the votes for each projected target once each tree in the forest forecasts the target value. Finally, as a final prediction, the random forest algorithm examines a heavily voted anticipated target. Figure 6 represents a random forest classifier's confusion matrix and Fig. 7 contains classification report.

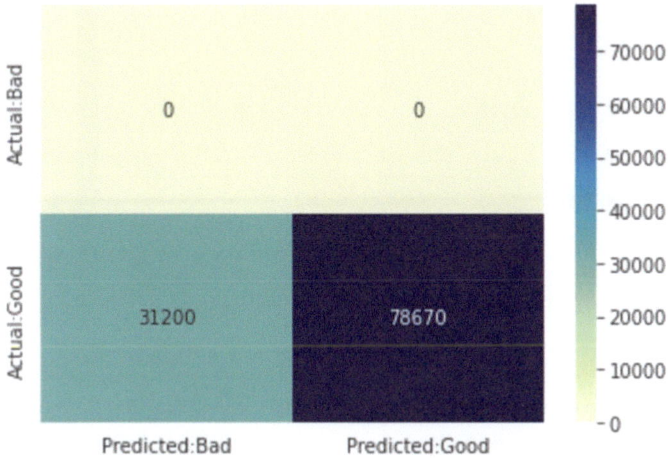

Fig. 6 Confusion matrix for random forest

	precision	recall	f1-score	support
Bad	0.00	0.00	0.00	0
Good	1.00	0.72	0.83	109870
accuracy			0.72	109870
macro avg	0.50	0.36	0.42	109870
weighted avg	1.00	0.72	0.83	109870

Fig. 7 Classification report for random forest

3.4.3 Naive Bayes Algorithm

Naive Bayes is a simple conditional probability-based probabilistic classifier. This classifier uses the fundamental Bayesian theorem and propagates a strong independence across attributes in addition to computing the posterior probability for each class and assigning the sample to the class with the highest probability. When processing the corpus, the algorithm estimates the probability (or frequency) of each feature within each class, treating every feature as independent of the others. It just has a single class label. In NB, it is considered that all neglected features are independent of one another. In sentiment analysis, spam filtering, recommendation systems, and other applications, Naive Bayes algorithms are frequently utilised. They are quick and simple to implement, but the necessity that predictors be independent is their major drawback. Figure 8 presents the confusion matrix and Fig. 9 the classification report for Naive Bayes classifier.

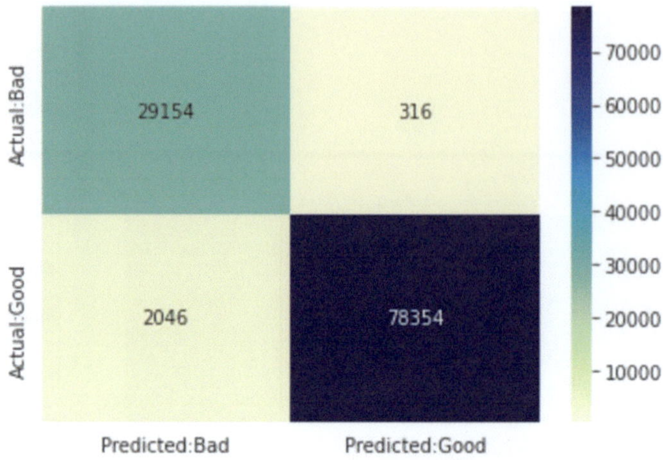

Fig. 8 Confusion matrix for Naïve Bayes

CLASSIFICATION REPORT

	precision	recall	f1-score	support
Bad	0.93	0.99	0.96	29470
Good	1.00	0.97	0.99	80400
accuracy			0.98	109870
macro avg	0.97	0.98	0.97	109870
weighted avg	0.98	0.98	0.98	109870

Fig. 9 Classification report for Naïve Bayes

3.4.4 Logistic Regression Algorithm

A statistical model is a logistic. It models a binary dependent variable with a logistic function. Our regression approach uses 80% tagged data to build a logistic function that can predict whether a website is legitimate or a phishing scam. When the data is linearly separable, logistic regression performs better. Because it is highly interpretable, it does not necessitate a lot of computational resources. Scaling the input features is simple and does not necessitate adjustment. Using logistic regression, creating and training a model is simple. It indicates the importance of a predictor (coefficient size) as well as the direction of relationship (positive or negative). Figure 10 presents the confusion matrix and Fig. 11 the classification report for logistic regression model.

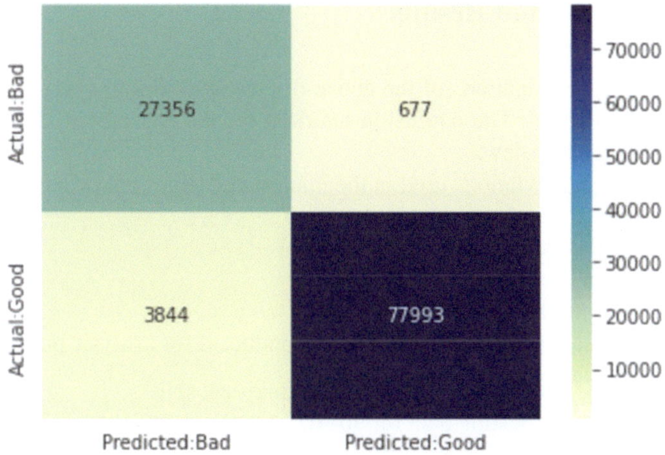

Fig. 10 Confusion matrix for logistic regression

CLASSIFICATION REPORT

	precision	recall	f1-score	support
Bad	0.88	0.98	0.92	28033
Good	0.99	0.95	0.97	81837
accuracy			0.96	109870
macro avg	0.93	0.96	0.95	109870
weighted avg	0.96	0.96	0.96	109870

Fig. 11 Classification report for logistic regression

3.4.5 Proposed Method Using Hybrid Classifiers

Hybrid classifiers were implemented by using ensembling of algorithms. The predictions of many different models are combined in ensemble learning, a general meta approach to machine learning, to enhance predictive performance. Ensemble models aim to reduce the generalisation error of the prediction. As long as the base models are varied and independent, the prediction error decreases when employing the ensemble approach.

4 Experiments and Results

All the possible combinations of the above-discussed algorithms were trained and tested with the dataset. The confusion matrices for the ensembled algorithms are plotted in the figures below.

Note: All the abbreviations used in the images and tables below are

LR—Logistic Regression.

NB—Naive Bayes.

DT—Decision Tree.

RF—Random Forest.

Following three types of experiments are conducted for analysis purpose:

1. Using training and testing data for 80–20 train-test split.
2. Using training and testing data for 90–10 train-test split.
3. Real-time testing for phishing website.

4.1 Experiments and Analysis for 80–20 Train-Test Split for Phishing Website Detection Using Hybrid Algorithms

The training and testing data was split into two ratios. First, training data comprised 80% of the testing data and 20% of the training data, respectively. Table 1 demonstrates the accuracy of classifiers based on this ratio.

The confusion matrices and their respective classification reports for all hybrid classifiers are represented in Figs. 12, 13, 14, 15, 16, 17, 18, 19, 20, 21, 22, 23, 24, and 25.

The Naive Bayes algorithm outperforms decision trees and other algorithms in terms of detection accuracy, with a rate of 97.8%. The results also suggest that as

Table 1 Classifiers and their corresponding accuracy scores for 80–20 split of training and testing data

Classifiers	Accuracy score
DT	80.01
RF	71.60
NB	97.85
LR	95.88
LR + NB	97.76
DT + RF	80.01
LR + DT	95.23
NB + RF	97.85
LR + RF	95.88
NB + DT	97.13
LR + DT + RF + NB	95.91

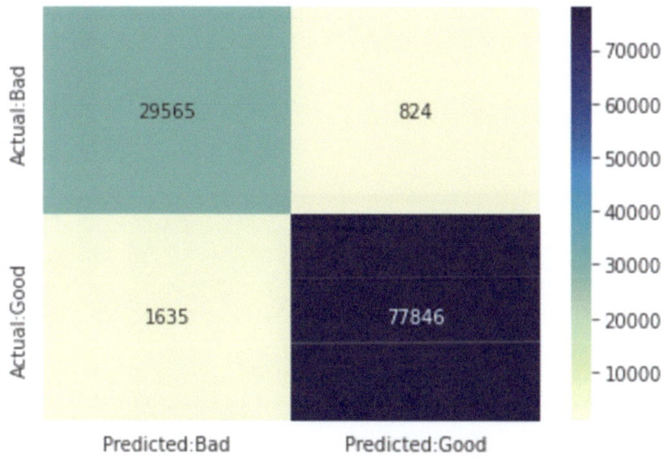

Fig. 12 Confusion matrix for LR + NB

CLASSIFICATION REPORT

	precision	recall	f1-score	support
Bad	0.95	0.97	0.96	30389
Good	0.99	0.98	0.98	79481
accuracy			0.98	109870
macro avg	0.97	0.98	0.97	109870
weighted avg	0.98	0.98	0.98	109870

Fig. 13 Classification report for LR + NB model

more datasets are utilised as training datasets, the detection accuracy of phishing websites improves. When 80% of the data is used as a training dataset, all classifiers perform well.

4.2 Experiments and Analysis for 90–10 Train-Test Split for Phishing Website Detection Using Hybrid Algorithms

Experiments are also conducted for second ratio with training and testing split of 90% and 10% and were also analysed. As expected, all the algorithms gave a greater accuracy as the training data was increased. Table 2 shows classifiers and their accuracies with this ratio.

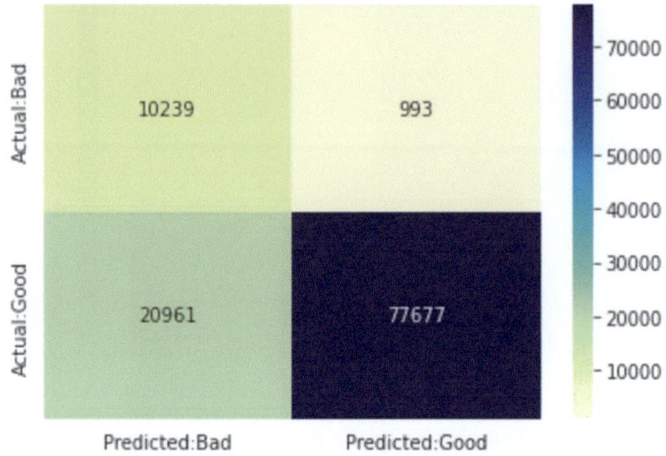

Fig. 14 Confusion matrix for DT + RF

CLASSIFICATION REPORT

	precision	recall	f1-score	support
Bad	0.33	0.91	0.48	11232
Good	0.99	0.79	0.88	98638
accuracy			0.80	109870
macro avg	0.66	0.85	0.68	109870
weighted avg	0.92	0.80	0.84	109870

Fig. 15 Classification report for DT + RF model

The Naive Bayes algorithm outperforms decision trees and other algorithms in terms of detection accuracy, with a rate of 97.85% when 80% data was used in training and 97.94% when 90% was used. The results also suggest that as more datasets are utilised as training datasets, the detection accuracy of phishing websites improves.

When 80% of the data is used as a training dataset, all classifiers perform well. When 90% of data is used as a training data only Naive Bayes and logistic regression perform well. The performance of random forest and decision tree decreases.

We can also draw the conclusion that accuracy grows as training data volume increases. The fact that the data is biased towards the training samples is one of the causes of this.

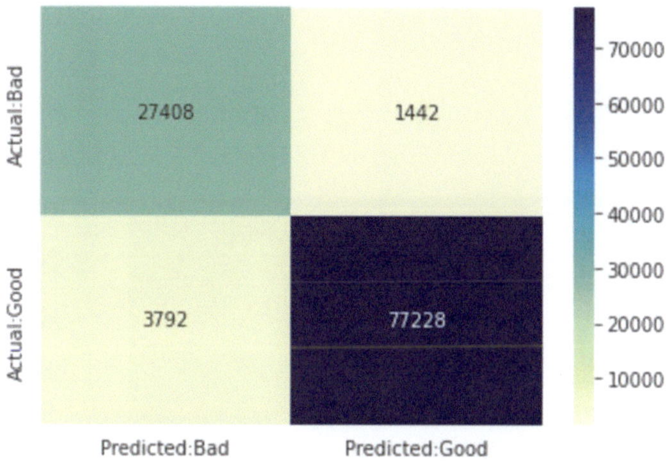

Fig. 16 Confusion matrix for LR + DT

CLASSIFICATION REPORT

	precision	recall	f1-score	support
Bad	0.88	0.95	0.91	28850
Good	0.98	0.95	0.97	81020
accuracy			0.95	109870
macro avg	0.93	0.95	0.94	109870
weighted avg	0.95	0.95	0.95	109870

Fig. 17 Classification report for LR + DT model

4.3 Real-Time Testing for Phishing Website

Analysis of real-time detection for phishing website is also included. Using the trained model, few websites are tested for analysis. The provided website URL can be classified in 'good' or 'bad' parameters. Figure 26 demonstrates the classification of website as good or bad using decision tree classifier.

Here in the figure above, decision tree model is used in the 'tree' variable for prediction. A URL is provided in 'x_predict' variable. The model is then analysed whether it is good or bad and then the result is printed.

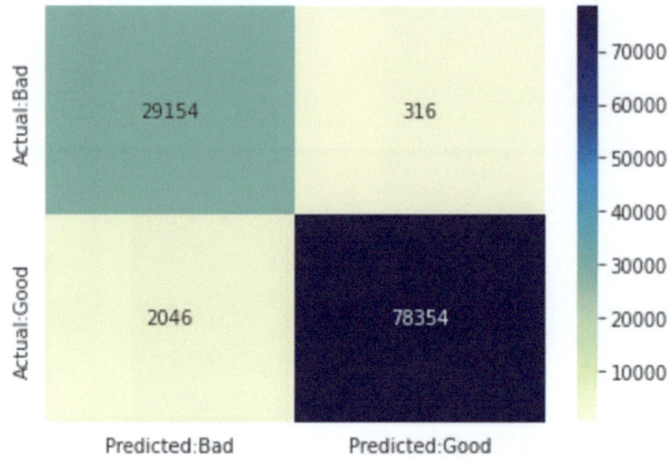

Fig. 18 Confusion matrix for NB + RF

CLASSIFICATION REPORT

	precision	recall	f1-score	support
Bad	0.93	0.99	0.96	29470
Good	1.00	0.97	0.99	80400
accuracy			0.98	109870
macro avg	0.97	0.98	0.97	109870
weighted avg	0.98	0.98	0.98	109870

Fig. 19 Classification report for NB + RF model

5 Conclusions

Phishing is a common way nowadays for obtaining sensitive information. The purpose of the research work is to compare and analyse the various machine learning approaches for detection of phishing websites. Decision tree, Random forest, Naive Bayes classifier, and different hybrid approaches are implemented and analysis is conducted based on the accuracy. From the analysis and experiments conducted, Naive Bayes approach demonstrates better accuracy for phishing website detection. Among the hybrid approaches implemented, the combination of Naive Bayes method and random forest approaches exhibits a comparable performance as compared to only Naive Bayes classifier but at the cost of increased classification time.

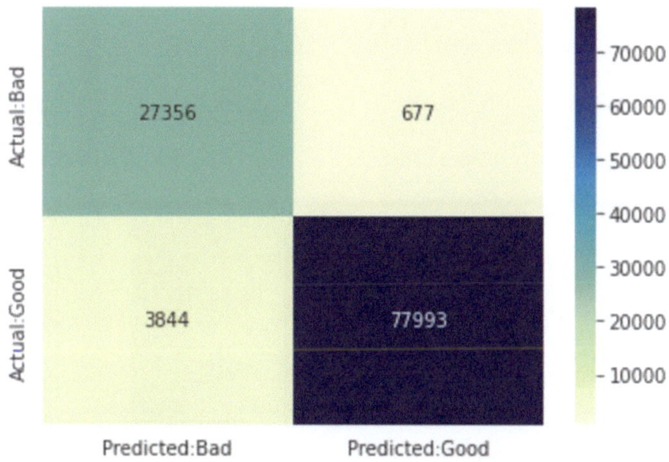

Fig. 20 Confusion matrix for LR + RF

CLASSIFICATION REPORT

	precision	recall	f1-score	support
Bad	0.88	0.98	0.92	28033
Good	0.99	0.95	0.97	81837
accuracy			0.96	109870
macro avg	0.93	0.96	0.95	109870
weighted avg	0.96	0.96	0.96	109870

Fig. 21 Classification report for LR + RF model

Therefore. we can conclude that Naive Bayes approach outperforms all other methods presented in the research work. One of the reasons for the same is Bayes classifier makes use of probabilities for the classification.

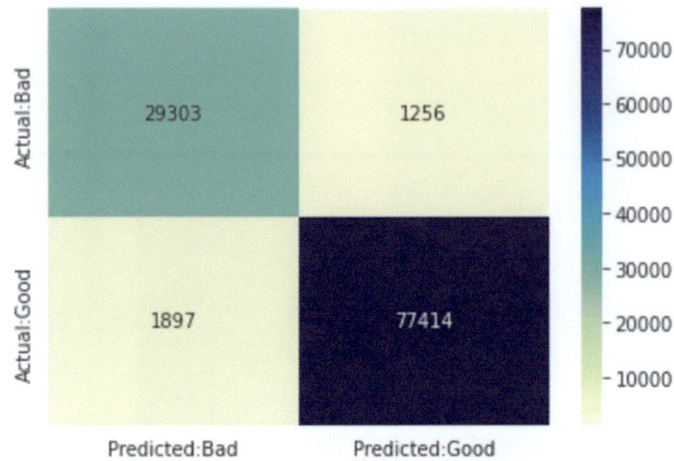

Fig. 22 Confusion matrix for NB + DT

CLASSIFICATION REPORT

	precision	recall	f1-score	support
Bad	0.94	0.96	0.95	30559
Good	0.98	0.98	0.98	79311
accuracy			0.97	109870
macro avg	0.96	0.97	0.96	109870
weighted avg	0.97	0.97	0.97	109870

Fig. 23 Classification report for NB + DT model

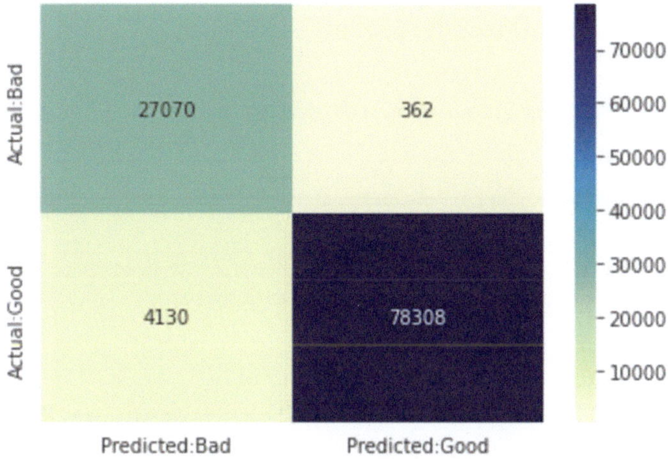

Fig. 24 Confusion matrix for LR + DT + RF + NB

CLASSIFICATION REPORT

	precision	recall	f1-score	support
Bad	0.87	0.99	0.92	27432
Good	1.00	0.95	0.97	82438
accuracy			0.96	109870
macro avg	0.93	0.97	0.95	109870
weighted avg	0.96	0.96	0.96	109870

Fig. 25 Classification report for LR + DT + RF + NB

Table 2 Classifiers and their corresponding accuracy scores for 90–10 split of training and testing data

Classifiers	Accuracy score
DT	79.83
RF	71.45
NB	97.94
LR	95.94
LR + NB	97.83
DT + RF	79.85
LR + DT	95.26
NB + RF	97.94
LR + RF	95.93
NB + DT	97.19
LR + DT + RF + NB	95.96

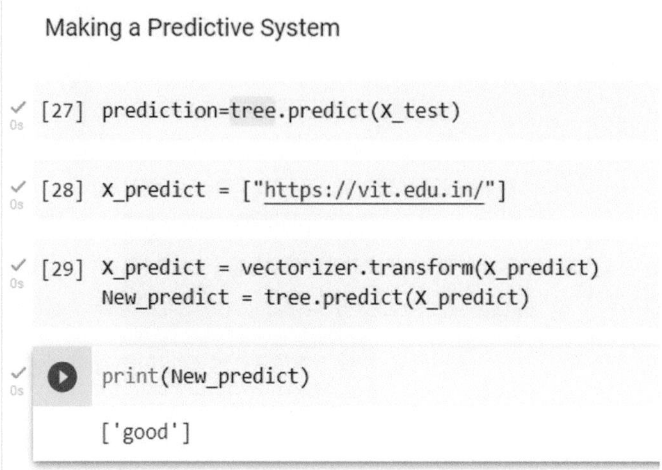

Fig. 26 Predictive system using decision tree

References

1. Chiew KL, Tan CL, Wong KS, Yong KSC, Tiong WK (2019) A new hybrid ensemble feature selection framework for machine learning-based phishing detection system. Inf Sci
2. Pan Y, Ding X (2018) Anomaly based web phishing page detection. In: Computer security applications conference, 2018. ACSAC'18. 22nd annual. IEEE, pp 381–392
3. Desai A, Jatakia J, Naik R, Raul N (2017) Malicious web content detection using machine learning. In: RTEICT 2017-2nd IEEE International conference recent trends electronic information communication technology proceedings
4. Fadheel W, Abusharkh M, Abdel-Qader I (2020) On feature selection for the prediction of phishing websites. In: 2020 IEEE 15th International conference dependable, automated security computing, 15th International conference pervasive intelligent computing, 3rd International conference big data intelligent computing cyber science technology, 2020
5. Mahajan R, Siddavatam I (2018) Phishing website detection using machine learning algorithms. Int J Comput Appl (0975–8887) 181(23)
6. Jain AK, Gupta BB (2018) PHISH-SAFE: URL features-based phishing detection system using machine learning. In: Cyber security. advances in intelligent systems and computing, vol 729
7. Abedin,NF, Bawm R, Sarwar T, Saifuddin M, Rahman MA, Hossain S (2020) Phishing attack detection using machine learning classification techniques. In: 2020 3rd International conference on intelligent sustainable systems (ICISS)
8. Rastogi M, Chhetri A, Singh DK, Gokul Rajan V (2021) Survey on detection and prevention of phishing websites using machine learning. In: 2021 International conference on advance computing and innovative technologies in engineering (ICACITE)
9. Rao RS, Pais AR (2019) Jail-Phish: an improved search engine based phishing detection system. Comput Secur 1(83):246–267
10. Aljofey A, Jiang Q, Qu Q, Huang M, Niyigena JP (2020) An effective phishing detection model based on character level convolutional neural network from URL. Electronics 9(9):1514
11. Almseidin M, Zuraiq AA, Al-kasassbeh M, Alnidami N (2019) Phishing detection based on machine learning and feature selection methods. Int J Interact Mobile Technol 13(12):171–183
12. Aljofey A, Jiang Q, Qu Q, Huang M, Niyigena JP (2020) An effective phishing detection model based on character level convolutional neural network from URL. Electronics

13. Sönmez Y, Tuncer T, Gökal H, Avci E (2018) Phishing web sites features classification based on extreme learning machine. In: 6th International symposium digital forensic security ISDFS 2018-proceeding, vol 2018
14. Abutaha M, Ababneh M, Mahmoud K, Al-Haj Baddar S (2021) URL phishing detection using machine learning techniques based on URLs lexical analysis. In: 2021 12th International conference on information and communication system (ICICS)
15. Desai A, Jatakia J, Naik R, Raul N (2018) Malicious web content detection using machine leaning. In: RTEICT 2017-2nd IEEE International conference recent trends electronic information communication technology proceeding, vol 2018
16. Parekh S, Parikh D, Kotak S, Sankhe PS (2018) A new method for detection of phishing websites: URL detection. In: 2018 Second international conference on inventive communication and computational technologies (ICICCT), 2018
17. Korkmaz M, Sahingoz OK, Diri B (2020) Detection of phishing websites by using machine learning-based URL analysis. In: 2020 11th International conference on computing, communication and networking technologies (ICCCNT), 2020: IEEE, pp 1–7
18. Shad J, Sharma S (2018) A novel machine learning approach to detect phishing websites Jaypee Institute of Information Technology, pp 425–430
19. Hong J, Kim T, Liu J, Park N, Kim SW (2020) Phishing URL detection with lexical features and blacklisted domains. In: Autonomous secure cyber systems. Springer
20. Purbay M, Kumar D (2021) Split behavior of supervised machine learning algorithms for phishing URL detection. In: Lecture notes in electrical engineering, vol 683

Security Assessment of Deep Learning Model for Intrusion Detection System Against Adversarial Example

Sushil Buriya and Neelam Sharma

Abstract Deep learning models have achieved so much attraction toward classification and prediction tasks with an impressive performance in the application of speech recognition, natural language processing, computer vision, and games playing applications with human brain-like precision. Deep learning models also proved their significance with impressive performance for cybersecurity applications, but these algorithms themselves are vulnerable to adversarial attacks. An adversary who has knowledge about the architecture of Deep Learning-based security tool can generate imperceptible noise to evade the decision. In this work, the vulnerability analysis of Deep Neural Network (DNN) architecture against evasion attacks for the implementation of intrusion detection systems is presented. We assess the impact of evasion attacks generated using the Fast Gradient Sign Method (FGSM) over the DNN model trained using two different datasets. The experimental results show the vulnerability of deep learning model against evasion attacks. It is possible to evade the detection of attack traffic flow as benign traffic flow by the intrusion detection system attacked by adversarial evasion attack.

Keywords Adversarial attacks · Evasion attacks · Deep learning · Deep neural network · Intrusion detection system

1 Introduction

Cybersecurity is a set of techniques, technologies, policies, and procedures intended to protect computers, information, networks, and programs connected through cyberspace in the form of the Internet against unauthorized access, malware, data

S. Buriya (✉) · N. Sharma
Department of Computer Science, Banasthali Vidyapith, Tonk, Rajasthan, India
e-mail: sushilburiya@banasthali.in

N. Sharma
e-mail: neelamsharma@banasthali.in

S. K. Biswas et al. (eds.), *Intelligent Computing Systems and Applications*, Lecture Notes in Networks and Systems 1307, https://doi.org/10.1007/978-981-96-3860-4_40

breaches, and any malicious activity that could cause harm, their confidentiality, integrity, and availability. Common cybersecurity threats and attacks are data breaches, phishing, DoS (denial of service), DDoS (distributed denial of service), malware and ransomware attacks, and zero-day attacks are also growing with the popularity of the Internet. The immense scale of cyberspace brings new technological challenges for intelligent analysis and actions for cybersecurity. There is an abundance of tools and techniques such as antivirus, antimalware, firewalls, IDSs (intrusion detection systems), and IPSs (intrusion protection systems) to protect cyberspace against threats and attacks [1].

The emergence of machine learning also provides a new weapon to intruders for the implementation of more sophisticated and intelligent zero-day cyberattacks. Self-organized and intelligent security management tools based on machine learning approaches to prevent and detect cybercrimes in a timely way are an urgent requirement for Internet security. Intrusion detection is a crucial part of the security system and it can be easily integrated with other security countermeasures such as access controls, encryption, and authentication systems. Cybersecurity intrusion detection has been changed into a new trend with impressive performance. Although most of the shallow learning techniques such as Neural Networks (NNs), Hidden Markov Model (HMM), and fuzzy models have achieved a significant performance in ML-based IDS, still they are suffering due to shallow architecture. Furthermore, the traditional shallow learning methods showed their limitations to solve complex classification problems. In contrast, Deep Learning (DL) methods showed an outstanding performance to solve complex classification problem with large-scale data analysis [2].

Deep learning models have gained so much attraction in the classification and prediction domain and proved themselves with an impressive performance in the application of speech recognition, natural language processing, computer vision, and games playing applications with human brain-like precision [3]. Deep learning networks are also purposed for cybersecurity applications like IDS and malware detection based on several architectures like Deep Neural Networks (DNNs), Recurrent Neural Networks (RNNs), Generative Adversarial Networks (GANs), Deep Belief Networks (DBNs), and Convolutional Neural Networks (CNNs). These architectures have shown significant success in security applications [4].

Deep learning systems mainly focus to learn models using a massive amount of data with high prediction and classification accuracy. Deep learning overly depends on training data, so the accuracy of the model also relies on the accuracy or correctness of training data. A small perturbation in training data can affect the accuracy of the model which can mislead the prediction of the unknown sample. It makes deep learning approaches quite brittle. In particular, an adversary can often synthesize imperceptible perturbations of the training and testing samples and persuade the learned model to take erroneous decisions with high confidence. These synthesized training data are known as adversarial examples. It is possible to break deep learning through adversarial examples [5].

In this paper, the vulnerability analysis of Deep Neural Network (DNN) architecture against evasion attacks for the implementation of intrusion detection systems is presented. We have analyzed the effect of evasion attacks over the DNN model trained using two different IDS datasets NSL-KDD and UNSW-NB15. Adversarial examples are generated using the Fast Gradient Sign Method (FGSM) at testing phase. The organization of the paper is as follows: Section 2 presents the recent work toward machine learning-based IDS and adversarial crafting against ML-based methods. Section 3 explains the adversarial threat model adopted for vulnerability assessment and the experimental setup for vulnerability analysis. In Sect. 4, results are discussed. Finally, conclusion is given in Sect. 5.

2 Related Work

Salama A. et al. proposed an intrusion detection hybrid framework that contains three main steps: pre-processing which performs normalization, DBN for dimensionality reduction which reduces the data from 41 features to 5, and last step classification which classifies the network traffic into two classes's legitimate traffic or attack using SVM. The proposed DBN-SVM hybrid framework is evaluated using the NSL-KDD dataset and results are compared with the individual performance of SVM and DBN. The results show that the accuracy of the proposed hybrid framework is more than 90% for several training settings [6].

Farahnakian F. et al. presented a Deep Learning (DL) model for intrusion detection system using a four-layer deep Autoencoder (AE) that was trained in an unsupervised learning way followed by a softmax classifier. The framework was evaluated on the KDD'99 dataset and all the numerical valued dimensions are normalized in 0 to 1 range and 117 dimensions were generated from the original 41 dimensions in the pre-processing phase. In performance evaluation, a maximum accuracy of 94.71% was reported with a 0.42% of false alarm rate [7].

Szegedy C. et al analyzed the robustness of artificial neural networks concerning small noise in the training dataset. They applied an imperceptible non-random perturbation with optimization of the input dataset to maximize the prediction error for a correctly classified input image and that arbitrarily changes the classifier's prediction for the same input image. Their work describes a way to generate perturbation in training data that can reduce the accuracy of a neural network and defines neural networks are vulnerable against adversarial examples [8].

Hu W. et al. proposed an algorithm named MalGAN to launch an adversarial attack against an ML-based malware detector with a black-box mechanism. A substitute detector has been implemented using neural network to replace the existing malware detector. A neural network-based adversarial example generator is learned to construct adversarial samples with the ability to fool the substitute detector. This approach can control the distribution of adversarial examples and frequent changes in probability distribution make the malware detector learning process unstable. This work has shown the potential of adversarial attacks to work in a black-box manner without prior knowledge about the machine learning algorithm [9].

Grosse K. et al. described a Jacobian matrix-based adversarial sample generation algorithm to mislead the neural network-based malware detection system. Therefore, the prediction of the classifier changed according to the adversary's goal. Since their algorithm can mislead malware detection systems based on neural networks, this adversarial example crafting algorithm has achieved a misclassification rate of 69% against the most recent models to achieve great classification performance. This work also advocates the potential of adversarial examples as a threat vector against machine learning-based computer security systems and suggests the defensive measure for machine learning [10].

Papernot N. et al. proposed a new class of methods for generating adversarial examples with help of knowledge about the architecture of Deep Neural Networks (DNN) and calculated the DNN's forward derivative with the help of the Jacobian matrix to construct an adversarial saliency map to uncover the input features from training data required by an adversary. Forward derivative and adversarial saliency map techniques followed by algorithms for adversarial examples crafting applied to a DNN used for handwritten digit recognition tasks. These algorithms achieved adversarial success rate of 97% toward DNN with only a 4.02% modified input features per sample. This work also exposes the vulnerability of DNN against adversarial examples [11].

Biggio B. et al. presented an evasion attack approach based on gradient-descent against neural networks (NNs) and support vector machines (SVMs) classification algorithms at the test phase. The effectiveness of the evasion attack is investigated using a handwritten digit classification task followed by a malware detection system for PDF files. The result shows that very popular classification algorithms (in particular, NNs and SVMs) have a high probability of evasion attack with limited knowledge of the input dataset. This work raises questions about the applicability of classification algorithms in security-sensitive systems [12].

Biggio B. et al. introduced a data poisoning attack against Support Vector Machine (SVMs) based on gradient ascent with the assumption that attackers have knowledge about training data. This attack is crucial from the practical point of view because an attacker doesn't have direct access to an existing training dataset but may provide a new training dataset to enhance the classifier efficiency. Most malware or intrusion detection systems enhance their knowledge base by collecting data from web-based repositories that can provide an opportunity to poison the training data by an attacker. This poisoning attack emphasized the need to consider robustness against adversarial training data for kernel-based learning algorithms [13].

Recent research work has exposed the vulnerability of deep learning architectures against adversarial evasion attacks, but most of the work focused on the task related to image classification. Currently, there is no significant research work available to analyze the performance of deep learning architectures for intrusion detection systems against adversarial evasion attacks.

3 Methodology

In this work, we have adopted the white-box threat model. An adversary gains the access of DNN architecture along with training and testing datasets in white-box class of adversarial attacks. The goal of adversary is to confuse the trained model and persuade to take erroneous decisions to consider attack traffic as benign.

3.1 Adversarial Attacks

An adversarial example is a training sample vector with small, intentional feature perturbations that affect the Deep Learning (DL) model to generate a false prediction. The presence of adversarial examples in the DL model makes it vulnerable to adversarial attacks. DL architectures depend on a large number of input data examples for their success which results in an incomplete exploration of data distribution of available training datasets which leads to loopholes in the training data manifold. Adversarial attacks exploit the unexplored subspace of the data manifold to mislead the DL model. This unexplored subspace of data distribution is also known as an adversarial subspace that contains adversarial examples generated using feature perturbation of legitimate training examples with wisely crafted synthetic noise.

Formally, let $x \epsilon R^N$, where x denotes a training feature vector from natural data manifold R^N and $x^{'} = x + \delta$, here $x^{'}$ denotes perturbation in x with $\delta \epsilon S$, here $S \subseteq R^N$ is a subspace of all possible perturbation. $x^{'}$ is called adversarial example of x under D and $f(x)$, if $D(x, x^{'}) \leq \epsilon$ and $f(x) \neq f(x^{'})$. Here D is a distance metric, ϵ is a constant parameter for adversarial example, and $f(x)$ is the learning task.

Most of the adversarial attacks proposed in research literature use l_p-norm-based distance metric D and optimize the loss function of learning task $f(x)$ to get the preferred δ. L_p-norm-based distance metric can be written as $\|x - x^{'}\|_p$ where p-norm $\| \, . \, \|_p$ in n-dimensional vector space is defined as

$$\|x - x'\|_p = \left(\sum_{i=1}^{n} | \, x_i - x_i^{'} \, |^p\right)^{\frac{1}{p}} \tag{1}$$

Three kinds of distance metrics are used in generating adversarial examples; all of these are based on L_p-norm [14]. These distance metrics are as follows:

1. L_0 distance measure gives the number of samples i such that $x_i \neq x_i^{'}$. It will return the number of training samples altered to generate adversarial examples.
2. L_1 distance is the standard Euclidean distance between original training example x and adversarial example $x^{'}$.

3. L_∞ distance measures the maximum change in any of the training samples.

$$\|x - x'\|_\infty = \max(| x_1 - x_1' |, | x_2 - x_2' |, \ldots\ldots\ldots, | x_n - x_n' |) \quad (2)$$

The general principle behind adversarial attacks is obtaining adversarial examples by perturbing the training or testing samples within the data manifold such that the perceptual quality of the learning model remains unchanged but the decision is misguided [15].

Based on the level of access or knowledge about the targeted machine learning model, adversarial attacks are categorized as black-box attacks and white-box attacks. The attacker doesn't have any knowledge about the targeted learning model and defense mechanism in black-box attacks. The attacking approach described in MalGAN [9] is a black-box-based adversarial attack. Another attack category is a white box which assumed that attackers gain access to targeted learning models.

Adversarial examples in existing research work are generated using optimization-based methods, Jacobian matrix-based methods, gradient-descent-based methods, or a combination of these or some other statistical approach to generate synthetic noise in the data manifold. White-box attacks are defined in the work of [8, 10–12].

Adversarial attacks may compromise machine learning algorithms during the training phase and testing phases. Taxonomy of adversarial threat categorized in Fig. 1. From the perspective of influence on machine learning security threats, they can be classified as causative attacks and exploratory attacks. In causative attacks, the adversary can change the distribution of training data of a learning algorithm which induces modified learning parameters when the model is retrained. Poison attacks come under the category of causative attacks. It is a security threat that affects the training phase of learning models [13].

Exploratory attacks don't modify already trained models but aim to reveal sensitive data related to training datasets and learning models. The security threats against the testing phase exploit the vulnerability of a trained model to explore the behavior and working of the learning model. Potential security threats in the testing phase

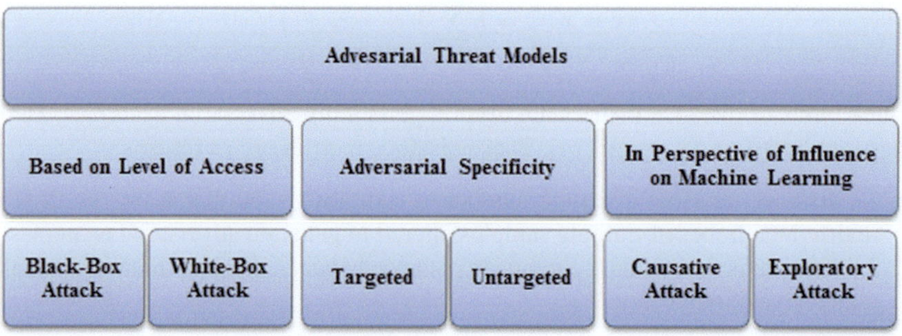

Fig. 1 Taxonomy of adversarial threat model

proposed in existing research work are evasion attacks, impersonate attacks, and inversion attacks. In the evasion attack, adversary generates adversarial samples using adversarial examples crafting algorithms to evade detection that compromises the learning model and misleads the output. Evasion attacks target the DL model used in cybersecurity systems to reduce the overall security of the targeted system [10, 12]. Our focus is on evasion attacks based on gradient-descent methods.

3.2 Evasion Attacks

It is the class of adversarial examples that targets the learning model during the inference phase to confuse the model to make accurate decisions. In evasion adversarial settings, malicious test samples with attack class are modified to evade detection, and it is misclassified as the benign sample. Figure 2 represents the evasion attack adversary in the DNN model testing phase. Evasion attacks can target both linear and nonlinear classifiers. An evasion attack is a white-box class attack, so an adversary needs to know the classification model to craft adversarial examples.

There are a lot of methods have been proposed to craft adversarial examples. The most popular and simplest technique to craft adversarial examples is the Fast Gradient Sign Method (FGSM). Goodfellow et al. [16], Mustafa et al. [17], Agarwal et al. [18] demonstrated the impact of FGSM on DNN architecture for image classification problems.

3.3 Fast Gradient Sign Method (FGSM)

The Fast Gradient Sign Method (FGSM) was given by Goodfellow et al. to generate adversarial images for image classification. FGSM aims to shift the input sample with a noise that leads to a diverted output by the DNN classifier. The logic behind

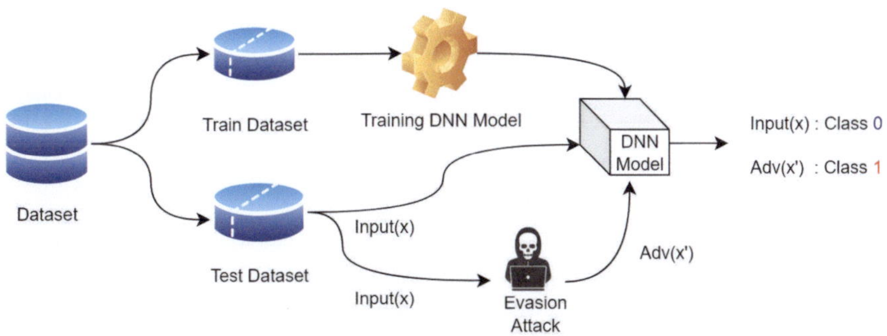

Fig. 2 Graphical representation of evasion attack

FGSM is to update the gradient of the loss function followed by its sign direction. The perturbation equation can be denoted by

$$x^{'} = x + \epsilon * \text{sign}(\Delta(J(\theta, x, y))) \quad (3)$$

Here $x^{'}$ denotes the adversarial example corresponding to input sample x, ϵ is the constant parameter, J is the loss function of the classifier, and θ denotes the learning parameter of the model for input sample x with target class label y.

FGSM works by taking an input sample that was classified correctly using the trained model, and then it computes the loss of the predicted class label. Thereafter, find out the gradients of the loss calculated concerning the input sample, and finally compute the sign of the gradient and multiply with a constant value. Perturbation calculated by FGSM is added to the input sample to find out the corresponding adversarial sample [16].

3.4 Experiment

In the experimental setup, we have trained the Deep Neural Network (DNN) model for binary classification of the network traffic instance as benign or attack based on statistical information on network flow. The DNN model first evaluated using the two benchmark datasets: NSL-KDD and UNSW-NB15, followed by the evaluation of the trained model under evasion attack crafted by FGSM method using the same datasets.

Dataset Description We have used two different datasets to evaluate the impact of evasion attacks over DNN for IDS implementation. NSL-KDD has flow-based network traffic and UNSW-NB15 has both packet-based and flow-based network traffic. Both the datasets have been preprocessed to remove missing values and normalization of numerical values. The categorical values are encoded using label encoder technique.

NSL-KDD NSL-KDD is the improved version of the KDD 99 dataset that addresses and overcomes from various problems of the KDD 99 dataset. This dataset contains five different attack classes. In this work, we have used the binary classes (attack and benign) only. Dataset describes 42 features, including class label. It has flow-based features. Each row in the NSL-KDD dataset is labeled as 0 for normal and 1 for attack records. In our experiments, we used 117478 records for training and 57863 for testing, including 56000 benign samples and 119341 attack samples [19].

UNSW-NB15 UNSW-NB15 dataset is a network flow generated by the IXIA perfecStorm tool in a lab setup under normal and attack conditions. TCPdump is used to capture the network traffic. This dataset contains nine different attacks that include: Fuzzers, Analysis, Backdoors, DoS, Exploits, Generic, Reconnaissance, Shellcode, and Worms. In our work, we have used the binary classes (attack and benign) only.

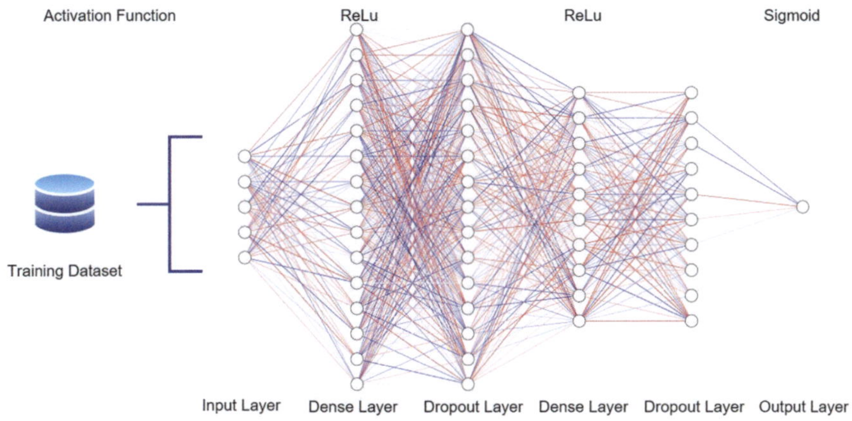

Fig. 3 DNN architecture for IDS implementation

Dataset describes 49 features, including class label. It has both packet-based and flow-based kind of features. Each row in the UNSW-NB15 dataset is labeled as 0 for normal and 1 for attack records. In our experiments, we used 175341 records for training and 82332 for testing [20].

DNN Model Implementation We have used the DNN architecture for the implementation of IDS as shown in Fig. 3 to experiment on both NSL-KDD and UNSW-NB15 datasets. The ADAM optimizer has been used and sliced the data into 64 batches repeated over the 20 epochs for training.

First, we have trained the model using the training dataset and analyzed the performance using the test dataset. Then we attacked the DNN testing phase with FGSM adversarial attacks and analyzed the performance under evasion attack.

3.5 Evaluation Metrics

Evaluation metrics are essential to measuring the performance of machine learning tasks. Our task is the binary classification of network traffic flow for that positive is attack traffic flow and negative is considered as normal traffic flow. Accuracy, precision rate (PR), recall rate (RR), F1-score, and AUC-ROC are used to analyze the performance of the DNN model under adversarial conditions and non-adversarial conditions [21]. Our model predicts a binary outcome as correct (True Positive, True Negative) or incorrect prediction (False Positive, False Negative). True Positive (TP) defines the number of attack network flow that are correctly predicted as an attack, True Negative (TN) states that the number of benign network flow is correctly predicted as benign, False Positives (FP) are those number of benign samples which are misclassified as attack, and False Negative (FN) are the number of attack samples which are misclassified as benign. Based on these values classification metrics are defined.

Accuracy: It describes how often the classifier predicts the correct class for an attack and benign sample. It is calculated as

$$\text{Accuracy} = \frac{TP + TN}{TP + TN + FP + FN} \tag{4}$$

Precision Rate: It explains how many of the attacks predicted traffic samples turned out to be positive.

$$\text{Precision Rate (PR)} = \frac{TP}{TP + FP} \tag{5}$$

Recall Rate: It defines how many of the actual attack traffic samples are predicted correctly with our model. The recall rate should be high for a network traffic analyzer.

$$\text{Recall Rate (RR)} = \frac{TP}{TP + FN} \tag{6}$$

F1-score: The F1-score is the harmonic mean of precision and recall. It is maximum when precision is equal to recall.

$$\text{F1 Score (F1)} = 2 * \frac{PR * RR}{PR + RR} \tag{7}$$

AUC-ROC: The Receiver Operator Characteristic (ROC) is a plot between TP (True Positive) and the FP (False Positive) at various threshold points. The Area Under the Curve (AUC) describes the ability of a classifier to distinguish between classes and reflects the model performance.

4 Result Evaluation and Discussions

We evaluate the performance of the trained DNN model for two different IDS datasets NSL-KDD and UNSW-NB15. DNN model performed well and achieved detection of attacks with 98.2% accuracy, 96.5% precision rate, 98.9% recall rate, and 97.6% F1-score for the NSL-KDD dataset as presented in Table 1 for non-adversarial settings. But the performance of the trained model degrades under the adversarial settings and the difference is presented graphically in Fig. 4. The accuracy of the trained model decreased to 87.3% but the evasion attack highly affects the precision and recall. Precision decreased to 45.6% and recall decreased to 29.3% with an F1-score of 35.6%.

AUC-ROC curve shown in Fig. 5 states that the confidence of the model to distinguish between attack and benign is 99% for non-adversarial settings but it degrades

Table 1 Performance comparison of DNN in the non-adversarial environment and under adversarial attack using the NSL-KDD dataset

	Accuracy (%)	Precision (%)	Recall (%)	F1-score (%)	AUC (%)
Non-adversarial environment	98.2	96.5	98.9	97.6	99.0
Under adversarial attack	87.3	45.6	29.3	35.6	59.2

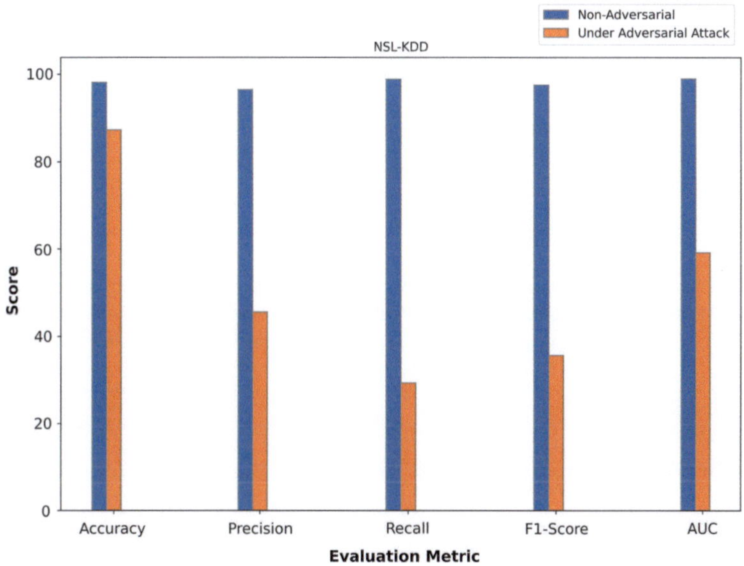

Fig. 4 Performance comparison of DNN in the non-adversarial environment and under adversarial attack using the NSL-KDD dataset

to 59.2% under evasion attack. So, we can state that model cannot distinguish the attack and benign traffic effectively under adversarial attacks.

For the UNSW-NB15 dataset, the DNN model also performed well and achieved detection of attacks with 97.8% accuracy, 94.7% precision rate, 98.4% recall rate, and 96.5% F1-score as presented in Table 2 for non-adversarial settings. But the performance of the trained model degrades under the adversarial settings and the difference is presented graphically in Fig. 6. The accuracy of the trained model decreased to 79.3% but evasion attack highly affects the precision and recall. Precision decreased to 32.5% and recall decreased to 24.4% with an F1-score of 27.9%.

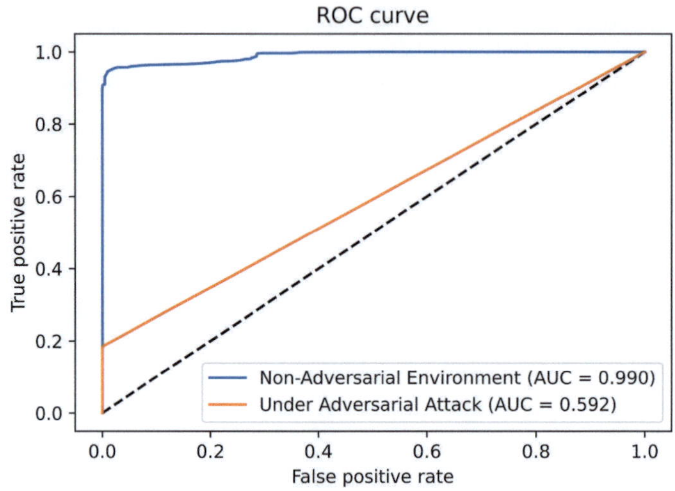

Fig. 5 ROC curve using NSL-KDD dataset

Table 2 Performance comparison of DNN in the non-adversarial environment and under adversarial attack using the UNSW-NB15 dataset

	Accuracy (%)	Precision (%)	Recall (%)	F1-score (%)	AUC (%)
Non-adversarial environment	97.8	94.7	98.4	96.5	98.2
Under adversarial attack	79.3	32.5	24.4	27.9	61.5

AUC-ROC curve shown in Fig. 7 states that the confidence of the model to distinguish between attack and benign is 98.2% for non-adversarial settings but it degrades to 61.5% under evasion attack. So, we can state that model is not able to distinguish the attack and benign traffic effectively under adversarial attacks.

We observe that the evasion attacks generated using FGSM highly affect the decision-making ability of DNN trained using IDS datasets. The precise goal of our experiment is to misclassify the attack traffic as benign traffic using an evasion attack to increase False Negatives (FN). A high rate of FN decreases the recall rate. Results show that recall and precision highly degrade under an adversarial environment. A lower rate of recall for IDS defines the inability of the system to detect the attack traffic. The results show that evasion attacks can misguide the IDS implemented using DNN and easily bypass the attack traffic without any detection.

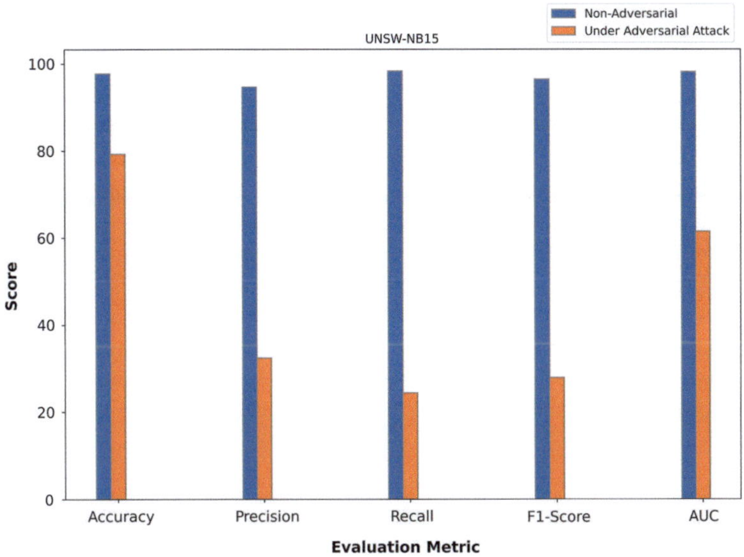

Fig. 6 Performance comparison of DNN in the non-adversarial environment and under adversarial attack using UNSW-NB15

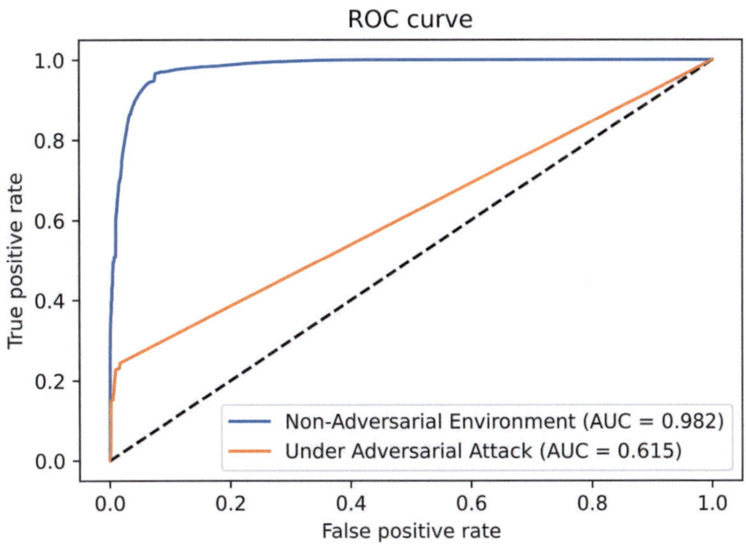

Fig. 7 ROC curve using UNSW-NB15 dataset

5 Conclusion

In this work, we have analyzed the impact of adversarial evasion attacks on the deep learning model for intrusion detection systems to present the robustness of DNN for the implementation of IDS. We have performed the evasion attack using FGSM and measured the performance of DNN under adversarial environments and non-adversarial environments with NSL-KDD and UNSW-NB15 datasets. Results show that the performance of the trained model degrades under evasion attacks for both datasets. Evasion attacks can persuade the DNN model to misclassification of the attack traffic by benign traffic to increase False Negatives (FN). Therefore, IDS based on deep learning models are vulnerable to evasion attacks and can be exploited to bypass the attack traffic without any detection. An adversarial security assessment is required for DNN-based IDS before adoption in cyberspace.

References

1. Sarker IH (2021) Deep cybersecurity: a comprehensive overview from neural network and deep learning perspective. SN Comput Sci 2(3):1–16
2. Mahdavifar S, Ghorbani AA (2019) Application of deep earning to cybersecurity: a survey. Neurocomputing 347:149–176
3. Liu Q, Li P, Zhao W, Cai W, Yu S, Leung V (2018) A survey on security threats and defensive techniques of machine learning: a data-driven view. IEEE Access 6:12103–12117
4. Berman DS, Buczak AL, Chavis JS, Corbett CL (2019) A survey of deep learning methods for cyber security. Information (Switzerland) 10(4). https://doi.org/10.3390/info10040122
5. Chakraborty A, Alam M, Dey V, Chattopadhyay A, Mukhopadhyay D (2018) Adversarial attacks and defences: a survey. x(x). http://arxiv.org/abs/1810.00069
6. Salama MA, Eid HF, Ramadan RA, Darwish A, Hassanien AE (2011) Hybrid intelligent intrusion etection scheme (2011). Soft computing in industrial applications. Springer, pp 293–303
7. Farahnakian F, Heikkonen J (2018) A deep auto-encoder based approach for intrusion detection system. In: 2018 20th International conference on advanced communication technology (ICACT). IEEE, 2018, pp 178–183
8. Szegedy C, Zaremba W, Sutskever I, Bruna J, Erhan D, Goodfellow I, Fergus R (2014) Intriguing properties of neural networks. In: 2nd International conference on learning representations, ICLR 2014-conference track proceedings, pp 1–10
9. Hu W, Tan Y (2017) Generating adversarial malware examples for black-box attacks based on GAN
10. Grosse K, Papernot N, Manoharan P, Backes M, Mcdaniel P (2017) Adversarial examples for malware detection, vol 1, pp 62–79
11. Papernot N, Mcdaniel P, Jha S, Fredrikson M, Celik ZB, Swami A (2016) The limitations of deep learning in adversarial settings. IEEE Eur Symp Secur Privacy (EuroS&P) 2016:372–387
12. Biggio B, Corona I, Maiorca D, Nelson B, Šrndic N, Laskov P, Giacinto G, Roli F (2013) Evasion attacks against machine learning at test time. In: Joint European conference on machine learning and knowledge discovery in databases, pp 387–402
13. Biggio B, Nelson B, Laskov P (2012) Poisoning attacks against support vector machines. In: Proceedings of the 29th international conference on machine learning, ICML 2012, vol 2, pp 1807–1814

14. Carlini N, Wagner D (2017) Towards evaluating the robustness of neural networks. In: Proceedings-IEEE symposium on security and privacy, pp 39–57. https://doi.org/10.1109/SP.2017.49
15. Kyatham V, Mishra M, Yadav TK, Mishra D, AP P (2019) Variational inference with latent space quantization for adversarial resilience
16. Goodfellow I, Pouget-Abadie J, Mirza M, Xu B, Warde-Farley D, Ozair S, Courville A, Bengio Y (2014) Generative adversarial nets. In: Advances in neural information processing systems, pp 2672–2680
17. Mustafa A, Khan SH, Hayat M, Goecke R, Shen J, Shao L (2020) Deeply supervised discriminative learning for adversarial defense. IEEE Trans Pattern Anal Mach Intell 8828(c):1–13. https://doi.org/10.1109/TPAMI.2020.2978474
18. Agarwal A, Singh R, Vatsa M, Ratha NK (2020) Image transformation based defense against adversarial perturbation on deep learning models. IEEE Trans Dependable Secure Comput 5971(c):1–14
19. Tavallaee M, Bagheri E, Lu W, Ghorbani A (2009) A detailed analysis of the KDD CUP 99 data set. In: Second IEEE symposium on computational intelligence for security and defense applications (CISDA)
20. Moustafa N, Slay J (2015) Unsw-nb15: a comprehensive data set for network intrusion detection systems (unsw-nb15 network data set) 2015 military communications and information systems conference (MilCIS), IEEE, pp 1–6
21. Goodfellow I, Bengio Y, Courville A (2016) Deep learning. MIT Press

Intelligent Embedded Systems

An Intelligent Swarm-Based Intrusion Detection Framework for Securing IoT Environment

Dukka Karun Kumar Reddy, Janmenjoy Nayak, and H. S. Behera

Abstract The increasing adoption of IoT technologies has led to a heightened emphasis on network security. However, due to limitations in resources and the intricate nature of algorithms, conventional ML intrusion detection methods face challenges when implemented within IoT networks. The leverage of the advancements of swarm algorithms has proven their efficiency in solving complex problems. This study aims to design, implement, and test an IDS placement strategy based on traditional ML algorithms with hyperparameter tuning. A population-based swarm algorithm called Firefly is introduced for extracting the pertinent parameters of ML algorithms through hyperparameter tuning. This framework demonstrates effective performance when applied to the TON_IoT dataset featuring new variables. Additionally, an analysis comparing the predictive outcomes of the proposed hyperparameter tuning method against those obtained using default parameters is conducted. The outcomes of the experiment validated that the suggested IDS framework designed for IoT successfully identifies genuine attacks, thereby bolstering the security within the IoT ecosystem. Additionally, fine-tuning the hyperparameters resulted in striking a balance between efficacy and resource utilization, leading to superior accuracy when contrasted with utilizing default ML parameters.

Keywords Internet of Things (IoT) · Swarm intelligence (SI) · Firefly (FF) · Intrusion detection system (IDS)

D. K. K. Reddy (✉) · H. S. Behera
Department of Information Technology, Veer Surendra Sai University of Technology, Burla, India
e-mail: karun.redy@gmail.com

J. Nayak
Department of Computer Science, Maharaja Sriram Chandra Bhanja Deo (MSCB) University, Mayurbhanj, Odisha, India

S. K. Biswas et al. (eds.), *Intelligent Computing Systems and Applications*, Lecture Notes in Networks and Systems 1307, https://doi.org/10.1007/978-981-96-3860-4_41

1 Introduction

The IoT has received significant consideration in recent years due to its unique support for a variety of applications areas, including industrial processes, health care, automation, smart environments, and so on. Although IoT offers an immense range of services, it is vulnerable to cyber-attacks. Because the IoT is a heterogeneity environment with an interoperability mechanism that does not support traditional security solutions. However, other aspects of IoT security, such as access controls, confidentiality, and data authentication are improved. These security mechanisms are designed in collaboration between the user and the IoT. Yet, they still have security flaws because security attack vectors have also evolved both in complexity and diversity in recent years. As a result, IoT threat analysis is crucial for sensing, identifying, and preventing new bouts in IoT networks. As a result, a specific module is required to assure IoT network security. IDS is an example of a notion that is already being used in wireless networks. Improving the IDS features of the wireless network will aid IoT in protecting the network from assaults and other vulnerabilities [1]. IDS is a second-line defensive system that monitors and detects malicious traffic in a node or network. IDS is critical for analyzing and exploring devices and user actions to detect signatures and detect intrusion activity. An IDS's objective is to keep monitoring and defending network nodes from various forms of intrusions and notify the user when intrusions are detected. The IoT network configures IDS devices for switches and routers based on low-power resource constraints and observes nodes and networks to detect various network incursions concerning the user's needs. The IoT landscape encompasses various protocols, applications, and platforms, providing a fertile ground for malicious actors to launch attacks unhindered.

An IoT security framework comprises three levels: perception, network, and application layer [2]. Perception nodes, such as sensors and actuators, play a vital role in gathering data within the perception layer. Key security concerns within this tier include ensuring secure communication among nodes and implementing lightweight authentication protocols. The network layer facilitates widespread access to the perception layer through ad-hoc networks, wireless networks, and similar means. Consequently, security threats like information leakage, network disruption, denial-of-service (DoS) attacks, and others emerge as prevalent challenges at this level. While designing an operative and responsive NIDS, there are primarily two obstacles. Choosing suitable characteristics from an IoT dataset poses challenges due to the dynamic and evolving nature of cyber threats. Features effective for detecting one type of attack may not be as effective for others. Additionally, the scarcity of labeled datasets from actual networks hampers the development of an IDS. Generating a tagged dataset from raw network traffic traces, whether gathered over time or in real-time, requires substantial dedication and labor. Furthermore, network administrators are hesitant to report any intrusion that may have occurred in their networks in order to protect the secrecy of the internal organizational network structure and the privacy of diverse users [3].

A consideration of the fact that nature-inspired metaheuristic will explore the search area effectively and efficiently for the success of building NP-hard problems [4]. The SI involves an intelligent exploration of the search space, meticulously examining regions for optimal solutions while venturing into uncharted territories. These processes are commonly referred to as intensification and diversification. Intensification involves utilizing acquired information by the metaheuristic within a defined timeframe, while diversification entails exploring inadequately addressed regions. The application of these two fundamental metaheuristic qualities can result in high-quality solutions for hyperparameter tuning. It is crucial to maintain a harmonious equilibrium between these two aspects for a metaheuristic's success because of swiftly finding areas with optimal solutions and avoids wasting time in areas with low-quality solutions or that have already been thoroughly investigated on the other [5].

SI-based techniques are an extremely sophisticated approach to assist in overcoming the problems of establishing an effective IDS to analyze data with a logical structure through nature-inspired biological systems. The SI is referred to as a collective behavior of self-organized systems with decentralized approach. SI approaches are characteristically composed of a population of simple agents cooperating locally in their environment. Although there is not a centralized control system dictating the behavior of individual agents, their localized interactions, which are partially random, lead to a collectively intelligent global behavior that remains unrecognized by the agents themselves. The characteristics exhibited by the agents enable independent learning through a combination of unsupervised and supervised training, allowing them to understand intricate and non-linear relationships without relying on pre-existing models. SI can address how to find the global minimum because these algorithms will not be trapped in the local minima. Due to the significant dynamical nonlinearity of positive feedback aids in the formation of convenient structures, whereas negative feedback counterbalances positive feedback and aids in the stabilization of the collective pattern. SI provides a balance between exploration and exploitation on function suitability with a valuable mean artificial creativity approach. SI also speeds up the convergence and enhances the performance due to multiple interactions of the swarm's agents rely on information from their neighbors. The information is disseminated over the network. This offer a way to consequently making IDSs parallel, self-organizing, adaptive, cost-efficient, and autonomous [6].

The primary significance of this paper is:

i. To investigate and evaluate the influential characteristics of the TON-IoT dataset to discern potential outcomes and enhance its capability to detect various types of attacks.
ii. Developing and executing a ML-driven IDS with hyperparameter tuning method to distinguish between abnormal and normal behaviors.
iii. The performance evaluation of the proposed work is verified by the utilization of various evaluation metrics.

In contrast to conventional ML classifiers, the recommended approach offers enduring benefits across diverse attack scenarios and enhances accuracy significantly.

The paper structure unfolds as follows: Sect. 2 scrutinizes pivotal prior research in this domain. Section 3 introduces the FF optimization method for hyperparameter tuning employing multiple ML techniques. Section 4 outlines the datasets, detailing their description and preprocessing. Section 5 delves into the experimental configuration and analysis of results. Lastly, Sect. 6 encapsulates the conclusions drawn from the study.

2 Related Works

Detecting anomalies in IoT environments stands as a pressing concern necessitating prompt attention within the IoT sphere. Numerous scholars have dedicated efforts towards fortifying IoT ecosystems through the employment of indiscriminate attack detection algorithms. This segment encompasses an evaluation of different scholars' methodologies using the TON_IoT datasets, a synopsis of their investigations, and concise insights into detecting attacks within IoT networks.

Moustafa [7] employed four ML-based algorithms: GB, RF, NB, and deep neural networks. A comparative analysis was performed between the TON_IoT and other relevant datasets to study the authentic and abnormal patterns. The work consists of a binary classification approach, achieving 99.98% of accuracy. Notably, the RF algorithm exhibited precision and recall rates of 99%, surpassing other ML algorithms.

Sarhan et al. [8] conducted a study focusing on IDS datasets, aiming to explore the effectiveness of combining feature reduction techniques with various ML approaches across diverse datasets. They employed six ML models: Deep feed forward, DT, CNN, LR, NB, and RNN. The study evaluated the uncovering accuracy of three characteristics mining methods: PCA, AE, and LDA across three benchmarks, which vary in terms of their feature sets, types of attacks, and network configurations. Notably, using DT with AE attribute selection, employing 20 attributes, resulted with 98.23% of accuracy and an F1-score of 99%. Specifically, DT exhibited strong performance in detecting attacks related to XSS, DoS, Password, and Backdoor classes.

Sarhan, Layeghy and Portmann [9] investigated optimal attributes sets for predicting network attacks, assessing their significance and predictive efficacy. They employed three attributes assortment algorithms: Chi^2, IG, and correlation to identify and rank data attributes. These features were then utilized with two machine learning classifiers, namely deep feed-forward and RF, to assess their accuracy in detecting attacks.

Abdallah et al. [10] made a detailed comparison study of various ML techniques to the test the vehicular ad hoc networks issues. For feature selection and class balance, the authors employed the Chi^2 method and the SMOTE. The multi-classification metric results with respect to normal and attack records shows that the XGBoost approach worked better than other ML algorithms. The Chi^2 method was employed to choose the features. It whittled down the features to 20. The XGBoost derives an accuracy of 98.3%, 98.2% using Chi^2, and 97.9% using the SMOTE technique for multi- classification for normal and attack records. The XGBoost using the Chi^2 technique and SMOTE technique derives an accuracy of 97.8% for multi-class classification. Idrissi et al. [11] projected a transfer learning technique by updating of DL-IDS that enables us to fine-tune pre-trained models on small datasets with new attack patterns. The suggested approach involves employing transfer learning with a CNN architecture, where the pre-trained convolutional base from IDS is utilized and kept fixed. This model is trained initially on the Bot-IoT dataset and subsequently fine-tuned with data from the TON-IoT dataset. Despite achieving an accuracy of 99.47% and a low loss of 0.23% on the validation set, it struggles to address the class imbalances effectively. Muna et al. [12] suggested a Threat Intelligence (TI) system based on DL methods that can find cyberthreats from different networks. The proposed framework comprises three integral components: a sophisticated pattern extractor and a method for identifying attack types based on TI analysis. The TI-driven detection leverages insights from the deep pattern extractor to uncover previously unseen patterns within IoT networks. Furthermore, the identification of attack types through TI analysis aids in promptly addressing security threats. Through evaluation on the TON_IoT dataset and employing a suitable defensive and mitigation approach, the proposed framework achieves remarkable performance metrics, boasting an accuracy of 99.42% and a detection rate of 99.45%.

3 Firefly Algorithm

Yang [13], drawing inspiration from the flashing behavior of fireflies, explored the potential applications of this natural phenomenon in addressing practical problems. Fireflies utilize their flashing property as a means of communication. This exploration is based on three key assumptions: first, that fireflies are unisex and are drawn to one another; second, that the brightness of their flashes is crucial in attracting others; and third, that fireflies will exhibit random movements when their brightness levels are identical.

FF Algorithm	
Step 1:	Initialize n, β,α, d, max_t constraints
Step 2:	Produce N initial set of solutions randomly
Step 3:	Till max_t is reached
	Step 3.1 Compute the fitness rate for individually firefly i.e, I_i at x_i is calculated by the fitness function $F(x_i)$
	Step 3.2 *for* $i \leftarrow 1\ to\ n-1$
	Step 3.2.1 *for* $j \leftarrow i+1\ to\ n$
	Step 3.2.2 *if* $I_i > I_j$
	Update the location of firefly i towards firefly j, as outlined in Eq. (1).
	Step 3.3 If no brighter firefly is selected, move randomly as given in Eq. (2)
Step 4:	Report the best solution

In this context, α defines the maximum range of a random step, while β governs the size of stepping towards an optimal solution. The parameter d represents the onlooker's distance from the source, and n dictates the size of population. Additionally, γ signifies the light absorption coefficient. The variable ε constitutes a random vector between (0, 1). The maximum iterations are determined by $\max_t$. Individual fireflies' positions are updated according to Eq. (1), moving firefly i towards firefly j, while Eq. (2) dictates that fireflies move randomly if a brighter one is not recognized.

$$x_i(t+1) = x_i(t) + \beta e^{\gamma r_{ij}^2}\left(x_j^t - x_i^t\right) + \alpha_t \in_i^t \tag{1}$$

$$x_i(t+1) = x_i(t) + \alpha\varepsilon_i \tag{2}$$

A set of parameters chosen at random from the parameter pool is used to initialize the hyperparameter space model of the FF population before it is executed. The algorithmic model calculates the fitness function after being trained from the hyperparameter space of the FF swarm. Updating of the individual best and overall best of FF's occurs until the maximum conditions are met. With the calibration of these characteristics related to the problem, the FFs adapt to the best location globally and produce an ideal outcome. The basic steps of the FF-based hyperparameter tuning method are depicted in Fig. 1.

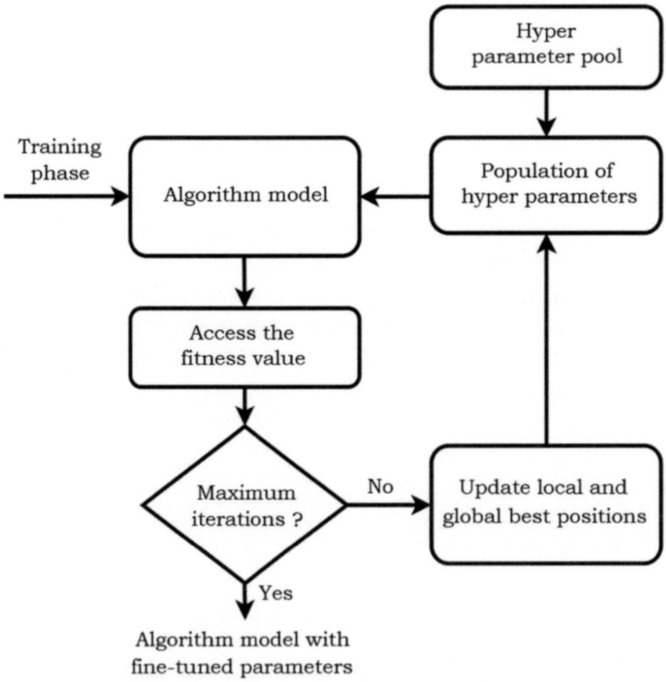

Fig. 1 FF-based hyperparameter tuning

An ML model constitutes a mathematical framework with numerous parameters requiring learning from data. Defining the model architecture offers flexibility, but determining the optimal parameters remains a challenge, demanding extensive exploration. Swarm-based parameter tuning, relies on agents exploring parameter space independently, guided by a fitness function derived from data. The selection of the optimal parameters hinges on evaluating this function. Various algorithms are employed based on the problem at hand, with the best-performing model selected based on performance metrics. Hyperparameters are user-specified variables influencing model construction. Hyperparameters precede parameters and play a crucial role in determining the ideal model configuration. The study introduces NatureInspiredSearchCV, a method leveraging nature-inspired algorithms to optimize hyperparameters and enhance model performance.

Hyperparameter tuning of NatureInspiredSearchCV	
Step :1	Set the algorithm constraints.
Step :2	Create an initial set of hyperparameters with a specified solution size:$X = \{x_i\}_i^n$
Step :3	Divide the dataset into segment folds for training and testing.
Step :4	Until criteria satisfied
	Step 4.1 Each firefly represents a unique mixture of parameters, which undergoes training as part of the optimization process. During this process, $f(X_{(t)})$ fitness function is computed for every firefly within the population across dataset folds.
	Step 4.2 Choose a portion of the most finely tuned parameters, $B_{(t)}$ from the fireflies of a specified size over a determined number of generations.
	Step 4.3 Construct an ML model, denoted as $ML_{(t)}$ utilizing the framework provided by $B_{(t)}$.
	Step 4.4 The model undergoes updates by sampling its current state, denoted as $ML_{(t)}$ to produce novel solutions $X'_{(t)}$ and then replacing these solutions back into the current state $X_{(t)}$.
	Step 4.5 The optimization runs are storing cached parameter combinations.
Step :5	Display the optimal combination of parameters for the chosen machine learning model.

4 Experimental Configuration and Dataset

4.1 System Environment

The experimentation took place on a Google Colab notebook. Data preprocessing was conducted using the NumPy framework, while hyperparameter optimization utilized the NatureInspiredSearchCV class from the Scikit-learn framework.

4.2 TON_IoT Dataset

Preparing data involves converting raw data into a format suitable for ML algorithms. This process includes addressing missing values, scaling data, and adjusting

Table 1 TON_IoT attack taxonomy with target attribute

S. no	Label	Type
1	Normal	Normal
2	Attacks	DDoS, Backdoor Injection, Password Scanning, Ransomware, XSS

Table 2 Label feature distribution

Label feature	Before pre-processing frequency count	After pre-processing		
		Frequency count	Percentage of total data	Normal class: Attack class
Normal	10,000	9874	62.5	1: 0.598
Attack	5980	5922	37.5	

the probability distribution of variables. Transformations may also be applied to modify the representation of historical data to align with the requirements of machine learning algorithms. However, there are instances where deviating from expectations or applying a transformation unrelated to the data could yield favorable or optimal outcomes. Therefore, FF is selected to transform the training data into a search or optimization problem aimed at effectively revealing the underlying structure of the data to ML algorithms.

The TON_IoT dataset represents a newly introduced set of Industrial and Internet of Things (IoT) data, specifically designed to evaluate the efficacy and dependability of diverse cybersecurity solutions. Termed as TON_IoT, these datasets amalgamate information from various sources, alongside publicly accessible IoT traffic datasets. Utilizing Performance Monitor Tool (PMT) dataset collectors on a Windows 7 machine, TON_IoT Windows datasets were captured, which differ from conventional network traces. This dataset comprises 134 attributes with 15,890 instances of data. The dataset's characteristics, along with feature descriptions and data types, are outlined by Moustafa [7]. The TON_IoT attack taxonomy, detailed in Table 1, includes 5,980 instances of anomaly classes and 10,000 instances of normal class. Table 2 indicates that the label class distribution of the dataset, and Table 3 which provides a comprehensive breakdown of attack distribution.

4.3 Data Pre-processing

The primary goal of a classification model is to effectively utilize information for classification purposes. Therefore, preprocessing is a vital in prioritizing and refining features, enhancing the classifier's performance. Primarily, data cleaning is conducted to handle inconsistencies such as missing data. In this specific dataset, approximately 184 instances contained missing values or NaN cells, which were subsequently removed to ensure data integrity. While these instances represented a

Table 3 Label feature distribution of attacks

Type attribute	Before pre-processing frequency count	After pre-processing		
		Frequency count	% of total data	% of attack data
DDoS	2134	2126	13.45	35.9
Backdoor	1779	1753	11.09	29.6
Injection	998	987	6.24	16.66
Password	757	755	4.78	12.75
Scanning	226	225	1.61	3.8
Ransomware	82	72	0.45	1.21
XSS	4	4	0.02	0.06

loss of significant data, they were deemed necessary for maintaining the integrity of the analysis. Additionally, columns with single constant values were identified and removed to streamline the dataset. A total of 51 features with zero values and 3 features with constant values of 10,000 and 1E + 09 were eliminated. No duplicate observations were found in the dataset. Lastly, object-type features were converted to float64 type to facilitate numerical computations. This comprehensive preprocessing ensures the dataset's suitability for subsequent analyses, particularly in the context of hyperparameter optimization for classifier accuracy enhancement.

5 Results

It is imperative to validate the technique to confirm the adequacy of dataset visualization and analysis. The analysis of simulation results on this dataset uncovers previously undetected anomaly cases. This section presents the analysis of results obtained from employing ML algorithms with FF optimization. The datasets encompass multi-classification tasks, and a fivefold cross-validation is conducted. Various evaluation measures, are illustrated to assess the presentation of the proposed work. Table 4 outlines the ML algorithms employed in the work along with the selected hyperparameters, which were optimized using FF optimization. Table 5 showcases the evaluation measures of ML algorithms with ideal hyperparameters for the TON_IoT dataset. Figure 2 illustrates various performance metrics for taken into consideration for the study. Furthermore, Table 6 provides a concise comparative analysis of various research efforts on the TON_IoT dataset using different evaluation metrics.

Table 4 ML classifiers with FF optimized hyperparameters

ML classifiers with selected hyperparameters		Hyperparameter optimization Firefly Algorithm
		For TON_IoT
Decision Tree (DT)	splitter = ["best", "random"], min_samples_split = range (2, 4, 1), max_features = ["auto","sqrt","log2"], criterion = ["gini","entropy"], max_depth = range (2, 16, 2)	splitter = random, min_ samples_split = 3, max_features = auto, criterion = gini, max_ depth = 12
Random Forest (RF)	max_depth = range (2, 16, 2), n_estimators = range (10, 120, 10), min_samples_split = range (2, 4, 1), max_features = ["auto", "sqrt", "log2"], criterion = ["gini","entropy"]	max_depth = 14, n_estimators = 40, min_samples_split = 2, max_features = auto, criterion = entropy
Multi-Layer Perceptron (MLP)	activation = ["identity","logistic","tanh", "relu"], hidden_layer_sizes = [70, 80, 100, 110, 120], solver = ["lbfgs", "sgd", "adam"], max_iter = range (50, 200, 50), learning_rate = ["constant","invscaling", "adaptive"]	activation = relu, hidden_layer_ sizes = 120, solver = adam, max_iter = 200, learning_rate = constant
K Nearest Neighbor (k-NN)	n_neighbors = range (2, 6, 1), weights = ["uniform", "distance"], algorithm = ["auto","ball_ tree","kd_tree", "brute"]	n_neighbors = 3, weights = distance, algorithm = brute
Extra Tree Classifier (ETC)	n_estimators = range (10, 120, 10), max_depth = range (2, 16, 2), min_samples_split = range (2, 4), max_features = ["auto","sqrt","log2"], criterion = ["gini","entropy"]	n_estimators = 50, max_depth = 14, min_samples_split = 2, max_features = auto, criterion = gini

(continued)

Table 4 (continued)

ML classifiers with selected hyperparameters		Hyperparameter optimization Firefly Algorithm For TON_IoT
AdaBoost (Ada)	n_estimators = range (10, 120, 10), algorithm = ["SAMME", "SAMME.R"], learning_rate = [0.5,1.0]	n_estimators = 30, algorithm = SAMME.R, learning_rate = 0.5
Gradient Boost Classifier (GB)	learning_rate = [0.5, 1.0], n_estimators = range (10, 120, 10), min_samples_split = range (2, 4), max_features = ["auto", "sqrt", "log2"], max_depth = range (2, 18, 2)	learning_rate = 0.5, n_estimators = 30, min_samples_split = 2, max_features = auto, max_depth = 18

6 Conclusion

Utilizing SI algorithms in conjunction with IoT datasets enables the detection of ongoing threats and validates the potential for identifying anomalous traffic for identification purposes. SI operates iteratively, continuously seeking to enhance a candidate solution in terms of a specified measure of fitness function quality. Swarm-based hyperparameter tuning goal is to discover best hyperparameters that optimizes a predefined loss function and give better results in the parameter search space. The primary focus of the paper lies in conducting a thorough evaluation of the characteristics of the TON_IoT dataset through the utilization of ML techniques, alongside hyperparameter tuning employing FF optimization. The regularized hyperparameters with FF optimization provides the ability to control the flexibility of the model. On new test data, this control prevents overfitting and a loss in predicting accuracy. The assault scenarios within each data sets, as well as their interrelationships, were studied with F1-score as fitness function in terms of the classification scheme's properties. The study effectively showcased its resilience through experimentation with the TON_IoT dataset. Empirical findings indicate that employing hyperparameter tuning with the FF algorithm notably enhances the performance of intrusion detection via machine learning algorithms. Particularly, the GB-FF algorithm exhibited exceptional accuracy in distinguishing between normal and abnormal activities, achieving a 100% accuracy rate on the TON_IoT dataset.

Table 5 Evaluation measures of ML classifiers with optimized hyperparameters

Model	Evaluation metrics	Backdoor	DDoS	Injection	Normal	Password	Ransomware	Scanning	XSS
RF	TPR	0.99	1.0	1.0	0.98	1.0	1.0	1.0	1.0
	FPR	0.0	0.0	0.0	0.0	0.0003	0.0	0.009	0.0
	Precision	1.0	1.0	1.0	1.0	0.99	1.0	0.6	1.0
	F1 score	0.99	1.0	1.0	0.99	0.99	1.0	0.75	1.0
	Roc curve	0.99	1.0	1.0	0.99	0.99	1.0	0.99	1.0
	ER	0.0003	0.0	0.0	0.009	0.0003	0.0	0.009	0.0
	Individual accuracy	0.99	1.0	1.0	0.99	0.99	1.0	0.99	1.0
	Overall accuracy	99.27%							
DT	TPR	1.0	0.96	0.98	0.98	0.97	1.0	1.0	1.0
	FPR	0.001	0.002	0.005	0.0	0.0	0.0	0.006	0.0003
	Precision	0.99	0.98	0.92	1.0	1.0	1.0	0.69	0.5
	F1 score	0.99	0.97	0.95	0.99	0.98	1.0	0.81	0.66
	Roc curve	0.99	0.98	0.98	0.99	0.98	1.0	0.99	0.99
	ER	0.0009	0.006	0.005	0.007	0.001	0.0	0.006	0.0003
	Individual accuracy	0.99	0.99	0.99	0.99	0.99	1.0	0.99	0.99
	Overall accuracy	98.54%							
ETC	TPR	1.0	1.0	1.0	0.87	1.0	1.0	1.0	1.0
	FPR	0.0	0.0	0.0	0.0	0.0	0.0	0.076	0.0
	Precision	1.0	1.0	1.0	1.0	1.0	1.0	0.158	1.0
	F1 score	1.0	1.0	1.0	0.93	1.0	1.0	0.27	1.0
	Roc curve	1.0	1.0	1.0	0.93	1.0	1.0	0.961	1.0
	ER	0.0	0.0	0.0	0.075	0.0	0.0	0.075	0.0

(continued)

Table 5 (continued)

Model	Evaluation metrics	Backdoor	DDoS	Injection	Normal	Password	Ransomware	Scanning	XSS
	Individual accuracy	1.0	1.0	1.0	0.92	1.0	1.0	0.924	1.0
	Overall accuracy	92.43%							
k-NN	TPR	0.99	1.0	1.0	0.98	1.0	1.0	1.0	1.0
	FPR	0.0	0.0	0.0	0.0	0.0003	0.0	0.009	0.0
	Precision	1.0	1.0	1.0	1.0	0.99	1.0	0.6	1.0
	F1 score	0.99	1.0	1.0	0.99	0.99	1.0	0.75	1.0
	Roc curve	0.99	1.0	1.0	0.99	0.99	1.0	0.99	1.0
	ER	0.0003	0.0	0.0	0.009	0.0003	0.0	0.009	0.0
	Individual accuracy	0.99	1.0	1.0	0.99	0.99	1.0	0.99	1.0
	Overall accuracy	99.01%							
MLP	TPR	0.98	0.84	1.0	1.0	0.93	0.86	1.0	0.0
	FPR	0.0	0.0	0.02	0.006	0.0003	0.0	0.0	0.003
	Precision	1.0	1.0	0.75	0.99	0.99	1.0	1.0	0.0
	F1 score	0.99	0.91	0.86	0.99	0.96	0.92	1.0	0.0
	Roc curve	0.99	0.92	0.98	0.99	0.96	0.93	1.0	0.49
	ER	0.001	0.02	0.02	0.002	0.003	0.0006	0.0	0.003
	Individual accuracy	0.99	0.92	0.98	0.99	0.96	0.93	1.0	0.49
	Overall accuracy	97.37%							
Ada	TPR	1.0	1.0	0.99	1.0	1.0	1.0	0.0	1.0
	FPR	0.0	0.0	0.0	0.038	0.0	0.0	0.0	0.0
	Precision	1.0	1.0	1.0	0.97	1.0	1.0	None	1.0
	F1 score	1.0	1.0	0.99	0.98	1.0	1.0	0.0	1.0

(continued)

Table 5 (continued)

Model	Evaluation metrics	Backdoor	DDoS	Injection	Normal	Password	Ransomware	Scanning	XSS
	Roc curve	1.0	1.0	0.99	0.98	1.0	1.0	0.5	1.0
	ER	0.0	0.0	0.0003	0.01	0.0	0.0	0.014	0.0
	Individual accuracy	1.0	1.0	0.99	0.98	1.0	1.0	0.98	1.0
	Overall accuracy	98.54%							
GB	TPR	1.0	1.0	1.0	1.0	1.0	1.0	1.0	1.0
	FPR	0.0	0.0	0.0	0.0	0.0	0.0	0.0	0.0
	Precision	1.0	1.0	1.0	1.0	1.0	1.0	1.0	1.0
	F1 score	1.0	1.0	1.0	1.0	1.0	1.0	1.0	1.0
	Roc curve	1.0	1.0	1.0	1.0	1.0	1.0	1.0	1.0
	ER	0.0	0.0	0.0	0.0	0.0	0.0	0.0	0.0
	Individual accuracy	1.0	1.0	1.0	1.0	1.0	1.0	1.0	1.0
	Overall accuracy	100%							

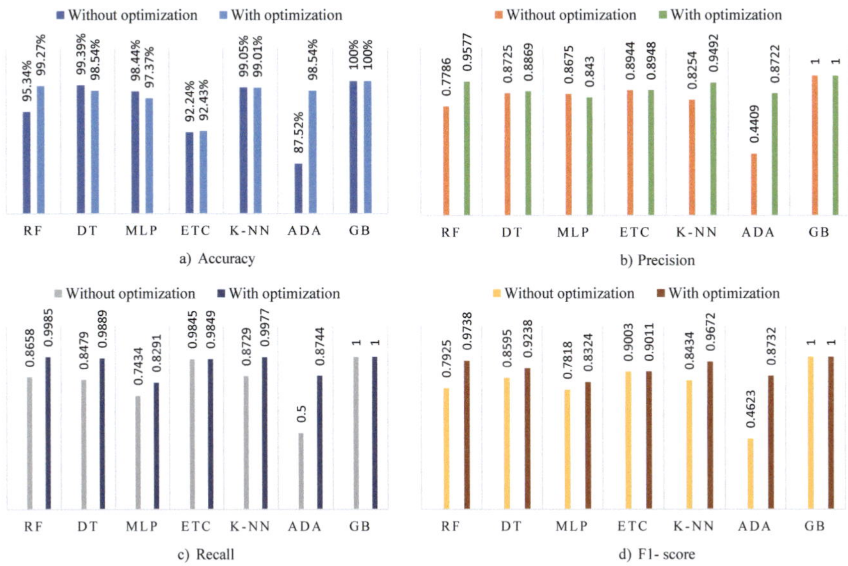

Fig. 2 Evaluation metrics of ML algorithms

Table 6 Comparative study for TON_IoT dataset

Proposed work	Various researcher proposed work	References
Accuracy (Acc) of **100%** for GB	GB (Acc = 93.83%, precision = 94.15%, recall = 87.75%, specificity = 97.09%), NB (AUC = 91.28%), DNN (Acc = 99.92%, precision = 99.9%, recall = 99.87%, specificity = 99.95%), and RF (Acc = 99.98%, precision = 99.95%, recall = 99.99%, specificity = 99.98%)	[7]
	DFF with 30 features (Acc = 96.93%, F1 = 0.98, DR = 98.25%, FAR = 40.86%, AUC = 0.801), CNN with 20 features (Acc = 96.78%, F1-score = 0.98, DR = 97.59%, FAR = 26.39%, AUC = 0.925), RNN with 5 features (Acc = 96.13%, F1-score = 0.98, DR = 96.9%, FAR = 26.02%, AUC = 0.9249), LR with 30 features (Acc = 95.46%, F1-score = 0.98, DR = 96.31%, FAR = 28.81%, AUC = 0.8375), DT with 20 features (Acc = 98.23%, F1-score = 0.99, DR = 98.28%, FAR = 3.21%, AUC = 0.9753), NB with 5 features (Acc = 97.94%, F1-score = 0.99, DR = 99.82%, FAR = 55.75%, AUC = 0.7203),	[8]
	RF for TON_IoT with 20 features (Acc = 97.49%, AUC = 0.97, F1-score = 0.99, DR = 97.51%, FAR = 2.9%), and RF for NF-TON_IoT with 6 features (Acc = 99.38%, AUC = 0.99, F1-score = 1.00, DR = 99.33%, FAR = 0.4%),	[9]
	XGBoost with 20 features (Acc = 97.8%, precision = 97.8%, recall = 97.8%, F1-score = 97.8%, false positive rate = 0.019%)	[10]
	DL using deep transfer learning (Acc = 99.43%, loss = 0.23%, F1-score = 99%)	[11]
	DL-enabled TI (Acc = 99.42%, DR = 99.45%, FPR = 0.6%, FNR = 0.55%)	[12]

References

1. Prakash KB, Nayak J, Madhav BTP, Padmanaban S, Balas VE (2021) Big data analytics and intelligent techniques for smart cities. CRC Press, Boca Raton
2. Jing Q, Vasilakos AV, Wan J, Lu J, Qiu D (2014) Security of the internet of things: perspectives and challenges. Wirel. Networks 20(8):2481–2501. https://doi.org/10.1007/s11276-014-0761-7
3. Javaid A, Niyaz Q, Sun W, Alam M (2016) A deep learning approach for network intrusion detection system. https://doi.org/10.4108/eai.3-12-2015.2262516
4. Blum C, Puchinger J, Raidl G, Roli A (2011) Hybrid metaheuristics. In: Hypar optimization and its applications, vol 45. Springer, pp 305–335.
5. Abd-Alsabour N (2016) Hybrid metaheuristics for classification problems. In: Pattern recognition—analysis and applications, vol 561. InTech, pp 57–72.
6. Zhang Y, Wang S, Ji G (2015) A comprehensive survey on particle swarm optimization algorithm and its applications. Math Probl Eng 2015:1–38. https://doi.org/10.1155/2015/931256
7. Moustafa N (2021) A new distributed architecture for evaluating AI-based security systems at the edge: Network TON_IoT datasets. Sustain Cities Soc. 72:102994. https://doi.org/10.1016/j.scs.2021.102994
8. Sarhan M, Layeghy S, Moustafa N, Gallagher M, Portmann M (2021) Feature extraction for machine learning-based intrusion detection in IoT networks. Available: http://arxiv.org/abs/2108.12722
9. Sarhan M, Layeghy S, Portmann M (2021) Feature analysis for ML-based IIoT intrusion detection. Available: http://arxiv.org/abs/2108.12732
10. Gad AR, Nashat AA, Barkat TM (2021) Intrusion detection system using machine learning for vehicular Ad Hoc networks based on ToN-IoT dataset. IEEE Access 9:142206–142217. https://doi.org/10.1109/ACCESS.2021.3120626
11. Idrissi I, Azizi M, Moussaoui O (2021) Accelerating the update of a DL-based IDS for IoT using deep transfer learning. Indones J Electr Eng Comput Sci 23(2):1059. https://doi.org/10.11591/ijeecs.v23.i2.pp1059-1067
12. Al-Hawawreh M, Moustafa N, Garg S, Hossain MS (2021) Deep learning-enabled threat intelligence scheme in the internet of things networks. IEEE Trans. Netw. Sci. Eng. 8(4):2968–2981. https://doi.org/10.1109/TNSE.2020.3032415
13. Yang, X-S, Karamanoglu M (2020) Nature-inspired computation and swarm intelligence: a state-of-the-art overview. In: Nature-inspired computation and swarm intelligence, vol. 927. Elsevier, pp 3–18

Optimum-Time Graph Theoretic Path Planning with Multi-sensor Fusion for AGV

Rapti Chaudhuri and Suman Deb

Abstract With a leap in the era of communication and sensor technology, Automated Guided Vehicle (AGV) has found its major applicability in point-to-point smooth locomotion. For accomplishing precise autonomous intelligent navigation, there is a need of environment perception which is basically achieved from combination of multiple sensors rather than through an individual sensor. Proper sensor calibration paves the way to achieve collision-free navigation in a complex environmental structure filled with obstacles. The concerned research work proposes a novel working architecture where the concept of sensor fusion for collecting data is incorporated with geometrical representation of grid-based graphical path planning strategy executed by a two active wheeled robot platform. Data from 2D LiDAR (light detection and ranging) and 3D sensor (Intel Realsense D455) are fused together for acquiring the required depth data of the surrounding environment to achieve collision-free optimized path. Specifically, GPS (global positioning system) denied indoor environment has been considered here for carrying out the experimental procedure. The paper concludes with a comparative performance analysis of different procedures adapted to collect environmental data with respect to precise obstacle detection and also a comparative performance analysis of the Dijkstra algorithm, a significant graph theoretic path-finding approach accompanied by the respective data collection methods executed by an automated customized differential drive. The multisensory path navigation achieved 94.4% accuracy in obstacle prediction with 162 s execution time of algorithmic execution which outperforms other compared techniques.

Keywords Automated Guided Vehicle (AGV) · Grid-based graphical path planning · Sensor calibration · GPS-denied indoor environment · Fused sensor · Differential drive

---

R. Chaudhuri (✉) · S. Deb
Department of Computer Science and Engineering, National Institute of Technology Agartala, Jirania, India
e-mail: rapti.ai@nita.ac.in

S. Deb
e-mail: sumandeb.cse@nita.ac.in

S. K. Biswas et al. (eds.), *Intelligent Computing Systems and Applications*, Lecture Notes in Networks and Systems 1307, https://doi.org/10.1007/978-981-96-3860-4_42

1 Introduction

AGVs equipped with intelligent machine learning technologies are able to achieve the solutions of complex path navigation-related problems. Basic path-searching techniques applied by a random mobile robot for point-to-point locomotion include A*, RRT, Dijkstra, genetic algorithm, particle swarm optimization, and many more. Real-time problems with constraints require relatively smooth and quick solvable path-searching algorithms. This significant requirement is near-perfectly achieved by Dijkstra algorithm [1]. Dijkstra is a grid-based path-finding approach and is suitable for both local and global path planning issues. The primary motive of the algorithm includes searching of the most suitable route accompanied by searching of the destinations in every possible directions. The respective visibility points which are generally referred to as collision-free points are added to the constructed path for producing the traversed map. The considered self-driving vehicle, with multi-depth sensors mounted on its top, are eligible to perceive the considered indoor or outdoor environment in a comparative precise manner resulting in quick detection of on-route obstacles optimizing the time required for achieving the goal [2]. Different variations of 2D LiDAR (light detection and ranging) and one RGB-D (red green blue-depth) sensor have been taken into use for collection of fused data [3]. The used sensors are preferred for their superiority in performance of measurements accumulation evident from extensive literature survey and experiments carried out, being significantly economical and for execution of active role in sensor fusion [4]. The probabilistic detection of the obstacles is achieved in the fused format of coordinate as well as 3D depth data. GPS-denied indoor environment with low light condition, presence of dynamic obstacles and reachability are also considered during the experiment. Primary contributions of this work are mentioned below:

- Creation and execution of a self-designed mobile robot working based on the principle of differential drive mechanism.
- Use of fused multi-sensors for relative collection of precise data of detected on-path obstacles.
- Achievement of graph theoretically optimized path navigation by the customized drive with comparative proof of analysis.

The visual reference of the scanned environment is collected from ROS (robot operating system) [5] visualizer [6].

Figure 1 presents the free space representation of movement principle followed by the used customized mobile agent. The next section of the paper gives an analytical overview of the reviewed related works followed by the generalized methodology section with schematic and geometric visualization of the procedure and comparative analysis of the experimented results of different applied techniques. The experiment table confirms the reliability of the proposed technique with reference to the real-life obstructed scenario.

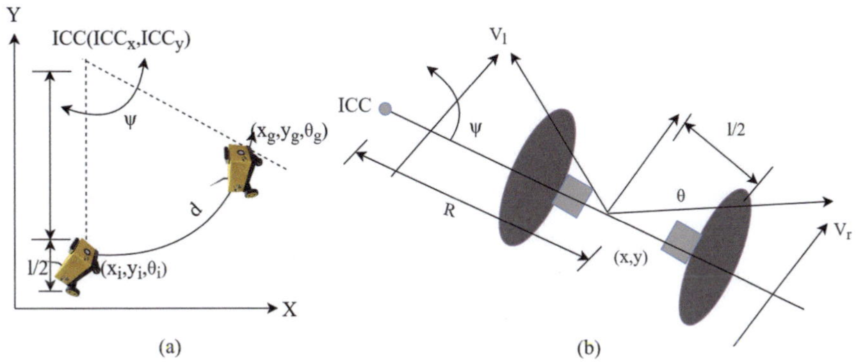

Fig. 1 **a** Transformation of customized mobile robot from one position to another relative to Instantaneous Center of Curvature (ICC), **b** Geometrical working architecture of differential drive mechanism followed by the customized robot platform

2 Related Work

Extensive studies have been done in the field of robot path planning as well as object detection using multi-sensor fusion techniques. Some works are analyzed to find out the research gaps and are tried to overcome with better results in this concerned paper work.

Dirik et al. in the paper [1] proposed a new approach of optimized path plan by combining grid-based Dijkstra's algorithm and sampling-based RRT (rapidly exploring Random Trees) algorithm and recorded the executed performance of a wheeled robot platform. In the work mentioned in the paper [7], Turcian et al. give a brief description of the significant usage of fused multi-sensor within the domain of speed control of a robot using integrated Arduino with video camera and ultrasonic sensor for perfect vehicle detection. Minhag et al. in the paper [8] analyzed the performance of Dijkstra's algorithm in obtaining optimized path by a mobile robot in different experimental and practical scenarios. The work presented in the paper [9] uses camera, radar, and LiDAR fusion in a particular robust dynamic obstacle detection in the field of autonomous vehicles. The significant error analysis in triangulation method of depth perception with comparative analysis of cost and efficiency has been briefly presented by Chaudhary et al. in the paper [10]. Robust camera-LiDAR fusion has been performed in the paper [11] along with adversarial training for 2D object detection.

The recognized research gaps in the aforesaid reviewed works include dynamic obstacle detection problem, achievement of smooth filtered out scanned environment as visual reference, and application of an optimized path planning approach accompanied with collection of input data by multi-sensor fusion. The present work tries to accomplish the desired goal by improving the identified gaps for getting optimum-time path navigation in a complex indoor environment by the mobile robot.

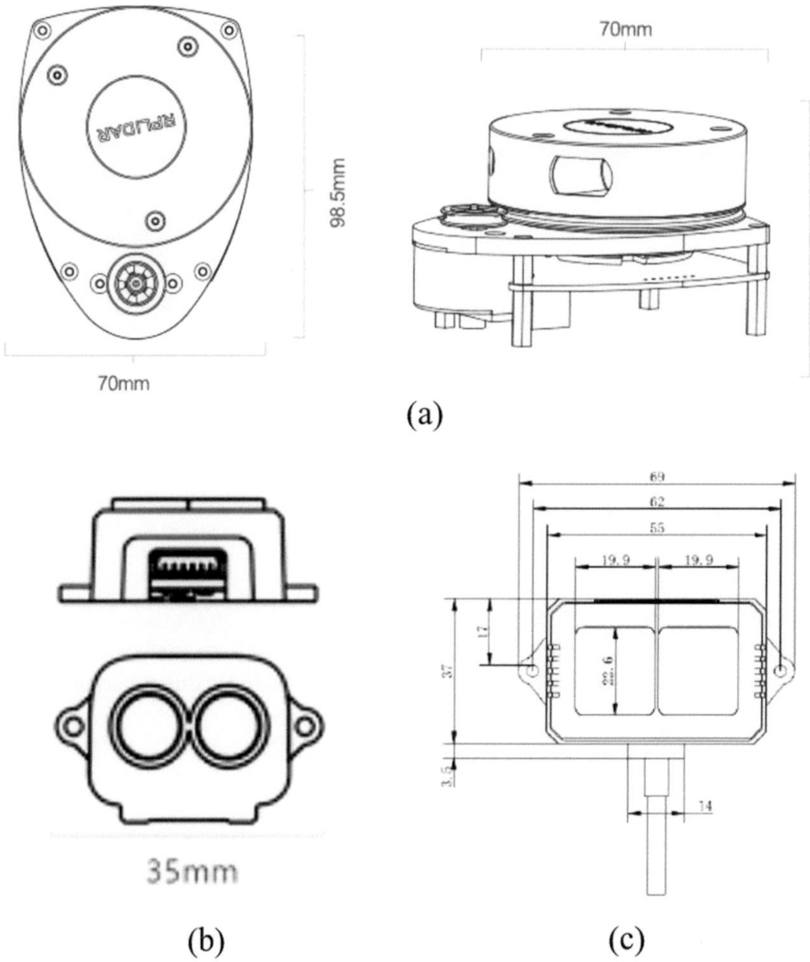

Fig. 2 **a** Internal structure of A1M8 series RPLiDAR for 2D scan, **b** TF micro-Luna LiDAR internal architecture used as auxiliary 2D scanning sensor, **c** TF02 pro Luna LiDAR internal structure used as supporting sensor for 2D percept of environment

3 Methodology

This section presents the schematic and geometrical analysis of working principle followed by the customized mobile robot platform and the fused sensors in carrying out the experiment. The schematic presentation is followed by algorithmic study of used graph theoretic technique to find out the optimized path from the start to desired goal point. The entire process has been structured schematically in Fig. 4.

3.1 LiDAR Sensor for 2D Scan

2D LiDAR has been preferred over other existing sensors for its precise obstacle periphery detection in one plane. Figure 2 presents all the variants of 2D LiDAR used in this research work. Three variants of 2D LiDAR taken for capturing various data are mentioned below:

RPLiDAR. A1M8 model of 2D RPLiDAR is for two-dimensional capture of indoor environment [12]. Periphery of the obstacles situated in same plane as the sensor is scanned by its limited triangular measurement. Maximum range of scanning is approximately 12m by radius.[1] A significant hindrance includes the disability of 2D LiDAR to sense obstacles in other plane rather than the sensor's which is covered up by RGB-D sensor.

TF02 pro Luna LiDAR. TF02-Pro is a single-point ranging LiDAR. It achieves stable, sensitive, and high-frequency range detection. The primary features include 40 m detection range, high frame rate, and ambient light resistance.[2]

TF-Luna Micro. It represents to be a single-point ranging LiDAR, based on the TOF (time of flight) principle. Its uniqueness depends on the optical and electrical design allowing stable, accurate, and highly sensitive range measurement. The range of detection is up to 8 m by radius, accompanied with voltage input and wide range of temperature adaptation. Both the variants of Luna LiDAR have been used in smoothing in environment scanning by detecting maximum possible nearest obstacle.[3]

3.2 RGB-Depth Sensor for 3D Scan

Three-dimensional depth data collection is accomplished by taking input from Realsense D455 [13]. Figure 3 depicts the internal architecture of Intel Realsense Depth sensor.[4]
Intel Realsense D455. Structural features of Intel Realsense D455 include depth module, IR projector, RGB color sensor, inbuilt IMU (inertial measurement unit) and vision processor. All the aforementioned 2D and 3D sensor data are fused together obtained in both coordinate and depth format and passed into the decision-making module to execute the optimized path-finding algorithm.

---

[1] https://www.alibaba.com/product-detail/Laser-Radar-Lidar-Rplidar-A1-2D_60763034393.html.

[2] https://www.mouser.in/new/benewake/benewake-tf02-i-lidar-sensors/.

[3] https://grobotronics.com/tf-luna-micro-single-point-lidar-module.html?sl=en.

[4] https://www.intel.com/content/dam/support/us/en/documents/emerging-technologies/intel-realsense-technology/Intel-RealSense-D400-Series-Datasheet.pdf.

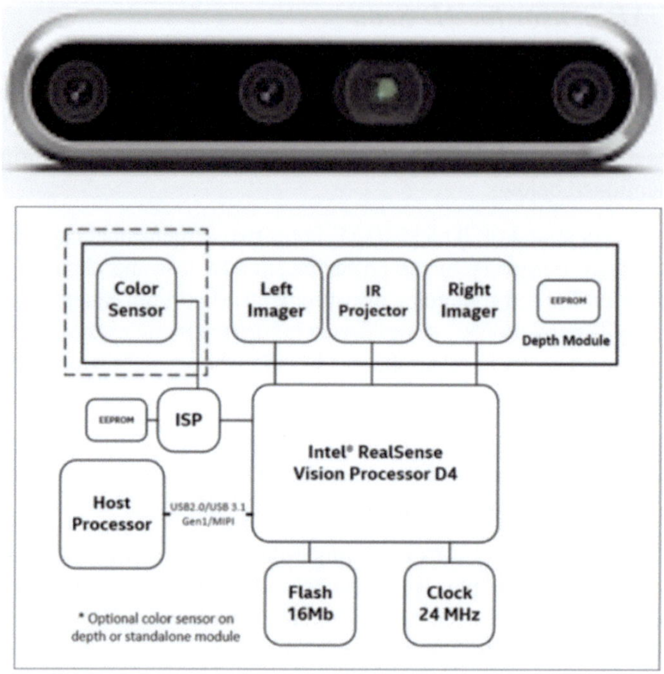

Fig. 3 Internal structure description of Intel Realsense D455 Depth sensor

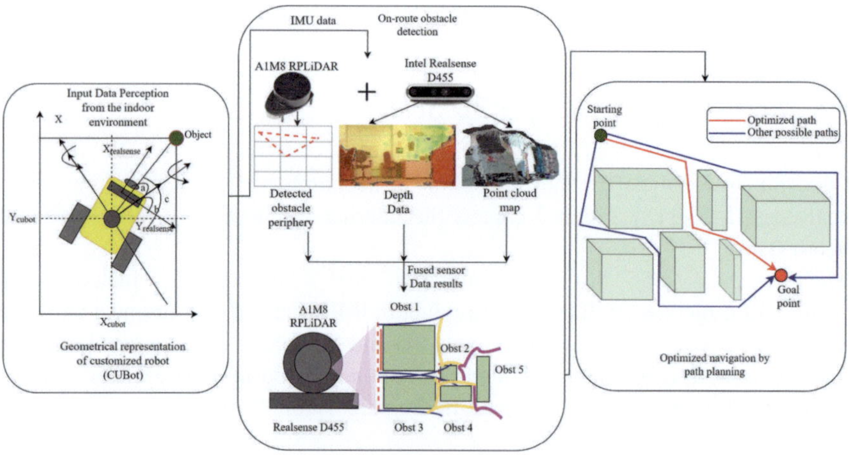

Fig. 4 Proposed model architecture presenting three structural module carried out during experimentation

3.3 Graphical Approach of Path Planning

Basically the performance of one of the shortest and quick path-finding procedures, Dijkstra's algorithm [14], has been analyzed by taking both individual primary sensors and fused multi-sensors for data collection in a complex indoor environmental stature. Geometrical presentation work as evidence of the process to accomplish graphical path navigation by the mobile robot within finite amount of time. The process has been carried out along with detection and avoidance of on-path static and dynamic obstacles which ensures the reachability of the concerned moving agent in a partially unknown obstructed indoor environment [15]. The working principle of Dijkstra's technique is similar to A* but unlike A* Dijkstra's algorithm covers up every possible grid for achieving near-optimum solution. In comparatively complex environment, performance of Dijkstra's algorithm has been appreciated rather than A*. The working process of this technique by taking input from multi-sensors has been illustrated in Algorithm 1.

Algorithm 1 Dijkstra's Algorithm with multisensor input for path plan

Input: - (G,V), E, a_n, f(s_1), f(s_2)// Graph with V vertex, Edges of the graph, set of vertices in V, Sensors data

Output: - (u,v)// optimum navigation graph

1: G = 0, V = 0
2: Q_{min} = {a_1, a_2, a_3, a_4....a_n}
3: **while** $Q_{min} \neq \phi$ and $f(s_1) \neq \phi$, $f(s_2) \neq \phi$ **do**
4: u = Min(Q_{min})
5: **for** V ∈ Q_{min}:(u,v)∈E **do**
6: Relax (u,v)

This algorithm has been proved to work better in case of maze robot path environment. Dijkstra's algorithm makes the mobile robot able to reach the goal even without searching the whole graph. The precise detection of on-path obstacles saves time to identify and detect followed by optimum-time path finding. Worst case of the algorithm performance is computed as O($|V|^2$) where V is the number of vertices in concerned graph. An auxiliary vector D is introduced initially. Shortest path in current start point from a_1 to each end of a_n is represented as $u[n]$. Q_{min} is the priority Queue. Initially $u[n]$ is infinity. So, length is expressed in Eq. (1):

$$u[m] = \min_n \quad \{u[n] | V \in Q_{min}\} \tag{1}$$

The basic formulation of the algorithm's working process is presented in Eq. (2).

$$dist[v] = \min(dist[v], dist[u] + cost[u][v]) \tag{2}$$

where v denotes the distance vertex adjacent to vertex u and it will get updated iff (dist[v], dist[u] + cost[u][v]) is less than dist[u] [16].

4 Experimental Analysis

This section specifically points out the significant sections carried out during the procedure. The first module includes individual and combined sensors' calibration. It is followed by application of optimized graphical approach executed by Dijkstra's algorithmic construction for decision-making in searching the possible optimum path accompanied by perfect obstacle detection by fused multi-sensors.

4.1 Sensor Calibration

2D LiDAR calibration. Estimation of rigid body transformation between Instantaneous Center of rotation (ICC) of actuator and ICC of LiDAR frame is dependent on calibrated LiDAR parameter [17]. Polar coordinate transformation of polar coordinate to Cartesian for 2D point achievement in LiDAR frame is presented in Eq. (3).

$$^{\Omega}x(\theta_i, p_i) = \begin{pmatrix} x_i \\ 0 \\ z_i \\ 1 \end{pmatrix} = \begin{pmatrix} \cos\theta_i \\ 0 \\ \sin\theta_i \\ 1 \end{pmatrix} p_i \tag{3}$$

where Ω denotes LiDAR frame and p denotes range measurement.

Proper triangulated measurement has been followed by 2D LiDAR to scan the nearest possible obstacle and the LiDAR frame is transformed to actuator frame by rotation along z-axis.

RGB-D sensor calibration. (a) Offline calibration. F_i and F_f are taken as initial and final frames, respectively. The extrinsic parameter matrix is presented in Eq. (4). P denotes point cloud.

$$\begin{bmatrix} P_f \\ 1 \end{bmatrix} = \begin{bmatrix} R_r^f & t_r^f \\ 0_{1\times 3} & 1 \end{bmatrix} \begin{bmatrix} P_r \\ 1 \end{bmatrix} = T_r^f \begin{bmatrix} C_r \\ 1 \end{bmatrix} \tag{4}$$

Final logical equation has been presented in Eq. (5).

$$P_{f-1}^0 = R_0^{(f)} R_f^0 P_q + R_0^{(f)} t_f^0 + t_0^{(f)} \tag{5}$$

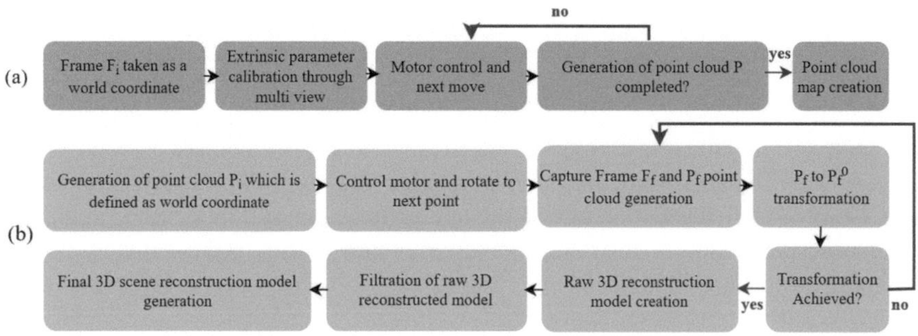

Fig. 5 **a** Stepwise offline calibration of RGB-D sensor **b** Stepwise online calibration of RGB-D sensor

(b) Online calibration. Transformation matrix obtained from offline calibration is preserved for online calibration [18]. Figure 5 denotes the calibration steps of RGB-Depth sensor. Point cloud P_1 of F_1 is transformed to F_0, and is presented in Eq. (6).

$$P_1^0 = T_1^0 P_1 \tag{6}$$

Reconstructed 3D scene model is presented as the set of point cloud presented in world coordinate w in Eq. (7).

$$P_w = \{P_0, P_1^0, P_2^0, \ldots., P_M^0\} \tag{7}$$

where M denotes camera views number.

4.2 Fused Sensor Calibration

Calibration of LiDAR with respect to Realsense has been performed based on certain parameterized constraints. World coordinate w is projected to image coordinate.

Equation (8) denotes w_g, the point for presentation of rigid transformation from camera coordinate system to LiDAR coordinate system.

$$w_g = \delta w + \phi \tag{8}$$

where δ presents 3×3 camera orientation orthonormal matrix relative to LiDAR and ϕ is the 3×1 vector with respect to its position.

4.3 Optimization in Graph Theoretic Path Planning Using Fused Sensor Data

The proposed method illustrates precise geometric shape detection of on-route obstacles resulting in reduced amount of time in achieving point-to-point smooth navigation. Constraint-rich situation affects the navigation strategy which has been numerically described below. Constrained penalty function process is presented by Eq. (9) for energy minimization.

$$E(k) = E(k_0) + g_0^T \delta k + 1/2\delta k^T H_0 \delta k \tag{9}$$

where g_0 = gradient $\frac{dE}{dk}$ at k_0, H_0 = Hessian $\frac{d^2E}{dk^2}$, $\delta k = k - k_0$ A remarkable process has been applied for optimization including addition of Lagrangian multipliers expressed in Eq. (10).

$$L(k) = E(k) + \sum_n \in \lambda_n C_n(k) \tag{10}$$

Conversion of the convergence has been depicted in Eqs. (11), (12):

$$\frac{\partial L(k)}{\partial \lambda_n} = g_0 + H_0 \delta k + \sum_n \in \lambda_n \frac{\partial C_n(\lambda)}{\partial k} = 0 \tag{11}$$

$$\frac{\partial L(k)}{\partial \lambda_n} = C_n(k) = 0 \tag{12}$$

Projection method is being applied by linear constraints where projector P eliminates the directions where displacement is zero, numerically expressed in Eqs. (13), (14):

$$P_{g_0} + P H_0 P \delta k + \beta(I - P) = 0 \tag{13}$$

$$P = I - \sum_n \in C_k C_k^T / |C_n|^2 \tag{14}$$

where C_n presents the set of orthogonal constraint vectors $\beta > 0$. Figure 6 depicts the sensor performance in detecting obstacles. An indoor lab scenario consisting of on-route 36 obstacles is taken. It is evident from LiDAR performance that 30 obstacles have been detected. RGB-D sensor has been able to detect 28 obstacles and the fused sensor results in detection of 34 obstacles.

The customized mobile robot scans and collects environmental data within the repulsive force range of the existed obstacles. After object detection with each applied sensor technology, the robot is made to make decided movement from point to point by applying Dijkstra's algorithm performance which is presented in Fig. 7. Data have been derived by collecting perceived input from the indoor environment through

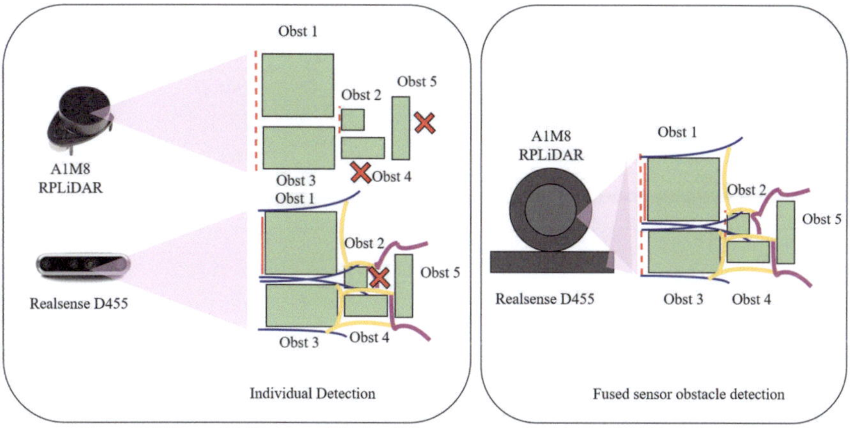

Fig. 6 Comparative visual analysis of individual and fused sensor performances

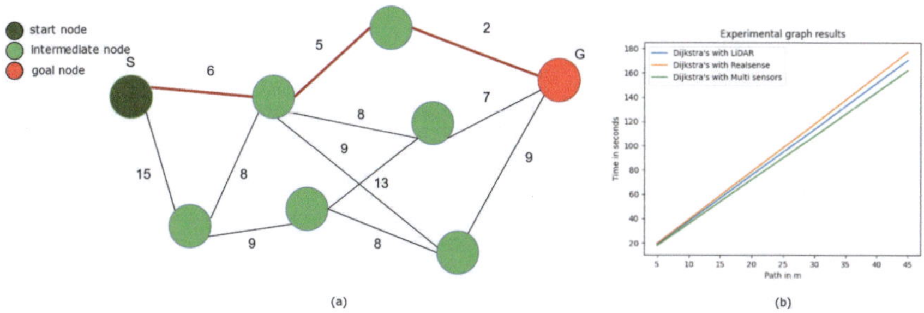

Fig. 7 **a** Path finding through Dijkstra's algorithm starting from start point and reaching to goal. **b** Graphical comparison of the performance of various individual and multi-sensor techniques followed by graphical path searching

Table 1 Experimental comparison of various sensor-mediated strategic procedures

Sensors	Obstacles present	Obstacles identified	Accuracy of obstacle detection
RP LiDAR	36	30	83.33% accuracy
Realsense	36	28	77.77% accuracy
Fused sensors	36	34	**94.44% accuracy**

sensors' fusion. Experimental results have been numerically analyzed and presented visually. The numerical analysis has been shown in Table 1 followed by Table 2.

Table 2 Experimental comparison of various sensor-mediated path planning strategy

Procedure adapted	Path in m	Time taken in seconds
2D LiDAR with Dijkstra's algorithm	45	170.37
3D RGB-D sensor with Dijkstra's algorithm	45	177
Fused multi-sensors with Dijkstra's	45	162

5 Conclusion

The experimental result obtained verifies the proposed methodology to be comparatively precise and efficient in detecting on-route obstacles. The simulated results confirm the successful detection of obstacles in any plane within the indoor environment followed by parameterized Dijkstra's algorithm for optimum-time smooth path finding. The performance of fused sensors and the performance of individual sensor in scanning the environment as well as reaching the desired goal in finite is noted time. An estimation of 94.44% accurate identification has been obtained by the usage of multi-sensor fusion. The proposed methodology and referential outputs would definitely serve for further research in partially unknown obstructed environment. The presented work would be a definite instance of reliability of the constructed technique in the field of collision-free optimum AGV navigation.

References

1. Dirik M, Kocamaz F (2020) Rrt-dijkstra: An improved path planning algorithm for mobile robots. J Soft Comput Artif Intell 1(2):69–77
2. Yeong DJ, Velasco-Hernandez G, Barry J, Walsh J et al (2021) Sensor and sensor fusion technology in autonomous vehicles: a review. Sensors 21(6):2140 (2021)
3. Sugiyama N, Sugiyama S, Catignani T, Chase ASZ, ernandez-Diaz JC (2021) Humans as geomorphic agents: Lidar detection of the past, present and future of the teotihuacan valley, Mexico. PloS one 16(9):e0257550 (2021)
4. dos Reis WPN, Morandin Junior O (2021) Sensors applied to automated guided vehicle position control: a systematic literature review. Int J Adv Manuf Technol 113(1):21–34
5. Mohamed Y, Lemaignan S (2020) Ros for human-robot interaction. In: 2021 IEEE/RSJ international conference on intelligent robots and systems (IROS). IEEE, pp 3020–3027
6. Majumder M, Hofmann H, Cook D (2013) Validation of visual statistical inference, applied to linear models. J Am Stat Assoc 108(503):942–956
7. Turcian D, Dolga V, Turcian D, Moldovan C (2020) Fusion sensors experiment for active cruise control. In: Joint international conference of the international conference on mechanisms and mechanical transmissions and the international conference on robotics. Springer, Berlin, pp 432–443
8. Zhou M, Gao N (2019) Research on optimal path based on dijkstra algorithms. In: 3rd international conference on mechatronics engineering and information technology (ICMEIT 2019). Atlantis Press, pp 884–892
9. Kocić J, Jovičić N, Drndarević V (2018) Sensors and sensor fusion in autonomous vehicles. In: 2018 26th telecommunications forum (TELFOR). IEEE, pp 420–425

10. Chaudhary P, Joshi NV, Paranjape R, Jadhav P (2021) An efficient system to quantify error in sensor fusion for cameras in adas. In: 2021 international conference on smart generation computing, communication and networking (SMART GENCON). IEEE, pp 1–4
11. Wang S, Wu T, Chakrabarti A, Vorobeychik Y (2022) Adversarial robustness of deep sensor fusion models. In: Proceedings of the IEEE/CVF winter conference on applications of computer vision, pp 2387–2396
12. Markom MA, Adom AH, Tan ESMM, Shukor SAA, Rahim NA, Shakaff AYM (2015) A mapping mobile robot using rp lidar scanner. In: 2015 IEEE international symposium on robotics and intelligent sensors (IRIS), pp 87–92
13. Tadic V, Odry A, Kecskes I, Burkus E, Kiraly Z, Odry P (2019) Application of intel realsense cameras for depth image generation in robotics. WSEAS Trans Comput 18:2224–2872
14. Karur K, Sharma N, Dharmatti C, Siegel JE (2021) A survey of path planning algorithms for mobile robots. Vehicles 3(3):448–468
15. Sharma CK (2015) Shortest path searching for road network using a* algorithm
16. Wang H, Yu Y, Yuan Q (2011) Application of dijkstra algorithm in robot path-planning. In: 2011 second international conference on mechanic automation and control engineering. IEEE, pp 1067–1069
17. Alismail H, Browning B (2015) Automatic calibration of spinning actuated lidar internal parameters. J Field Robot 32(5):723–747
18. Tsai C-Y, Huang C-H (2017) Indoor scene point cloud registration algorithm based on rgb-d camera calibration. Sensors 17(8):1874

IOT-Based Agricultural System with Standalone Solar Power Management

P. V. Nithara, N. Rakesh, N. Surendra, A. M. Rahul, and K. S. Ranjith

Abstract In last few years, over the time, the rainfall patterns and weather conditions have been changing and cause simultaneous threat to crop cultivation in farms. Due to change in environment conditions smart agriculture systems have come in to existence and many farmers are slowly adapting to these agriculture systems due to its advantages. It is a method where manual procedure is automated with the help of information technology linked with IOT. IOT has wide range of scope, especially in areas of wireless environments. The smart sensors with this technology have led to sustainable farming even in different climatic conditions. The whole system can be controlled by wireless communication with internet and can be monitored also. The main objective of smart agriculture is to automate irrigation system using IOT technologies and powered by solar system. This paper focused on the off-grid renewable agricultural methods as well. The proposed system used to provide the information such as soil moisture, water level, humidity, and temperature to the farmers through IoT system. From that information, smart agriculture can be achieved. Because farmer can able to access real-time data in a user-friendly and easily accessible manner. Here smartphone can be used to monitor the crop field condition and it will help to conduct the required action based on current circumstances. Also, it helps to achieve the solar system power management.

Keywords Agriculture · Power management · IoT · Smart irrigation

1 Introduction

Agriculture in India is one of the major sources of livelihood especially in rural areas. In the last few decades, it is known that crop development in agriculture sector is shrinking. The monitoring of crops depends on water supply to the crops, amount of sunlight, and level of humidity. In recent years, the rise of smart phones, tablets

P. V. Nithara (✉) · N. Rakesh · N. Surendra · A. M. Rahul · K. S. Ranjith
Department of EEE, CMR Institute of Technology Bengaluru, Bengaluru, Karnataka, India
e-mail: nithara.p@cmrit.ac.in

S. K. Biswas et al. (eds.), *Intelligent Computing Systems and Applications*, Lecture Notes in Networks and Systems 1307, https://doi.org/10.1007/978-981-96-3860-4_43

have brought automation in almost many major fields. These technologies can bring development in agriculture sector even in long-term perspective.

Conventional methods used for farming has more complexity such as required more hard work, experienced workers, variation in weather condition, and so on. Due to this, the outcome of the farming has been reduced. So, the use of Internet of Things (IoT) devices for monitoring of soil fertility, moisture content, and vegetation are one of the possible strategies for achieving better agriculture. IoT devices are tiny, power-efficient embedded electronics with the capacity to send data via a network. When devices connect with one another to cooperate in attaining a shared objective, this is frequently referred to as an Internet of Things (IoT) network. The IoT technology has played an important role in many sectors and the agriculture sector is also one among them.

Nowadays, importance of renewable energy sources has been increased. So, the utilization of solar energy for agriculture purpose has investigated in this work. As per literature review, the gaps identified in the agriculture field is lack of technology usage, availability of economical methods, variable weather condition, etc. Those problems have rectified by using solar energy for the standalone areas. So, the solar energy has been utilized for agriculture purpose as a source. So, in this work solar panel used as a single energy source to meet the energy demand in the isolated areas.

In this work, we offer a wireless Internet of Things (IoT)-based smart agricultural [2] monitoring system that is solar-powered. The planned network may measure the soil conditions, temperature, and humidity in an effort to aid agricultural production. In many ways, technology and IoT have the ability to change the agriculture. Such as smart agricultural sensors have collected a tone of data, which will be used by AI and ML for huge improve judgements in future. Improved product quality, volume and high profitability, better internal process management, which lowers production risks. Cost management and waste reduction due to better production control and increasing business efficiency through adaptable procedures.

Figure 1 explains the structure of smart agriculture, that has proposed in this work. Here supply is given by the solar panel, that is connected to the farm for irrigation, lightning and so on. So, with the help of the IoT the agriculture become smart. With the help of the proposed system, energy management also achieved through this work.

The remainder of this work is structured as follows: The relevant research on smart agricultural is reviewed in Sect. 2. An overview of the intended system is presented in Sect. 3, and the result and discussion is described in Sect. 4. Section 5 concludes this work.

2 Literature Survey

In paper [3], proposed work signifies on every aspect of the smart agriculture. This model is robust and highly efficient. But it has limitations such as more costly, complex, and it is an outdated model. Different types of sensors utilized in the paper

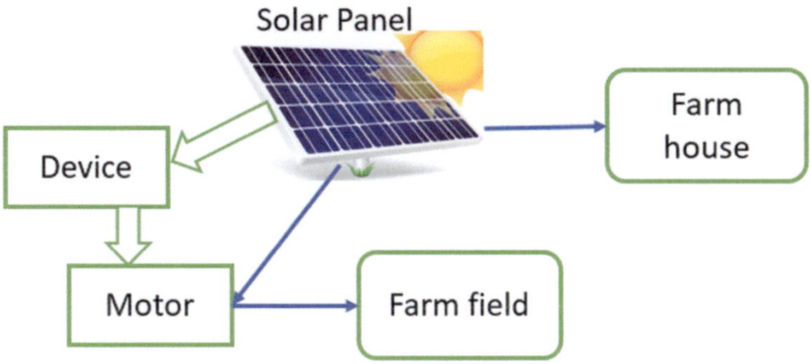

Fig. 1 Solar energy-based agriculture system [1]

and how to use them to interface with the field has been analyzed in this paper [4]. Those sensors given wide variety of options to smart agriculture. But lack of knowledge of the compatibility of various technologies mentioned in that paper. This paper [5] demonstrates how to create the hardware and the Internet of Things (IoT) for monitoring the seed flow into the soil for the seed plantation. This method is a smart analyzer for seed flow rate detection. But IR rays reduce the quality of crops. The proposed model in [6] explains the information of different sensors and how they used to achieve the whole smart processing systems. Here variety of sensors used to give an accurate reading. But this model is unable to handle a huge range of areas and it is an obsolete Model.

3 Methodology

A variety of hardware elements and methods were chosen for the proposed system in order to support a smart agriculture. Main hardware components have explained as follows.

3.1 Sensors

Farm field may have different crops growing at the same time at different parts of the agricultural land. In the farm field, mainly 2 main sensors are fitted and being used in the project, namely DHT22 sensor which is the temperature sensor and humidity sensor for getting environmental condition and moisture sensor measurements into the real-time input to the backend server. The sensors are directly connected to the main microcontroller ESP82. ESP32 microcontroller is low cost and low power which includes integrated Wi-Fi for additional functionalities and better usage of microcontroller. The real-time sensor values are fetched and uploaded to the backend server which can be further processed for useful decisions and processes.

3.2 Solar Panel

Additionally, a solar panel is connected to the farm field which will help capture the solar energy and convert to electrical energy which can be used to power up the whole motor for complete irrigation and can be helpful in providing electricity for the whole farm house [7].

3.3 Blink Backend and Frontend

The clients are given access to the system with a user-friendly and easy to operate application or a website through which the user can completely see all previous inputs from the sensors and completely control the farm field. The clients are given the frontend which is given by blink server which is a very reliable and secured third-party framework which supports all kinds of major frontend languages like HTML, CSS, and all kinds of latest frameworks.

3.4 Data Processing and Future Use

The real-time data are accessed from the app servers and are being processed and presented in a very meaningful pictorial and graphical standard which is easier to understand the condition of the crops [8]. The user can take the input from the sensors and he can keep the sensor readings for future use for deciding which crops to grow at which season and he can make maximum utilization of the smart agriculture to book maximum profits with minimum amount of loss or negligible amount of loss.

3.5 Smart Electricity Usage

The user can directly access all the input values and can turn on the motor for irrigation at any given time and the user need not to worry about the electricity for the motor since the electricity is being provided by solar energy the user will be relieved with all electricity charges [9].

The basic flowchart of the proposed system is shown in Fig. 2. There is different sensor such as moisture sensor, humidity, and temperature sensor. These sensors are required to give the information regarding the requirement of water availability. As per the data from the sensor control action from the NODE MCU will sent to the pumps through relays. Like this with the help of proposed model irrigation, lightening so on can be controlled [10]. Through this and by using supply from the solar panel, a better power management also can be achieved.

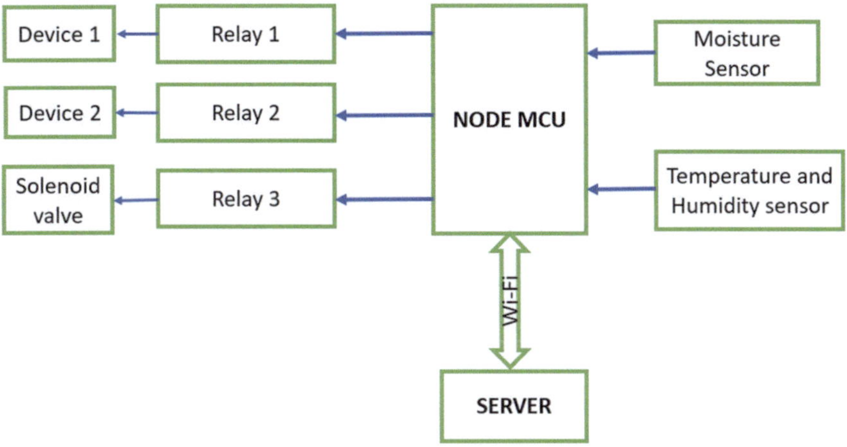

Fig. 2 Flowchart of the working model [1]

4 Result and Discussion

NODEMCU (ESP8266) is a microcontroller which is secured, easy to operate, and used in many prototypic concepts and sometimes in industrial standards. The main peculiarity of the proposed work is as follows:

1. The user is given a blink application in his smartphone which is user-friendly and easy to operate through which the whole Farm can be monitored.
2. The sensor values can be sent to user in terms of notification so that user can take decision on what to be done [13].
3. The whole process can be divided into two different modes [11], one as manual and one is automatic, over the manual the irrigation can be done manually based on the real-time values of the sensors or in the automated method the irrigation system will be automatically powered on when the particular threshold value is crossed.
4. In automatic mode, the complete control is given to the controller itself so much less responsibility is removed from users.
5. From the app, the user can have access to other devices in or out of the farmhouse, e.g., lights and actuators [12].

Circuit diagram of the proposed model is shown in Fig. 3. The main components used in the prototype are Node MCU (ESP8266), Soil-Moisture sensor with probe, DHT11 sensor, two 5V relay modules, water pump, LED lights with resistor, 12V supply, servo motor, two LDR sensors, MQTT solar charge controller, and Raspberry pi [14]. In Fig. 4, main components such as Humidity Analyzer, Soil-Moisture Sensor, and Arduino are mentioned.

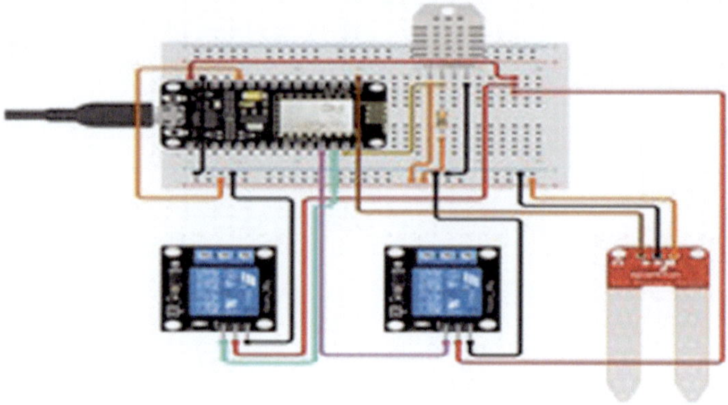

Fig. 3 Circuit diagram of the proposed model

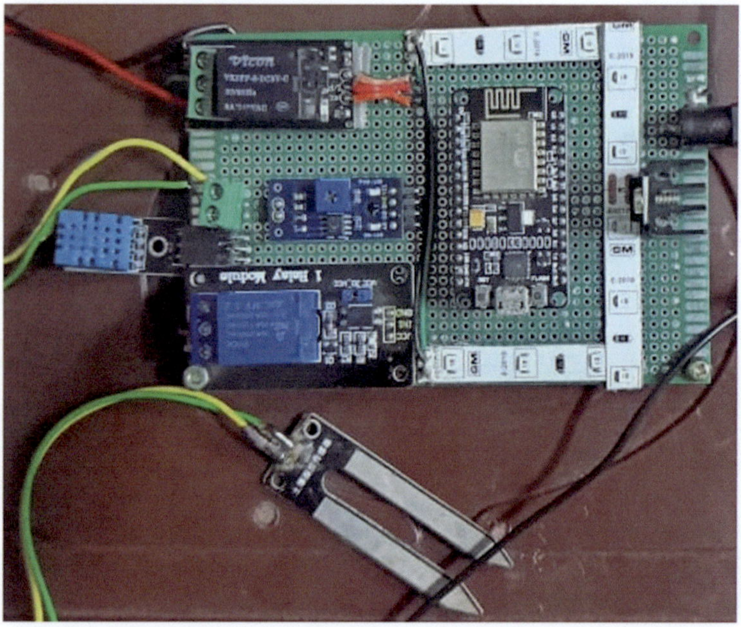

Fig. 4 Humidity analyzer, soil-moisture sensor, and Arduino

Fig. 5 Model of working prototype

Proposed IoT-based smart agriculture model has been implemented and tested in real-life application. Figure 5 represents the working model of the proposed system. And Fig. 6 represents the experimental results fetched from sensors. As per this data, user can understand the condition of soil and action to be taken either automatically or manually [14].

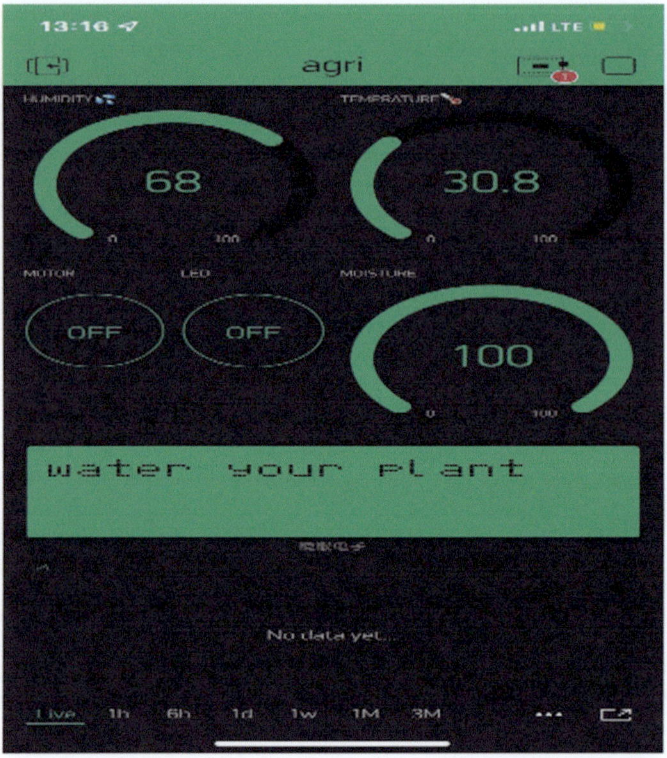

Fig. 6 Experimental results fetched from sensors

5 Conclusion

The main contribution of this paper is to develop possible model which can help remote villages to meet all the requirements of farmers in Agriculture field. This paper proposes a smart agricultural model which is integrated with IoT. IoT has always been the top priority and importance in the field of agriculture because it plays vital role in development of agriculture field. Farmers have grown a lot of crops over a decade's due to which the farmers had a nice idea about all conditions to make maximum profit from crops but due to rapid and undesirable climatic changes and soil changes have made a tough challenge for the farmers to grow crops in present situation. Through the approach given in this paper, the new adoption of smart agriculture will help farmers to adapt to the undesirable changes and to get maximum benefit out of the crops. This is a challenging task which requires a lot of knowledge to predict the climatic changes in different contexts and over different farm lands. This model of IOT-based irrigation system will help the farmers in every aspect to get maximum yield with smart electricity and smart irrigation with less human introversion.

References

1. Islam MS, Dey GK (2019) Precision agriculture: renewable energy based smart crop field monitoring and management system using WSN via IoT. In: 2019 International conference on sustainable technologies for industry 4.0 (STI). IEEE, pp 1–6
2. Sharma H, Haque A, Jaffery ZA (2019) Maximization of wireless sensor network lifetime using solar energy harvesting for smart agriculture monitoring. Ad Hoc Netw 1(94):101966
3. Ayaz M, Ammad-Uddin M, Sharif Z, Mansour A, Aggoune EHM (2019) Internet-of-Things (IoT)-based smart agriculture: toward making the fields talk. IEEE Access 7:129551–129583
4. Villarrubia G, Paz JFD, Iglesia DH, Bajo J (2017) Combining multi-agent systems and wireless sensor networks for monitoring crop irrigation. Sensors 17(8):1775
5. Karimi H, Navid H, Besharati B, Behfar H, Eskandari I (2017) A practical approach to comparative design of non-contact sensing techniques for seed flow rate detection. Comput Electron Agric 142:165–172
6. Khanna A, Kaur S (2019) Evolution of Internet of Things (IoT) and its significant impact in the field of precision agriculture. Comput Electron Agric 157:218–231
7. Sudharshan N, Karthik AK, Kiran JS, Geetha S (2019) Renewable energy based smart irrigation system. Procedia Comput Sci 1(165):615–623
8. Bhattacharyya P, Pathak H, Pal S (2020) Energy management for climate-smart agriculture. In: Climate smart agriculture. Springer, Singapore, pp 73–84
9. Liu J, Chai Y, Xiang Y, Zhang X, Gou S, Liu Y (2018) Clean energy consumption of power systems towards smart agriculture: roadmap, bottlenecks and technologies. CSEE J Power Energy Syst 4(3):273–282
10. Mahto R, Sharma D, John R, Putcha C (2021) Agrivoltaics: a climate-smart agriculture approach for Indian farmers. Land 10(11):1277
11. Nithara PV, Nayak F, Shashank R (2022) Energy management using Atmega2560 MCU for residential infrastructure, pp 311–315
12. Nithara PV (2021) Voltage controlled inverter for islanded mode of solar panel system. In: 2021 7th International conference on electrical energy systems (ICEES), pp 15–18
13. Abbasi AZ, Islam N, Shaikh ZA (2014) A review of wireless sensors and networks' applications in agriculture. Comput Stand Interfaces 36:263–270
14. Nikolidakis SA, Kandris D, Vergados DD, Douligeris C (2015) Energy efficient automated control of irrigation in agriculture by using wireless sensor networks. Comput Electron Agric 113:154–163

An Efficient Probability Attestation Scheme for Multipath Routing and Performance Enhancement in IoT

J. V. N. Raghava Deepthi, Ajoy Kumar Khan, and Tapodhir Acharjee

Abstract The interconnection of different wireless communication devices is known as the Internet of Things (IoT). This allows for the ubiquitous accessibility of devices and their built-in intelligence capacity. The benefits of these networks are numerous and few of them are no infrastructure support, deployed spontaneously, cost-effective setup, and so on. In order to improve service quality, the Internet of Things (IoT) facilitates communication between devices that have different types of connections, which allows for the creation of interfaces and connections. Many risks impact the efficiency of the network and the data transfer between IoT devices. But still, security mechanisms are challenging. Remote Attestation (RA) is one tool that efficiently identifies a malicious entity in the network. This research offers a probability-based attestation scheme (PROB_ATT) for improving IoT network routing and performance. Depending on the estimated probability value of the journey, the attestation value is included in the network with the data. With the application of trickle time, attestation is verified and data transmission is performed. After the estimation of probability within the network, data transmission is performed based on the trust mechanism. The PROB_ATT scheme was found to decrease energy consumption, packet loss ratio, and delay in the simulation analysis. Also, the proposed PROB_ATT exhibits improved efficiency, packet delivery rate, and throughput than existing techniques.

Keywords Remote attestation · Trust mechanism · Packet delivery ratio · Throughput

J. V. N. Raghava Deepthi (✉) · T. Acharjee
Assam University, Silchar, Assam, India
e-mail: deepthijonnalagadda1@gmail.com

A. K. Khan
Mizoram University, Aizawl, India

S. K. Biswas et al. (eds.), *Intelligent Computing Systems and Applications*, Lecture Notes in Networks and Systems 1307, https://doi.org/10.1007/978-981-96-3860-4_44

1 Introduction

The Internet of Things (IoT) links a vast array of physical devices. Internet of Things (IoT) objects can be either real or virtual; they can make their own decisions and perform their functions; they can communicate with other devices; and, because they are interoperable, they can operate interactively. In addition, the Internet of Things allows for versatile, always-on, anywhere-service interaction with objects [1]. In recent years, there has been a tremendous growth in IoT. Many different types of heterogeneous Internet of Things (IoT) devices are interconnected for a wide range of uses, including health and environmental monitoring, smart city planning, building automation, and military operations. These simplify, intelligently, and automatically make life much easier. But now there are dangers, and hackers have a new cyberspace to operate in. Researchers are motivated to develop security and safety mechanisms after attacks like the Mirai botnet and the smart TV hack [2]. Users are vulnerable to a multitude of attacks due to devices' lack of security [3–7]. Many times, these types of attacks, such as man-in-the-middle [8] attacks, smart attacks [9], and Denial-of-Service (DoS) attacks [10], result in monetary loss or even worse consequences [11]. It is possible to detect these harmful devices by performing inexpensive attestation. However, a cheap, unskilled device-to-device remote attestation solution has issues with communication overhead and long attestation times [12]. Plus, systems that use a cluster of dynamic technology devices, such as ITS and oil and gas search robots, do not scale [13]. New, reliable attestation solutions that are both scalable and innovative are thus required to guarantee the safe operation of IoT devices on networks [14, 15]. The devices connected to the infrastructure of IoT are attacked by attackers where the protection mechanism is easily broken [16]; these circumstances motivated the researchers to develop various protection methods that provided network security [17]. In addition to security, other challenges are also faced in IoT networks. Security solutions are a need for an expanding Internet of Things (IoT) network, but they add complexity, computational overhead, and expense as the network's node count grows. This highlights the need for the development of a scalable, secure, and lightweight protocol that can adapt to the ever-changing requirements of networks [18]. To make Internet of Things (IoT) communications more secure, this study developed a probability-based attestation scheme (PROB_ATT). Additionally, this study employs a trust mechanism to efficiently route data within the network. In the proposed PROB_ATT scheme, within the network, the node generates and sends its attestation. Once the attestation is created, the proposed PROB_ATT calculates the different paths and transmits data across the network. If the node responds with the correct attestation within the tickle time, data is transmitted in the network; otherwise, transmission is terminated. The last step in efficient data routing within the network is the application of a trust mechanism. According to the results of the simulations, the proposed PROB ATT is far superior to the existing methods.

2 Related Works

Security at every layer of the IoT-based applications is the most significant feature to be considered. In an IoT system, implementation is successful if security is concentrated more. Hence, security is addressed while designing applications based on IoT networks [18]. Jonathan Tournier et al. [19] have addressed a common approach with dual challenges, and Yousefnezhad et al. [20] provided a promising solution to address security in IoT. The studies focus on security and its taxonomy on product lifestyle, security solutions for every stage in the lifecycle, and issues in these stages that introduce challenges, AntarShaddadH. Abdul-Qawy et al. [21] attempted to provide a common and widespread taxonomy with complete classification covering almost all the major energy-saving approaches of IoT-based constrained WSNs. Aarika et al. [22] examined various solutions for IoT attacks at perception, network, and application layers. Karunanithy and Velusamy [23] implemented a CEEDG protocol to enhance lifetime and throughput. Elappila et al. [24] created the Survivability Aware Channel Allocation (SACA) fuzzy-based method, in which the nodes in the network could express their preferences for which channels to used. When compared to other methods, SACA significantly improved network reliability, PDR, and throughput. Lee et al. [25] reviewed the RPL's performance after making some adjustments. Step one involves disabling the timer of the vacation spot integrated outlet trickle due to its unsuitability for use in periodic topology changes. In Tian et al. [26], by focusing on the node's location, the GI-RPL scheme improves upon the RPL in VANETs and makes it possible to deal with periodic changes to the topology. Martocci et al. [27] offer an analysis of RPL using the most recent revision. Examples include possibilities shown by RPL, point-to-point RBL (P2P-RBL), assessment, performance, challenges, and research resistance. Brandit et al. [28] planned Goal Function, the new feature based on fuzzy logic (OF-FL), which allows RPL to advance beyond the standard objective capability restriction by taking into account crucial nodes.

3 Overview of Remote Attestation

To determine whether an adversary is present in a device, Remote Attestation (RA) is an appropriate technique. An "untrusted" device's integrity is confirmed by a trusted entity (Vrf), which is especially important in RA. (Prv) by checking, in the absence of the adversary, that the device is using the updated software version. Figure 1 depicts a challenge sent by *Vrf* to *Prv*. When the challenge is received, *Prv* performs the necessary operation and responds back to *Vrf*. Depending on this response, Vrf verifies the device's health. Even though RA efficiently validates the health of the device, implementing it on huge networks is challenging as it is a 1–1 verification approach. RA gains cost in many factors like computing, storing, communicating overhead, and power of batteries. Low-end embedded devices suffer

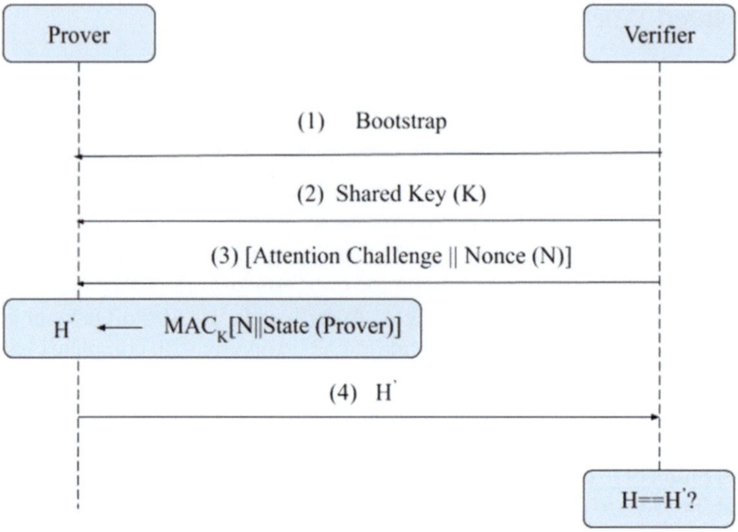

Fig. 1 Steps in remote attestation

from computational cost, hence several IoT applications do not involve high security mechanisms. To develop low-cost secure networks, RPL-integrated RA is suitable due to their interoperability. Thus, with the RPL framework, the attestation process is made lightweight and efficient. In PROB_ATT, the DAO packet structure is used for communicating attestation results.

Moreover, conventional attestation methods send the attestation request through *Vrf* to the network, but here drip time is used which automatically initiates the process of attestation. Hence, it overcomes the communication overhead and tackles several network attacks. In Fig. 1, conventional remote attestation scheme is presented.

Operating under the principle of Destination-Oriented Directed Acyclic Graph (DODAG), RPL [17] is a virtual routing topology. In DODAG, while each node can have multiple parents, the objective functions and routing metrics determine which parent each node is assigned to. The parent node forwards the data packets. The DODAG structure supports N-1 communication in RPL. Rank ID is provided to every node based on the distance from the node to the root. DODAG Information Objects (DIO) with IC-MPv6 control packets create and maintain DODAG. In RPL, every node distributes DIO packets which include a link, node metrics, and an objective function which selects the preferred parent of a node among its neighbors. Node metrics include residual energy and expected transmission count (ETX).

4 Proposed Prob_Att Scheme

The RPL protocol is the current standard for Internet of Things (IoT) infrastructure implementation, serving as the network's de facto routing mechanism. The PROB_ATT design exemplifies a safe and effective means of communication. In the multi-path routing method, different paths are considered for each node in the network that is meant to be a destination. Higher frequency nodes in the network necessitate a higher energy level for data transmission costs. Using the network's probabilistic value, routing decisions are observed through the incorporation of the probability model. In Eq. (1), connection probability evaluation of proposed PROB_ATT is provided:

$$G(P) = \begin{cases} 1 \; p \leq p_0 \\ 0 \; p > p_0 \end{cases} \tag{1}$$

If the connection was successful, the value will be 1, and if it was unsuccessful, the value will be 0. In this, the probability of transmitting a packet is p. The communication radio environment can be better understood with the help of probability routing analysis. In order to maximize channel performance, this multi-path optimization method evaluates end-to-end user communication using a game-theory approach.

The average path for the IoT communication is represented in Eq. (2):

$$f(v_i) = \frac{1}{\sigma\sqrt{2\prod}} e^{-(p_i - p_{\mathrm{avg}})^2/2\sigma^2} \tag{2}$$

We can estimate the effective path for information communication on the IoT network using Eq. (3):

$$E(N) = \rho d\lambda \tag{3}$$

within this number of nodes, transmitted packets, and other factors are introduced.

For attestation, this approach employs RPL DAO IC-MPv6 control messages with improved header fields. The updated data structures are sent in the following formats: (i) a 4-bit flag field that indicates the node ID; (ii) an 8-bit reserved field that includes a timestamp and several bits (00 for a GOOD node and 11 for a BAD node) that represent the results of the self-attestation; and (iii) a 32-bit option field that allows for the transmission of an encrypted hash of DAOcrypt in plaintext along with the attestation results. To show how recent a time-bound item is, the outcome includes the hash value of the device's software and the attestation time stamp. The attestation message is as follows in Eq. (4):

$$A_k^{\mathrm{Att}} = (\mathrm{Hash}||\mathrm{Timestamp})\mathrm{Root}_{pk} \tag{4}$$

$Root_{pk}$ denotes the public key of the root node. Only the root node can decrypt the attestation report, which is encrypted with this public key. Before encrypting with the root node's public key, nodes perform a hash on the modified DAO IC-MPv6 message during self-attestation. An optional 32-bit field in the DAO stores both the encrypted result and the plaintext self-check report. The security of DAO messages and self-attestation reports in DAO messages relies on encryption and hashing. An adversary modifies the root path attestation report, for instance. By comparing the plaintext attestation report's hash with the decrypted hash string, changes of that kind can be found in PROB_ATT.

4.1 Encryption Using PROB_ATT

The proposed PROB_ATT is adopted for minimizing computational cost and improved security. However, the attestation result is forwarded to the root node in PROB_ATT by exploiting DAO message packets. The software's public key for root and attestation can be found in Traffic-based Energy Estimation (TEE). The proposed PROB_ATT is adopted for minimizing computational cost and improving security. However, in PROB_ATT, the root node receives the authentication result through the DAO message packet. Attestation and root software public keys are stored in Traffic Energy Estimation (TEE). Since encryption is done using its public key, decryption can only be done at the root.

Figure 2 presents the overall architecture of PROB_ATT for security improvement. Based on probability estimation and Tickle-Timer data are transmitted within the network. This approach utilizes the non-storing mode of RPL as MOP2, which is ideal for devices with limited resources, as it supports computational requirements and minimum memory. Another feature of MOP2 is that all network nodes communicate with the root via control messages. At regular intervals (referred to as a "Trickle-Timer"), the DAO message in PROB_ATT transmits the device attestation report to the root. This timer monitors the rate at which beacon messages are generated [18]. With the timer set to relay DAOcrypt messages to the root, the root can learn the network's health status and use that information to redirect RoamingAdv-derived threats as shown in Fig. 2. The overall architecture of the proposed PROB_ATT scheme for WSN-IoT security is presented in Fig. 2.

For analyzing IoT path and the performance of data transmission, the following assumptions are coined and are stated below:

PROB_ATT has Verifier (S_{rf}) and Prover *(Qrv)* as basic stakeholders. Algorithm 1 provides the pseudocode incorporated in the encryption process adopted in PROB_ATT for security improvement in WSN-IoT is provided.

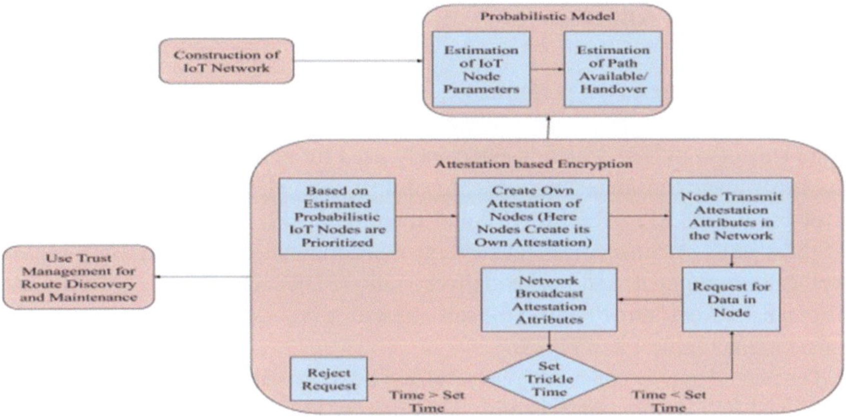

Fig. 2 Proposed model for flow of PROB_ATT

Algorithm1: Encryption by PROB_ATT

1. *Initialize the Set device id as* N_{id}
2. *Set cryptographic details of node as* E_{cry}
3. *Define attestation details of each node in the network* N_{att}
4. *Set trickle time for attestation* $T\ T_{att}$
5. *Network Cryptography* $D_{crypt}=0$
6. *While true do*

   Formation();
7. *If* $(T\ T_{att} == True)$ *then*

   Perform N_{att}

   $D_{att} \leftarrow$ *Encryption of Message*

   $D_{crypt} \leftarrow D_{att}$

   Transmit the obtained D_{crypt}

   else

   Perform normal operation

   end

   end

   R_{msg} = *Received attestation values*

   E_{node}- *Elimination of faulty nodes*

   While True do

   Check ()←flag

   If check () ==0 then

   Verify D_{crypt}

   Perform E_{node}

   else

   Keep those nodes

   Perform normal Operation

   end

   end

5 Simulation Analysis

This section illustrates the performance of PROB_ATT with the simulation results obtained and is compared with traditional RPL in various situations. PROB_ATT is implemented in NS2 which is extensively used for network deployment. Hence, this experiment considered the increase of attacker nodes related to node density as of traditional RPL. PROB_ATT and traditional RPL data are compared with the parameters calculated, loss rate, energy consumption, energy efficiency, routing overhead, throughput, and packet delivery ratio (PDR). This is necessary because different attackers harm PDR and the self-attestation process has an impact on energy consumption (Table 1 and Fig. 3).

The overall energy consumption calculated by proposed PROB_ATT scheme and traditional RPL is presented in Fig. 3. Certainly, PROB_ATT requires minimum

Table 1 Energy consumption versus number of nodes

No. of nodes	Existing (RPL)	Proposed (PROB_ATT)
0	0	0
20	14.87	9.57
40	21.85	15.67
60	28.53	21.63
80	37.83	29.50
100	38.66	30.76

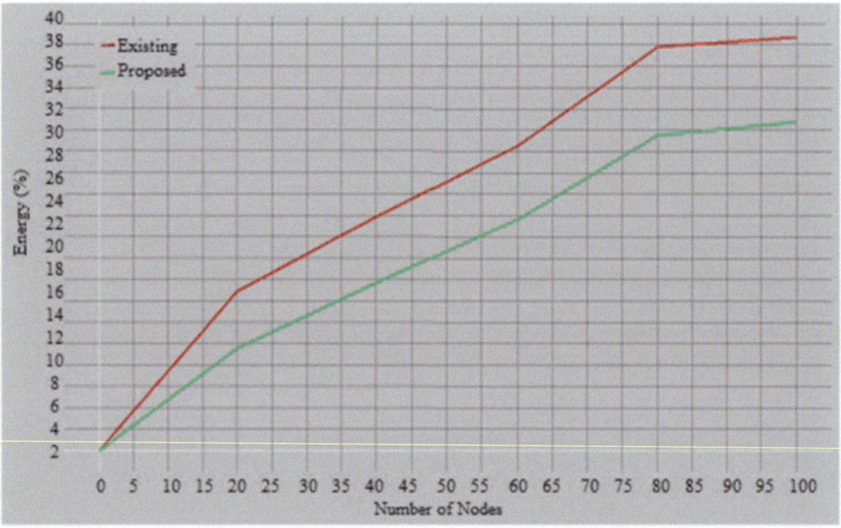

Fig. 3 Comparison of energy consumption

energy than traditional RPL. As complex hashing is utilized to provide encrypted attestation results, PROB_ATT achieves a lower energy consumption than the traditional RPL. In Fig. 4, the energy efficiency of the proposed PROB_ATT with traditional RPL is presented (Table 2).

Simulation results prove that with the increase in simulation time, energy required for PROB_ATT drastically decreases which increases energy efficiency. As malicious nodes are detected at the early stages, healthy nodes are prevented from forwarding iterative messages. In Fig. 5, packet loss rate (PLR) of the proposed PROB_ATT is comparatively examined with the traditional RPL scheme (Table 3).

From Fig. 5, the delivery probability $PTMG_{(S,D)}$ is estimated by varying the probability which is randomly defined in [10%; 20%; 30%] for each node. For the assigned probability value, loss rate of the packet is examined. From Fig. 4, the suggested PROB_ATT scheme outperforms the conventional RPL in terms of energy efficiency. This, in turn, decreases the packet loss rate compared with the existing traditional

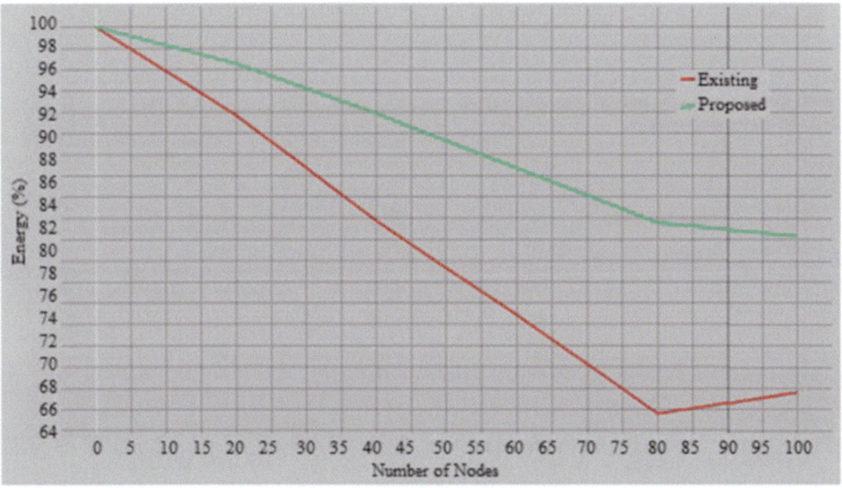

Fig. 4 Comparison of energy efficiency

Table 2 Energy efficiency versus number of nodes

No. of nodes	Existing (RPL)	Proposed (PROB_ATT)
0	0	0
20	91.67	96.53
40	81.78	91.87
60	72.86	86.68
80	63.57	81.59
100	65.66	80.33

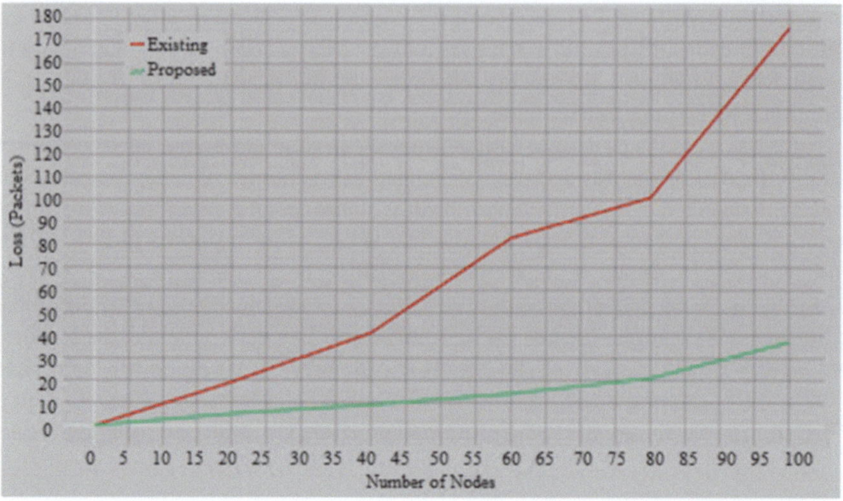

Fig. 5 Comparison of PLR

Table 3 Loss rate versus number of nodes

No. of nodes	Existing (RPL)	Proposed (PROB_ATT)
0	0	0
20	19	5
40	41	9
60	83	14
80	101	21
100	176	37

RPL scheme. In Fig. 6, the Packet Delivery Rate (PDR) of the proposed PROB_ATT with the traditional RPL is presented (Table 4).

The results shown in Fig. 6 provide better PDR for the proposed PROB_ATT approach. This approach utilizes a multi-path disjoint routing method and when the route discovery mechanism is considered, each trusted node with high PDR discovers a valid secure link. Hence, it is concluded that the proposed approach is better for larger networks like Mobile Ad Hoc Networks (MANET). In Fig. 7, throughput measurement of the proposed PROB_ATT scheme with existing RPL scheme is provided (Table 5).

In Fig. 7, throughput measurement of the proposed PROB_ATT scheme with existing traditional RPL is presented. The analysis stated that an increase in PDR and reduced PLR increases the network throughput. In Fig. 8, delay measurements for varying nodes are presented. In Table 2, delay measured for proposed PROB_ATT scheme for security improvement is presented (Table 6).

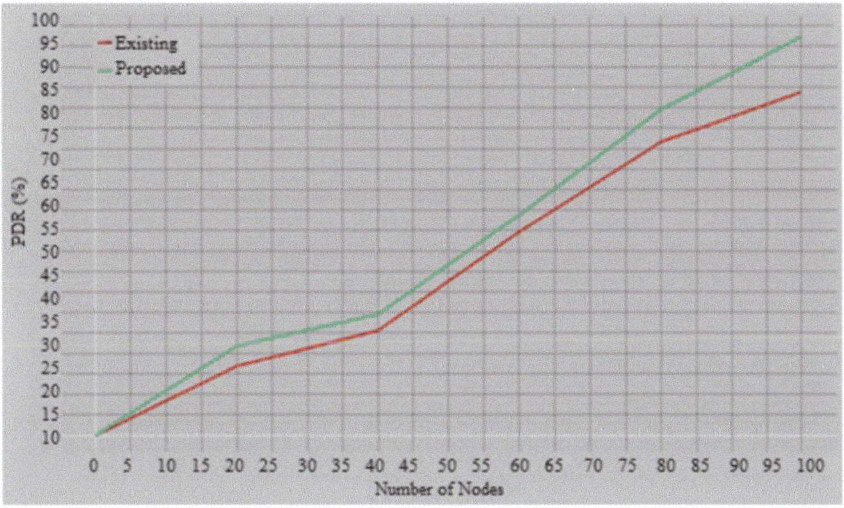

Fig. 6 Comparison of PDR

Table 4 Packet delivery ratio versus number of nodes

No. of nodes	Existing (RPL)	Proposed (PROB_ATT)
0	0	0
20	16.89	21.87
40	25.62	29.57
60	49.57	53.67
80	71.57	79.52
100	83.78	97.23

Figure 8 shows the delay in the network. Network size and control packets play a role in delay while transmitting messages to the root because of multiple hop distances. In RPL, DAO messages maintain the downward topology of the network. By DAO messages, the root obtains the network view globally which controls the DODAG topology. In Fig. 9, routing overhead measurement for the proposed PROB_ATT scheme and traditional RPL is presented (Table 7).

In Fig. 9, PROB_ATT, from routing overhead, exploits the DAO packet to transmit the attestation report to the root. Attestation and encryption with SHA-256 are expensive. Still, it is proved that PROB_ATT does not delay in transmitting DAO messages which is cost-effective for security. Particularly, PROB_ATT obtains network security without any additional delays and the root gets updated with the status of the network without any network delay.

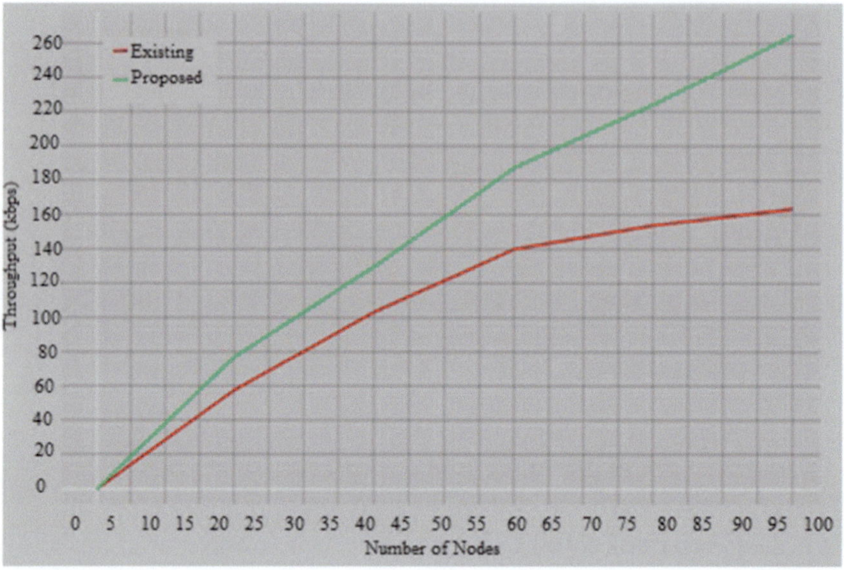

Fig. 7 Comparison of throughput

Table 5 Throughput versus number of nodes

No. of nodes	Existing (RPL)	Proposed (PROB_ATT)
0	0	0
20	57.26	76.67
40	103.28	129.57
60	139.78	187.57
80	153.57	224.29
100	163.57	264.29

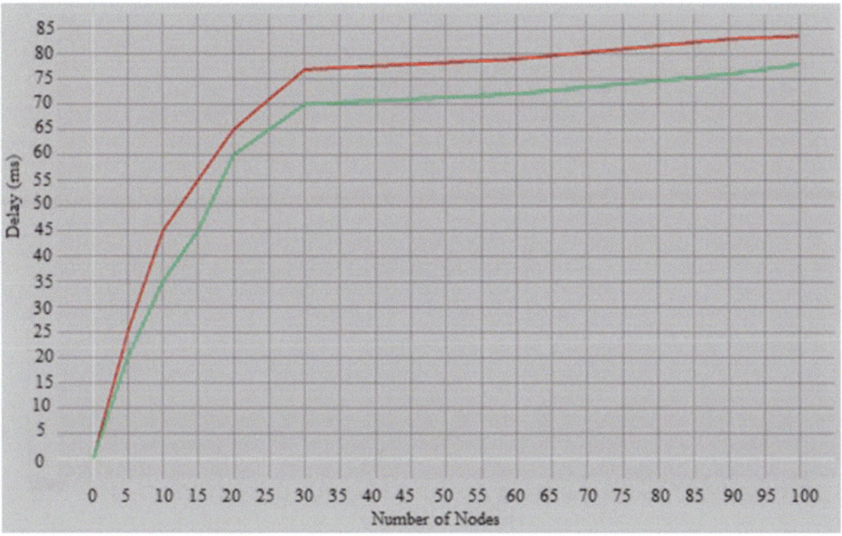

Fig. 8 Comparison of delay

Table 6 Delay versus number of nodes

No. of nodes	Existing (RPL)	Proposed (PROB_ATT)
0	0	0
5	0.025	0.020
10	0.045	0.035
15	0.055	0.045
20	0.065	0.060
30	0.077	0.070
60	0.079	0.072
90	0.083	0.076
100	0.0835	0.078

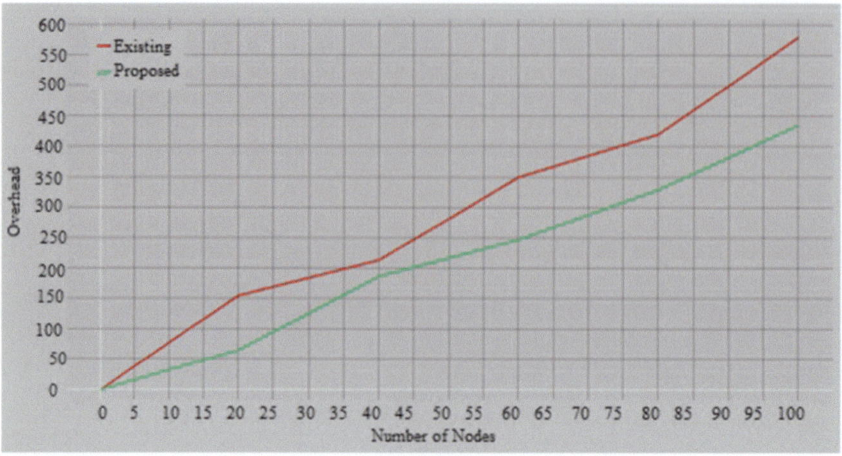

Fig. 9 Comparison of routing overhead

Table 7 Routing overhead versus number of nodes

No. of nodes	Existing (RPL)	Proposed (PROB_ATT)
0	0	0
5	0.025	0.020
10	0.045	0.035
15	0.055	0.045
20	0.065	0.060
30	0.077	0.070
60	0.079	0.072
90	0.083	0.076
100	0.0835	0.078

6 Conclusion

IoT infrastructure is based on an ad hoc network with numerous benefits like no infrastructure, deploying spontaneously, and a chance to set up cost-effective temporary and dynamic networks. Still, few characteristics, like unreliable wireless links, dynamic topology, and energy and power constraint objects, develop new protocols that ignore the security of applications for satisfying quality of service (QoS). The purpose of this study was to enhance the efficiency of Internet of Things (IoT) communication by creating a probability-based attestation scheme (PROB ATT). Also, for effective routing of data within the network this research utilizes a trust mechanism. The node in the proposed PROB_ATT scheme generates and sends its attestation across the network. The proposed PROB_ATT creates an attestation, then estimates the paths through the network and transmits data. All of the PROB_ATT evaluations include up to 100 nodes, and the suggested method yields satisfactory results. Simulation analysis stated that the PROB_ATTT scheme reduces energy consumption, packet loss ratio, and delay by 20%, 10%, and 5%, respectively. In terms of energy efficiency, throughput, and packet delivery rate, performance improved around 20% compared to the existing method. In the future, we can extend this work by implementing the variable nodes up to 500.

References

1. Airehrour D, Gutierrez JA, Ray SK (2019) SecTrust-RPL: a secure trust-aware RPL routing protocol for Internet of Things. Futur Gener Comput Syst 93:860–876
2. Meneghello F, Calore M, Zucchetto D, Polese M, Zanella A (2019) Iot: internet of threats? a survey of practical security vulnerabilities in real iot devices. IEEE Internet Things J 6(5):8182–8201
3. Tomić I, McCann JA (2017) A survey of potential security issues in existing wireless sensor network protocols. IEEE Internet Things J 4(6):1910–1923
4. Qiu T, Chen N, Li K, Qiao D, Fu Z (2017) Heterogeneous adhoc networks: architectures, advances and challenges. AdHoc Networks 55:143–152
5. Deebak BD, Al-Turjman F (2020) A hybrid secure routing and monitoring mechanism in IoT-based wireless sensor networks. Ad Hoc Netw 97:102–122
6. Haseeb K, Islam N, Almogren A, Din IU, Almajed HN, Guizani N (2019) Secret sharing-based energy-aware and multi-hop routing protocol for IoT based WSNs. IEEE Access 7:79980–79988
7. Neshenko N, Bou-Harb E, Crichigno J, Kaddoum G, Ghani N (2019) Demystifying IoT security: an exhaustive survey on IoT vulnerabilities and a first empirical look on internet-scale IoT exploitations. IEEE Commun Surv Tutor 21(3):2702–2733
8. Conti M, Kaliyar P, Rabbani MM, Ranise (2020) Attestation-enabled secure and scalable routing protocol for IoT networks. Ad Hoc Networks 98
9. Conti M, Kaliyar P, Rabbani MM, Ranise S (2018) SPLIT: a secure and scalable RPL routing protocol for Internet of Things. In: 14th International conference on wireless and mobile computing, networking and communications (WiMob), pp 1–8
10. Feng W, Qin Y, Zhao S, Feng D (2018) AAoT: Lightweight attestation and authentication of low-resource things in IoT and CPS. Comput Netw 13(4):167–182
11. Tournier J, Lesueur F, Le Mouël F, Guyon L, Ben-Hassine HA (2020) Survey of IoT protocols and their security issues through the lens of a generic IoT stack. Internet Things 100–264
12. Yousefnezhad N, Malhi A, Främling K (2020) Security in product lifecycle of IoT devices, a survey, pp 102–779
13. Abdul-Qawy ASH, Almurisi NMS, Tadisetty S (2020) Classification of energy saving techniques for IoT-based heterogeneous wireless nodes. Procedia Comput Sci 17(1):2590–2599
14. Aarika K, Bouhlal M, Abdelouahid RA, Elfilali S, Benlahmar E (2020) Perception layer security in the internet of things. Procedia Comput Sci 17(5):591–596
15. Elappila M, Chinara S, Parhi DR (2020) Survivability aware channel allocation in WSN for IoT applications. Pervasive Mob Comput 6(1):101–107
16. Karunanithy K, Velusamy B (2020) Cluster-tree based energy efficient data gathering protocol for industrial automation using WSNs and IoT. J Ind Inf Integr 100–156
17. Tao M, Li X, Yuan H, Wei W (2020) UAV-aided trustworthy data collection in federated-WSN-enabled IoT Applications. Inf Sci
18. Rabbani MM, Vliegen J, Winderickx J, Conti M, Mentensa N (2019) Scalable heterogeneous layered attestation. IEEE Internet Things J 6(6):10240–10250

19. Libert B, Ling S, Mouhartem F, Nguyen K, Wang H (2016) Signature schemes with efficient protocols and dynamic group signatures from lattice assumption, pp 373–403
20. Yousefnezhad N, Malhi AK, Främling K (2020) Security in product lifecycle of IoT devices: a survey. J Netw Comput Appl 171:102779
21. Abdul-Qawy AS, Almurisi NM, Tadisetty S (2020) Classification of energy saving techniques for IoT-based heterogeneous wireless nodes. Procedia Comput Sci 171:2590–2599
22. Aarika K, Bouhlal M, AitAbdelouahid R, Elfilalia S, Benlahmar E (2020) Perception layer security in the internet of things. Procedia Comput Sci 175:591–596
23. Karunanithy K, Bhanumathi V (2020) Cluster-tree based energy efficient data gathering protocol for industrial automation using WSNs and IoT. J Ind Inf Integr
24. Elappila M, Chinara S, Parhi DR (2019) Survivability aware channel allocation in WSN for IoT applications. Pervasive and Mobile Computing
25. Lee KC, Sudhaakar R, Dai L, Addepalli S, Gerla M (2012) RPL under mobility. In: Proceedings of 2012 IEEE consumer communications and networking conference (CCNC), Las Vegas, NV, USA
26. Tian B, Hou KM, Shi H, Liu X, Diao X (2013) Application of modied RPL under VANETWSN communication architecture. In: Fifth (ICCIS), pp 1467–1470
27. Martocci EJ, Mil PD, Riou N, Vermeylen W (2010) Building automation routing requirements in low-power and lossy networks. RFC:5867
28. Brandt A, Buron J, Porcu G (2010) Home automation routing requirements in low-power and lossy networks

Resource Management of Mobile Edge Computing (MEC) Enabled 5G IoT Networks

Jayalakshmi G. Nargund, Chandrashekhar V. Yediurmath, M. Vijayalaxmi, and Vishwanath P. Baligar

Abstract The dramatic growth of mobile devices and the use of user data will result in exceptional difficulties to manage in the current mobile network. It may decrease data quality and service quality of the network. Next generation network 5G and Mobile Edge Computing (MEC) technologies are the desired solutions to manage the resources of mobile networks. Quality of Services (QoS) of end to end mobile network can be provided by combining MEC network with 5G. MEC is an intermediate network between end devices and back bone network. The end device network may include varity of Internet of Things (IoT) enabled 5G applications. MEC servers are processing huge amount data received from thc users. This overburdens the MEC servers. Resources like Band Width (BW) management is necessary. So balancing the load of servers is essential for efficiency usage of BW. Service efficiency, system performance, and operating costs have been improved by using appropriate load balancing algorithm on MEC servers. The authors have introduced mobility load balancing in MEC network with 5G to prevent service interruptions due to unequal load. The proposed algorithm Load Balancing Efficiency Factor (LBEF) considers the load on MEC server connected to different neighbouring cells. Edge user equipment of each of the overloaded cells are considered to determine the load efficiency factor. It specifies the sequence of overloaded cells and starts offloading process in them. The proposed method of load balancing on MEC servers in 5G increases the throughput of network with minimal standard deviation.

Keywords Load balancing · MEC server · 5G Enabled IOT networks load balancing factor

---

J. G. Nargund · C. V. Yediurmath (✉) · M. Vijayalaxmi · V. P. Baligar
KLE Technological University, Hubballi, India
e-mail: chandruvy143@gmail.com

S. K. Biswas et al. (eds.), *Intelligent Computing Systems and Applications*, Lecture Notes in Networks and Systems 1307, https://doi.org/10.1007/978-981-96-3860-4_45

1 Introduction

5G is the most recent iteration of cellular technology, designed to boost the speed and responsiveness of wireless networks. Data carried through wireless broadband connections in 5G is having peak speed as high as 20 Gbps [1]. Due to higher accessible bandwidth and enhanced antenna technology, 5G will enable a significant increase in the volume of data delivered across wireless systems.

MEC brings resources closer to the end user. Data is processed and stored at edge of the networks [9], rather than at a remote data center. Thus it reduces latency to support the real-time applications. The edge computing has been designated as one of the core technologies in 5G [8] to provide low latency, mission critical and and future IoT services. The MEC is built from scratch to support edge computing in an efficient and flexible manner [2]. The combination of MEC, IoT, and 5G technologies can drastically reduce transmission latency and computational activities to provide services. Edge computing is the next step in the evolution of cloud computing [4].

Load balancing has received a lot of attention as a promising approach to improve resource management, system performance, and cost reduction of system. It is an effective solution for cellular networks to manage the spatial-temporal fluctuation problem of mobile data traffic [4]. Existing load balancing techniques rely on channel adoption from neighbouring cells and it cannot be directly applicable to 5G wireless networks [2]. Device-to-device (D2D) communication is recommended to assist load balancing without the use of additional spectrum in an Orthogonal Frequency Division Multiple Access (OFDMA) ultra-dense tiny cell network [1].

Figure 1 depicts the small dense cells in 5G network. Authors investigate the uplink of an OFDMA ultra-dense small cell network with one Small cell Base Station (SBS).

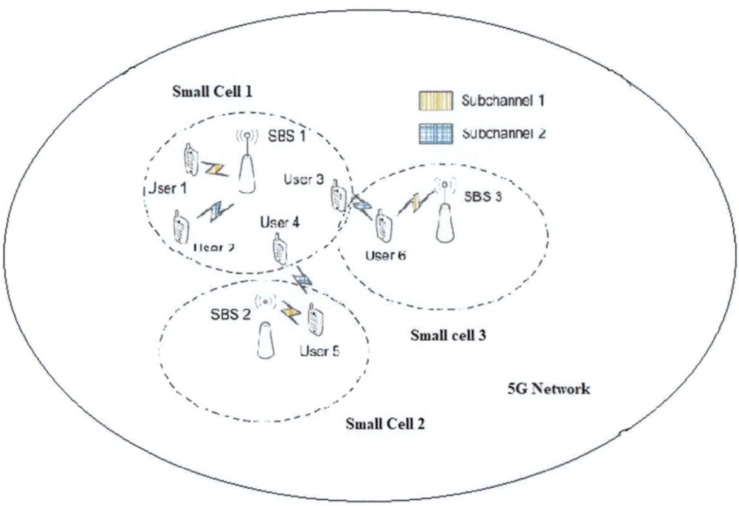

Fig. 1 Cell architecture in 5G

It is assumed that only users have their own data to transmit, and they correspond to a data flow in the system. To balance the load in each small cell, users can either send their data through the SBS in their own cell or use multi-hop D2D communications to transport their data to a neighbouring SBS. In Fig. 1. User 4 of Small cell 1 normally transmit the data through its base station SBS1. Suppose Small Cell 1 is overloaded, then User 4 can send the data through SBS2 of neighboring cell.

1.1 MEC in 5G Enabled IoT Networks

Figure 2 shows the adoption of MEC with other technologies in the proposed system. Thus the MEC—IoT integration poses new challenges to service and content providers [6]. The data generated through IoT devices is directly stored in the local server and it can export data to excel file. Authors of this article are visualize and analyze the collected data using machine learning models. MEC is located in between core network and applications as shown in Fig. 2. Thus, it reduces the physical and virtual communication distance between users and network. MEC server analyses, processes and stores the data near the end users. This method improves the performance of high bandwidth applications such as smart campus augmented reality applications by reducing the latency [10].

Technology gives an opportunity for new services and it leads to a new revenue for the industry [4]. MEC also helps to reduce the cloud data storage and transport cast. The main features of MEC are proximity, low latency, high bandwidth and virtualization [8]. The distributed cloud or data centers are also resided along with MEC servers at the edge of network as shown in Fig. 2. The data that can be accessed anywhere is stored in centralized cloud. The aggregated data is stored in distributed cloud at edge network. The machine learning/deep learning techniques are used by

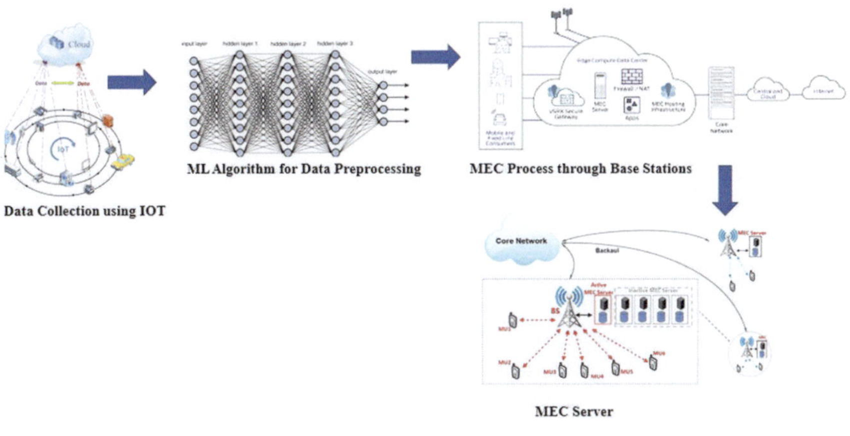

Fig. 2 MEC in 5G enabled IoT networks

the MEC server to analyse the data. The predictions of the model are communicated to users by MEC.

The proposed Load Balancing Efficiency Factor (LBEF) algorithm on MEC servers considers the load of overloaded cells and under loaded cells, which are connected to sever for calculating Load Balancing (LB) factor.

The rest of this paper is organized as follows: In Sect. 2, authors describe the related works on load balancing techniques for MEC servers enabled with 5G networks. Section 3 provides the system model, resource allocation and D2D routing sub-problems. Section 4, describes the methodology to implement MEC on 5g networks. Evaluation of results is depicted in Sect. 5 and paper is concluded in Sect. 6.

2 Related Work

The Wang et al. in [2] present an analysis of the number of ants (i.e. edge nodes) in ant colony system algorithm. The study focuses on the effect of changing the number of ants in the algorithm behavior rather than find the optimum number. The factors investigate in this study are algorithm execution time, best solution, pheromones accumulative, pheromone dispersion, and the number of new solutions found by the ants. The results show that the number of ants changes the algorithm behavior intensely. Therefore, tuning the parameter number of ants in ant colony system could be easier by applying the min and max number of ants recommended in this study.

The Nguyen, et al in [1] propose a blockchain-empowered framework for implementing resource trading and task assignment in edge computing. The smart contracts tool can provide reliable resource transactions and immutable records in the block-chain. This tool can also be adopted for facilitating a vehicular edge consortium block-chain. It enables secure communication to share their computation resources with service requester. To support security and privacy for cognitive edge computing, spatial-temporal smart contracts are used with intensive mechanisms to accelerate the economy sharing in smart campuses.

The Addali et al. in [1] present mobility load balancing to prevent quality of service degradation due to unbalanced load across small network cells. It considers the load of neighboring cells and edge user equipment for each congested cell to determine the load balancing efficiency factor. The performance of this system is analyzed using system-level simulations. The results show that the algorithm minimizes the standard deviation and increases the throughput of the network compared to existing algorithms from the literature.

In 5G Era data center is the essential device for data exchange and its internal network traffic shows a sharp increase in [8] of Sun Yanzhi et al. This focuses on the coexistence of bandwidth-sensitive traffic and delay-sensitive traffic in a power data center network. This paper proposes a Multiple Factor Load Balancing (MFLB) algorithm. Considering the multidimensional statistical characteristics of links, the algorithm integrates six parameters such as four dimensions of hops, path delay, average link delay, maximum link delay, maximum residual bandwidth, and minimum residual bandwidth for path filtering. The simulation environment is built on a mininet. Compared with existing algorithms, the performance of the MFLB algorithm improves the average throughput, standardized total throughput, and link delay.

The survey motivates the authors of this paper to propose load balancing algorithm for MEC servers using mobility of users.

3 Proposed Work

In this section explore the proposed system model, load balancing process in 5G and LBEF algorithm.

3.1 System Model

The MEC 5G enabled IoT networks integration poses new challenges for propose a novel techniques to improve resource management, load balancing and performance in delay-sensitive applications. The authors have explored IoT MEC and 5G concepts to dnevelop the system as shown in Fig. 3.

In Phase I the smart campus application has developed using IoT components by researchers to monitor vehicle/people movement, air and sound pollution etc. This application is generating data and provides interaction between user and network. 5G network test bed is simulated in NS3 [13] and it makes use of data generated by the application. Different sensors such as pH, temperature, moisture sensors and more are used to collect the data. The generated data by the application is stored in the local server. Thus Phase I of the research helps to monitor the pollution in the campus.

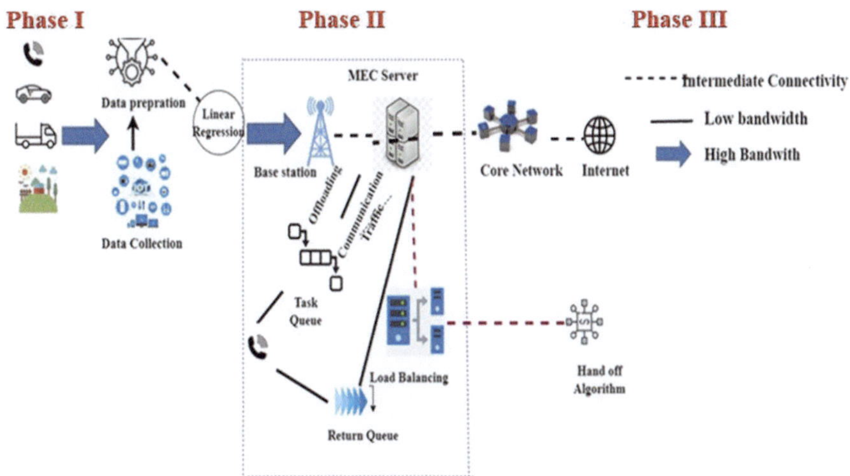

Fig. 3 Proposed system model

The collected data from different IoT devices is integrated and data set is prepared with required fields. In Phase II the machine learning techniques are used to analyse and pre-process the data. The MEC unit is prepared using NS3 and it receives the trained data from IoT network or apps in Phase I. The MEC servers are deployed with proposed load balancing algorithm LBEF to provide speedy services to users. Hence MEC system improves response time of the applications.

The aggregated data is stored in distributed cloud at core network as depicted in Phase III. It helps to acess data globally in the internet.

3.2 Load Balancing in 5G

The load balancing process in MEC is shown in Fig. 4 depicts the load balancing procedure in MEC. The authors of this paper modify the handover parameters across a small cell network to propose an LBEF algorithm. The process of assessing the load of user equipment (UEs) and the remaining capacity at neighboring cells yields the load balancing efficiency factor. The list of cells needs to be updated following each cell offloading procedure because the load on the cells varies during the operation. The same internet service provider owns the tiny cells in 5G.

Additionally, small cells use the X2 interface to communicate with one another. To carry out the handover of UEs from one cell to another, they exchange information pertaining to handover over the X2 interface. Through X2 interfaces, load control

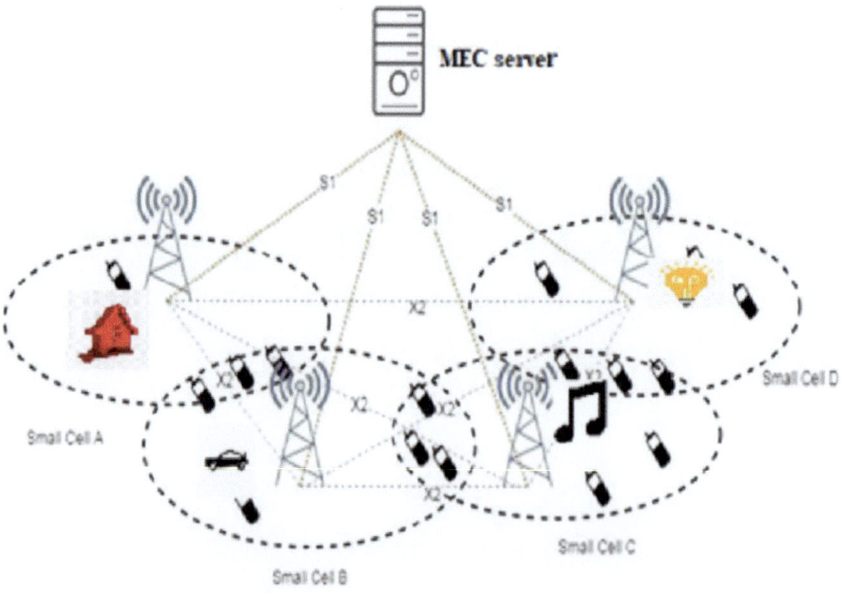

Fig. 4 Radio access network with a MEC server

functions such as resource status and traffic load can be supplied in addition to mobility management [6]. Information is frequently gathered by the MEC server. To balance the load across the network, optimize and update the handover parameters of small cells if any overloaded cells are found.

In Fig. 4, small cell A is under-loaded with a load of 50%, whereas small cells B, C, and D are overloaded and have load values of 70, 75, 80 and 85%, respectively. If the fixed and adaptive-threshold load balancing algorithms are adopted, they sequentially take the overloaded cells from the list [B, C, D]. The load balancing algorithm considers order of cells loaded and offload their excessive load to the lightly loaded neighbour cells. Since B is a neighbour of A cell depending on adaptive-threshold excessive load of B is transferred to A. The off loading process is repeated periodically.

3.3 LBEF Algorithm

LBEF algorithm is depicted in Algorithm 1 and it is used for MEC servers in 5G networks considers load of all cells attached to the MEC sever the variables used in the algorithm are denoted in Table 1. It sorts the cells in increasing order according to their load in terms BW utilization. The algorithm compares maximum loaded cell s_max with TH_Max. If s_max is greater than TH_Max then load balancing process starts. LBEF generates overloaded cell set OVR as depicted from line number 3 to 7. LB_factor is calculated in line number 9. Its computation depends on load of edge User Equipment (UE) and non-edge UEs. Overloaded cells are sorted in decreasing order depending on LB_factor. Transfer of load of overloaded cell is shown from line number 11 to 16. After adjusting the load of overloaded cell, it is removed from

Table 1 Variable description of LBEF algorithm

Sl no.	Variable	Meaning
1	Avg_NL	Average network load, 5G minimum BW = 24 GHz
2	N	No. Of requests messages, OVR: 70% (0.7)
3	TH_Max	Maximum Threshold
4	TH_UL	Threshold of Under loaded cell
5	OVR	Set of overloaded cells
6	QL	Queue length
7	LB_factor	Load balancing efficiency factor for overloaded cells
8	TN	Set of neighbor cells
9	PE	Estimated load after handover at edge UE in under loaded cells
10	P	Probability of load of UEs in the cell
11	1–P	Probability of load of non edge UEs
12	TH_LG	Threshold of load gap

OVR set. Re-computation of LB_factor is done again. Thus the LBEF algorithm balances the load of small cell, which are attached to a MEC server.

Algorithm 1 Load balancing Efficiency Factor (LBEF)

1. Input: Small cell information in set SCI [i.e. BW assigned to SBS]
2. Sort elements in SCI in increasing order. i.e SCI=[s1,s2,..,s_max]
3. **If** s_max ≥ TH_Max **then**
4. TH_UL = max(Avg_NL, TH_Max)
5. **for** $s\_i\ in\ SCI\ do$ $//i = 1\ to\ s\_m\_a\_x$
6. **If** $s\_i \geq$ TH_UL
7. Include $s\_i\ in\ OVR$
8. **endif**
9. **endfor**
10. $LB\_Factor = \sum_{I\ in\ TN} Min(PE_i, (1 - P_i))$
11. $Resort\ cells\ in\ OVR\ decreasing\ order\ according\ to\ LB\_Factor$
12. **for** $J\ in\ OVR$ **do**
13. $Find\ TN\ of\ J$ // $under\ loaded\ neighbors\ of\ J$
14. **for** $K\ in\ TN$ **do**
15. **if** $(Load(J) - Load(K) > TH\_LG)$
16. $Transfer\ excess\ Load\ of\ J\ to\ K$
17. **endif**
18. **endfor**
19. $Remove\ J\ from\ OVR$
20. $Recompute\ the\ LB\_factor.$
21. **endfor**
22. **endif**

4 Simulation Results

To study the impact of the load balancing efficiency factor, the authors have designed a algorithms which can be simulated using network simulators NS3. A small cell network consisting of 50 small cells with 80 UEs. Authors have used 24 GHz bandwidth for each small cell. Resource Based (RB) utilization is simulated using a Pseudo Random Binary Sequence (PRB). The transmission power is set to 24 dBm, path loss is modeled as a non-line of sight propagation loss. Channel and QoS Aware (CQA) scheduler is used to allocate resource to the UEs. Authors set the initial overload threshold to 0.75 for the adaptive and proposed LBEF algorithm.

50 small cells are divided into 3 clusters. Regarding the initial UEs distribution for each of the clusters is 90% and they are modeled with a Circular Way (CW) and randomly distributed over the network coverage area. The rest 10% of UEs are static non-uniformly distributed over the overlapping area of the small cells. Hence the small cells will be able to assign the required resources to the UEs, which results in increased network throughput. Some overloaded cells will not be able to offload their excess load if their neighbour cells are congested. LBEF algorithm on MEC computes load balancing efficiency factor as specified in Algorithm 1. It finds list of

Table 2 Description of simulation parameters

Sl no.	Parameter	Values
1	System bandwidth	20 GHZ
2	Transmission power	24 dBm
3	Number of small cells	50
4	Antenna mode	Isotropic
5	Number of users	80
6	Path loss	146.4 + 43.4 log10(R)
7	Fading	Log-normal
8	Scheduler	CQA

overloaded cells, which are to be balanced. Each overloaded cells are processed and offloaded their data to under loaded neighbors.

Simulation environment parameters are specified in Table 2. 5G network environment is simulated in NS3 (Version 3.27) using "5G New Radio access configuration Modules". As specified in Sect. 3, Phase I of the research work is used to collect the user data from smart campus application using IoT network. The collected data is trained using Linear regression machine learning model. The processed data is in the form of excel file and it is exported to NS3. The excel file is copied to ns3 base folder and using build.py python script real data is taken as input to simulate MEC with 5G environment. The build.py script includes configuration of MEC server and tracePath, which depicts the path of IoT enabled real time data.

4.1 Results and Discussion

Performance of LBEF Algorithm is analyzed using standrad deviataion, Resource Based (RB) utilization and Throughput against adaptive threshold. The further part of this sub section describe the comprative study.

The authors have contrasted the effectiveness of the suggested LBEF algorithms, adaptive threshold, and load balancing scenarios. One metric used to assess how loads are distributed throughout the network is the standard deviation. At any one time, the amount of load balancing in the network is determined by monitoring a standard deviation. The data traffic for every UE was configured by the authors to have a Guaranteed Bit Rate (GBR) of 2 Mbps.

The 5G MEC network without load balancing algorithm shows high standard deviation of the load and it remain constant throughout the simulation. The threshold of load is fixed in the network and it is considered as adaptive threshold. Load in the network is balanced depending on this adaptive threshold. The proposed LBEF algorithm balance the load in the network based on load balancing factor as described in Algorithm 1. LBEF algorithm gives a smaller standard deviation compared with the other approaches, as shown in Fig. 5. Due to resource (BW) is properly utilized

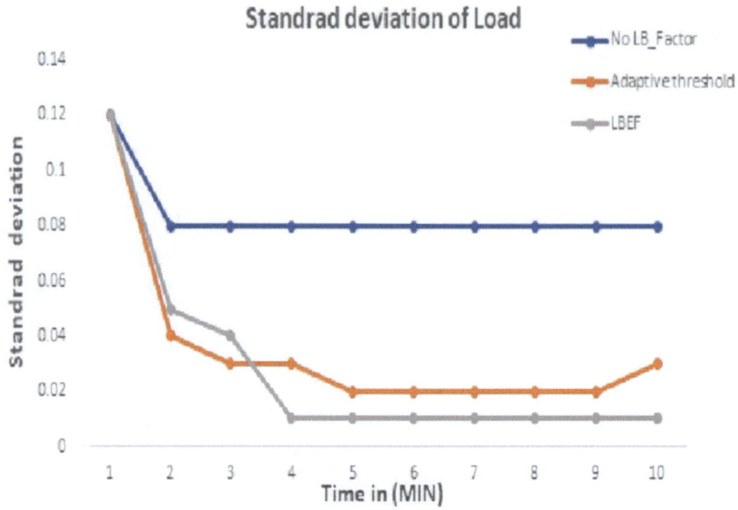

Fig. 5 Comprative study of LBEF using standrad deviation

by the severs of the small cell, the variance in load among small cells become low, and therefore the system is more balanced. The comparative study shows that LBEF algorithm improves standard deviation by 20% than the adaptive threshold algorithm.

Load balancing modifies how the network uses its resources. The average resource use for the two load balancing strategies is displayed in Fig. 6. The algorithms always result in a higher network resource use rate. In the event that there is no load balancing

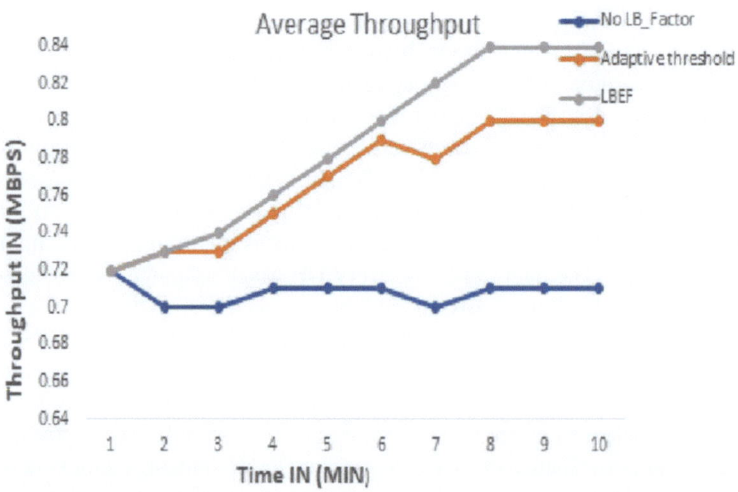

Fig. 6 Comprative study of LBEF using throughput

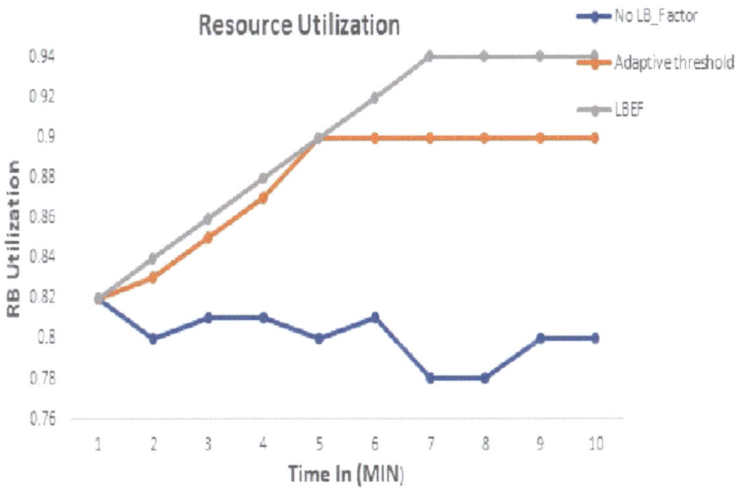

Fig. 7 Comprative study of LBEF using RB utilization

in the network, the overloaded cells cannot assign the necessary BW to the UEs due to a lack of resources. The lightly loaded cells are still not being used in the meantime. By distributing UEs from heavily loaded tiny cells to neighboring cells that are lightly loaded, the LBEF algorithm can relieve resources at the overloaded cells. Hence, UEs can be allocated the required resources from the lightly-loaded cells and thus the overall resource utilization increases. The simulation without load balancing shows least utilization resources. The LBEF algorithm improves the resources utilization by 30% than adaptive threshold algorithm.

Eventually, the authors have evaluated network performance in terms of throughput. Figure 7 shows the average cell throughput with two load balancing approaches. The UEs cannot be served with the required data rate because the overloaded small cells cannot provide the UEs with the needed resources. Hence, the UE in an overloaded cell of a network without load balancing will experience a limited throughput. The load across the small cells is evenly distributed when the network is adopted load balancing algorithms. The small cells will be able to assign the required resources to the UEs, which results in increased network throughput. The proposed LBEF algorithm increases the average throughput than adaptive threshold algorithm.

5 Conculsion

MEC server with 5G enabled IoT networks are useful technologies for real world applications. Authors of this article conduct the research in three phases. In the Phase I smart campus application is developed using IoT networks. The collected data from the application is trained using linear regression machining learning model.

The trained data is uploaded to the MEC servers. The proposed LBEF algorithm is running on MEC severs to balance the load in cells connected to the server in Phase II. The aggregated data is uploaded to the cloud in Phase III. The results show that LBEF algorithm performs better than adaptive threshold algorithm. As a future work authors want to extend their work to balance the load between cluster of MEC servers.

References

1. Addali KM, Bani Melhem SY, Khamayseh Y, Zhang Z, Kadoch M (2019) Dynamic mobility load balancing for 5G small-cell networks based on utility functions. IEEE Access 7:126998–127011 (2019). https://doi.org/10.1109/ACCESS.2019.2939936
2. Addali K, Kadoch M (2019) Enhanced mobility load balancing algorithm for 5G small cell networks. IEEE Can Conf Electr Comput Eng (CCECE) 2019:1–5. https://doi.org/10.1109/CCECE.2019.8861598
3. Hasan MM, Kwon S, Na J (2018) Adaptive mobility load balancing algorithm for LTE small-cell networks. IEEE Trans Wirel Commun 17:2205–2217
4. Shahid SM, Seyoum YT, Won SH, Kwon S (2020) Load balancing for 5G integrated satellite-terrestrial networks. IEEE Access 8:132144–132156. https://doi.org/10.1109/ACCESS.2020.3010059
5. Drozdova V, Akhpashev R (2020) The usage of load intensity balance approach For 5G MAC protocol chatacteristics investigation. In: Ural symposium on biomedical engineering. Radioelectronics and information technology (USBEREIT), pp 292–294. https://doi.org/10.1109/USBEREIT48449.2020.9117725
6. Giuseppi A, Maaz Shahid S, De Santis E, Ho Won S, Kwon S, Choi T (2020) Design and simulation of the multi-RAT load-balancing algorithms for 5G-ALLSTAR systems. In: 2020 international conference on information and communication technology convergence (ICTC), pp 594–596. https://doi.org/10.1109/ICTC49870.2020.9289485
7. Zhang H, Song L, Zhang YJ (2018) Load balancing for 5G ultra-dense networks using device-to-device communications. IEEE Trans Wirel Commun 17(6):4039–4050. https://doi.org/10.1109/TWC.2018.2819648. June
8. Mei C, Xia X, Liu J, Yang H (2020) Load balancing oriented deployment policy for 5G core network slices. In: IEEE international symposium on broadband multimedia systems and broadcasting (BMSB), pp 1–6. https://doi.org/10.1109/BMSB49480.2020.9379563
9. Yanzhi S, Yuming L, Quan W, Cheng F, Liqiong J (2021) Multi-factor load balancing algorithm for 5G power network data center. In: 2021 3rd international conference on advances in computer technology, information science and communication (CTISC), pp 188–193. https://doi.org/10.1109/CTISC52352.2021.00042
10. Hatipoğlu A, Başaran M, Yazici MA, Durak-Ata L (2020) Handover-based load balancing algorithm for 5G and beyond heterogeneous networks. In: 2020 12th international congress on ultra modern telecommunications and control systems and workshops (ICUMT), pp 7–12. https://doi.org/10.1109/ICUMT51630.2020.9222456
11. Park J-B, Kim KS (2018) Efficient load balancing scheme in 5G heterogeneous cellular networks. In: 2018 IEEE 88th vehicular technology conference (VTC-Fall), pp 1–2. https://doi.org/10.1109/VTCFall.2018.8690808
12. Zhang Q, Xu X, Zhang J, Tao X, Liu C (2020) Dynamic load adjustments for small cells in heterogeneous ultra-dense networks. In: 2020 IEEE wireless communications and networking conference (WCNC), pp 1–6. https://doi.org/10.1109/WCNC45663.2020.9120688
13. ns-3 manual, Release ns-3-dev, ns-3 project September 8 2022. Available at https://www.nsnam.org/releases/ns-3-27/

PID and Fuzzy PID Controller Design for Quadruple Tank System

S. Srinivasulu Raju, Jalalu Guntur, K. Guravaiah, and I. Samyuktha

Abstract The majority of process industry systems needed to regulate multiple variables by manipulating several controlled variables. An MIMO system is one that has multiple inputs and outputs. The process exhibits a nonlinear response as a result of the presence of disturbances in the multivariable system; these effects must be reduced in order to linearize the system. The mathematical formulation of an MIMO system and its linearization process around a desired operating point are both demonstrated in this work. In this article, we'll examine the behaviour of a multivariable nonlinear model, and attempt to linearize and regulate the system with a PID controller using the MATLAB software. And also concluded that, the controller effectively leads the system to reach the target set point based on various circumstances around the operating region. This paper presents a MATLAB-based implementation of a proportional–integral–derivative (PID) controller. To validate the functionality of the model, PID is applied to the quadruple tank process. MATLAB aids in the development of an ease of handling designed primarily for research scientists.

Keywords Multi-input–multi-output · Nonlinear system · PID · MATLAB · Absolute error

S. S. Raju (✉) · J. Guntur
IEEE Member, Department of Electronics and Instrumentation Engineering, VR Siddhartha Engineering College, Vijayawada, India
e-mail: srinu85raju@gmail.com

K. Guravaiah
Department of Computer Science and Engineering, Indian Institute of Information Technology Kottayam, Kottayam, India
e-mail: kguravaiah@iiitkottayam.ac.in

I. Samyuktha
Department of Electronics and Instrumentation Engineering, VR Siddhartha Engineering College, Vijayawada, India

S. K. Biswas et al. (eds.), *Intelligent Computing Systems and Applications*, Lecture Notes in Networks and Systems 1307, https://doi.org/10.1007/978-981-96-3860-4_46

1 Introduction

An MIMO system has numerous inputs and outputs in contrast to an SISO system's single input and output. An example of an MIMO system is the quadruple tank design. This kind of system is ideal for analysing various control strategies. It is more difficult to regulate the entire process and achieve the intended results when dealing with industrial control challenges since there are more controlled and manipulated variables. Controllers are added to the system to help with smooth functioning and improved operator control on the system to get the desired response. This helps to overcome these systemic issues. Fuzzy controllers, adaptive controllers, PID and predictive controllers, and many more are examples of different types of controllers with various controlling techniques. The controllers are chosen based on the operator requirements and the complexity of the system. The PID controller and its impact on multivariable systems like quadruple tanks to manage the level in the tanks at various set points using LabVIEW software will be the main focus of this experimental analysis. The process sector has been using PID (proportional–integral–derivative) controllers as their primary control technology for many years. There are many various control techniques available everywhere, but they are complicated, expensive, and virtually ineffective from the perspective of industrial need. Finding a mathematical model of the system is essential before using the PID control technique in a plant because the controller's whole tuning range is derived directly from the system. There are many methods described to accomplish the control of level of QT tank systems utilizing PID controller. Harsha Vardhan et al. [1, 2] implemented and demonstrated the fuzzy PID controller for tracking the DC motor position. Fuzzy logic, PID, and fuzzy PID controllers are used for tracking the position of DC motor and the performance analysis of an open-loop system was carried out to conclude that fuzzy PID provides better tracking as compared to PID and Fuzzy Logic controllers. MPC and adaptive PID controllers were used to demonstrate level control of nonlinear conical tank system by Pushpaveni et al. [3, 5, 6]. The authors used relay feedback method to tune the PID parameters, and MPC algorithm designed and simulated in MATLAB for multi-set point tracking of level control. Authors concluded that MPC tracks the servo and regulatory response without overshoot and within less time. Vinoth Kumar et al. [4] used the tuning methods—Cohen Coon and Ziegler Nicholas for tuning the controller parameters of P, PI, and PID for level control of nonlinear structures used in pharmaceutical manufacturing. The open-loop performance analysis was carried out to conclude that the PID controller was able to track the level within specified limits of nonlinear system. Srinivasulu et al. [7, 8] designed and implemented the DMC control algorithm for concentration control of CSTR process in MATLAB simulation environment. The control algorithms PID and DMC are analysed and found dynamic matrix control provided better response for tracking the concentration of CSTR system. Fuzzy PID controller was used to track the temperature set point of a continuous stirred tank reactor process by Ravi Kiran et al. [9] A relay feedback technique is used to find the PID parameters using ultimate gain and period. Fuzzy PID was implemented for the process using data process technique by authors and concluded that advanced controller like Fuzzy PID enhances the performance of temperature control in CSTR process. The effectiveness of the controller's operation is directly correlated with the model's correctness. A good controller will have less overshoot, offsets, and time to accurately reach the destination as a result of adequate tuning. The QTP is a multi-input, multi-output system that can be used to investigate various control strategies. It is referred to as a two-tank process.

2 Quadruple Tank System

The quadruple tank method is a research laboratory procedure using two pumps and four interconnected tanks. The inputs are $u1$ and $u2$ and outputs of QTP are considered as $y1$ and $y2$. Main aim of the work is to obtain the set point tracking of level of two tanks by considering the inlet flow via pump. QT processes have four tanks, each tank is connected with reservoirs. Tanks 2 and 4 are interconnected as same as tanks 1 and 3 as shown in Fig. 1. Model identification process has been carried out by using RLS identification method and RGA analysis has been done to arrive at the relation between control loops of multivariable system. Tank 4's output is fed into tank 1, while tank 1's output is sent to tank 2. These connections provide a tight connection among the tanks, allowing the system to be controlled in multiple ways. For a control system engineer, controlling and identifying errors in the QTP system is a demanding undertaking.

3 Modelling of QT System

The most crucial phase in designing of control system is modelling. Theoretical and experimental methods are employed. The mathematical representation of the QT process is obtained by using fundamental equations, which represent the dynamic nature of the process. The solution of the equations carried out using ODE 45 solver in MATLAB software. Each tank process is represented with its mass balance equation in interacting mode to arrive at the level control of tanks 1 and 3. Different characteristics of multivariable control system can be explained using the multivariable quadruple tank method. The mathematical modelling of the quadruple tank system is as follows:

$$A_i \frac{dh_i}{dt} = q_{\text{in}i} - q_{\text{out}i} \tag{1}$$

where $i = 1, 2, 3, 4$;

Ai represents the cross-sectional area; $q_{\text{in}1}$, $q_{\text{in}2}$ are the tank inlet and outlet flow; and h_i is level height in tank.

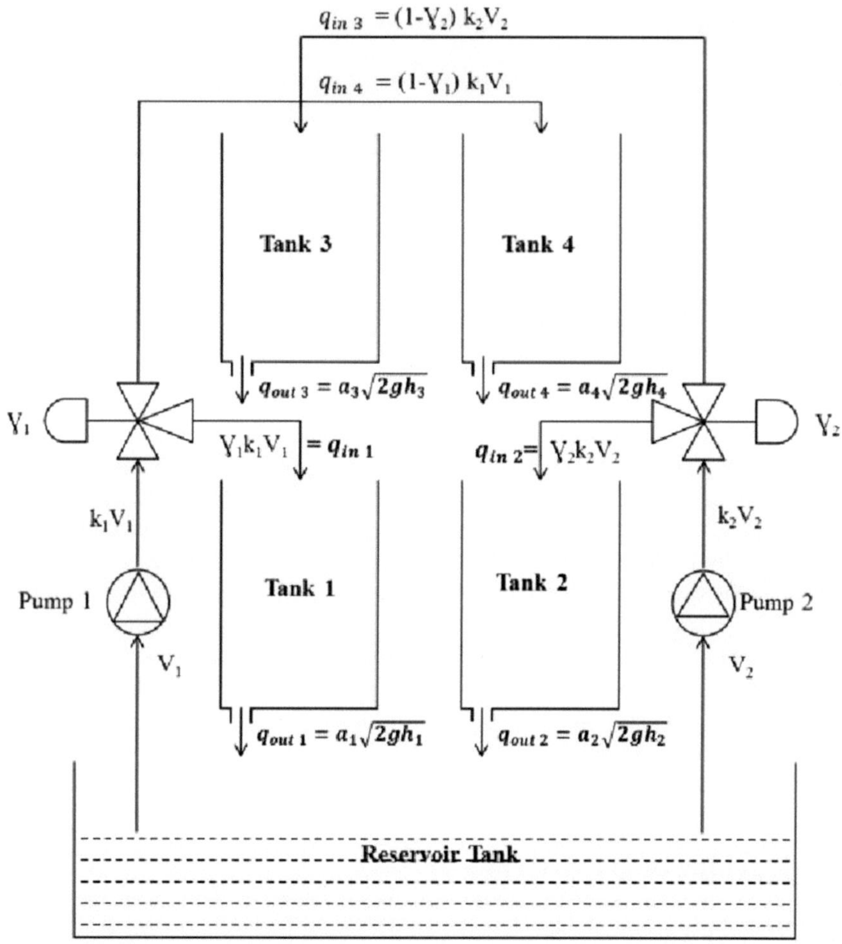

Fig. 1 Block diagram of quadruple tank system

Hence

$$q_{\text{in}1} = \gamma_1 k_1 v \tag{2}$$

$$q_{\text{in}2} = \gamma_2 k_2 v_2 \tag{3}$$

$$q_{\text{in}3} = k_2 v_2 (1 - \gamma v_2) \tag{4}$$

$$q_{\text{in}4} = k_1 v_1 (1 - \gamma v_1) \tag{5}$$

where $k1$, $k2$ are constants, $v1$, $v2$ are flow velocity of water, and $\gamma 1$ and $\gamma 2$ are the ratio of control valve.

$$q_{\text{out}i} = ai\sqrt{2ghi} \tag{6}$$

Mass balance equation of QT process is given as follows:

$$\frac{dh_1}{dt} = \frac{1}{A_1}\left[\gamma_1 k_1 v_1 + a_3\sqrt{2gh_3} - a_1\sqrt{2gh_1}\right] \tag{7}$$

$$\frac{dh_2}{dt} = \frac{1}{A_2}\left[\gamma_2 k_2 v_2 + a_2\sqrt{2gh_2} - a_4\sqrt{2gh_4}\right] \tag{8}$$

$$\frac{dh_3}{dt} = \frac{1}{A_3}\left[-a_3\sqrt{2gh_3} + (1-\gamma_2)k_2 v_2\right] \tag{9}$$

$$\frac{dh_4}{dt} = \frac{1}{A_4}\left[-a_4\sqrt{2gh_4} + (1-\gamma_1)k_1 v_1\right] \tag{10}$$

To derive a state-space representation from QTP, Eqs. (7)–(10) are solved using the Taylor series and Jacobian conversion matrix.
Time constant, $T = \frac{Ai}{ai}\sqrt{2gh}\, i$.
Jacobian and Taylor series are used to arrive the A, B, and C values:

$$A = \begin{bmatrix} \frac{-1}{T1} & 0 & \frac{A3}{A1T3} & 0 \\ 0 & \frac{-1}{T2} & 0 & \frac{A4}{A2T4} \\ 0 & 0 & \frac{-1}{T3} & 0 \end{bmatrix} \tag{11}$$

$$B = \begin{bmatrix} \frac{\gamma 1K1}{A1} & 0 \\ 0 & \frac{\gamma 2K2}{A2} \\ 0 & \frac{(1-\gamma 2)K2}{A3} \end{bmatrix} \tag{12}$$

$$C = \begin{pmatrix} K_c & 0 & 0\,0 \\ 0 & K_C & 0\,0 \end{pmatrix} \tag{13}$$

The system TF is derived by converting the state space in S to time domain using

$$G(s) = C(sI - A)^{-1}B + D \tag{14}$$

Hence

$$\text{G(s)} = \begin{pmatrix} \frac{c1\gamma 1}{(sT1+1)} & \frac{C1(1-\gamma 2)}{(sT1+1)(sT3+1)} \\ \frac{c2(1-\gamma 1)}{(sT2+1)(sT4+1)} & \frac{c2\gamma 2}{(1+sT2)} \end{pmatrix} \tag{15}$$

where $C1 = \frac{T_1 K_c K_1}{A_1}$ & $C2 = \frac{T_2 K_c K_2}{A_2}$.

As a result, the state-space equation is used to create Eq. (15) which indicates the TF of QT system. Because the zero lies on left half of s-plane, the QT system represents in minimum phase. And the procedure was established as a minimal step on

$$1 < \gamma_1 + \gamma_2 < 2$$

The operating conditions are selected from the laboratory setup.

Process control parameter symbols	Minimum phase
$A1$ & $A3$, $A2$ & $A4$	28, 32 [cm^3]
$a1$ & $a3$, $a2$ & $a4$	0.071, 0.057 [cm^3]
$T1$, $T3$	62, 90
$T2$, $T4$	23, 30
$y1$, $y2$	0.7, 0.6
$k1$, $k2$ (cm^3/vs)	3.33, 3.35
G	9.81

In the QTP system, the control parameter variables are applied, and the transfer function of the system is obtained as

$$G(s) = \begin{pmatrix} \frac{2.6}{(62s+1)} & \frac{1.5}{(23s+1)(62s+1)} \\ \frac{1.4}{((30s+1)(90s+1))} & \frac{2.8}{(90s+1)} \end{pmatrix} \tag{16}$$

4 Controller Design for QT System

PID Controller

The fundamentally utilized regular controllers are conventional controllers. Figure 1 depicts PID, which is an additive mix of proportional, integral, and derivative operations. The PID tuning parameters are obtained using Ziegler–Nicholas method and employed in the MATLAB simulation to analyse the performance of open-loop and closed-loop systems.

Fuzzy Logic Controller

Design and implementation of intelligent control systems became an area of interest in field of process control. AI plays an important role and provides the effective methodology in designing the control systems for robotic vehicle. Fuzzy logic controller mostly used to track the desired response mostly for control problems

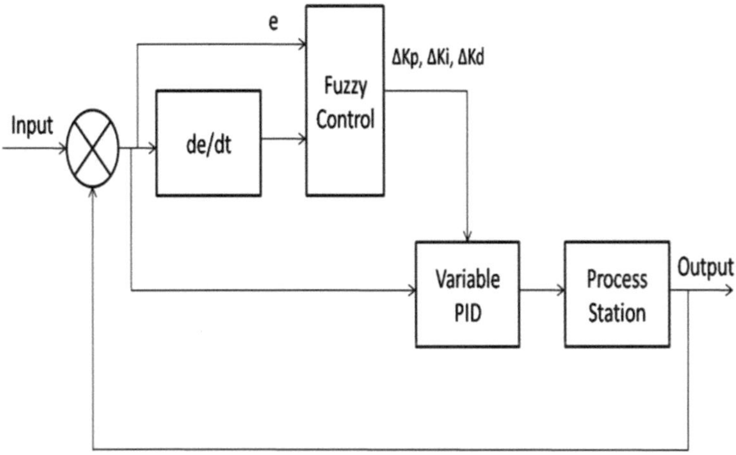

Fig. 2 Structure of fuzzy adaptive PID

in manufacturing and product-based industries. A simple rule-based "IF X AND Y THEN Z" method is used to frame the rules based on the membership functions for solving the control engineering problems instead of striving to analytical model of a system. Fuzzy logic is mostly used in analysis and design of the control system for engineering problems, which are very complex in nature and it is difficult to attain the model of a system. The expert knowledge and his experience about the process will be required to frame the rule base, which is effectively used for setpoint tracking of the robot vehicle system. The basic architecture of fuzzy logic is shown in Fig. 2. From Fig. 2, the four principal techniques are involved in the fuzzy logic are knowledge base block, decision-making block, fuzzification, and defuzzification units. Membership functions are used to frame the rules using knowledge base and inference engine implies fuzzy operations on rules, and finally the fuzzified control output transforms to numeric value to the process into the system.

Fuzzy PID Control Algorithm

The formation of fuzzy PID algorithm is presented in Fig. 2, which comprises two sections: PID is one and the second one is fuzzy controller. The proposed work consists of three outputs and two input fuzzy PID controllers for tracking the level of the tanks 1 and 3. Change in error and error are considered as input to fuzzy PID, and the output of controller is to find tuning parameters of the PID. The PID controller tuning parameters are adapted in online to meet the desires specified by the designer to achieve the good stability.

A fuzzy system could reflect extensive language knowledge, such as that provided by a human expert who follows rules.

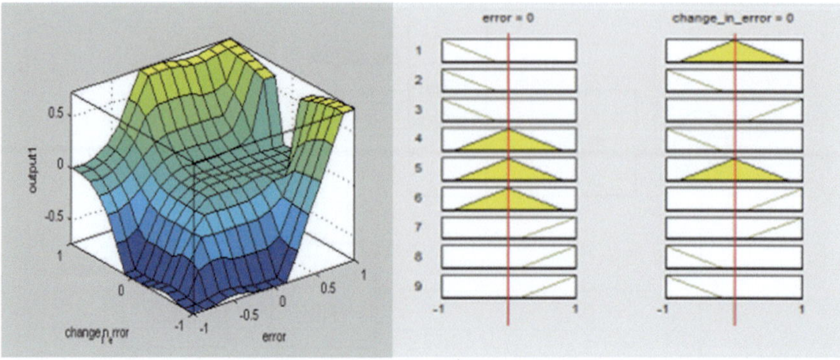

Fig. 3 Rule base overlay of fuzzy PID and rule viewer

Model of Fuzzy PID controller

The input precise magnitude to the fuzzy quantity is created by the fuzzy controller. The flow process base's experienced expertise is housed in the knowledge base. Each linguistic variable's membership function is included in fuzzy data. The controlling rules are from the fuzzy database. The fuzzy quantity converts to non-fuzzy quantity using defuzzification method, in which the centroid method is mostly used to convert into precise quantity.

Membership Function

Triangular and Gaussian membership function is used by fuzzy controller to frame the linguistic variables. The input ranges from -7 to $+7$, and the fuzzy subsets are NB, NM, NS, ZO, PS, PM, and PB. In the implementation of the fuzzy controller, the scaling and quantization factor is extremely important.

The fuzzy PID ruler view and surface view are shown in Fig. 3.

Essentially, the choice is made using the fuzzy system's rule, and the PID detects the error between the measured and set value. For the fuzzy controller, only nine rules are derived. The defuzzification module converts the rules' output into crisp variables. The fuzzy PID simulated output is shown below in lines.

5 Performance Analysis and Simulation Results

The open- and closed-loop studies have been done for the level control of QT system to achieve desired level setpoint. MATLAB software (R2020a) is applied to obtain the analytical results and analysed the performing criterion. The parameters of PID are designed using Ziegler–Nichols closed-loop method and used to analyse the closed-loop performance of the system with controller (Fig. 4). The simulation studies are analysed in MATLAB (R2020a) and Figs. 5, 6, 7, and 8 show the respective results.

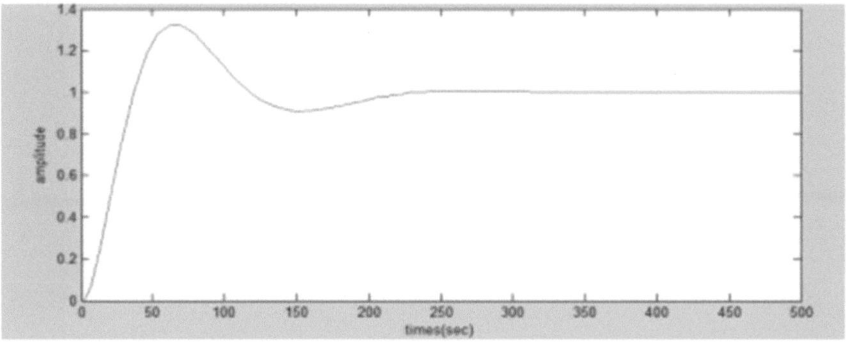

Fig. 4 PID controller response for level in tank 1

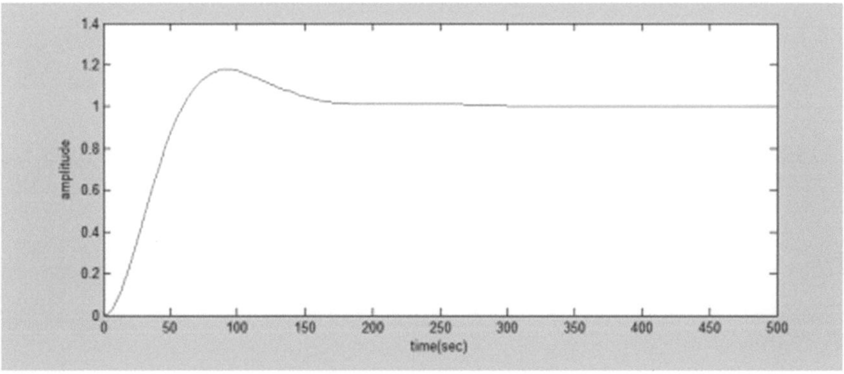

Fig. 5 The response of PID controller for level in tank 3

The response of first principal model is equated with fuzzy PID output and displayed in Fig. 8. The use of a fuzzy PID overcomes the difficulties of employing a PID controller for level tracking, to tune the parameters of PID adaptively with respect to change in the plant dynamics. Fuzzy logic is used to tune the control parameters to follow the set point with limitations in a discrete domain. The fuzzy algorithm for desired level of level is designed and simulated in MATLAB software. The system would be processed into discrete domain before being sent on to controller, which will adjust the process parameter in proportion to the output of fuzzy PID is shown in Fig. 5. From the results, finally, we may say that the fuzzy PID is effective. Stabilizes the process, level reaches the setpoint. Hence, we determine that fuzzy PID is able to track the level setpoint of QT system without any overshoot, with less settling time and zero rise time. Even if there are disturbances, the system will be stable.

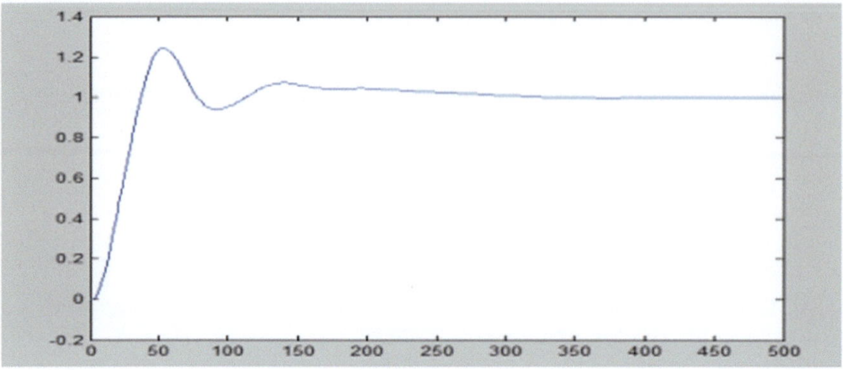

Fig. 6 Fuzzy PID controller response for level in tank 1

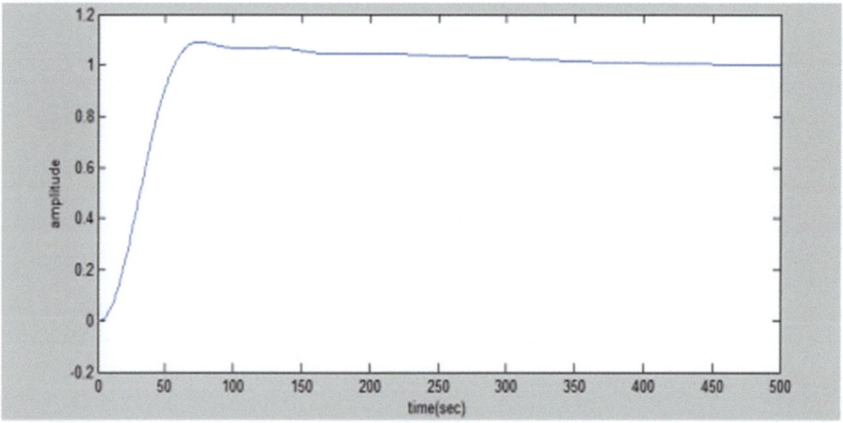

Fig. 7 Fuzzy PID controller response for level in tank 3

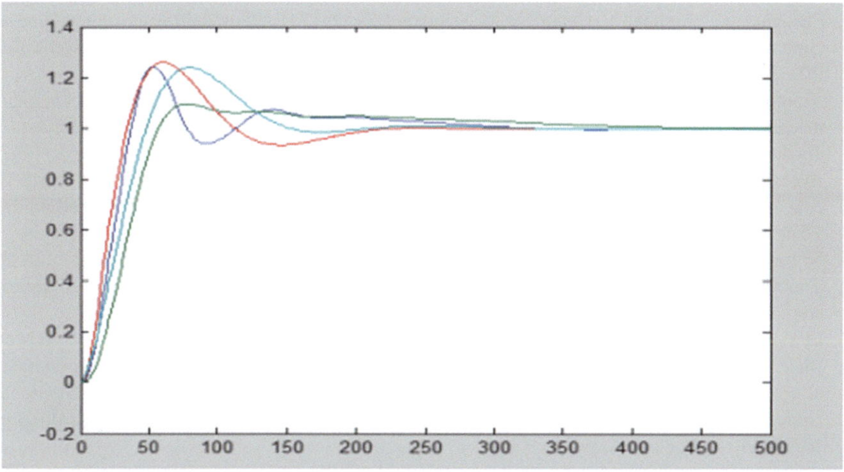

Fig. 8 Validation response of level in tanks 1 and 3 using PID and fuzzy PID

6 Conclusion

In this work, the authors considered the level control of QT system. Conventional controller PID and fuzzy PID are used to track the level setpoint and reject the perturbations at output level. Ziegler–Nichols method is used to determine the PID parameters using ultimate gain and period. The tuning parameters are imposed in MATLAB environment for tracking the level of QP system tanks 1 and 3. When a dependable model is not accessible in the QTP system, designing a controller with the intended target becomes more challenging. By comparing the performance and reaction of QT interaction system, control strategy is acquired utilizing different controllers. According to the simulation, the fuzzy PID controller outperforms the PID. Peak overshoot is minimized with fuzzy PID, and a soft response is achieved. This strategy is straightforward because the fuzzy PID uses membership functions to improve responsiveness. Model-based controllers with healthy functioning can be designed using the models developed thus far. The performance of PID and fuzzy PID is satisfied for tracking the level of QT system at laboratory setup with regulatory and servo response.

References

1. Karl Henrik Johansson "The Quadruple-Tank Process: A Multivariable Laboratory Process with an Adjustable Zero", Member, IEEE (2000).
2. Harsha Vardhan A, Sai Bharadwaj A, Srinivasulu Raju S, Archana N (2013) Implementation of fuzzy Pid controller and performance comparison with Pid for position control of Dc motor. Int J Eng Res Dev 7(9):78–84

3. Pushpaveni T, Archana N, Chandana M, Srinivasulu S (2013) Raju modelling and controlling of conical tank system using adaptive controllers and performance comparison with conventional PID. Int J Sci Eng Res 4(5)
4. Vinothkumar C, Esakkiappan C (2019) Level control of nonlinear hopper tank process in pharmaceutical industry using Ziegler Nicholas and Cohen Coon tuning techniques. Int J Innov Technol Explor Eng (IJITEE) 8(10)
5. Ayeesha Siddiqa M, Ravi Kiran TKS, Viswanath M, Srinivasulu Raju S (2013) Control of concentration in CSTR using DMC and conventional PID based on relay feedback method. Int J Eng Sci Technol (IJEST) 7(4)
6. Nithya S, Vijayarekha K, Sivakumaran N, Balasubramanian T, Anantharaman N (2008) Controllers implementation based on softcomputing for non-linear process. In: 34th Annual conference of IEEE in industrial electronics, pp126–132
7. Li Y, Ang KH, Chong G (2008) PID control system analysis and design. IEEE Control Syst Mag
8. Raju SS, Swamy GN, Bharath Y, Nandini CN (2020) Simulation and performance analysis of autonomous underwater vehicle using advanced control algorithms. In: 2020 International conference on communication and signal processing (ICCSP), pp 0905–0909. https://doi.org/10.1109/ICCSP48568.2020.9182176
9. Anandanatarajan R, Chidambaram M, Jayasingh T (2005) Design of controller using variable transformations for a nonlinear process with dead time. ISA Trans 44(1):81–91

Event Track: TextSumEval'2022

TextSumEval-2022: Overview of the Textual Summary Evaluation Task

Tamwo, Pankaj Dadure, Prottay Kumar Adhikary, Partha Pakray, Benoit Favre, Sivaji Bandyopadhyay, and Thierry Artires

Abstract TextSumEval aim to provide a platform for the researcher working the field of text summarization and evaluation. In this task, we have provided the system-generated summaries and gold summaries to the task participants. Participants need to design a summary evaluation technique that compares system-generated summaries with gold summaries and generated their semantic resemblance score. To facilitate the above-mentioned task, we have provided a total of 6000 summaries which are selected from the existing databases, namely, CNNDM, MSMO, and XSUM. We have calculated the correlation scores between the BLEU score, ROUGE score, and our baseline system's semantic score with participants submitted semantic scores using a Pearson correlation coefficient to verify the participants' scores. Moreover, based on the mean of the correlation scores, we have highlighted the methodology of the best-performed team.

Keywords Automatic evaluation · Text summary · Summarization · Semantic similarity · Pearson correlation

1 Introduction

Text summarization [10] is one of the well-known applications of Natural Language Processing (NLP) and has come under the limelight since a few decades. In the advancement and subsequent effect of the digital world on the burgeoning publication sector, time to sincerely read an article, document or book is no more feasible, particularly in view of the lack of time. Furthermore, as the number of articles published and traditional print publications is digitalized, it has become pretty hard to keep track of increasing web publications. In this case, a summary of the text can serve

Tamwo · P. Dadure · P. K. Adhikary · P. Pakray (✉) · S. Bandyopadhyay
National Institute of Technology Silchar, Assam, India
e-mail: partha@cse.nits.ac.in

B. Favre · T. Artires
Aix-Marseille University, CNRS, LIS, Marseille, France

S. K. Biswas et al. (eds.), *Intelligent Computing Systems and Applications*, Lecture Notes in Networks and Systems 1307, https://doi.org/10.1007/978-981-96-3860-4_47

to minimize the complexity of the texts. Summarization requires a comprehensive understanding of input documents, identification of relevant content, and generation of a synthetic perspective of the document, often subject to a length constraint [9]. Building better summarization systems requires progress in summarization evaluation metrics, which are used to assess the quality of the generated summaries. Generally, there are two ways to evaluate the performance of any summarization system [13]: manual evaluation and automatic evaluation. The manual evaluation consists of ranking summaries or parts of summaries according to linguistic fluency and perseverance of the semantic meaning of the documents. The automatic evaluation is a system generated metric produced by comparing a set of summaries. Manual evaluation is more accurate but much more costly than automatic evaluation, and it is often not actionable in a machine learning environment (systems require rapid and repeated evaluation of their output in order to learn how to summarize). However, the existing methods of automatic evaluation are effective, and it requires advancement to preserve the linguistic fluency and semantic meaning of the original documents.

This paper provides the overview of the TextSumEval[1] task which is organized by the Department of Computer Science and Engineering department of the National Institute of Technology Silchar. The TextSumEval aim to provide the platform to the researcher working the field of text summarization and evaluation. TextSumEval aims to design an automatic evaluation system that compares the system generated summaries with gold summaries and estimates their relevance in terms of semantic score. Herein, 6000 system generated summaries and gold summaries have been provided to the participants, selected from the existing three datasets, namely CNNDM, MSMO, and XSUM. As a result, the participants have submitted semantic scores which depicts the semantic resemblance between system generated summaries and gold summaries. To verify the participants score, we have calculated the correlation scores between BLEU score, ROUGE score, our baseline system's semantic score with participants submitted semantic score using a pearson correlation method. Moreover, based on the mean of the correlation scores, we have highlighted the methodology of best performed team.

The following sections make up the paper: Sect. 2 discusses relevant work, Sect. 3 discusses dataset statistics, Sect. 4 discusses our suggested techniques of our baseline system, Sect. 5 contains the procedures the teams has followed, the evaluation method of the proposed work has been reflected on Sect. 6, Sect. 7 discusses experimental results, and Sect. 8 wraps up the paper and suggests future research directions.

2 Background

Semantic Textual Similarity [16] is a Natural Language Processing (NLP) task that uses a predetermined metric to score the relationship between texts or documents [7]. Measuring semantic similarity scores between texts is useful in many applications in

[1] http://icicsa.nits.ac.in/textsumeval22/index.html.

NLP and related fields. Text summarization [6, 25], extractive summarization [23], Word Sense Disambiguation (WSD) [14], relevance feedback and text classification [22] and Question Answering (QA) [19] are just a few of the applications that have already reported the use similarity score of text. Many semantic web applications, such as community ontology development, entity disambiguation and extraction, rely on Semantic Textual Similarity (STS). It's also useful for the Twitter Search program [23], which necessitates the capacity to reliably quantify semantic relatedness between concepts or things. One of the primary difficulties in information retrieval (IR) is retrieving a set of documents and images by captions semantically connected to a given user query in an online search engine [8]. Since 2012, the goal of semantic similarity has included not only identifying similarities between two texts but also generating a similarity score from 0 to 5 using various SemEval tasks1 [2–4]. In this assignment, a score of 0 indicates perfect semantic equivalence while a score of 5 indicates complete unrelatedness. Since 2012, the SemEval Semantic Textual Similarity (STS) shared tasks [1–5] in the general English domain have been arranged to encourage the natural language processing community to develop automated approaches for this criterion. Many NLP applications must determine how similar two short texts are in terms of meaning. For example, search engines must simulate the relevance of a page to a query that goes beyond the overlap in words. Question-and-answer sites like Quora, for example, must identify whether a question has already been asked. This type of text similarity is frequently calculated by embedding two short paragraphs and then computing their cosine similarity.

Since 2012, the purpose of semantic similarity has been to provide a similarity score from 0 to 5 utilising various SemEval tasks. A score of 5 implies perfect semantic equivalence, whereas a score of 0 shows complete unrelatedness in this assignment. Word Sense Disambiguation, relevance feedback and text categorization, and Question Answering are only a few of the applications that have already documented the usage similarity score of text. Although popular semantic evaluation techniques such as word embeddings like Word2Vec and GloVe have established common methodologies for determining semantic similarity between two words, there is less agreement on how to compute sentence embeddings.

3 Dataset

In the TextSumEval-2022 task, we have provided 6000 gold summaries to the participant teams, which are prepared from the openly available dataset, namely CNNDM,[2] MSMO,[3] and XSUM.[4] We have extracted a total of 2000 gold summaries from each of the datasets. In addition to the gold summary, we have also provided the predicted summaries, which have been generated by our designed baseline system [26]. In the

[2] https://github.com/abisee/cnn-dailymail.

[3] https://aclanthology.org/D18-1448/.

[4] https://github.com/EdinburghNLP/XSum/tree/master/XSum-Dataset.

Table 1 Statistics of the dataset

Dataset	Gold summary	Predicted summary
CNNDM	2000	2000
MSMO	2000	2000
XSUM	2000	2000

provided predicted summaries, we have included 50% of those summaries whose semantic score is high with respect to gold summaries, and the remaining 50% summaries include those summaries which are partially similar to the gold or completely diverse from the gold summary. The selection of these summaries does not depend on the semantic score of our baseline system. We have selected these summaries manually by carefully analyzing the gold summary and baseline system's predicted summary. After careful analysis and based on human judgment, the selection has been performed. The statistics of the dataset is given in Table 1.

4 Baseline System

The designed baseline system utilizes two of the five modules from [15]; (1) measuring the similarity and interpretability score, (2) instead of a classification algorithm, we use a regression algorithm that works sequentially to delivered the semantic textual similarity. A pair to sentence to be evaluated are pre-processed for feature extraction where the ten text similarity measures are extracted between the two texts. In the second phase, the extracted features are used to train a linear regression model and the trained model is employed to generate a semantic similarity on unknown pairs of texts. The baseline system generates a interpretability score between two texts. The workflow of the baseline system is illustrated in Fig. 1 and discussed in the following subsection.

4.1 Baseline System Architecture

The baseline system originally extract two feature groups where the two aligned chunks used to train the regression model and return the feature values range from 0 to 1. The first one is string similarity measurements which includes Jaccard similarity coefficient, cosine similarity, and common word ratio. Herein, by counting the number of shared and distinct words across chunks, and generates the alignment score. The second one uses the word vectors as distributional representations of words to calculate similarity. By using above-mention feature group, the baseline system

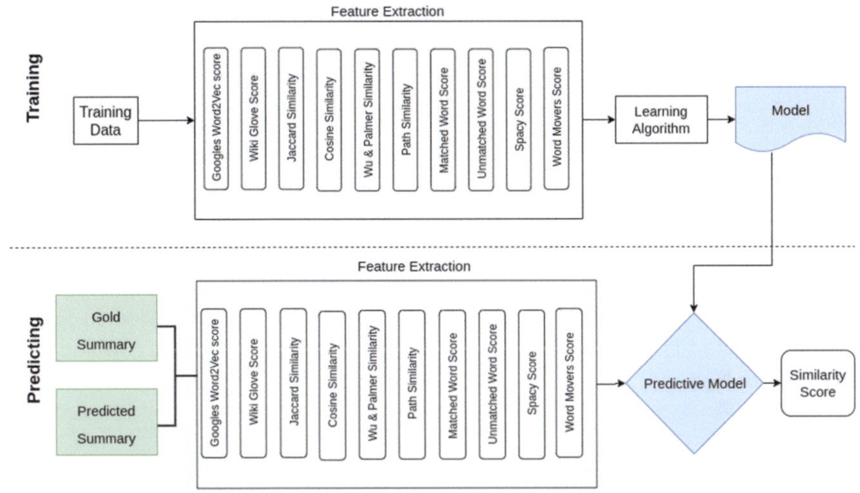

Fig. 1 Work flow of the baseline system

extracts total of ten features: Word2Vec [20], Wiki GloVe [20], SpaCy similarity, Word Mover's Score, Jaccard similarity [21], Wu and Palmer similarity [11], Path similarity and Common word ratio.

- Word embeddings: The baseline systems extract the various word similarity score using four techniques, i.e., Word2Vec, GloVe, Word Mover's and SpaCy similarity. Word embedding is a phrase for the representation of words in text analysis, often in the form of a real-valued vector that encodes the meaning of the word and predicts that words adjacent in the vector space have similar meanings.
- Jaccard similarity: The Jaccard similarity is a statistical method for determining sample set similarity and variety. The Jaccard similarity is determined by the fraction of number of observations in both sets and the number of observations in each set.
- Cosine similarity: A statistic that is used to determine how similar data objects are regardless of their size. It captures the cosine angle between the two text fragments.
- Wu and Palmer similarity: It determines similarity based on how similar the word senses are and where the Synsets appear in the hypernym tree in relation to one another.
- Path similarity: Path similarity is also a similarity measure that determines the distance between two synsets as the length of the shortest path.
- Common word ratio: The ratio of shared vocabulary between the two sentences are also used as a feature.

The above-mentioned methodology of the baseline system has been tested on the dataset mentioned Sect. 3. The experimental results of the baseline system in depicted in Table 2. Herein, we have estimated BLEU and ROUGE score, and compared with semantic score obtained from designed baseline system. The obtained experimental

Table 2 Experimental results of baseline system

Dataset	BLEU	ROUGE	Semantic score
CNNDM	36.12	36.13	42.85
MSMO	42.05	40.76	44.46
XSUM	35.98	34.65	46.22

results revels that the designed baseline system delivered higher semantic score than BLEU and ROUGE score. These obtained results has been used to judged the participants submitted score to verify their authenticity and their correlation as discussed in Sect. 6.

5 Participated Teams

In TextSumEval task, total of 3 teams have been participated.

5.1 NITS_APSS

The NITS_APSS [17] is one of the team participated in TextSumEval from National Institute Technology Silchar, Assam, India. Team NITS_APSS has used pre-trained language model namely MiniLM [27] which generates the 384 dimensional numerical vector based embedding of the gold and system generated summary. After the embedding generation, cosine similarities have been calculated and reported as the similarity scores between the gold and the predicted summaries.

5.2 NITS_DMA

The NITS_DMA [12] is one of the team participated in TextSumEval from National Institute Technology Silchar, Assam, India. Team NITS_DMA has uses the DistilRoBERTa pre-trained model [24] for creating the 768-dimension embeddings of the gold and system generated summaries. Afterward, the cosine similarity between each pair of gold and predicted summaries has been calculate which further consider as semantic score between the summaries.

5.3 SC-NEHU

Team SC-NEHU [18] is a research team from North-Eastern Hill University, Shillong, Meghalaya, India. Team SC-NEHU has submitted total of 2 runs where each of which follows the different working phenomena. The frist run rely on the concept of sentence embedding. Herein, team SC-NEHU has uses the Sentence-BERT (S-BERT) model which utilizes pre-trained BERT and RoBERTA networks and fine-tunes them to generate fixed-size sentence embeddings. To construct a fixed-sized sentence embedding, S-BERT adds a pooling operation to the output of a BERT/RoBERTA model. Pooling method is used for calculating the mean of all output vectors. After the model has been trained, S-BERT employs a regressive objective function for inference within a siamese network, identical to the one used for fine-tuning. Afterwards, the cosine similarity between two sentence embeddings (gold summary and system generated summary) are computed as a score between [–1... 1]. The cosine similarity measure, in this case, is given as semantic score between the summaries.

The second methodology uses by the SC-NEHU for run 2 is Singular Value Decomposition (SVD) from linear algebra. In this method the same sentence embeddings obtained in the first methodology are used as the vectors of the matrix M. The dimensionality of M is 2×384 where 2 corresponds to the number of summaries that need to be compared and 384 is the length of the sentence vectors. SVD is applied to matrix M and the dot product of the two components of the two summaries is taken. This process transforms the summaries into a topic encoded data whereby the topics are latent. The vectors in the result of the dot product indicate how much a summaries (gold and system generated) is expressed in each of the 10 topics.

6 Evaluation Method

The participants performance has been assessed using a method called the correlation coefficient, specifically the Pearson correlation coefficient. This coefficient is commonly used to measure the strength and direction of a linear relationship between two variables. It's a descriptive statistic that summarizes the characteristics of a dataset, indicating how much and in which direction two quantitative variables are related. The Pearson correlation coefficient is a number ranging from –1 to 1. A value between 0 and 1 signifies a positive correlation (both variables change in the same direction), 0 indicates no correlation, and a value between 0 and –1 represents a negative correlation (variables change in opposite directions). The interpretation of relationship strength, also called effect size, may vary across different disciplines. For instance, when the correlation coefficient is above 0.5, both variables have a strong and positive impact. If it falls between 0.3 and 0.5, the impact is considered moderate and positive. A range of 0 to 0.3 signifies a weak but positive impact. A correlation coefficient of 0 indicates no relationship between the variables. On the other hand, if the coefficient is between 0 and –0.3, it suggests a weak negative impact. Ranging from –0.3 to –0.5 implies a moderate negative impact, and below –0.5 suggests a strong negative impact. Here's a mathematical representation for estimating the Pearson correlation coefficient (r):

$$r = \frac{\sum(x_i - \bar{x})(y_i - \bar{y})}{\sqrt{\sum(x_i - \bar{x})^2 \sum(y_i - \bar{y})^2}} \tag{1}$$

where, r = correlation coefficient, x_i = values of the x-variable in a sample, $\bar{x}$ = mean of the values of the x-variable, y_i = values of the y-variable in a sample, $\bar{y}$ = mean of the values of the y-variable.

Using the above-mentioned formula, we have calculated the correlation coefficient between the BLEU, ROUGE, baseline semantic score with participants score. Subsequently, the mean of the correlation score of the BLEU, ROUGE, baseline semantic score with participants score has been considered as a final result value of the participant. The Table 3 shows the obtained correlation scores of the each participant where team NITS_APSS has achiveved the highest score. Moreover, the first run of the team SC-NEHU has shown the very good performance by competing with team NITS_APSS. Team NITS_APSS and team SC-NEHU-R1 has achieved the same mean semantic score. However, in correlation coefficient, semantic score of the team NITS_APSS is more favorable. However, team NITS_DMA achieved the highest semantic score but it shown the less correlation with BLEU, ROUGE, baseline semantic score. The second run of the team SC-NEHU shown the 50% of semantic similarity between the gold and system generated summary, but it shows very less correlation with BLEU, ROUGE, baseline semantic score.

7 Conclusions and Future Scope

TextSumEval provides the platform for researchers working in summary generation and evaluation to contribute toward the design of effective summary assessment techniques. Herein, participants need to design a summary evaluation technique that compares system-generated summaries with gold summaries and generates their semantic resemblance score. To achieve this objective, a total of 6000 summaries have been provided, which are selected from the existing databases, namely, CNNDM, MSMO, and XSUM. In this task, three teams participated. The validation of the participants submitted results have been done using the Pearson correlation coefficient method. In the obtained results, team NITS_DMA achieved the highest semantic score, whereas team NITS_APSS achieved the highest correlation score. In the future, the dataset will be improved by incorporating multiple system generated summaries for each gold summary. Also, several new techniques will be used to validate the participant results.

Table 3 TextSumEval task results of participated teams

Team name	Mean semantic score	Correlation	Correlation score	Average correlation score
NITS_APSS	0.6518	BLEU - NITS_APSS Score	0.6239	0.6332
		ROUGE - NITS_APSS Score	0.6887	
		Baseline Semantic Score - NITS_APSS Score	0.5869	
NITS_DMA	0.6579	BLEU - NITS_DMA Score	0.6081	0.6104
		ROUGE - NITS_DMA Score	0.6635	
		Baseline Semantic Score - NITS_DMA Score	0.5596	
SC-NEHU-R1	0.6518	BLEU - SC-NEHU-R1 Score	0.6238	0.6331
		ROUGE - SC-NEHU-R1 Score	0.6885	
		Baseline Semantic Score - SC-NEHU-R1 Score	0.5870	
SC-NEHU-R2	0.50293	BLEU - SC-NEHU-R2 Score	0.01446	0.01209
		ROUGE - SC-NEHU-R2 Score	0.02077	
		Baseline Semantic Score - SC-NEHU-R2 Score	0.00106	

Acknowledgements The organized shared task falls under the Research Project Grant No. IFC/4130/DST-CNRS/2018-19/IT25 (DST-CNRS targeted program). The organizer would like to express gratitude to the Centre for Natural Language Processing and Artificial Intelligence Lab, Department of Computer Science and Engineering, National Institute of Technology Silchar, India for providing infrastructural facilities and support.

References

1. Agirre E, Banea C, Cardie C, Cer D, Diab M, Gonzalez-Agirre A, Guo W, Lopez-Gazpio I, Maritxalar M, Mihalcea R et al (2015) Semeval-2015 task 2: semantic textual similarity, english, spanish and pilot on interpretability. In: Proceedings of the 9th international workshop on semantic evaluation (SemEval 2015), pp 252–263
2. Agirre E, Banea C, Cardie C, Cer DM, Diab MT, Gonzalez-Agirre A, Guo W, Mihalcea R, Rigau G, Wiebe J (2014) Semeval-2014 task 10: Multilingual semantic textual similarity. In: SemEval@ COLING, pp 81–91
3. Agirre E, Banea C, Cer D, Diab M, Gonzalez Agirre A, Mihalcea R, Rigau Claramunt G, Wiebe J (2016) Semeval-2016 task 1: Semantic textual similarity, monolingual and cross-lingual evaluation. In: SemEval-2016. 10th international workshop on semantic evaluation; San Diego, CA. Stroudsburg (PA): ACL; 2016. pp 497–511. ACL (Association for Computational Linguistics)
4. Agirre E, Cer D, Diab M, Gonzalez-Agirre A (2012) Semeval-2012 task 6: A pilot on semantic textual similarity. In: * SEM 2012: the first joint conference on lexical and computational semantics–volume 1: proceedings of the main conference and the shared task, and volue 2: proceedings of the sixth international workshop on semantic evaluation (SemEval 2012), pp 385–393
5. Agirre E, Cer D, Diab M, Gonzalez-Agirre A, Guo W (2013) * sem 2013 shared task: Semantic textual similarity. In: Second joint conference on lexical and computational semantics (* SEM), volume 1: proceedings of the Main conference and the shared task: semantic textual similarity, pp 32–43
6. Alguliev R, Aliguliyev R et al (2009) Evolutionary algorithm for extractive text summarization. Intell Inf Manag 1(02):128
7. Cer D, Diab M, Agirre E, Lopez-Gazpio I, Specia L (2017) Semeval-2017 task 1: Semantic textual similarity-multilingual and cross-lingual focused evaluation. arXiv:1708.00055
8. Coelho TA, Calado PP, Souza LV, Ribeiro-Neto B, Muntz R (2004) Image retrieval using multiple evidence ranking. IEEE Trans Knowl Data Eng 16(4):408–417
9. El-Kassas WS, Salama CR, Rafea AA, Mohamed HK (2021) Automatic text summarization: a comprehensive survey. Expert Syst Appl 165:113679
10. Fattah MA, Ren F (2008) Automatic text summarization. World Acad Sci Eng Technol 37(2):192
11. Guessoum D, Miraoui M, Tadj C (2016) A modification of wu and palmer semantic similarity measure. In: The tenth international conference on mobile ubiquitous computing, systems, services and technologies, pp 42–46
12. Jain D, Borah MD, Biswas A (2022) Performing semantically rich summarization evaluation using pretrained distilroberta embeddings. In: International conference on intelligent computing systems and applications. Springer
13. Khilji AFUR, Sinha U, Singh P, Ali A, Pakray P (2021) Abstractive text summarization approaches with analysis of evaluation techniques. In: International conference on computational intelligence in communications and business analytics. Springer, Berlin, pp 243–258
14. Li C, Sun A, Datta A (2011) A generalized method for word sense disambiguation based on wikipedia. In: European conference on information retrieval. Springer, Berlin, pp 653–664

15. Majumder G, Pakray P, Das R, Pinto D (2021) Interpretable semantic textual similarity of sentences using alignment of chunks with classification and regression. Appl Intell 51(10):7322–7349
16. Majumder G, Pakray P, Gelbukh A, Pinto D (2016) Semantic textual similarity methods, tools, and applications: a survey. Computación y Sistemas 20(4):647–665
17. Mishra A, Roy P, Bandyopadhyay S, Das S (2022) Automatic summarization evaluation: a pre-trained sentence transformer based semantic similarity calculation approach. In: International conference on intelligent computing systems and applications. Springer, Berlin
18. Mitri AM, Thabah NDJ, Maji AK (2022) Automatic evaluation of text summaries. In: International conference on intelligent computing systems and applications. Springer, Berlin
19. Mohler M, Bunescu R, Mihalcea R (2011) Learning to grade short answer questions using semantic similarity measures and dependency graph alignments. In: Proceedings of the 49th annual meeting of the association for computational linguistics: Human language technologies, pp 752–762
20. Nematzadeh A, Meylan SC, Griffiths TL (2017) Evaluating vector-space models of word representation, or, the unreasonable effectiveness of counting words near other words. In: CogSci
21. Niwattanakul S, Singthongchai J, Naenudorn E, Wanapu S (2013) Using of jaccard coefficient for keywords similarity. In: Proceedings of the international multiconference of engineers and computer scientists, vol 1, pp 380–384
22. Rocchio, J (1971) Relevance feedback in information retrieval. In: The Smart retrieval system-experiments in automatic document processing, pp 313–323
23. Salton G, Singhal A, Mitra M, Buckley C (1997) Automatic text structuring and summarization. Inf Process Manag 33(2):193–207
24. Sanh V, Debut L, Chaumond J, Wolf T (2019) Distilbert, a distilled version of bert: smaller, faster, cheaper and lighter. arXiv:1910.01108
25. Steinberger J, Jezek K et al (2004) Using latent semantic analysis in text summarization and summary evaluation. Proc ISIM 4(93–100):8
26. Tawmo T, Bohra M, Dadure P, Pakray P (2022) Comparative analysis of t5 model for abstractive text summarization on different datasets. In: International conference on computational intelligence and innovative technologies
27. Wang W, Wei F, Dong L, Bao H, Yang N, Zhou M (2020) Minilm: Deep self-attention distillation for task-agnostic compression of pre-trained transformers. Adv Neural Inf Process Syst 33:5776–5788

Automatic Evaluation of Text Summaries

Aiom Minnette Mitri, N. Donald Jefferson Thabah, Arnab Kumar Maji, Goutam Saha, and Bipul Shyam Purkayastha

Abstract This paper is an attempt to automate the evaluation of the performance of an automatic summarization system by comparing the golden summary with the predicted summary. The paper proposes two methods for evaluation and both of them score the predicted summary based on how semantically similar it is to its corresponding golden summary. The first method uses S-BERT's pre-trained sentence embeddings to generate sentence feature vectors for the predicted and golden summary correspondingly. Cosine similarity is used to determine the similarity between the two sentence Vectors. The second method uses Singular Value Decomposition (SVD) on the embeddings of both the gold and predicted summary to generate the latent topics. Angular cosine similarity is computed between the corresponding vectors of the gold and the predicted summary, taking into consideration only the top 2 latent topics.

Keywords Embeddings · S-BERT · Siamese network · Singular value decomposition

1 Introduction

The task of automatic text summarization is to produce a concise yet fluent summary of a given text while preserving key information and overall meaning. Automatic summarization task is challenging because computers lack the capability of understanding language like humans do. But all the more challenging and non-trivial is the task of evaluating an automatically generated summary. It is challenging because a human generated golden summary can take different forms depending on the summarizer. So there is not one acceptable standard summary. In this work, however,

A. M. Mitri (✉) · A. K. Maji · G. Saha
North Eastern Hill University, Umshing Mawkynroh, Shillong, Meghalaya, India
e-mail: mitriaiom@gmail.com

N. D. J. Thabah · B. S. Purkayastha
Assam University, Silchar, Assam, India

S. K. Biswas et al. (eds.), *Intelligent Computing Systems and Applications*, Lecture Notes in Networks and Systems 1307, https://doi.org/10.1007/978-981-96-3860-4_48

what has been given is a large dataset that consists of golden summary and predicted summary pairs [11]. The purpose of this work is to evaluate how semantically similar is the predicted summary to its corresponding golden summary and quantifying this into a number between 0 and 1. Both the methods proposed here utilize embeddings and in particular, sentence embeddings obtained using S-BERT. An elaborate description of the two methods viz., evaluation using sentence similarity with S-BERT and evaluation using Singular value decomposition follows in the sections below.

2 Related Work

In 2019, a model, S-BERT (Sentence BERT) is developed by Reimers et al. [3] by using the pre-trained BERT [1] and RoBERTA networks [2] and fine-tuning them to generate fixed-size meaningful sentence embeddings which is achieved using a Siamese network structure [3]. In the field of Natural Language Processing, Singular Value Decomposition or SVD [10] is the underlying method for Latent Semantics Analysis (LSA) of text data. SVD is a matrix factorization technique in linear algebra which can also be used in dimensionality reduction of vector spaces and has been used extensively in the task of *extractive summarization* by researchers in the area [4].

3 Method 1: Evaluation Using S-BERT

In the first method, S-BERT (Sentence BERT) has been used by leveraging on pre-trained embeddings. Embeddings are vector representations of texts that encapsulate semantics. Generation of sentence embeddings is a document processing technique that involves mapping sentences to real valued vectors that not only encapsulate meaning but also render them fit for processing in machine learning.

To construct a fixed-sized sentence embedding, S-BERT adds a pooling operation to the output of a BERT/RoBERTA model. Pooling method is used for calculating the mean of all output vectors, and the mean, obtained, is the embedding of a sentence. In Fig. 1, u and v are the sentence embeddings for Sentence A and Sentence B respectively. Prior to the softmax classifier, the concatenation step, denoted by $(u, v, |u - v|)$, has been shown to be the optimal representation for fine-tuning. $|u - v|$ is an element-wise difference of u and v vectors. Cross-entropy loss is used to optimise the classification objective function, which may be represented by the following equation.

Objective function: $o = Softmax\,(W_t(u, v, |u - v|))$, Where $W_t \in \mathbb{R}^{3n \times k|}$

u = Sentence embedding of u

v = Sentence embedding of v

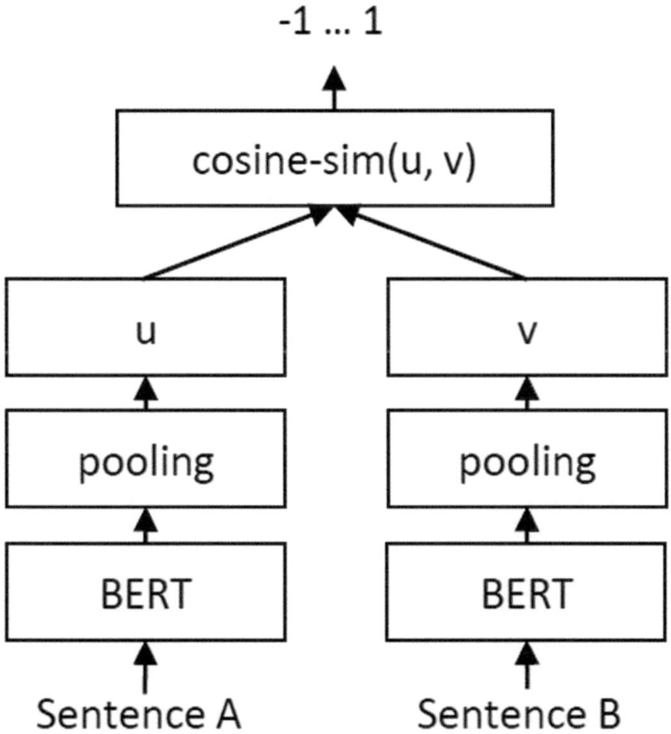

Fig. 1 A regressive objective function is used with the S-BERT architecture for inference [3]

n = Dimensions of sentence
k = number of labels.

Inference with S-BERT After the model has been trained, S-BERT employs a regressive objective function for inference within a siamese network, identical to the one used for fine-tuning. For this semantic evaluation the u and v dimensions of S-BERT are 384. As shown in Fig. 1, the cosine similarity between two sentence embeddings (u and v) are computed as a score between [–1 to 1] (Fig. 1). Generally, two extremely similar vectors, have a cosine similarity of 1, while two statistically independent vectors have a similarity of 0 and two opposite vectors have a similarity of –1. In this evaluation, the cosine similarity result is in a positive space that is neatly bounded by the interval [0, 1].

4 Method 2: Evaluation Using Singular Value Decomposition

Using Singular Value Decomposition, a matrix M with dimensions mxn is factorised as follows:

$M_{m\times n} = U_{m\times r} S_{r\times r} (V_{r\times n})^T$

where,

U is an mxr matrix such that $U^T U = I_{m\times m}$ where I is the identity matrix. The columns of U indicate left singular vectors.

S is a diagonal rxr matrix with positive real numbers on the diagonal of the matrix, and V^T is an rxn matrix such that $VV^T = I_{n\times n}$, where I is the identity matrix. The rows of V indicate right singular vector. For the task of evaluation of the summaries, the same sentence embeddings obtained in the first method have been used to represent the sentences that make up the gold and corresponding predicted summary. The vectors have been combined to make up rows of a matrix, analogous to the matrix M above. Since the embedding dimension is 384, the matrix dimension is 2×364. The first dimension is 2 because the given golden and corresponding predicted summary happen to be, rather lengthy, single line sentences devoid of any sentence delimiter. So the matrix M, will contain two rows of embedding vectors, the first row makes up the golden summary embeddings and the second row, the predicted summary embeddings. SVD is then performed on S to get the factors U, S and V^T.

From the perspective of text analytics, the matrix M can be viewed as a document by features matrix, where the document here happens to be the gold and predicted summaries. The matrix, U, is a mapping of the document into some concepts, the matrix S is a diagonal singular matrix whose values indicate the strength of a concept and the matrix V^T can be considered as features by concept mapping. After applying SVD to matrix M, the dot product of S and U is taken. This process transforms the original data into concept or topic encoded data. The vectors in the result of the dot product indicate how much a summary document (either gold or predicted) is expressed in each of the topics. In this work, top 2 topics have been selected for comparison using the angular distance measure, bounded between [0, 1], as

$$S_\theta = 1 - D_\theta$$

where,

$$D_\theta = \text{angular distance} = \frac{2.arccos(\text{cosine similarity})}{\pi}$$

The number of topics chosen depends on the size of the document. Since the document size is relatively very small (the golden as well as the predicted summary are each one line long), there is not much diversity in the contents of the document and two latent topics are considered for the similarity measure.

5 Results and Observations

Different schemes have been incorporated into the core methods for the three sets of test data [11]. The result of doing so is as shown in Table 1. It is observed that, performing the preprocessing steps of removal of stopwords and lemmatization before feeding the texts to S-BERT, gives a better score in cases where the sentences are, relatively, complex and difficult to comprehend as in the sample shown in Table 2. This may be because such sentences are synthesised during the preprocessing steps. In general, however, preprocessing does not have a significant impact on the semantic scores obtained. This may be due to the fact that removing stopwords could break the overall sentence semantics.

Table 1 Mean semantic score

Methods	Test1	Test2	Test3
LSA with preprocessing	0.259	0.446	0.383
LSA with SBERT embeddings	0.506	0.502	0.500
SBERT with preprocessing	0.572	0.644	0.652
SBERT	0.607	0.667	0.681

Table 2 Samples where semantic score are better with text preprocessing

Text preprocessing	Gold summary	Predicted summary	Semantic score
No	Brighton boss chris hughton urges his team to sign off for international break on a high with a win over swansea	Brighton boss chris hughton urges his side to fight for every result ahead of swansea clash	0.676
Yes	Brighton boss chris hughton urge team sign international break high win swansea	Brighton boss chris hughton urge fight result ahead swansea clash	0.739
No	Six people taken hostage in a kosher market siege say media outlet endangered their lives they hid in a cold room during the attack in paris by gunman amedy coulibaly	Six survivors of the paris kosher supermarket siege are suing a french media outlet they call the hostage taking dangerous live broadcasting bfmtv is accused of endangering the lives of hostages hiding in a cold room	0.758
Yes	People take hostage kosher market siege medium outlet endanger life hide cold room attack paris gunman amedy coulibaly	Survivor paris kosher supermarket siege sue french medium outlet hostage take dangerous live broadcasting bfmtv accuse endangering live hostage hide cold room	0.811

Table 3 Results of Run1 and Run 2

Team name	Mean semantic score	Correlation	Correlation score	Average correlation score
Anonymous—R1 (S-BERT)	0.6518	BLEU—Participant score	0.6238	0.6331
		ROUGE—Participant score	0.6885	
		TextSumEval score—Participant score	0.5870	
Anonymous—R2 (SVD)	0.50293	BLEU—Participant score	0.01446	0.01209
		ROUGE—Participant score	0.02077	
		TextSumEval score—Participant score	0.00106	

Table 3 shows the results for Run1 and Run2. Run1 corresponds to the results obtained using the first method as described in Sect. 3 and Run2 corresponds to the results obtained using the second method described in Sect. 4. As in the table, the correlation scores namely, BLEU, ROUGE and semantic score, of Team TextSumEval'2022 [11] and our resultant semantic score is shown in the fourth column. The average corelation score for Run1 is found to be 0.6331 and the average corelation score of Run2 is 0.01209.

6 Conclusion

The score evaluation methods (Run1 and Run2) employ sentence embeddings. These embeddings have been trained to detect semantic similarities between sentences and hence the similarity score, thus obtained, is a semantic similarity score. It is seen from Table 1 that the average corelation score using S-BERT sentence similarity is higher than the one using SVD. It may be mentioned that the BLEU score measures the extent by which two groups of words (n grams) are identical in the documents concerned. In effect, it gives higher performance for overlapping words. However, if the words are not identical but have similar meanings or are semantically similar, the BLEU score will not take that into account. Sentence embeddings, on the other hand, detect sentence similarity if the vector embedded sentence have a similar projection in the vector space.

References

1. Devlin J, Chang MW, Lee K, Toutanova K (2018) Bert: Pre-training of deep bidirectional transformers for language understanding. https://doi.org/10.48550/ARXIV.1810.04805
2. Liu Y, Ott M, Goyal N, Du J, Joshi M, Chen D, Levy O, Lewis M, Zettlemoyer L, Stoyanov V (2019) RoBERTa: a robustly optimized bert pretraining approach. https://doi.org/10.48550/ARXIV.1907.11692
3. Reimers N, Gurevych I (2019) Sentence-BERT: sentence embeddings using siamese BERT-networks. https://doi.org/10.48550/ARXIV.1908.10084
4. Smith TF, Waterman MS (1981) Identification of common molecular subsequences. J Mol Biol 147:195–197. https://doi.org/10.1016/0022-2836(81)90087-5 @INPROCEEDINGS9142938, author=Manjari, K Usha and Rousha, Syed and Sumanth, Dasi and Sirisha Devi, J, booktitle=2020 4th International Conference on Trends in Electronics and Informatics (ICOEI)(48184), title=Extractive Text Summarization from Web pages using Selenium and TF-IDF algorithm, year=2020, volume=, number=, pages=648-652, doi=10.1109/ICOEI48184.2020.9142938
5. May P, Ehrlich H-C, Steinke T (2006) ZIB structure prediction pipeline: composing a complex biological workflow through web services. In: Nagel WE, Walter WV, Lehner W (eds) Euro-Par 2006. LNCS, vol 4128. Springer, Heidelberg, pp 1148–1158. https://doi.org/10.1007/11823285_121
6. Foster I, Kesselman C (1999) The grid: blueprint for a new computing infrastructure. Morgan Kaufmann, San Francisco
7. Czajkowski K, Fitzgerald S, Foster I, Kesselman C (2001) Grid information services for distributed resource sharing. In: 10th IEEE international symposium on high performance distributed computing. IEEE Press, New York, pp 181–184. https://doi.org/10.1109/HPDC.2001.945188
8. Foster I, Kesselman C, Nick J, Tuecke S (2002) The physiology of the grid: an open grid services architecture for distributed systems integration. Technical report, Global Grid Forum
9. National Center for Biotechnology Information. http://www.ncbi.nlm.nih.gov
10. Hendrickson B (2007) Latent semantic analysis and fiedler retrieval. Linear Algebra Appl 421(2):345–355. Automatic Evaluation of Text Summaries 3 https://www.sciencedirect.com/science/article/pii/S0024379506004393, special Issue in honor of Miroslav Fiedler, https://doi.org/10.1016/j.laa.2006.09.026
11. Tamwo PD, Adhikary PK, Pakray P, Favre B, Bandyopadhyay S, Artières T (2022) TextSumEval-2022: overview of the textual summary evaluation task. In: International conference on intelligent computing systems and applications

Automatic Summarization Evaluation: A Pretrained Sentence Transformer Based Semantic Similarity Calculation Approach

Arnab Kumar Mishra, Pinki Roy, Sivaji Bnadyopadhyay, and Sujit Kumar Das

Abstract Most of the summarization evaluation metrics that are typically utilized to assess text summarization lack semantic similarity calculation. In this regard, a pretrained sentence transformer based semantic similarity calculation approach is proposed in this working note, as a submission for the TextSumEval 2022 shared task. The main idea of this work is to utilize the MiniLM based high-dimensional semantic embeddings of gold and predicted summaries, for performing semantic similarity calculations. From experimental evaluation on the training data, we have found that such an approach has 0.43 and 0.55 Pearson's correlation values with respect to the BLEU and ROUGE metrics. Moreover, the evaluation of test dataset reveals that the proposed approach has a correlation of 0.6239, 0.6887 and 0.5869 compared to BLEU, ROUGE and Organizers' semantic score measures. Such kind of positive correlation shows that the proposed approach can give evaluations that are similar to that of the popularly used metrics.

Keywords Summarization evaluation · Semantic similarity · Pretrained sentence transformer · Text similarity

1 Introduction

Automatic text summarization is an emerging field in the broader area of Natural Language Processing (NLP). There are two main approaches utilized for generating automatic summaries-Extractive Summarization and Abstractive Summarization. The idea of extractive summarization is to extract summary worthy sentences from the document itself; whereas the idea of abstractive summarization is to perform natural language generation so that the main semantic summary can be produced

---

A. K. Mishra (✉) · P. Roy · S. Bnadyopadhyay · S. K. Das
Department of CSE, National Institute of Technology Silchar, Silchar, India
e-mail: arnab.mishra.1992@gmail.com

P. Roy
e-mail: pinki@cse.nits.ac.in

S. K. Biswas et al. (eds.), *Intelligent Computing Systems and Applications*, Lecture Notes in Networks and Systems 1307, https://doi.org/10.1007/978-981-96-3860-4_49

from a longer document. In order to evaluate the performance of a system generated summary, text similarity calculation needs to be performed with respect to the ground truth gold summaries. Currently most of the evaluation metrics that are considered for evaluating automatic summarization techniques focus on the matching of ngrams and word sequences. Such kinds of word matching approaches do not consider the semantic similarity aspect of the two texts that are being compared. In this TextSumEval 2022 [1] shared task submission, our team (NITS_APSS) proposes a semantically rich similarity calculation approach that firstly represents the reference and predicted summaries as vectors in high-dimensional semantic embedding space, and then cosine similarity based vector similarity computation is performed to arrive at the final similarity score between the two texts.

The traditionally utilized metrics like BLEU [2] and ROUGE [3] scores fail to capture the semantic similarity component between two pieces of text. However, in case of most of the text generation based approaches that fall in the class of abstractive summarization, the even though the predicted summaries might not match the reference summaries word-by-word, they might still have the same meaning. In such kind of scenarios, the utilization of the traditional word matching based metrics might not give accurate similarity measure. This is where the importance of appropriate semantic representation of the texts come into the picture.

Recently, several transformer based approaches have been proposed in the NLP literature (BERT [4] and its variants [5, 6]), which have shown significant performances with respect to many NLP tasks. Such language models have been extensively utilized to perform high-dimensional representation of text data, in a semantically rich embedding space. These types of models can be utilized for the task of text similarity calculation also, since they can help represent both the reference and the predicted summary texts in a semantically rich manner before performing the actual comparison between them. With this motivation, in this work we have utilized the MiniLM model [10] to represent both the texts in higher-dimensional embedding space, followed by a cosine similarity based scoring.

The rest of the working note is organized as follows- firstly we discuss the data that is provided for the shared task in Sect. 2, following which we discuss the overall methodology of the proposed approach in Sect. 3. After this, the experimental results and discussion are presented in Sect. 4, and finally the working note is concluded in Sect. 5 with a summary of the key aspects of the work, findings and potential future research directions.

2 Data Description

In the TextSumEval 2022 shared task, there are two datasets provided by the organizers, one consisting of 100 training samples and the other containing 6000 testing samples. For the 100 training samples, the reference/gold summaries, predicted summaries, BLEU and ROUGE scores are provided, so that the initial model design can be performed. For the 6000 testing samples, three subsets are provided of 2000 samples each, where the subsets are consisting summaries from the popular text summarization datasets like CNN Daily Mail [7], XSUM [9] and MSMO [8]. For each of the testing sample, the gold and predicted summaries are provided. The goal of this shared task is to design automatic summarization evaluation approaches for finding out the text similarity scores between the gold and predicted summaries.

3 Methodology

Our working model is illustrated in Fig. 1. The main idea in this work is to utilize a pretrained sentence transformer model (MiniLM) to represent the reference and predicted summaries in terms of 384-dimensional vectors. These high dimensional vectors are expected to capture the semantic information present in the two summaries, so that the comparison between them can be performed in a semantically richer fashion. In this work, we use pretrained model MiniLM [10] which is a compressed version of transformers model. Specifically, the model made use of deep self-attention distillation framework. It means that the model is designed in such a way that it deeply mimics the self attention modules of the transformer based teacher and student models. The model is used to create 384-dimensional vector. In this way, we get embeddings of size 384, having maximum sequence length of 128 tokens, for each of the summaries.

After obtaining the 384-dimensional representation of the two summaries, the next task is to perform a comparison between the two vectorized representations. For this purpose we have used the popularly utilized Cosine Similarity metric, which is defined in Eq. 1, for the reference summary vector (V_{ref}) and predicted summary vector (V_{pred}).

$$Cosine(V_{ref}, V_{pred}) = \frac{V_{ref} \cdot V_{pred}}{||V_{ref}|| \; ||V_{pred}||} \tag{1}$$

Once the Cosine Similarity measure is computed, a post-processing step is also applied on the similarity scores. In this post-processing step, we handle negative Cosine Similarity values, where we truncate the negative valued similarity scores to a constant value of 0, representing that the two summaries have no similarity whatsoever. This step ensures that the final score generated via the proposed approach is always in the range [0, 1].

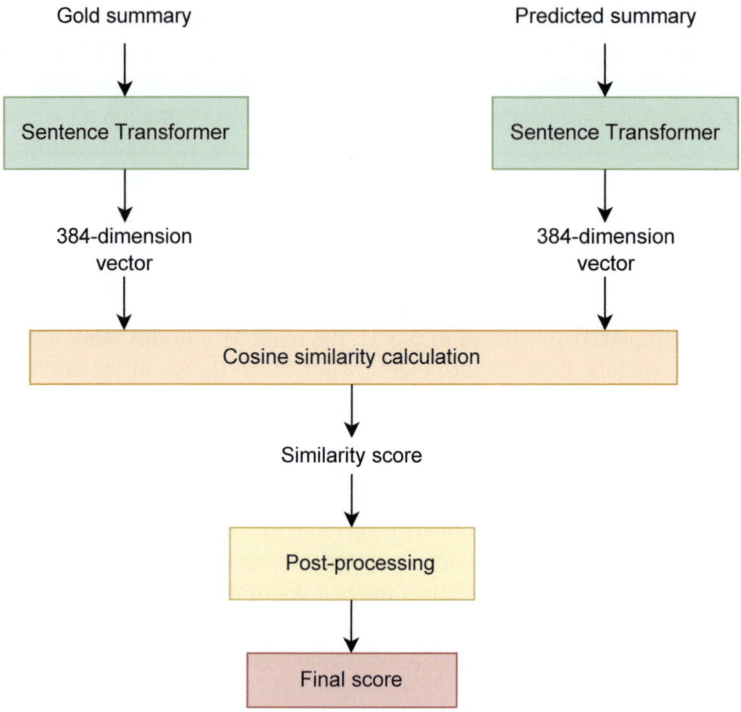

Fig. 1 Illustration of assigning semantic score

4 Experimental Results and Discussion

4.1 Experimental Setup

All the experimentations in this work is carried out in the Google Colaboratory environment. The sentence transformer model implementation is performed with the help of the Python package sentence-transformers version 2.2.0 [12]. Apart from this, other popular Python packages like Numpy [13] and Pandas [14] were also utilized during the implementation of the proposed approach.

4.2 Training Set Results

In this subsection, we present the experimental results with respect to the 100 sample training set, in terms of two aspects-correlation heatmap with respect to other popular metrics like BLEU and ROUGE scores (Fig. 2) and sample-wise score line chart (Fig. 3).

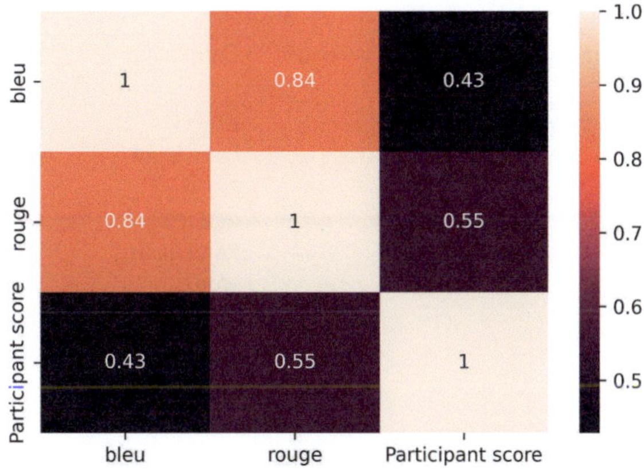

Fig. 2 Correlation heatmap of NITS_APSS Team Scores against BLEU and ROUGE metrics on training data

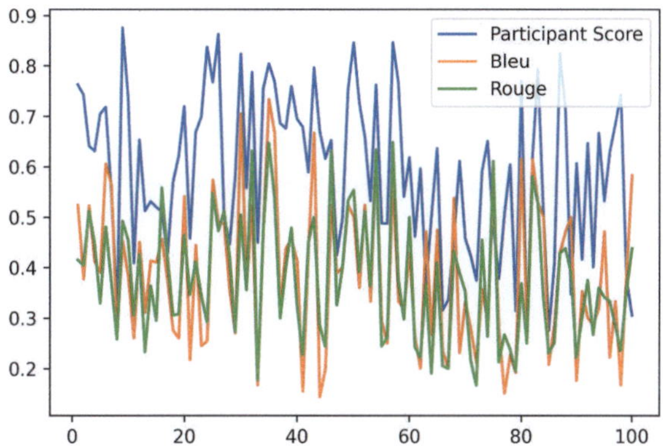

Fig. 3 Sample-wise similarity score comparison via line chart

From the correlation heatmap, we can observe that the proposed scoring technique has a positive correlation value of 0.43 and 0.55 with respect to the BLEU and ROUGE metrics for the training dataset. Such kind of positive correlation shows that the proposed approach is capable of capturing the similarity between the texts in a manner that is somewhat similar to the BLEU and ROUGE metrics. Along with this, since the proposed approach doesn't work on the token dimension and works on the high-dimensional semantic embedding space, the semantic similarity of the two texts are also captured in this measure.

From the sample-wise score comparison presented in Fig. 3, we can observe that in general the proposed scoring approach is generating scores that are overestimates with respect to the BLEU and ROUGE metrics. This can be attributed to the fact that the predicted summaries are abstractive in nature and in such scenarios semantic similarity measures can perform better evaluation of the summarization, as opposed to the token matching based measures like BLEU and ROUGE.

Table 1 Test set results

Team name	Mean semantic score	Correlation	Correlation score	Average correlation score
NITS_APSS	0.6518	BLEU—Participant score	0.6239	0.6332
		ROUGE—Participant score	0.6887	
		Organizer semantic score—Participant score	0.5869	

4.3 Test Set Results

The experimental results on the test set is depicted in Table 1. From the test set results it can be observed that the proposed approach is able to obtain a mean semantic score of 0.6518, and an average correlation score of 0.6332. This suggests that there is a high degree of positive correlation between the scores generated by the proposed scoring technique and other popular text similarity scoring measures.

5 Conclusion and Future Work

In this working note we have presented our TextSumEval shared task submission system. The main idea of our approach was to utilize MiniLM based high-dimensional embeddings of the texts to be compared, so that semantically rich summarization evaluation could be performed. From the training set results we can observe that our proposed approach has a high degree of positive correlation with other popular token-matching based evaluation metrics. Also, the test set results suggest that the proposed approach can perform effective summarization evaluation, both with respect to the mean semantic score and average correlation scores. Another key benefit of this proposed approach is that since it utilized MiniLM based representations, a compressed transformer based technique is used to generate the text embeddings, resulting in 384-dimensional vectors. This kind of representations could be faster than many BERT-like model representations that are typically double the size in terms of the embedding dimensions.

Future research works can be carried out by considering the combination of word-matching based metrics along with semantic similarity metrics. Such kind of approaches could prove to be even more effective. Moreover, domain specific evaluation metric development could be explored, so that more specific summarization evaluation can be possible.

References

1. Tamwo PD, Adhikary PK, Pakray P, Favre B, Bandyopadhyay S, Artières T (2022) TextSumEval-2022: overview of the textual summary evaluation task. In: International conference on intelligent computing systems and applications
2. Papineni K, Roukos S, Ward T, Zhu Q-J (2002) Bleu: a method for automatic evaluation of machine translation. In: Proceedings of the 40th annual meeting of the Association for Computational Linguistics, pp 311–318
3. Lin C-Y (2004) Rouge: a package for automatic evaluation of summaries. In: Text summarization branches out, pp 74–81
4. Devlin J, Chang MW, Lee K, Toutanova K (2018) Bert: pre-training of deep bidirectional transformers for language understanding. arXiv:1810.04805
5. Lan Z, Chen M, Goodman S, Gimpel K, Sharma P, Soricut R (2019) Albert: a lite bert for self-supervised learning of language representations. arXiv:1909.11942
6. Liu Y, Ott M, Goyal N, Du J, Joshi M, Chen D, Levy O, Lewis M, Zettlemoyer L, Stoyanov V (2019) Roberta: a robustly optimized bert pretraining approach. arXiv:1907.11692
7. Nallapati R, Zhou B, Gulcehre C, Xiang B (2016) Abstractive text summarization using sequence-to-sequence rnns and beyond. arXiv:1602.06023
8. Zhu J, Li H, Liu T, Zhou Y, Zhang J, Zong C (2018) MSMO: multimodal summarization with multimodal output. In: Proceedings of the 2018 conference on empirical methods in natural language processing. Association for Computational Linguistics, Brussels, Belgium, pp 4154–4164
9. XSUM dataset, Hugging Face. https://huggingface.co/datasets/xsum
10. Wang W, Wei F, Dong L, Bao H, Yang N, Zhou M (2020) Minilm: Deep self-attention distillation for task-agnostic compression of pre-trained transformers. Adv Neural Inf Process Syst 33:5776–5788
11. Google Colaboratory Environment. https://colab.research.google.com
12. Sentence Transformers Python Package. https://pypi.org/project/sentence-transformers/
13. Harris CR, Millman KJ, Van Der Walt SJ, Gommers R, Virtanen P, Cournapeau D, Wieser E, Taylor J, Berg S, Smith NJ, Kern R (2020) Array programming with NumPy. Nature 585(7825):357–362
14. McKinney W, Others (2010) Data structures for statistical computing in python. In: Proceedings of the 9th python in science conference, vol 445, pp 51–56

Performing Semantically Rich Summarization Evaluation Using Pretrained DistilRoBERTa Embeddings

Deepali Jain, Malaya Dutta Borah, and Anupam Biswas

Abstract Traditional summarization evaluation metrics like ROUGE, BLEU, etc. have a serious limitation due to the fact that they essentially use word-matching based calculations while comparing a predicted summary against a reference summary, and ignore the semantic similarity completely. To deal with this issue, in this TextSumEval 2022 shared task submission, we propose a semantically rich summarization evaluation approach. The main idea is that both the summaries are firstly represented as a DistilRoBERTa based high-dimensional vector, and then cosine similarity based vector semantic calculation is carried out. The proposed approach achieves Pearson's correlation values of 0.44 and 0.62 respectively w.r.t the BLEU and ROUGE metrics for the training set, and 0.6081 and 0.6635 for the test set. Also, the proposed approach obtains a mean semantic score of 0.6579 on the test set. With the proposed approach, we can generate positively correlated scores like traditional metrics, while capturing semantic similarity.

Keywords Semantic similarity · Summarization evaluation · DistilRoBERTa embeddings · Text similarity

1 Introduction

Summarization Evaluation is an important research problem in the specific area of automatic text summarization and also in terms of the more general text similarity assessment problem. The main problem setup is that the reference/gold summary and system generated summary will be given, and the goal is to generate a similarity

D. Jain (✉) · M. D. Borah · A. Biswas
Department of CSE, National Institute of Technology Silchar, Assam, India
e-mail: deepali_rs@cse.nits.ac.in

M. D. Borah
e-mail: malayaduttaborah@cse.nits.ac.in

A. Biswas
e-mail: anupam@cse.nits.ac.in

S. K. Biswas et al. (eds.), *Intelligent Computing Systems and Applications*, Lecture Notes in Networks and Systems 1307, https://doi.org/10.1007/978-981-96-3860-4_50

score in a specific range like [0, 1], where values near 1 represent very high similarity and 0 represents complete dissimilarity between the two texts. When it comes to the automatic text summarization problem, there are two primary approaches that are typically utilized in the Natural Language Processing (NLP) literature—Extractive Summarization and Abstractive Summarization. In Extractive Summarization, the main idea is that if important summary worthy sentences are picked from the document and combined to form the summary, then the key information present in the document can be identified. Whereas, in Abstractive Summarization, the main idea is that based on the content of the document we want to generate shorter length texts using Natural Language Generation (NLG) techniques, which summarizes the important information effectively. Most of the popular summarization evaluation techniques utilized in the literature, like ROUGE [1] variants or BLEU [2] metrics, perform word-matching based similarity evaluation. These evaluation techniques cannot capture the semantic similarity between the reference and the system generated summaries. In order to deal with this issue, in this TextSumEval 2022 [3, 4] shared task submission, our team (NITS_DMA) proposes a DistilRoBERTa [5] based semantic similarity evaluation metric, which can be effectively applied as a summarization evaluation approach.

The traditionally utilized summarization evaluation metrics like ROUGE-variants and BLEU perform a very primitive text similarity calculation, which is completely based on word and word-sequence matching based ideas. But a system generated summary might be produced in such a way that the key information present in the document is captured succinctly with the help of novel words. In such kind of scenarios, using metrics like ROUGE or BLEU will not give a very similarity score, even though semantically both the predicted as well as the reference/gold summary might be very similar. In order to deal with this problem, effective summarization evaluation measures need to be developed that can capture the semantic similarities between two pieces of texts.

Recently, in the NLP domain, transformer based pretrained embedding approaches have enabled significant improvements in terms of several NLP tasks (BERT [6] and its variants [7, 8]). Such kind of language models can generate very effective high-dimensional semantic embeddings for a given piece of text. The proper utilization of such techniques can enable better text representations that can be helpful for any downstream predictive modeling task. Motivated by this, in this shared task submission we have utilized the DistilRoBERTa language model based semantic embedding representations for the reference and the predicted summaries. These vector representations finally compared with the help of the Cosine Similarity measure, to generate the final similarity scores.

The rest of the working note is organized as follows—the datasets utilized in this work are discussed in Sect. 2, followed by a description of the proposed approach in Sect. 3. The experimentation results and discussion are presented in Sect. 4 and finally the working note is concluded in Sect. 5 by summarizing the key aspects of the work and also highlighting potential future research directions.

2 Data Description

A demo summary sample has been provided by the organizers which includes gold summary, predicted summary, BLEU similarity score, and ROUGE similarity score. These has been provided so that a method can be designed for testing or calculating relevance between the gold and predicted summary. Whereas, the test set consists of 6000 testing samples, three subsets are provided of 2000 samples each, where the subsets are consisting summaries from the popular text summarization datasets like CNN Daily Mail [9], XSUM [11] and MSMO [10]. For each of the testing sample, the gold and predicted summaries are provided. The primary goal is to design an automatic summarization evaluation approach for finding out how similar are the gold and predicted summaries.

3 Methodology

Our working model is illustrated in Fig. 1, where the main idea is that we firstly find the DistilRoBERTa based embeddings for both the reference and predicted summaries, following which a Cosine Similarity based vector similarity calculation is performed to arrive at the final similarity scores.

In this work, we use a pretrained model, namely, DistilRoBERTa, the base version. The model follows the same training procedure as that of [5]. The term 'distillation' means that the student model (or the compressed model) is trained in such a way that it mimics the teacher model. DistilRoBERTa was pretrained on OpenWebTextCorpus, a reproduction of OpenAI's WebText dataset (it is ~4 times less training data than the teacher RoBERTa). The model produces a 768 dimension vector. In this work, we utilize this model to create 768-dimension vectors for each of the gold as well as predicted summaries.

Once the 768-dimensional vector representation of the gold summary (V_G) and predicted summary (V_P) are obtained, the next step is to perform vector similarity calculation using the Cosine Similarity measure. The Cosine Similarity calculation is performed as per Eq. (1).

$$\text{Cosine}(V_G, V_P) = \frac{V_G \cdot V_P}{||V_G||\ ||V_P||} \tag{1}$$

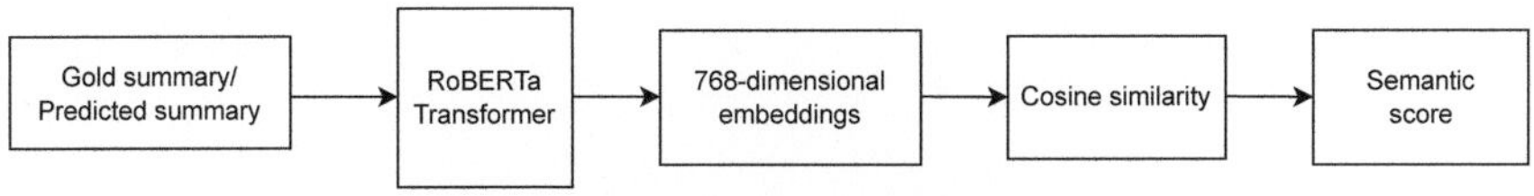

Fig. 1 Illustration of computing semantic score using contextualized embeddings

The Cosine Similarity measure calculated via Eq. (1) is finally reported as the similarity score between the two summaries.

4 Experimental Results and Discussion

4.1 Experimental Setup

For the purpose of experimentation, we have utilized the Google Colaboratory environment which can provide efficient compute support. The DistilRoBERTa based pre-trained model utilization is performed with the help of the Python package sentence-transformers version 2.2.0 [12]. Apart from this, other popular Python packages like Pandas [13] and Numpy [14] were also utilized for the purpose of data reading and further processing.

4.2 Training Set Results

The score correlation of our proposed approach and other popular approaches like ROUGE and BLEU on the training set are shown pictorially, with the help of a heatmap diagram in Fig. 2.

From the correlation heatmap, we can observe that the proposed similarity scoring approach is having a high degree of positive correlation with other popular word-matching based approaches. With respect to the BLEU metric, the correlation coefficient value is 0.44, whereas with respect to the ROUGE metric the correlation value is 0.62. Such results suggest that the word-matching based similarity aspect is also captured by the proposed semantic similarity calculation approach.

Fig. 2 Score correlation heatmap on training data

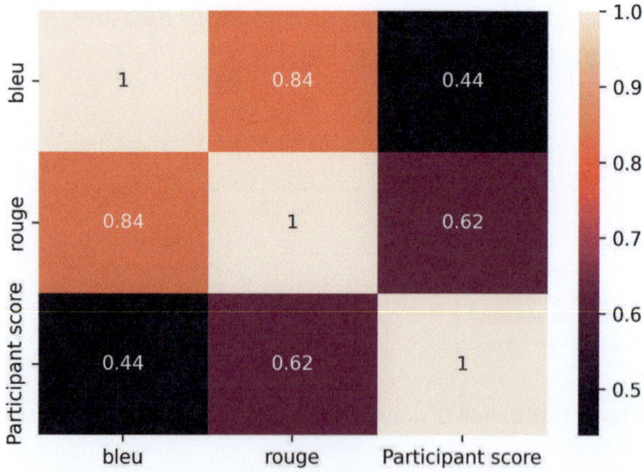

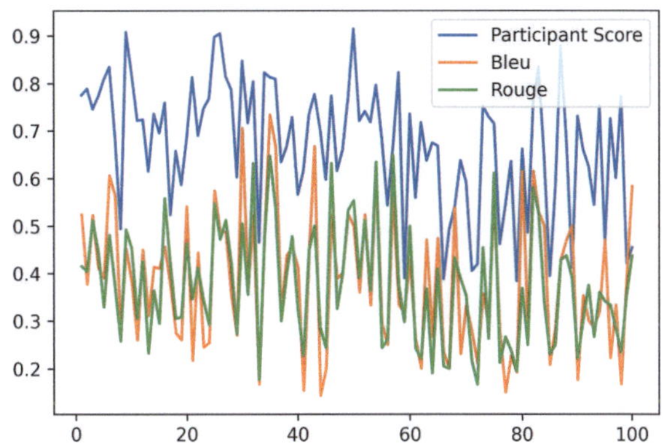

Fig. 3 Sample-wise similarity score comparison on training set

A sample-wise score comparison is also presented with the help of a line-chart diagram in Fig. 3. In this score comparison diagram, we can observe that compared to the BLEU and ROUGE metrics, the proposed approach overestimates the similarity scores up to some extent. This can be due to the fact that semantically similar predictions having different words in reference and predicted summaries are given higher scores by the proposed approach as compared to the traditional word-matching based metrics.

4.3 Test Set Results

The test set results are presented in Table 1, where our proposed approach is represented as Participant score. From the table, we can observe that the proposed approach is able to generate evaluation scores that are having high degree of positive correlation with traditional scoring approaches like BLEU and ROUGE. Moreover,

Table 1 Test set results

Team name	Mean semantic score	Correlation	Correlation score	Average correlation score
NITS_DMA	0.6579	BLEU—Participant score	0.6081	0.6104
		ROUGE—Participant score	0.6635	
		Organizer semantic score—Participant score	0.5596	

the proposed approach is able to achieve a mean semantic score of 0.6579 and an average correlation score of 0.6104, with a 0.56 correlation score compared to Organizer's semantic scores. Such kind of results suggest that the proposed approach is able to generate effective similarity scores, while capturing the semantic similarities between the gold and predicted summaries.

5 Conclusion

In this work, the problem of designing an effective summarization evaluation approach is undertaken, as part of the TextSumEval 2022 shared task submission. In order the overcome the limitations of current evaluation approaches like ROUGE and BLEU, we have proposed a semantically rich summarization evaluation technique in this work. The main idea is to represent the reference and predicted summaries in terms of semantically rich DistilRoBERTa based high-dimensional embeddings, and then to perform Cosine Similarity based calculations. With this approach we have observed that scores that are highly correlated to the traditional word-matching based metrics can be obtained, while capturing the semantic similarity as well. Such an approach can overcome the limitations of word-matching based summarization evaluation metrics by introducing the semantic component.

As part of the future work, a score combination based approach can be explored where both word-matching as well as semantic similarity matching based approaches can be fused to arrive at much better scoring techniques.

References

1. Lin CY (2004) Rouge: a package for automatic evaluation of summaries. In: Text summarization branches out, pp 74–81
2. Papineni K, Roukos S, Ward T, Zhu WJ (2002) Bleu: a method for automatic evaluation of machine translation. In: Proceedings of the 40th annual meeting of the association for computational linguistics, pp 311–318
3. TextSumEval 2022 Shared Task. http://icicsa.nits.ac.in/textsumeval22/index.html
4. Tamwo PD, Adhikary PK, Pakray P, Favre B, Bandyopadhyay S, Artières T (2022) TextSumEval-2022: overview of the textual summary evaluation task. In: International conference on intelligent computing systems and applications, 23–24 September 2022
5. Sanh V, Debut L, Chaumond J, Wolf T (2019) DistilBERT, a distilled version of BERT: smaller, faster, cheaper and lighter. arXiv preprint arXiv:1910.01108
6. Devlin J, Chang MW, Lee K, Toutanova K (2018) Bert: pre-training of deep bidirectional transformers for language understanding. arXiv preprint arXiv:1810.04805
7. Lan Z, Chen M, Goodman S, Gimpel K, Sharma P, Soricut R (2019) Albert: a lite Bert for self-supervised learning of language representations. arXiv preprint arXiv:1909.11942
8. Liu Y, Ott M, Goyal N, Du J, Joshi M, Chen D, Levy O, Lewis M, Zettlemoyer L, Stoyanov V (2019) Roberta: a robustly optimized Bert pretraining approach. arXiv preprint arXiv:1907.11692
9. Nallapati R, Zhou B, Gulcehre C, Xiang B (2016) Abstractive text summarization using sequence-to-sequence RNNs and beyond. arXiv preprint arXiv:1602.06023

10. Zhu J, Li H, Liu T, Zhou Y, Zhang J, Zong C (2018) MSMO: multimodal summarization with multimodal output. In: Proceedings of the 2018 conference on empirical methods in natural language processing, Brussels, Belgium. Association for Computational Linguistics, pp 4154–4164
11. XSUM dataset, Hugging face. https://huggingface.co/datasets/xsum
12. Sentence transformers Python package. https://pypi.org/project/sentence-transformers/
13. McKinney W, others (2010) Data structures for statistical computing in python. In: Proceedings of the 9th Python in science conference, vol 445, pp 51–56
14. Harris CR, Millman KJ, Van Der Walt SJ, Gommers R, Virtanen P, Cournapeau D, Wieser E, Taylor J, Berg S, Smith NJ, Kern R (2020) Array programming with NumPy. Nature 585(7825):357–362

MIX
Papier aus verantwortungsvollen Quellen
Paper from responsible sources
FSC® C105338

If you have any concerns about our products,
you can contact us on
ProductSafety@springernature.com

In case Publisher is established outside the EU,
the EU authorized representative is:
Springer Nature Customer Service Center GmbH
Europaplatz 3, 69115 Heidelberg, Germany

Printed by Libri Plureos GmbH
in Hamburg, Germany